T0321745

Strategies and Opportunities for Technology in the Metaverse World

P.C. Lai
University of Malaya, Malaysia

A volume in the Advances in Web Technologies and Engineering (AWTE) Book Series

Published in the United States of America by
IGI Global
Engineering Science Reference (an imprint of IGI Global)
701 E. Chocolate Avenue
Hershey PA, USA 17033
Tel: 717-533-8845
Fax: 717-533-8661
E-mail: cust@igi-global.com
Web site: http://www.igi-global.com

Library of Congress Cataloging-in-Publication Data

Names: Lai, P. C., 1974- editor.
Title: Strategies and opportunities for technology in the metaverse world / P.C. Lai, editor.
Description: Hershey, PA : Engineering Science Reference, [2022] | Includes bibliographical references and index. | Summary: "The rise of metaverse technologies with social impact will value add to the ecosystem and this book explores the opportunities and challenges of metaverse enterprises and outlines possible avenues for the growth of this sector"-- Provided by publisher.
Identifiers: LCCN 2022034136 (print) | LCCN 2022034137 (ebook) | ISBN 9781668457320 (h/c) | ISBN 9781668457337 (s/c) | ISBN 9781668457344 (eISBN)
Subjects: LCSH: Metaverse--Social aspects. | Metaverse--Economic aspects. | Internet in education.
Classification: LCC QA76.9.M47 S77 2022 (print) | LCC QA76.9.M47 (ebook) | DDC 005.4/35--dc23/eng/20220825
LC record available at https://lccn.loc.gov/2022034136
LC ebook record available at https://lccn.loc.gov/2022034137

This book is published in the IGI Global book series Advances in Web Technologies and Engineering (AWTE) (ISSN: 2328-2762; eISSN: 2328-2754)

British Cataloguing in Publication Data
A Cataloguing in Publication record for this book is available from the British Library.

Advances in Web Technologies and Engineering (AWTE) Book Series

Ghazi I. Alkhatib
The Hashemite University, Jordan
David C. Rine
George Mason University, USA

ISSN:2328-2762
EISSN:2328-2754

Mission

The **Advances in Web Technologies and Engineering (AWTE) Book Series** aims to provide a platform for research in the area of Information Technology (IT) concepts, tools, methodologies, and ethnography, in the contexts of global communication systems and Web engineered applications. Organizations are continuously overwhelmed by a variety of new information technologies, many are Web based. These new technologies are capitalizing on the widespread use of network and communication technologies for seamless integration of various issues in information and knowledge sharing within and among organizations. This emphasis on integrated approaches is unique to this book series and dictates cross platform and multidisciplinary strategy to research and practice.

The **Advances in Web Technologies and Engineering (AWTE) Book Series** seeks to create a stage where comprehensive publications are distributed for the objective of bettering and expanding the field of web systems, knowledge capture, and communication technologies. The series will provide researchers and practitioners with solutions for improving how technology is utilized for the purpose of a growing awareness of the importance of web applications and engineering.

Coverage

- Integrated user profile, provisioning, and context-based processing
- Information filtering and display adaptation techniques for wireless devices
- Knowledge structure, classification, and search algorithms or engines
- Web user interfaces design, development, and usability engineering studies
- Web Systems Architectures, Including Distributed, Grid Computer, and Communication Systems Processing
- Ontology and semantic Web studies
- Security, integrity, privacy, and policy issues
- Data and knowledge validation and verification
- Software agent-based applications
- Case studies validating Web-based IT solutions

IGI Global is currently accepting manuscripts for publication within this series. To submit a proposal for a volume in this series, please contact our Acquisition Editors at Acquisitions@igi-global.com or visit: http://www.igi-global.com/publish/.

The Advances in Web Technologies and Engineering (AWTE) Book Series (ISSN 2328-2762) is published by IGI Global, 701 E. Chocolate Avenue, Hershey, PA 17033-1240, USA, www.igi-global.com. This series is composed of titles available for purchase individually; each title is edited to be contextually exclusive from any other title within the series. For pricing and ordering information please visit http://www.igi-global.com/book-series/advances-web-technologies-engineering/37158. Postmaster: Send all address changes to above address.

Titles in this Series

For a list of additional titles in this series, please visit: http://www.igi-global.com/book-series/advances-web-technologies-engineering/37158

3D Modeling Using Autodesk 3ds Max With Rendering View
Debabrata Samanta (CHRIST University, India)
Engineering Science Reference • © 2022 • 291pp • H/C (ISBN: 9781668441398) • US $270.00

Handbook of Research on Gamification Dynamics and User Experience Design
Oscar Bernardes (ISCAP, ISEP, Polytechnic Institute of Porto, Portugal & University of Aveiro, Portugal) Vanessa Amorim (ISCAP, Polytechnic Institute of Porto, Portugal) and Antonio Carrizo Moreira (University of Aveiro, Portugal)
Engineering Science Reference • © 2022 • 516pp • H/C (ISBN: 9781668442913) • US $380.00

Advanced Practical Approaches to Web Mining Techniques and Application
Ahmed J. Obaid (University of Kufa, Iraq) Zdzislaw Polkowski (Wroclaw University of Economics, Poland) and Bharat Bhushan (Sharda University, India)
Engineering Science Reference • © 2022 • 357pp • H/C (ISBN: 9781799894261) • US $270.00

Handbook of Research on Opinion Mining and Text Analytics on Literary Works and Social Media
Pantea Keikhosrokiani (School of Computer Sciences, Universiti Sains Malaysia, Malaysia) and Moussa Pourya Asl (School of Humanities, Universiti Sains Malaysia, Malaysia)
Engineering Science Reference • © 2022 • 462pp • H/C (ISBN: 9781799895947) • US $380.00

Security, Data Analytics, and Energy-Aware Solutions in the IoT
Xiali Hei (University of Louisiana at Lafayette, USA)
Engineering Science Reference • © 2022 • 218pp • H/C (ISBN: 9781799873235) • US $250.00

Emerging Trends in IoT and Integration with Data Science, Cloud Computing, and Big Data Analytics
Pelin Yildirim Taser (Izmir Bakircay University, Turkey)
Information Science Reference • © 2022 • 334pp • H/C (ISBN: 9781799841869) • US $250.00

App and Website Accessibility Developments and Compliance Strategies
Yakup Akgül (Alanya Alaaddin Keykubat University, Turkey)
Engineering Science Reference • © 2022 • 322pp • H/C (ISBN: 9781799878483) • US $250.00

701 East Chocolate Avenue, Hershey, PA 17033, USA
Tel: 717-533-8845 x100 • Fax: 717-533-8661
E-Mail: cust@igi-global.com • www.igi-global.com

Table of Contents

Detailed Table of Contents

Humanity has entered a new digitalization formation with the invention of the computer. The formation of digitalization has passed through many phases, but even today it shows its effect to a great extent. The establishment of network connections and provision of the internet for the use of humanity, wireless network connections, cloud computing, and cryptocurrencies comes at the very beginning of these phases. Today, the formation of digitalization has taken on a completely different concept from the metaverse and has taken the whole world under its influence. This chapter would provide detailed information about the metaverse universe with an overview on its advantages and disadvantages. Additionally, the possibility of taking the role of the real world will be evaluated by comparing the metaverse universe with the real world.

The advancement of SMEs is accelerated by technological expansions using blockchain technology in the Industrial Revolution (IR) 4.0 era. Based on current trends in AI and blockchain technology, this study proposes that the distance between entrepreneurs all over the world and their potential workers may be greatly decreased to virtually real-time. A secondary literature review is carried out in order to identify the key developments in IR 4.0 technologies in the SMEs industry, as well as the potential trend that will lead the business sector. The adoption of AI and blockchain technology in the IR 4.0 technologies is projected to make seeking treatments overseas more reasonable, accessible, and health records readily available on a real-time and protected basis. However, it is necessary to highlight that the expansion of SMEs raises the eyebrows of society from the security, social, and economic viewpoints.

Technology is playing a very important role in transforming the industry. Currently, the advancement has been to Web3 technology, where we are using mutual interaction online in order to communicate with each other, or to organize the meetings or any social gathering with the help of internet. In the evolution of technology from radio to internet, metaverse is the current and upcoming technology supporting web3. Metaverse is expanding exponentially, the implementation of new technology is necessary for the progression to deliver the best experience using AR (augmented reality), VR (virtual reality), XR (extended reality). This chapter discusses the incredible potentials in which one can achieve outstanding results for implementing Metaverse technology, or to establish connection and connect with people in more live manner (in virtual spaces) and to achieve real-time experience.

This chapter aims to analyze the business models' opportunity represented by the Metaverse (Foundation, n.d.), proposing a framework composed of four interdependent blocks; the "illusions" (Slater, 2003). The word Metaverse has been used to describe a sort of thing, mainly related to virtual reality (VR) (Lanier, 1992) and the non-fungible tokens (NFTs). The immersive experience of virtual reality is characterized by three illusions: the illusion of place, the illusion of embodiment, and the illusion of plausibility (Slater et al., 2009). This chapter describes the use of blockchain technology, namely the NFTs, as the origin of a fourth illusion.

The growth of the metaverse is an imminent phenomenon for the world to tap its economic opportunities. Non-fungible token (NFT) is the choice of a new form of financial technology in metaverse to authenticate digital asset ownership. The objective of this study was to examine the impacts of NFT on the metaverse communities. Based on the unified theory of acceptance and use of technology model, the sample was financial users in the metaverse who are familiar with NFT. The findings validated that UTAUT constructs influence the intention to use NFT payments. Also, self-efficacy was confirmed as the antecedent to explain the need for managing the processes while payments were made. This study provides an understanding of the use of NFT by examining the theoretical aspect of the current usage scenario in the metaverse. Furthermore, the required personal ability of the users in determining the benefits of NFT offers practical consideration for wider adoption. Hence, it contributes to the research on the behavioral finance aspect of NFT usage.

At the end of 2019, individuals' outdoor activities were restricted due to the emergence of COVID-19. As a result of this phenomenon, interest in online activities and interaction in the metaverse environment has increased. Online games have exploded in popularity with the young generation in Metaverse where they can earn money through the platforms. Thus, it is desirable to investigate emerging technology and analyse how to invest using techniques, such as sentiment analysis and machine learning (ML), to predict crypto trends. This study analysed time series data for crypto price and text, where information like news, articles, and feedback from social media can use the input to generate the sentiment score to understand the crypto trends. FinBERT is a sentiment model that was used for this study to generate the result. The AI investing framework is built to incorporate both sentiment analysis technique and predictive model for this chapter, to address the research questions and enable one to make more informed decisions.

This chapter overviewed several existing works being conducted on the metaverse in immersive education. Most of the reviews have clearly stated that the metaverse would be a promising asset and tool in education besides the conventional method. The metaverse gives space where real-world specifications can be replicated via digital twin technology. This enables smarter problem-solving in industries like construction, architecture, healthcare, life sciences, and more. Reviewed studies on the effectiveness of those technologies for education found moderate evidence for the metaverse to assist students in their daily learning. The application of the metaverse to the education sector has been well studied and widely accepted as an engaging and effective method of delivering learning experiences. Hence, the metaverse might be a promising tool to provide immersive learning experiences to users nowadays. An application named wonders of the world (WoW) has been developed using XR technology to provide immersive learning experiences to the users.

According to the action plan for improving the quality of vocational education (2020-2023), the vocational school evaluation system includes professional ethics, professional quality, technical skills, employment quality, and entrepreneurial ability as important content areas for assessing the quality of talent training that should be strengthened. This plan was released by the Ministry of Education and nine other government departments. An example higher vocational college in Guangdong is used to illustrate the importance of quality talent training in the new era, and a method for improving talent training quality in vocational colleges is proposed. This chapter is intended to serve as a model for other higher vocational colleges seeking to improve the quality of talent training while also promoting the modernization and internationalization of high-quality vocational education.

Metaverse is a new learning spectrum for Malaysians where it proffers a parallel virtual universe to academics and students. The metaverse campus (MC) is an idea of converting learning from physical classes into virtual worlds. The initiative of MC allows students worldwide to partake in classes and events in the same virtual world. A study was conducted among 25 Malaysian academicians from local and private tertiary instructions using the fuzzy Delphi method (FDM). This study is aimed to get consensus from the academicians to discover the social perspectives towards MC from accessibility, diversity, equality, and humanity perspective.

This paper studies the purchasing behavior of college students using the webcast platform; understanding the factors that college students use the webcast platform and understanding the main factors that promote their consumption so that school managers can effectively manage and guide students to consume rationally. At the national economic level, we need to develop live e-commerce, and at the national education level, we need to ensure that students are not affected by external transition interference. The research uses the technology acceptance model (TAM). Several key factors include webcast platform design, webcast marketing, webcast celebrity webcast, perceived usefulness, and perceived ease of use. The survey data are collected by using the five-point scale, the goodness of fit of the measurement model is used for statistics, and the data are analyzed by using the standardized regression weight of the structural model.

Classroom interaction is one of the most commonly used teaching methodologies that can be applied to develop linguistic competencies in second language instruction. Interaction in the classroom refers to the conversation between teachers and students, as well as interactions between the students. Active participation and interaction are both critical themes and vastly invaluable concerning the learning of students and second language acquisition. This paper examines how classroom interaction helps students to learn and acquire a second language. Using the findings of previous studies as a baseline, this study reveals some factors that affect how instructors should use this teaching methodology in different situations. In addition, this paper mentions applying innovative ideas for second language acquisition with the support of new scenarios and developed applications (e.g., role-play activities in personalized settings, etc.), aimed at exploring these possibilities for language teaching.

Despite massive support for blockchain startups, countries in Southeast Asia have yet to fully launch themselves into metaverse. The appeal of playing lucrative crypto games such as Axie Infinity, where active players can earn up to US$15 per day, may provide these countries the incentive to kickstart many of the business opportunities in metaverse gaming. Metaverse is a potential hype. The international audience has started riding on the back of metaverse's success. Countries in Southeast Asia (SEA) are at with their respective development within the metaverse space. It is important to recognise the level of development of metaverse in these countries. Among those who felt negative about the Metaverse, the most common reason was their concerns about data security and privacy. This research intends to understand the development of metaverse in countries in SEA – Vietnam, Indonesia and Malaysia. Through understanding the development of metaverse in these countries, these countries can properly position themselves to reap the most benefit out of this hype.

This chapter examines how Turkish educational researchers used the Metaverse concept during the Covid-19 pandemic. A thorough examination of the term metaverse is conducted from its emergence, evolution, past and current usage within the academic circles, along with the terms edufication and gamification. Document analysis method is chosen to do a systematic literature review on academic journals. Search was conducted with added keywords to the Metaverse within the literature in Turkish and English. Although the term is a relatively new entrant to the academic research, it is continuously evolving along with the technological advancements its applications to the industry and social life; hence, the concept and its usage will be a continuous subject of further research. The result of the analysis showed that Turkish researchers mainly prefer 'three-dimensional virtual world' instead of the term Metaverse, and the ones found to use Metaverse mainly misuse the term out of its internationally accepted definition.

The rapid development of technology has penetrated almost all sectors of the society and makes any form of resistance almost impossible. The incorporation of digital technology into adolescent's daily life, as well as its impact on their cognitive, emotional, and social development, is growing by the day. They can use technology to play, explore, and learn in a variety of ways. This is because their brains are so adaptable, these learning opportunities represent a vital growth stage throughout this time period, which helps and encourages them to improve their communication skills and knowledge. No one will

Preface

Every time technology changes, it creates threats to established ways of doing business and opportunities for new ways to offer products and services. The metaverse is the driving force bringing the current technology together in a unified, immersive experience. One indication that the final stage has arrived is when the terminology becomes part of our common culture. This is where the metaverse world is the hottest topic today. The topic of the metaverse world now appears regularly in national media and becomes a point of discussion during business meetings and during social gatherings. Have you embarked on the metaverse world?

OPPORTUNITIES IN THE METAVERSE WORLD

The metaverse has a lot of technologies underlying its foundation and is probably the next iteration of the internet and mobile platform. Let's explore the key technologies that drive the metaverse world. The availability of the 5G and 6G infrastructure level will be the driver of the metaverse world. These will provide the infrastructure speed to deploy devices mainly referring to Augmented Reality (AR) or Virtual Reality (VR), smart wearable (SM), and Omnidirectional treadmill (ODT) fields. VR interactive equipment, including virtual reality interaction technology, motion capture technology, and ergonomics research development as well as deployment has driven the growth of the metaverse world. Participants in the Metaverse will be able to have more immersive experiences that converge reality and the virtual world by utilizing mixed reality and AR/VR technologies. Blockchain, NFT, and virtual currency, as the core technologies of the meta-universe, can provide scarcity for the virtual world. These technologies are the basis for the establishment of social systems, legal systems, and business systems in the Metaverse ecosystem. Blockchain financial technology focuses on the innovation of blockchain technology and products. It has been applied to digital assets, equity bonds, supply chain traceability, joint credit investigation, and data security areas. In the field of NFT, at present, NFT companies and start-ups are mainly concentrated in three fields: digital collections, game assets, and the virtual world. The driving force sectors of the Metaverse world will be games, smart medicine, industrial design, and smart education.

According to GrayScale (2021), the metaverse will likely infiltrate every sector in some way in the coming years, with the market opportunity estimated at over $1 trillion in yearly revenues. How attractive is the metaverse world? Based on MatthewBall (2021), the metaverse offers opportunities to spend on virtual goods at $54 billion per year, almost double the amount spent buying music. The GDP for Second Life was about $650M in 2021 with nearly $80M USD paid to creators (Zdnet, 2022) and not to mention even brick-and-mortar shop like Walmart is preparing to enter the metaverse (CNBC, 2022).

Obviously, not to miss out on the gaming industry. In-game ad spending is set to reach $18.41 billion by 2027 (IPS, 2021). Thus, there is the possibility of the metaverse massively expanding access to the marketplace for consumers from emerging and frontier economies. Furthermore, there is a shift toward online technologies during Covid has been convoyed by a rapid uptake of technologies like metaverse and others (Agur, Peria, and Rochon, 2020; Lai and Tong, 2022). There is a need for more structured research and development as well as strategies to support these growing industries. (Pisano, 2015, Stam and Wennberg, 2019; Lai, 2020; Lai and Liew, 2021).

STRATEGIES IN THE METAVERSE WORLD

Organizations are prioritizing metaverse hiring, customer research, and upskilling as much or even more than their investments in metaverse-related technology. It's a good strategy since technology is primarily an enabler. Even countries are implementing favorable metaverse strategies. Let's explore the development of the metaverse in China and the country's strategy as well. More recently, the Guangzhou Huangpu District and Guangzhou Development Zone jointly released Measures for Promoting the Innovation and Development of the Metaverse (also referred to as the "10 Metaverse Measures"), the first policy of its kind in the Guangdong-Hong Kong-Macao Greater Bay Area. Some of the highlights include:

- Promoting innovation agglomeration and encouraging enterprises to form clusters and become "highly specialized";
- Supporting technological leadership, where institutions and businesses should work hand in hand to solve technical bottlenecks.
- Strengthening the protection of intellectual property rights.
- Increasing talent acquisition; and
- Establishing a "Metaverse Industrial Fund" to attract social capital.

A new wave of innovation, strategies, development, and deployment across consumer industries has been seen globally with the rapid development of the metaverse world. This pandemic has changed the way people live, work, and socialize, accelerating demand for innovation, as retailers, consumer goods, and travel companies shift from reacting to the crisis to reinventing products and services according to findings of a new global survey from Accenture. We can see the huge demand for technologies from the popular zoom (virtual meeting), online movie streaming like Netflix, video-sharing (TikTok), delivery services like Grubhub, and others. Similarly, this pandemic has transformed online shopping from a nice-to-have to a must-have around the world. The pandemic has accentuated the trend toward greater adoption of social media (e.g: Facebook, Instagram, etc) and growth in sales through e-commerce websites. Shifts in consumption habits have also been observed, driven by the need for sourcing essential items. Social media and owning e-commerce shops are important sales channels for e-commerce companies. Both channels have witnessed higher growth since the beginning of the pandemic due to the lockdown. Organizations are differentiating themselves by adding new technologies like AR, VR, virtual coins, etc to attract customers to their platforms.

The organization's decision-makers are eyeing Augmented Reality, Virtual Reality, Artificial Intelligent, Blockchain, E-payment, and cutting-edge computing technologies in the metaverse world. It is eminent that these technologies have significant implications in the metaverse world that are stalwart in inspiring social impact. This has opened new areas of opportunities for organizations to strategies themselves toward the metaverse world. Businesses primarily focused on the creation of wealth previously while education and technology have taken the centre stage during this Covid pandemic, pressure to behave more ethically and responsibly has prompted organizations to consider investments that serve Environment, Social, and Governance (ESG) goals. These encompass strategies and practices that feature the management of economic, social, and environmental performance to optimize benefits for both business and society with metaverse technologies. Therefore, metaverse technologies do revolutionize how the world works and the impact on the society and environment. Furthermore, this has been attracting more and more metaverse academicians, researchers, developers, and professionals across the globe to facilitate the ecosystem for a better world.

ORGANIZATION OF THE BOOK

This book offers multi-disciplinary Strategies and Opportunities for Technology in the Metaverse World coming from 6 continents. The book is organized into seventeen chapters. A brief description of each of the chapters follows:

The first section of the book highlights contributions sharing the analysis, strategies, opportunities, and future direction of the metaverse world.

Chapter 1 starts with the "ANALYSIS OF METAVERSE TECHNOLOGY, IS IT REAL OR VIRTUAL?". Today, the formation of digitalization has taken on a completely different concept from the metaverse and has taken the whole world under its influence. This chapter would provide detailed information about the metaverse universe with an overview of its advantages and disadvantages. Additionally, the possibility of taking the role of the real world will be evaluated by comparing the metaverse universe with the real world.

Chapter 2 is the up-and-coming trend of the "Artificial Intelligence and Blockchain Technology in the 4.0IR Metaverse Era: Implications, Opportunities, and Future Directions". Based on current trends in AI and blockchain technology, this study proposes that the distance between entrepreneurs all over the world and their potential workers may be greatly decreased to virtually real-time.

Chapter 3 is entitled "Exploring The Incredible Potentials And Opportunities Of Metaverse World." Metaverse is expanding exponentially, implementing new technology for the progression to deliver the best experience using AR (Augmented reality), VR (Virtual reality), and XR (Extended reality). This chapter discusses the incredible potential by which we can achieve outstanding results for implementing Metaverse technology or establishing a connection and connecting with people in a more live manner (in virtual spaces) and achieving real-time experience.

Chapter 4 is about the metaverse new economy titled "The Fourth Illusion: How a New Economy of Consumption is Been Created in The Metaverse". The immersive experience of Virtual Reality is characterized by three illusions: the illusion of place, the illusion of embodiment and the illusion of plausibility (Slater, 2009) (Slater et al., 2009). This chapter describes the use of Blockchain Technology, namely the NFTs, as the origin of a fourth illusion.

Chapter 5 will dive into the title "User Acceptance Towards Non-fungible Token (NFT) as the Fintech for Investment Management in the Metaverse". Non-fungible token (NFT) is the choice of a new form of financial technology in the metaverse to authenticate digital asset ownership. The objective of this study was to examine the impacts of NFT on the metaverse communities.

Chapter 6 is the not to be missed opportunity the "Metaverse in Investment using Sentiment Analysis and Machine Learning." Non-fungible token (NFT) is the choice of a new form of financial technology in the metaverse to authenticate digital asset ownership. The objective of this study was to examine the impacts of NFT on the metaverse communities.

In the next section, the metaverse will focus on the education section.

Chapter 7 will start with the "Wonders of the World: Metaverse for Education Delivery". This chapter overviewed several existing works being conducted on metaverse in immersive education. The application of metaverse to the education sector has been well studied and widely accepted as an engaging and effective method of delivering learning experiences.

Chapter 8 then look at the quality of education with the metaverse "Research on Improving the Quality of Talent Training in Higher Vocational Colleges in China". This Chapter is intended to serve as a model for other higher vocational colleges seeking to improve the quality of talent training while also promoting the modernization and internationalization of high-quality vocational education.

Chapter 9 will focus on the "Delphi study on Metaverse Campus Social Perspectives". The Metaverse Campus (MC) is an idea of converting learning from physical classes into virtual worlds. The initiative of MC allows students worldwide to partake in classes and events in the same world.

Chapter 10 focuses on the metaverse payment potential through the "Research on College Students' purchasing using a webcast platform". This chapter studies the purchasing behavior of college students using the webcast platform and understands the factors that college students use the webcast platform in the metaverse world.

Chapter 11 is on how education can use metaverse to support language learning with the title "Classroom Interaction and Second Language Acquisition in the Metaverse World." This chapter mentions applying innovative ideas using metaverse for second language acquisition with the support of new scenarios and developed applications (e.g., role-play activities in personalized settings, etc.), aimed at exploring these possibilities for language teaching.

The last section of the book will investigate a few different spectrums from legal to the adoption of a metaverse in a few countries.

Chapter 12 will start exploring the "ADOPTION OF METAVERSE IN SOUTHEAST ASIA: - VIETNAM, INDONESIA, MALAYSIA" in ASEAN. Countries in Southeast Asia (SEA) are at with their respective development within the metaverse space. It is important to recognize the level of development of the metaverse in these countries.

Chapter 13 explores the edification with the title "Metaverse or Not Metaverse: A content analysis of Turkish Scholars' Approach to Edufication in the Metaverse. This chapter examines how Turkish educational researchers used Metaverse concept during the Covid-19 pandemic. A thorough examination of the term metaverse is conducted from its emergence, evolution, past and current usage within the academic circles along with the terms edufication and gamification

Chapter 14 investigates the "Effects of Digital Technologies on Academic Performance of Nigerian Adolescents. The rapid development of technology has penetrated almost all sectors of society and makes any form of resistance almost impossible. The incorporation of digital technology into adolescent's daily life, as well as its impact on their cognitive, emotional, and social development, is growing by the day. They can use technology to play, explore, and learn in a variety of ways.

Chapter 15 covers the "A hybrid SEM-ANN approach for intention to adopt metaverse using C-TAM-TPB and IDT in China". The study aims to identify factors affecting university students' intention to adopt metaverse due to the 4th industrial revolution and the COVID-19 pandemic in China based on the integration of C-TAM-TPB model and IDT theory

Chapter 16 will focus on the satisfaction of shopping in the metaverse world with the title "Customer e-satisfaction Towards Online Grocery Sites in the Metaverse world". The main objective of this study is to establish the dimensions of customer satisfaction which is hypothesized to have a relationship with customer intention to purchase from online grocery sites and this will provide organizations insights into customer satisfaction in e-commerce in the metaverse.

Chapter 17 ends with the legal area title "The Role of Legal Governance Framework in the Metaverse World." The book chapter analyses how the current legal framework of self-governance is weak in regulating the metaverse world. The rule of law, as a discourse that emphasizes the legitimacy of governance and appropriate limits on the exercise of power, provides a useful framework as a first step to reconceptualizing and evaluating these tensions in communities at the intersection of the real and the virtual, the social and the economic, and the public and the private.

P. C. Lai
University of Malaya, Malaysia

REFERENCES

Agur, I. Peria, S. M., & Rochon, C. (2020). Digital Financial Services and the Pandemic: Opportunities and Risks for Emerging and Developing Economies. *International Monetary Fund (MF).*

Alpha Inc. (2022). Engage. *IPS News.* http://ipsnews.net/business/2021/10/04/in-game-advertising-market-to-reach-usd-18-41-billion-by-2027-is-going-to-boom-withrapidfire-inc-playwire-media-llc-atlas-alpha-inc-engage/.

Ball, M. (2021). Payments, Payment Rails, and Blockchains, and the Metaverse. https://www.matthewball.vc/all/metaversepayments.

CNBC. (2022). Walmart is quietly preparing to enter the metaverse. *CNBC.* https://www.cnbc.com/2022/01/16/walmart-is-quietly-preparing-to-enter-the-metaverse.html.

Grider, D. (2021). The Metaverse Web 3.0 Virtual Cloud Economies. *Grayscale Research.* https://grayscale.com/wp-content/uploads/2021/11/Grayscale_Metaverse_Report_Nov2021.pdf.

Icrowdnewswire. (2021). In-game advertising market to reach usd 18.41 billion by 2027, is going to boom with rapidfire inc., playwire media llc, atlas alpha inc, engage. ISP News.

Lai, P. C. (2020). *Intention to use a drug reminder app: a case study of diabetics and high blood pressure patients*. SAGE Research Methods Cases., doi:10.4135/9781529744767

Lai, P. C., & Liew, E. J. Y. (2021). *Towards a Cashless Society: The Effects of Perceived Convenience and Security on Gamified E-Payment Platform Adoption*. Australasian.

Lai, P. C., & Tong, D. L. (2022). *An Artificial Intelligence-Based Approach to Model User Behavior on the Adoption of E-Payment. Handbook of Research on Social Impacts of E-Payment and Blockchain.* IGI Global.

industry and Information Technology Bureau. (2022). Measures for Promoting the Innovation and Development of the Metaverse in Guangzhou Huangpu District and Guangzhou Development Zone. *People's Government of Huangpu District.* http://www.hp.gov.cn/gzjg/qzfgwhgzbm/qgyhxxhj/xxgk/content/post_8171935.html

Murray, A. (2022). High Fidelity invests in Second Life to expand the virtual world. *ZD Net.* https://www.zdnet.com/article/high-fidelity-invests-in-second-life-to-expand-virtual-world/.

Pisano, G. P. (2015). You Need an Innovation Strategy. *Harvard Business Review*.

Stam, E., & Wennberg, K. (2009). The roles of R&D in new firm growth. *Small Business Economics, 33*(1), 77–89. doi:10.100711187-009-9183-9

Wright, O., Standish, J., & Weiss, E. (2011). *How consumer companies can innovate to grow?* Accenture.

Acknowledgment

I'm grateful for the many blessings from the Almighty God. A special thanks to IGI Global publisher for publishing this book and the support from IGI Global team. The editor would like to acknowledge the help of all the people involved in this project and, more specifically, to the authors and reviewers that took part in the review process. Without their support, this book would not have become a reality. First, the editor would like to thank each one of the authors for their contributions. Our sincere gratitude goes to the chapter's authors who contributed their time and expertise to this book. Second, the editor wishes to acknowledge the valuable contributions of the reviewers regarding the improvement of quality, coherence, and content presentation of chapters. Most of the authors also served as referees; we highly appreciate their double tasks. Here the editor likes to especially thank those who are not authors and offered their valuable time to review the book chapters coming from both academic and industry from five (5) Continents. Last but not least, all those who read the book chapters and will take the research further, thank you and best wishes.

P. C. Lai
University of Malaya, Malaysia

Chapter 1
Analysis of Metaverse Technology:
Is It Real or Virtual?

Ersin Caglar
https://orcid.org/0000-0002-2175-5141
European University of Lefke, Turkey

ABSTRACT

Humanity has entered a new digitalization formation with the invention of the computer. The formation of digitalization has passed through many phases, but even today it shows its effect to a great extent. The establishment of network connections and provision of the internet for the use of humanity, wireless network connections, cloud computing, and cryptocurrencies comes at the very beginning of these phases. Today, the formation of digitalization has taken on a completely different concept from the metaverse and has taken the whole world under its influence. This chapter would provide detailed information about the metaverse universe with an overview on its advantages and disadvantages. Additionally, the possibility of taking the role of the real world will be evaluated by comparing the metaverse universe with the real world.

INTRODUCTION

Digitalization, which has become a part of our lives with computer science, affects humanity every day through its opportunities (Elmassah and Hassanein, 2022), which are the internet, cloud computing, internet of things, and blockchains as the leading technologies with such effects. With these opportunities, communication, data storage, and device controls have become much easier. Thus, all sectors started to keep up with this digitalization and adapted their sectoral activities to digital platforms (Mustapha et al., 2021; Nižetić et al. 2020; Oliveira et al., 2020). Some of the leading sectors are enterprises, educational institutions, government institutions, entertainment, and banks. New industry models and brand-new concepts have entered our lives due to the digitalization such as e-government, e-commerce, and online education.

DOI: 10.4018/978-1-6684-5732-0.ch001

As mentioned above, digitization, with its significance influence on our lives, has gained a completely different dimension with the concept of the metaverse. Metaverse not only affected the world that we live in but also changed the world by turning it into a completely different one.

In fact, the metaverse may seem like a new concept, but it's not. Although far from its current meaning, the metaverse word appeared in 1992 in a fiction novel called "Snow Crash" by science fiction writer Neal Stephenson (Stephenson, 2003). Novel, explains metaverse as a massive virtual environment parallel to the real one (Joshua, 2017). In the novel, the metaverse is called "Oasis" and users interact through digital avatars like today's metaverse (Damar, 2021).

The definition of metaverse is "the concept of a fully immersive virtual world where people gather to socialize, play, and work" (Laeeq, 2022). Shortly, the metaverse is a "layer between humanity and reality".

Stephenson, Metaverse is a virtual space, combining virtual reality (VR), augmented reality (AR) and the internet. (Murray, 2020). Addition to AR, VR, internet, today's metaverse concept consists of blockchain, NFT, and Web 3.0 (Mystakidis, 2022).

As mentioned above, today's metaverse is 3D virtual world platform where users can do any activity with the help of AR and VR services (Yonhap News Agency, 2021). Such platforms became popular in the last decade and people started to shift their activities to online platforms. Especially, COVID-19 which entered into our lives in the beginning of the 2020, online platforms are much more popular with the pandemic curfew (Gaubert, 2021; Harapan et al., 2020).

Even COVID19 couldn't make the metaverse popular but the statements of Facebook's CEO Mark Zuckerberg at the Facebook connect conference made the metaverse popular. According to Zuckerberg's description, the metaverse is "a virtual environment where any user can present himself/herself using the possibilities of digital spaces. You can think of it as an embodied Internet." Besides this, Zuckerberg described that "metaverse will be the next generation of the internet after the rise of mobile web and the smart phones." According to this claim, Facebook is rebranding itself as a Meta. After this statement, the eyes and attention of the whole world were turned to the metaverse (Zuckerberg, 2021; Díez, 2021).

METAVERSE

Metaverse which is "Snow Crash" in 1992, is still a complex and incomprehensible concept. In the novel, the writer explains this virtual world as a shared that the mixture of VR, AR, and the internet. Today's concept of metaverse has a structure similar to the novel (Stephenson, 2003; Damar, 2021).

Metaverse is the combination of the prefix "meta" (Greek prefix meaning post, after, or beyond) with the word "universe", which describes a hypothetical synthetic environment linked to the world. In other words, the Metaverse is a permanent and persistent multi-user environment, which is a post-reality universe that mixes digital virtuality and physical reality. (Mystakidis, 2022).

Shortly, metaverse is a virtual environment that blends physical and digital with the convergence between Internet technology and Extended Reality (XR). According to researchers (Milgram et al., 1995), XR integrates digital and physical to various degrees, such as; augmented reality (AR), mixed reality (MR), and virtual reality (VR).

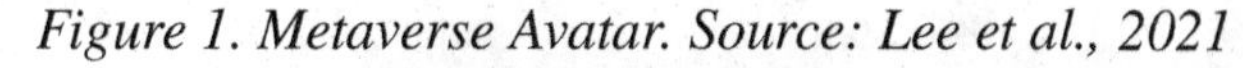
Figure 1. Metaverse Avatar. Source: Lee et al., 2021

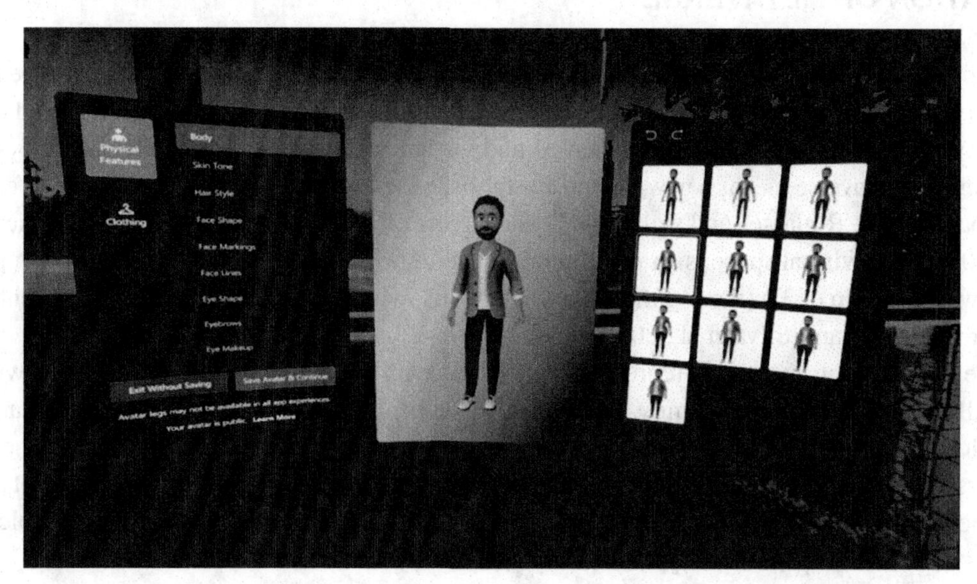

Similarly in Snow Crash, the metaverse reflects a replica of digital environments and the duality of the real world. So, all individual users have their own avatar (Figure 1), experiencing an alternative life in virtuality, which is a metaphor for the user's real world, by imitating the user's physical self. (Lee et al., 2021).

- **Extended Reality (XR):** XR is some kind of key term that contains a range of immersive technologies such as electronic, digital media in which data is represented and reflected. XR includes VR, AR and Mixed Reality (MR). In all the XR aspects mentioned above, people observe and interact in a fully or partially synthetic digital area created by technology. (Milgram et al., 1995).
- **Virtual Reality (VR):** VR is digitally generated environment which is artificial. Users feel that they feel immersed in a different world in virtual reality, working in similar ways as in their physical environment (Slater & Sanchez-Vives, 2016). The specialized multi-sensory equipment such as immersion helmets, VR headsets and versatile treadmills, this experience is amplified by sight, sound, touch, movement and natural ways of interacting with virtuality. (Pellas, et al., 2020; 2021).
- **Augmented Reality (AR):** AR adopts a different approach towards physical spaces where it embeds digital inputs, and virtual elements into the physical environment to enhance it (Ibáñez and Delgado-Kloos, 2018). It spatially merges the physical with the virtual world (Klopfer, 2008). The end outcome is a spatially projected layer of digital artifacts mediated by devices, e.g., smartphones, tablets and etc… (Mystakidis, et al., 2021).
- **Mixed Reality (MR):** MR is a more complex concept than other terms. The definition of MR, which has changed over time, reflects contemporary technological and dominant linguistic meanings and narratives. MR is assumed to be an advanced iteration of AR in the sense that the physical environment interacts with digital data accepted in real time. (El Beheiry et al., 2019). For example, a non-player character written in an MR game recognizes the physical environment and hides behind a table or sofa. Similar to VR, special glasses are required for MR. (Milgram et al., 1995).

UTILIZATION OF METAVERSE

As mentioned above, the Metaverse, a short form of meta-universe, is a virtual world where the real and virtual mix in a science-fiction vision, allowing people to move between different devices and communicate. In practical terms, it refers to augmented and virtual reality products and services (Mystakidis, 2022). Hence, Metaverse means a new universe established in a virtual space. It is possible to enter this virtual space with VR glasses and walk around the streets of a virtual city without any physical effort (Narin, 2021). The virtual space is an environment where the user can do many things that is possible on earth with the help of 3D (three-dimensional) elements, with holograms and user's representative as an Avatar (Metahuman) (Silva et al., 2018).

People control their avatars to represent themselves with avatars, people communicate with each other, and virtually create a community. In the metaverse the digital currency is used to buy many items such as clothes, cinema tickets, land, etc... Users can travel the metaverse using virtual reality headset or controllers. (Kiong, 2022). Users don't necessarily need their virtual image called Avatar to get a virtual land for now. However, producing digital copy with the help of Avatar Creator produced by Epic Games while it is still free is a predictive approach (Collins, 2021).

Metaverse technology is currently a concept supported by several different companies. The most important company that produces metaverse products is Decentraland. In other words, when users buy the Decentraland Land token, have virtual land on the company's game platform. In this virtual land, it is possible to set up the virtual environments which users want with 3D models. Moreover, although Facebook and Microsoft have not yet revealed their products, they are planning to enter the market in a short time (Associated Press, 2021; Damar, 2021).

As mentioned above, metaverse technologies produced by game companies such as Decentraland and Axie Infinity are a small part of the iceberg. It is unclear how much potential these firms have to improve existing technology. Therefore, it is very risky to make very large investments before well-established companies enter the sector. The point of how the system will rise will be the launch of the Facebook investment in a short time (Duan et al., 2021).

Despite these risky options, it is possible to buy and sell products and make small or large investments in the metaverse. Users need to use tokens (cryptocurrencies) to do these transactions in the metaverse (Jeon et al., 2022). The easiest way to obtain these tokens is to go to crypto exchanges like WazirX or CoinDCX and use your currency-loaded wallet to buy the tokens directly. Among the most well-known metastore tokens are MANA, the native currency of the Decentraland metastore; The currency of the Sandbox metastore is SAND and the local currency of the Axie Infinity metastore is AXS (Kim, 2021; Caulfield, 2021).

Users need to create an account in the metaverse that user wants to buy and link their crypto wallet to their account like signing to Decentraland for virtual real estate, Axie Infinity for characters and plots of land, and Sandbox to buy/sell artistic creations. For convenience, users can create an account with OpenSea if a user wants to access all NFTs in one common market (Caulfield, 2021).

NFT (Non-Fungible Token) is a crypto asset open for purchase and sale, which is formed by encrypting a digital file or idea over the blockchain or metaverse. NFT, which means the sale of digital products with crypto money, will play a big role in the future of the metaverse (Jeon et al., 2022). The best part is, that there are NFT-making sites and platforms (NFT Market) where everyone can do the job of producing and selling NFTs for free (Ante, 2021).

As mentioned above, the encryption and sale of a commercially available product with a cryptosystem are called NFT. At this point, it is possible to sell an image of a soccer ball, which is considered valuable for people, as NFT. NFT manufacturers design collections with content files such as music, images, videos, 3D modeling etc. Then these designs turn into crypto assets with open-source codes. After coding, the sales process begins (Wang et al., 2021).

Users must have an account on cryptocurrency exchanges for buying, selling, and barter transactions. At this point, most of the current sales are made with Ethereum. User needs a platform to encode digital content that is deemed as important to people. Rarible.com and OpenSea are currently the most popular platforms (Yoder, 2022).

ADVANTAGES

Despite being an old technology, the newly popular metaverse arouses great curiosity for humanity. With this curiosity, the metaverse has many opportunities for humanity.

- Connecting the world and negating physical distance: Metaverse professionals for improving communication can deliver engaging and interactive experiences for all users. (Damar, 2021).
- New opportunities for brands: Brands could create and showcase digital versions of products and solutions in shared virtual worlds. Users can gain near real-time experience viewing and controlling product features. (Knox, 2022).
- Improvements to the work environment: The work environment has always benefited from new technologies. (Knox, 2022).
- Overcoming obstacles: Information technology such as the internet, cloud, and cryptocurrency, helps humanity to make life easier. But, with metaverse people who have a disability can do anything easily (Lee et al., 2021).
- Traveling the world without moving: Any metaverse user can travel all around the world with minimum effort (Collins, 2021).
- More possibilities in education: With the metadata store, learning will be more accessible than ever before. For example, the physical location of the class no longer needs to be considered (Duan et al., 2021).

DISADVANTAGES

Apart from the benefits mentioned above, the metaverse has disadvantages as well. Although the metaverse is new, it also has its disadvantages. Since the metaverse is new, these disadvantages may change and perhaps increase in the future.

- Required advanced equipment: Metaverse brings with it many new and advanced technologies. However, not every individual has access to advanced technologies. (Lee et al., 2021).
- Privacy and security implications (Cybercrime / Cybersecurity): Cybercrime is a serious problem since the internet existence. All metaverse data which is the representation of each user virtually, under the risk. (Di Pietro and Cresci, 2021).

However, as a new, the metaverse still does not take advantage of these sophisticated levels of cyber-security that make it extremely vulnerable to all kinds of illegal activities. (Wang et al. 2022).

- Other problems: Addiction can become a problem. Some argue that there is a greater risk of addiction with the metaverse, as you are completely immersed in a virtuality. You don't need to leave your VR setup aside from basic physical needs like eating and sleeping. The group of children is most at risk, as experts suggest that individuals under the age of 18 spend too much time in the metaverse would severely damage their development. (Dutilleux and Chang, 2022).
- Losing connection with the physical world: Metaverse aims to provide users with immersive experiences by blurring the gap between the real and virtual worlds. Metaverse disadvantages also bring the possibility of metaverse, which affects how people perceive real relationships and interactions. (Dutilleux and Chang, 2022).
- Mental health issues: There is a serious mental health risk to metaverse users. Although virtual reality is used in controlled settings to help patients with schizophrenic symptoms, user cannot count on the metaverse to be controlled or generated to help people with these diseases.

Psychological research has also shown that immersion in this digital world and disconnecting ourselves from the real world will increase the likelihood of permanent separation from reality and may even lead to symptoms close to psychosis.

- Virtual bullying: Malicious behavior towards others can be found all over the internet, whether on social media or in games.

Despite numerous campaigns to prevent or limit these behaviors, taking the internet into a 3D environment and making users feel completely immersed in the metaverse will give these bad guys more power. This leads to greater opportunities to attack other users and can make victims more vulnerable as there is no other universe anymore.

Bullying, harassment and personal attacks against anyone will inevitably become one of the dark sides of the meta universe, and unfortunately there is little that can be done to combat it. (Di Pietro and Cresci, 2021).

WILL THE METAVERSE REPLACE THE REAL WORLD?

Despite of its uncertainty and risk, many famous companies established a brand or make an attempt to establish. Besides from famous companies, the university was established in the metaverse. It is still too risky to establish a brand in the virtual world even for a famous company, but there are exist in the virtual world.

- **Facebook:** As mentioned above, Mark Zuckerberg claimed that metaverse will be the next big computing platform at the Facebook Connect conference. After this claim, Zuckerberg introduced that Facebook shifted all companies to metaverse as a meta. This claim attracted the attention of all companies and increased the interest of the metaverse (Zuckerberg, 2021).
- **Nike**: Hire virtual wear designers and virtual footwear, among other things. So, Nike bought RTFKT which is a famous metaverse brand with a shoe collection (Hollensen et al., 2022).

- **Adidas:** According to CNBC report, Adidas purchased land on Sandbox VR, a virtual real estate firm (Altun, 2021).
- **Gucci:** Collaborated with Roblox to market and sell their products in the metaverse (Park and Kim, 2022).
- **Balenciaga:** This company created virtual boutiques to sell and market their produts (Averbek and Türkyilmaz, 2022).
- **Walmart:** is retail corporation that operates a chain of hypermarkets aim to sell things in metaverse (Openkov and Tetenkov, 2021).

Besides for these famous company, many company established their brand in metaverse, such as (Grider and Maximo, 2021):

- **Art Galleries**
- **Business Offices**
- **Games & Casinos**
- **Advertising**
- **Sponsored Content**
- **Music Venues**

Other than these famous companies, even education is slowly shifting to the metaverse. Communication 166/266 Virtual People, taught by Professor Jeremy Bailenson, deals with emerging VR and use cases. It covers the expanding impact of VR in many different fields. Related research shows that how VR can be used effectively in educational settings. Related with course structure, students will use their class time to engage in VR experiences, either alone or as a group, including classroom discussions. In 2021, over 20 weeks and two courses, 263 students, all with their own VR headset, spent more than 3,500 shared hours together on Metaverse. Thus, as a result of the research, a classroom was set up in a virtual reality environment using Facebook's Oculus Quest 2 headsets and training began. (Hirsh-Pasek et al., 2022).

Figure 2. Concert in Metaverse

Even governments have begun to switch their activities to the metaverse. Seoul state that, the city will be the first major city government to enter the metaverse. On 3/11/2021, the South Korean capital announced its plan to launch various public services and cultural events on the metaverse, an immersive internet based on virtual reality. if the goal is successful, Seoul residents can visit a virtual city hall to do everything from visiting a historic site to filing a civil complaint by wearing virtual reality glasses. (Um et al., 2022; Ramesh et al., 2022; Damar, 2021).

Apart from these companies which established brand in metaverse, some activities like gaming and concerts are performed in metaverse as well. Virtual activities have the potential to be more profitable than physical activities. A massive concert held in Fortnite was watched by 45 million people and grossed nearly $20 million, including product sales. The biggest advantage of this event is that those who cannot attend such events due to geography or cost can easily participate. (Moy and Gadgil, 2022).

Except all these attempts related with metaverse, Bloomberg company claimed that "The Metaverse creates a new economy model where wealth is produced, traded and augmented in a different currency, but linked to real world money. Soon we will see businesses set up, office buildings built, meetings held for remote workers, and job interviews in the metaverse world."

The Metaverse universe, which has many functions and uses, is known by many people, although it is currently in its infancy. Therefore, although it is not known exactly what will happen in the future, it is quite certain that the concept of Metaverse will be popular. However, it would be wrong to restrict this digital world only to games and entertainment areas. We will be able to perform many functions virtually in this universe. Any user will be able to attend meetings and conduct business negotiations with our personalized avatars. Any user will go to digital concerts with our friends and have as much fun as we want without getting tired. Although these events may seem remote or inaccessible, giant technology companies investing in the digital universe predict that they can happen in 10-15 years.

In short, everything we have today, our daily activities, hobbies and much more will find a place in the digital world. In fact, such a life is not far from us. The whole world was shaken by a global pandemic in early 2020 and its effects still continue. In this process, works, schools, meetings and artistic activities were done completely online. So, people have experienced the Metaverse universe to some extent. But there are huge differences between working remotely and working in the Metaverse universe. For example, not being able to see people's facial expressions, hand and body movements in online meetings was a huge problem. But thanks to VR technology, this is no longer a problem. Thanks to 3D systems, people can work as if they were together. While this provides endless opportunities for employees, it also relieves employers economically.

Although Metaverse is seen as a science fiction product in today's conditions, it is called the future of the internet by many. It is quite exciting to enter such a digital environment and to do almost all our work from there. Giant technology companies are also expected to invest more in this area and increase the possibilities in the digital universe. Maybe in the future, there will be developments that we cannot even predict at the moment, and these developments will become a part of our lives. Just like the phones that did not exist 50-60 years ago, do not fall into our hands now.

CONCLUSION

With technological developments such as wireless networks, cloud computing, internet of things, and cryptocurrencies, the effect of digitalization has begun to affect the world rapidly. Through such technological developments, businesses, educational institutions, government institutions and many other sectors have been under the influence of digitalization. Digitalization, under the influence of the whole world, has gained a new dimension that directly affects human life with the metaverse.

It seems that the world of the Metaverse is now clearly coming. This world can harbor both benefits and harms as expected. It is not possible to predict the impact of the metaverse today as it is still in the testing phase, yet just as other technologies, metaverse has good and bad sides. But regardless of the testing process, many companies established brand and many activities were done in metaverse.

Back in the 1970s, people couldn't exactly predict the future of the internet. Internet use was available, but limited. We can say that today's Metaverse is what the internet was then. Although it cannot be predicted exactly what it will cause in the future, it is thought that its use will become widespread in about 10-15 years. For this reason, Facebook founder Mark Zuckerberg wanted to be among the Metaverse companies by changing the name of his company. It has now proven that it exists not only in social media, but also in virtual reality under the name Meta. In addition to the Meta company, platforms such as Microsoft, Fortnite and Roblox are also investing in this universe. They think that the Metaverse will replace the internet in the future and become a part of our lives.

REFERENCES

Altun, D. (2021). *Sanal Ve Artirilmiş Gerçeklikle Dönüşen Yeni Nesil Sosyal Medya Mecrasi: Metaverse* [The Next Generation Social Media Channel Transforming with Virtual and Augmented Reality: Metaverse.]. Uluslararası İşletme ve Pazarlama Kongresi.

Ante, L. (2021). Non-fungible token (NFT) markets on the Ethereum blockchain: Temporal development, cointegration and interrelations.

Averbek, G. S., & Türkyilmaz, C. A. (2022). Sanal Evrende Markalarin Geleceği: Yeni İnternet Dünyasi Metaverse Ve Marka Uygulamalari. [The Future of Brands in the Virtual Universe: The New Internet World Metaverse and Brand Applications.] Sosyal Bilimlerde Multidisipliner Çalışmalar Teori, Uygulama Ve Analizler, 99.

Caulfield, B. (2021). *What is the Metaverse?* The Official NVIDIA Blog.

Collins, B. (2021). The Metaverse: How to Build a Massive Virtual World. *Forbes Magazine*. https://www. forbes. com/sites/barrycollins/2021/09/25/the-metaverse-how-tobuild-a-massive-virtual-world

Damar, M. (2021). Metaverse Shape of Your Life for Future: A bibliometric snapshot. *Journal of Metaverse*, *1*(1), 1–8.

Di Pietro, R., & Cresci, S. (2021). Metaverse: Security and Privacy Issues. In *Third IEEE International Conference on Trust, Privacy and Security in Intelligent Systems and Applications (TPS-ISA)*. IEEE. 281-288.

Díez, J. L. (2021). Metaverse: Year One. Mark Zuckerberg's video keynote on Meta in the context of previous and prospective studies on metaverses. *Pensar la publicidad: revista internacional de investigaciones publicitarias, 15*(2), 299-303.

Duan, H., Li, J., Fan, S., Lin, Z., Wu, X., & Cai, W. (2021). Metaverse for social good: A university campus prototype. In *Proceedings of the 29th ACM International Conference on Multimedia,* (pp.153-161). 10.1145/3474085.3479238

Dutilleux, M., & Chang, K. M. (2022). Future Addiction Concerned for Human-Being. [IMJST]. *International Multilingual Journal of Science and Technology*, *7*(2), 4724–4732.

El Beheiry, M., Doutreligne, S., Caporal.,, C., Ostertag, C., Dahan, M., & Masson, J. B. (2019). Virtual reality: Beyond visualization. *Journal of Molecular Biology*, *431*(7), 1315–1321. doi:10.1016/j.jmb.2019.01.033 PMID:30738026

Elmassah, S., & Hassanein, E. A. (2022). Digitalization and subjective wellbeing in Europe. Digital Policy, Regulation, and Governance.

Gaubert, J. (2021). Seoul to become the first city to enter the metaverse. What will it look like? *Euronews*. https://www.euronews.com/next/2021/11/10/seoul-to-become-the first-city-to-enter-the-metaverse-what-will-it-look-like

Gillieron, L. (2021). Facebook wants to lean into the metaverse. Here's what it is and how it will work. NPR. https://www.npr.org/2021/10/28/1050280500/what-metaverse-isa nd-how-it-will-work

Grider, D., & Maximo, M. (2021). *The metaverse: Web 3.0 virtual cloud economies*. Grayscale Research.

Harapan, H., Itoh, N., Yufika, A., Winardi, W., Keam, S., Te, H., Megawati, D., Hayati, Z., Wagner, A. L., & Mudatsir, M. (2020). Coronavirus disease 2019 (COVID-19): A literature review. *Journal of Infection and Public Health*, *13*(5), 667–673. doi:10.1016/j.jiph.2020.03.019 PMID:32340833

Hirsh-Pasek, K., Zosh, J., Hadani, H. S., Golinkoff, R. M., Clark, K., Donohue, C., & Wartella, E. (2022). *A whole new world: Education meets the metaverse*. Policy.

Hollensen, S., Kotler, P., & Opresnik, M. O. (2022). Metaverse–the new marketing universe. *The Journal of Business Strategy*. doi:10.1108/JBS-01-2022-0014

Ibáñez, M. B., & Delgado-Kloos, C. (2018). Augmented reality for STEM learning: A systematic review. *Computers & Education*, *123*, 109–123. doi:10.1016/j.compedu.2018.05.002

Jeon, H. J., Youn, H. C., Ko, S. M., & Kim, T. H. (2022). Blockchain and AI Meet in the Metaverse. *Advances in the Convergence of Blockchain and Artificial Intelligence, 73*. 10.5772/intechopen.99114

Joshua, J. (2017). Information Bodies: Computational Anxiety in Neal Stephenson's Snow Crash. *Interdisciplinary Literary Studies*, *19*(1), 17–47. doi:10.5325/intelitestud.19.1.0017

Kim, J. (2021). Advertising in the Metaverse: Research Agenda. *Journal of Interactive Advertising*, *21*(3), 141–144. doi:10.1080/15252019.2021.2001273

Kiong, L. V. (2022). *Metaverse Made Easy: A Beginner's Guide to the Metaverse: Everything you need to know about Metaverse*. NFT and GameFi.

Klopfer, E. (2008). *Augmented learning: Research and design of mobile educational games*. MIT press. doi:10.7551/mitpress/9780262113151.001.0001

Knox, J. (2022). The Metaverse, or the Serious Business of Tech Frontiers. *Postdigital Science and Education*, *4*(2), 207–215. doi:10.100742438-022-00300-9

Laeeq, K. (2022). *Metaverse: Why*. How, and What.

Lee, L.H., Braud, T., Zhou, P., Wang, L., Xu, D., Lin, Z., Kumar, A., Bermejo, C. & Hui, P. (2021). All one needs to know about metaverse: A complete survey on technological singularity, virtual ecosystem, and research agenda.

Milgram, P., Takemura, H., Utsumi, A., & Kishino, F. (1995). Augmented reality: A class of displays on the reality-virtuality continuum. In *Telemanipulator and telepresence technologies* (Vol. 2351, pp. 282–292). International Society for Optics and Photonics. doi:10.1117/12.197321

Moy, C., & Gadgil, A. (2022). Opportunities in the Metaverse: How Businesses Can Explore the Metaverse and Navigate the Hype vs. Reality. *Opportunities in the metaverse, 23*(02).

Murray, J. H. (2020). Virtual/reality: How to tell the difference. *Journal of Visual Culture*, *19*(1), 11–27. doi:10.1177/1470412920906253

Mustapha, U. F., Alhassan, A. W., Jiang, D. N., & Li, G. L. (2021). Sustainable aquaculture development: A review on the roles of cloud computing, internet of things and artificial intelligence (CIA). *Reviews in Aquaculture*, *13*(4), 2076–2091. doi:10.1111/raq.12559

Mystakidis, S. (2022). Metaverse. *Metaverse. Encyclopedia*, *2*(1), 486–497. doi:10.3390/encyclopedia2010031

Mystakidis, S., Christopoulos, A., & Pellas, N. (2021). A systematic mapping review of augmented reality applications to support STEM learning in higher education. *Education and Information Technologies*, 1–45.

Narin, N. G. (2021). A Content Analysis of the Metaverse Articles. *Journal of Metaverse*, *1*(1), 17–24.

Nižetić, S., Šolić, P., González-de, D. L. D. I., & Patrono, L. (2020). Internet of Things (IoT): Opportunities, issues, and challenges towards a smart and sustainable future. *Journal of Cleaner Production*, *274*, 122877. doi:10.1016/j.jclepro.2020.122877 PMID:32834567

Oliveira, T. A., Oliver, M., & Ramalhinho, H. (2020). Challenges for connecting citizens and smart cities: ICT, e-governance and blockchain. *Sustainability*, *12*(7), 2926. doi:10.3390u12072926

Openkov, M. Y., & Tetenkov, N. B. (2021). Digital Empires And The Annexation Of The Metaverse. 86-89.

Park, S., & Kim, S. (2022). Identifying World Types to Deliver Gameful Experiences for Sustainable Learning in the Metaverse. *Sustainability*, *14*(3), 1361. doi:10.3390u14031361

Pellas, N., Dengel, A., & Christopoulos, A. (2020). A scoping review of immersive virtual reality in STEM education. *IEEE Transactions on Learning Technologies*, *13*(4), 748–761. doi:10.1109/TLT.2020.3019405

Pellas, N., Mystakidis, S., & Kazanidis, I. (2021). Immersive Virtual Reality in K-12 and Higher Education: A systematic review of the last decade scientific literature. *Virtual Reality (Waltham Cross)*, *25*(3), 835–861. doi:10.100710055-020-00489-9

Ramesh, U. V., Harini, A., Gowri, C. S. D., Durga, K. V., Druvitha, P., & Kumar, K. S. (2022). International Journal of Research Publication and Review. Metaverse. *Future of the Internet.*, *3*(2), 93–97.

Silva, H., Resende, R., & Breternitz, M. (2018). Mixed reality application to support infrastructure maintenance. In *International Young Engineers Forum (YEF-ECE),* (pp. 50-54). IEEE. 10.1109/YEF-ECE.2018.8368938

Slater, M., & Sanchez-Vives, M. V. (2016). Enhancing our lives with immersive virtual reality. *Frontiers in Robotics and AI*, *3*, 74. doi:10.3389/frobt.2016.00074

Stephenson, N. (2003). *Snow crash: A novel*. Spectra.

Um, T., Kim, H., Kim, H., Lee, J., Koo, C., & Chung, N. (2022). Travel Incheon as a Metaverse: Smart Tourism Cities Development Case in Korea. In *ENTER22 e-Tourism Conference,* (pp. 226–231). Springer. doi:10.1007/978-3-030-94751-4_20

Wang, Q., Li, R., Wang, Q., & Chen, S. (2021). Non-fungible token (NFT): Overview, evaluation, opportunities and challenges.

Wang, Y., Su, Z., Zhang, N., Liu, D., Xing, R., Luan, T. H., & Shen, X. (2022). A survey on metaverse: Fundamentals, security, and privacy.

Yoder, M. (2022). An" OpenSea" of Infringement: The Intellectual Property Implications of NFTs. *The University of Cincinnati Intellectual Property and Computer Law Journal.*, *6*(2), 4.

Yonhap News Agency. (2021). Seoul to offer new concept administrative services via metaverse platform. *Korea Herald*. http://www.koreaherald.com/view.php?ud=20211103000692

Zuckerberg, M. (2021). *Connect 2021 Keynoate: Our Vision for the Metaverse*. Facebook.

Chapter 2
Artificial Intelligence and Blockchain Technology in the 4.0 IR Metaverse Era:
Implications, Opportunities, and Future Directions

Mohammad Rashed Hasan Polas
Sonargaon University, Bangladesh

Bulbul Ahamed
Sonargaon University, Bangladesh

Md. Masud Rana
Sonargaon University, Bangladesh

ABSTRACT

The advancement of SMEs is accelerated by technological expansions using blockchain technology in the Industrial Revolution (IR) 4.0 era. Based on current trends in AI and blockchain technology, this study proposes that the distance between entrepreneurs all over the world and their potential workers may be greatly decreased to virtually real-time. A secondary literature review is carried out in order to identify the key developments in IR 4.0 technologies in the SMEs industry, as well as the potential trend that will lead the business sector. The adoption of AI and blockchain technology in the IR 4.0 technologies is projected to make seeking treatments overseas more reasonable, accessible, and health records readily available on a real-time and protected basis. However, it is necessary to highlight that the expansion of SMEs raises the eyebrows of society from the security, social, and economic viewpoints.

DOI: 10.4018/978-1-6684-5732-0.ch002

INTRODUCTION

The latest industrial revolution known as "Industry 4.0" is the result of recent advancements in production techniques and automation. Data management, industrial competitiveness, production processes, and efficiency may all be included under the umbrella term "industry 4.0" (Alaloul et al., 2020). Cyber physical systems, the Internet of Things, artificial intelligence, big data analytics, and digital twins are just a few of the key enabling technologies that are referred to as part of the "Industry 4.0" movement. These technologies are seen as the main drivers of automated and digital manufacturing environments. The United Nations (UN) Sustainability 2030 agenda highlights sustainability as the cornerstone of corporate strategy and calls for smart manufacturing, energy-efficient construction, and low-impact industrialisation (Rakshit et al., 2022). Industry 4.0 technology aid in making company processes more sustainable. Globalization, mass customisation, and a competitive business climate are pressuring "traditional" sectors to embrace new business models and move toward Industry 4.0. Industry 4.0 technologies are a recent revolution in manufacturing that seeks to maximize productivity and efficiency while utilizing the least amount of resources possible. Industry 4.0 technologies have ushered in a new manufacturing trend in sectors that intended to maximize production through efficient resource usage. Industry 4.0's "smart manufacturing" or "digital manufacturing" can be seen as its fundamental component since it enables businesses to undertake flexible production processes with mass customisation (Alcácer & Cruz-Machado, 2019).

Artificial intelligence (AI) and blockchain technology, for example, have made significant contributions to economic growth and civilizational advancement. AI and blockchain technology are two technological developments with significant growth potential since they are altering the way economic transactions are conducted. Artificial intelligence and blockchain technology are already in widespread usage; it's just an issue of scale and perspective (Ariffin & Ahmad, 2021). After limiting their reach to local partners for so long, SMEs are now expanding their reach, particularly through digital technology, into a global strategy that is spreading in all directions. Individuals and small companies are increasingly being targeted by enterprises, and consumers and customers are becoming "cocreators" and actively engaging. Due to the limitations of blockchain technology, it must recreate itself on the basis of renewed trust and ethical concepts. Given that AI and BT are ubiquitous, it's interesting to observe how AI and blockchain technology interact with 4.0 IR, and one of the most intriguing features for academics is its ability to revolutionize innovation processes. One of the primary purposes of AI and blockchain technology is to assist researchers in rethinking their company and competitive strategies. Researchers can get closer to their aim by combining AI and BT with VR and AR (Merugula et al., 2021).

The current century is known as the Industrial Revolution 4.0. I.R. 4.0, which includes concepts such as the Internet of Things (IoT), blockchain technology, and artificial intelligence (AI), plays an important role and is widely recognized as a successful model. This chapter discusses the role of AI in advancing manufacturing technologies in metaverse-based SMEs. This new AI technology, which includes VR and AR, is having an influence not only on the communication and computing environment, but also on the industry's evolving employment skill requirements. Without a doubt, industrial demands and academic fields are intricately related. Robotics and artificial intelligence have an impact on nanotechnology, drones, sensor technologies, and computer vision, and as a result, SMEs are becoming more active (Tsang & Lee, 2022; Brodny & Tutak, 2022).

To begin, an overview of the AI and blockchain technology enablers, as well as the four analytics capabilities, is offered. Finally, upcoming research and development subjects are explored, such as making AI and blockchain technology available to SMEs, as well as the accompanying future trends and difficulties. Artificial intelligence (AI) and blockchain technology are only two of the technologies that comprise the fourth industrial revolution, often known as Industry 4.0. For years, both research and industry have researched Industry 4.0, with various firms and academic organizations seeking to categorize the technologies and methodology of the fourth industrial revolution. Businesses that operate in difficult environments are continuously seeking for methods to gain a competitive advantage, which usually takes the form of technical benefits, for example, the advantage of AR and VR. Small and medium-sized enterprises (SMEs) might apply artificial intelligence technology to reduce business risk to gain a competitive edge. To do so successfully, businesses must first have a better grasp of the technology, followed by a better understanding of the risk management framework and the incorporation of AI solutions into their business processes (Jamwal et al., 2021).

This chapter is split into two parts. The first portion includes an introduction, a history of the Industrial Revolution, and a description of the stages of artificial intelligence. It also provides an assessment of how far certain technologies have advanced in the business sector. Every country on the planet recognizes the impact of the industrial revolution on industries like mobile networks and internet-based social media. This chapter also shows how the AI-based algorithm boosted and targeted the sensitive brain, as well as its sympathetic expression at each level.

AUGMENTED AND VIRTUAL REALITY IN RELATION TO AI AND BLOCKCHAIN TECHNOLOGY

The phrase "augmented reality" refers to a set of technologies that allow people to perceive their real-world environment through computer visuals that have been altered or supplemented. A number of devices can be used to improve the visual qualities of the system's physical entities. Although there are several applications for augmented reality in Industry 4.0 to assist BT and AI, experts are primarily interested in the manufacturing sector. From design through prototype, maintenance operations, and assembly operations, real-time information is necessary at all phases of the product life cycle (Tambare et al., 2021). Augmented reality may be utilized to alleviate many of these stages' concerns through simulation, minimizing downtime for organizations and simplifying procedures. Simulation tools, smart machines, data analytics, and high connectivity among workstations have all created new potential for businesses to develop creative production units. These changes to industrial buildings increased industry production capacity and allowed it to adjust to changing market demands. Lee (2019) provided a real-time material flow model and architecture to address industrial difficulties. The proposed framework blends AR and manufacturing methods to address production challenges in large firms that use AI and BT. In terms of AI and BT, the study's findings revealed that simulation and augmented reality offered more accurate results for production modeling.

Jamwal et al. (2021) observed that VR technologies lower the amount of energy utilized during manufacturing in Industry 4.0 during their review of VR applications for energy-efficient manufacturing. Tambare et al. (2021) analyzed the global aviation manufacturing headquarters in the United States using surveys and interviews to provide a preliminary assessment of the organizational impact of AR and VR technologies when BT and AI are used. Javaid et al. developed an AR-based lesson assistance system for industrial applications. The proposed technique is effective for handling digital data while minimizing the need for human training. Brodny and Tutak (2022) examined some of the principal industrial uses of augmented reality (AR) technologies in the manufacturing industry, including assembly and maintenance operations. The major enablers for smart manufacturing processes have been identified as AR and VR technology. Malik et al.'s (2020) inquiry on the development and enhancement of VR technologies shows for human-centered manufacturing systems. The paper proposes a simulation-based human-centered virtual reality (VR) framework for manufacturing to reduce cycle time, robot control, layout optimization, and process plan creation. As part of an Industry 4.0 model-based manufacturing system, Mandic (2020) proposed merging AR and rapid technologies to design and construct a sheet metal process.

The Linkage between AI and Blockchain Technology with AR and VR

When discussing AI with blockchain technology with AR and VR, the term "coopetition" comes to mind. Indeed, with contemporary data science, classification algorithms, and artificial intelligence technologies, it is clear that rivals must pool their data in order to capitalize on a market opportunity (Tang & Veelenturf, 2019). To allow such sharing, a new governance model based on the notion of a shared data repository is required. Full blockchain technology is not appropriate for all use cases, is difficult to implement, and will be impossible to achieve if the network is subjected to intense competitive forces. Academics, on the other hand, believe that when blockchain technology, BT, and other intelligent open-source technologies are combined, the most transformative value may be generated. Transparency and privacy in the service relationship between humans and technology underpin the dynamic of individuals and organizations being trust-free in blockchain business services. Because every blockchain node stores transaction history eternally, AI and BT, with the aid of AR and VR, enable anybody to access the records of every transaction they do. Furthermore, because blockchain transactions are recorded using public and private keys (long sequences of characters that no one can read), individuals may stay anonymous while allowing other parties to authenticate their identities (Li et al., 2021).

INDUSTRIAL REVOLUTION 4.0: AN OVERVIEW

Many SMEs throughout the world, like Microsoft, Mac, Facebook, Opera, Dell, and others, are now integrating AI-enabled solutions to improve business operations and customer service. For example, the smart grid project is one of the pillars for integrating AI in SMEs; it enables spatial navigation through interactive and automated technologies that combine augmented reality and virtual reality to display the dynamics of the SMEs grid. Because of digitalization, SMEs are now better able to target individual demands, which has increased output and economic performance. Therefore, AI presents a chance for improved city administration; AI concepts and technology may affect and enhance how the SMEs functions. SMEs have clearly benefited much from using AI to develop with the aid of AR and VR and execute SMEs management techniques so far. However, it is believed that there is still a knowledge gap, particularly in how the public views the application of such technologies and how they feel about the widespread use of AI and BT in their SMEs. Policymakers might better comprehend how the general public feels about various parts of AI and BT by having a strong understanding of how people in their cities perceive these concepts and technology. As a result, governing bodies would be more ready to incorporate blockchain and AI technologies and to respond to public needs (Kalsoom et al., 2020).

Numerous essential characteristics of Industry 4.0, such as AR, VR, networking, and big data, have an impact on AI. The most crucial factor is connectivity. If the devices are not linked and data is obtained from multiple sensor systems present in controllers, robot arms, and end-of-arm tools, it is hard to gain any value. It is also critical to act in reaction to these statistics. Engineers might respond to past complaints and pattern thresholds on purpose to encourage maintenance activities. Modern computer models, such as networked mobile and interactive systems, will be utilised in tomorrow's intelligent factories. Artificial intelligence, cloud computing, and data processing may all assist industrial robots. The purpose of Industry 4.0-enabled robots is to maximize performance while minimizing downtime. More sensor-equipped robots are less sensitive to interference and are digitally linked. Unplanned plant downtime is one of the most important causes of today's production inefficiency. Industry 4.0 is expanding the automation of formerly manual operations via the use of a built-in set of real-time AI and computer networking. Industry 4.0 is coming together to help businesses maximize their investment returns (Stâncioiu, 2017).

Human input and customisation are crucial for industrial power automation, and Industry 4.0 facilitates this. Supervisors can direct robotic tasks via portable devices such as cell phones and tablets, bypassing traditional technology. It enables producers to stay adaptive in a fast-paced processing world when they were previously linked to a computer or fixed data infrastructure. Industry 4.0 is a big stride forward in equipment development and growth, which has already seen pioneering productivity advances in living quality for hundreds of years. Industry welcomes digital transformation by expanding intelligent manufacturing, intelligent factories, and Internet of Things applications. Machines now have more online access and are linked to the whole supply chain visualization system. Rapid improvements in automation technologies contribute in increased production.

Recent advances in AI technology have aided Industry 4.0. In industrial operations, AI technology combining AR and VR is used to detect anomalous behavior. This assumes that machine problems may be identified and addressed before a system fails. Failures spare plant operators and administrators the stress and cost troubles that come with them. As a consequence of predictable maintenance, equipment repair engineers' duties have also developed, allowing them to visit a spot before purchasing the essential components. They will collect sensor data from any area and classify the issue into the components needed to make their work more efficient and successful.

In the twenty-first century, industry is undergoing a revolution known as Industrial Revolution 4.0. Every century, at least one industrial revolution happens, altering technology. That technology has had a tremendous influence on human culture, skill sets, and economies throughout the world, resulting in the rise of new sectors and increasing economic opportunities. Since the commencement of the industrial revolution, steam power, coal, electricity, and a range of machinery, such as the spinning industry, have all proven to be useful to people and in the development of jobs. Initially dependant on animals for transportation, such as horses and buffalo, humans have gradually transitioned to train, airplane, and car means of transportation. Every industrial revolution reduces the gap between humans and machines.

I.R. 4.0 is on the verge of superhuman replication. These technologies are propelling smart cities, smart homes, and secure transactions such as blockchain technology and deep learning. AI technology pervades not just modern industry, but also a large element of the community's lifestyle. I.R. 4.0 seamlessly integrates AI, machine learning, and deep learning in all fields ranging from medicine to automobiles, just like I.R. 3.0 did with current concepts such as computer science, information technology, and telecommunications.

With a significant change within the mobility enhancement idea, IR 4.0 is generating a stir in the SME company. Thanks to mobile care, customer services will be more affordable and accessible, and health records will be more accessible and secure. Some instances of mobile devices and infrastructure are aiding in the development of the future. It has assisted many patients and physicians by optimizing processes, synthesizing data, and delivering real-time updates. The most recent technical breakthroughs will provide you a huge competitive advantage, especially when it comes to developing faster and more effective therapies.

THE EMERGENCE OF AI IN DEVELOPING SMART SMES

The aim of building smart SMEs is to minimize expenses, eliminate waste, and raise living conditions for residents by integrating numerous technological components into the system. The multiple disciplines of AI and IoT together provide a greater possibility for transforming traditional SMEs into smart ones. Various AI disciplines may be used to investigate how company owners use SMEs. AI-developed pattern recognition technology may be used to assess a significant amount of raw data, patterns, and hidden correlations collected from multiple sensors installed across the city for various reasons such as traffic monitoring, weather monitoring, and so on. Using AI technology to evaluate sensor data enables the development of prediction models that can be used to run apps and send commands to IoT devices. Deep learning can be used to monitor automobile speeds, read license plates, count the number of vehicles passing by and pedestrians walking along the street, and track vehicles parked in parking lots. A smart city built with AI and IoT can be used for a variety of purposes, such as smart environments, smart buildings, smart traffic management, smart grid, smart farming, smart waste management, smart health sectors, smart security systems, smart transportation, smart parking, drone delivery, smart postal services, smart disaster and calamity management, and more.

AI enhances interactions between professionals and consumers at a distance by offering necessary product or service ideas (Zigiene et al., 2019), making it a vital communication and engagement tool between customers and professionals on the other side of the planet. As part of its efforts to position Abu Dhabi as a unique health tourism destination, the United Arab Emirates (UAE) has taken AI seriously in regulating the development of its healthcare industry, introducing the first AI policy, postponing the UAE Artificial Intelligence Strategy, and appointing the world's first Minister of State for Artificial Intelligence.

The Revolution of Blockchain Technology in the SMEs

Blockchain technology is drawing the interest of a diverse variety of businesses, including energy firms, startups, technology developers, and financial institutions, as well as governments and academic institutions throughout the world, as an emerging technology. According to several industry insiders, blockchain technology has the potential to deliver significant benefits and innovation. Blockchain technology was inspired by decentralized digital currency, for example. The benefits of blockchain technology include decentralization and stability in distributed systems, as well as safe data storage and zero exchange transaction costs. A blockchain is a distributed ledger network in which nodes communicate with one another in order to exchange data and transactions. A blockchain must have both decentralization and immutability.

Blockchain technology may help the energy economy's operations, markets, and customers, making it more environmentally friendly as a result. Green energy can help to minimize greenhouse gas emissions and CO2 emissions generated by current use. Firms such as banks, manufacturers, small and medium-sized businesses, and agricultural product producers are already using blockchains to support green innovation that involves automation and remote sensors. By building a safe and private mesh network, blockchain technology has provided Bangladeshi green innovation with a new framework that eliminates the risks associated with central server design (Treiblmaier & Pan, 2022). Blockchain technology, which eliminates the need for traditional centralized cloud servers, enables a safe and low-power architecture for remote monitoring of physical activity. As a result, the ecosystem of the Internet of Things becomes more ecologically friendly.

THE IMPORTANCE OF BLOCKCHAIN TECHNOLOGY AND AI TOWARDS GREEN INNOVATION IN THE 4.0 IR

Due to their lower environmental effect than fossil fuels, renewable energy sources have recently emerged as a critical aspect for blockchain technology adoption. Renewable energy sources such as wind, geothermal heat, sunshine, and rain may all be utilized to create power (Bradu et al., 2022). The usage of these resources results in the production of clean, renewable energy. People may now create their own power without relying on expensive fossil fuels thanks to advancements in renewable energy technology [34]. This is where the smart grid comes in, allowing us to tap into all of the free and plentiful green energy sources in our immediate vicinity. Information technology and analogue or digital communications are used in the smart grid, which is a modernized electrical system (Tseng et al., 2021).

Furthermore, AI has significant challenges in terms of security, interoperability, and data transport. Renewable energy sources' variable and unpredictable nature needs the integration of complicated technologies into the present network (Hassoun et al., 2022). To encourage the use of renewable resources and build a better environment, a blockchain-based AI is being employed. The suggested paradigm is based on the inversion of the blockchain's recently disclosed forking attack. A central authority and dispersed peers represent stakeholders in the energy grid idea. This AI is promising, and it might be used to develop ethical smart grid systems in order to create a more environmentally friendly world (Dantas et al., 2021).

AI and blockchain technologies might be used to implement green credentials in an industrial operating system. The use of AI and blockchain to study the viability of peer-to-peer energy transfers was investigated. A blockchain-based AI system is thought to be capable of correctly assessing and monitoring the environmental effect of energy-related assets (Feng et al., 2022). As a result of blockchain's ability to enhance the energy supply chain, it is conceivable to go green. Various authors have advocated for new market models and the democratization of energy sources as a result of AI and Blockchain technologies enabling energy-sharing economy applications. When blockchains and smart contracts are joined, they offer transparent, tamper-proof, and secure platforms capable of delivering innovative, environmentally friendly business solutions (Rymarczyk, 2020).

Blockchain Technology in Enhancing Smart SMEs

Despite the fact that utilizing blockchain to construct smart SMEs is a relatively new concept, it helps to connect the services of smart SMEs while also boosting security and transparency. Blockchain technology protects and anonymizes data by encrypting and storing it in the cloud. This also reduces paper use, pollution, and waste. The blockchain network may be used in SMEs to store information required in the production system. This facilitates the retrieval of relevant information. It can also aid in billing and transaction processing, as well as smart grid energy sharing.

Integration of IR 4.0 technologies in the general SMEs industry can boost corporate development by revolutionizing customer service options and patterns, notably during the first check and subsequent follow-up phases (Li et al., 2021). This benefit is especially important for the elderly and those who may have difficulty moving about. Over the last several years, investment in digitalized and linked services has expanded, as has the broad usage of technology-enabled customer service, making the concept of the "Smart SME" a reality (Chen et al., 2021). In recent years, the push for the adoption of blockchain technology to improve the safety and effective use of consumer data within the SME business has gained traction (Khan et al., 2021). Customers' face-to-face consultations may be freed up by the use of technology, particularly for follow-up sessions after they return to their home country, resulting in a more convenient and cost-effective experience (Wang et al., 2019).

ARTIFICIAL INTELLIGENCE'S KEY BENEFITS/ADVANTAGES

AI may do many functions at the same time. Facility administrators can maximize industrial floor space while lowering hardware footprint by substituting a large multi-core processing capacity with an existing programmable logic controller. AI has been implemented into major food processing plants. The AI conducts a number of development tasks using vision technologies, cameras, and AI. They can cut, measure, pack, and palletize almost anything. Sensor-rich industries that use AR and VR will track machinery and manufacturing processes in real time to avoid inaccurate output and services. Machine vision robots can perform complicated optical tasks with pinpoint accuracy. Microscopic structural flaws or minor color changes may be identified to preserve performance quality (Kaushik, 2022).

Automation is becoming far more dependable than human labor. Robots are meticulously built such that production and services can be sustained indefinitely without the risk of human mistake. For decades, the business has used robots on floors. Lasers and cameras are included into the robots, allowing for high-precision welding (Belfiore et al., 2020). Automation's dependability allows manufacturers to decrease overall waste in the production process. When robotic systems may be improved or reallocated as the market model evolves to fulfill new operations, replacement costs are reduced. Industrial robots, although appearing to be an expensive initial investment, rapidly pay for themselves through decreased labor costs and quicker production cycles. Long-term operating and maintenance expenses would be cheaper than if an employee performed the same tasks (Ferreira et al., 2021).

The attachment of a smart camera on the end of a robotic arm opens up a wide range of applications since the arm will travel over the component being checked to monitor many characteristics. For robot-based inspection and guiding, machine vision includes a variety of advances to give practical outcomes from picture capture and interpretation (Al-Smadi et al., 2019). Machine vision inspection is a fast-growing technology that assesses surface defects, color, and presence/absence. The automotive sector has been a significant driver of industrial robotics, using the majority of today's robots. Body welding has been the most popular robotic operation in the automotive industry. At one stage during the process, two metal pieces are fused together to generate a combined fusion of both components. A strong electrical current is conveyed at a low voltage (Chahkoutahi & Khashei, 2017; Georgantzinos et al., 2021; Paschek et al., 2022).

In a variety of sectors, AI is often used to apply sealing material cables or adhesives. The material that has to be applied is a liquid or a paste. It's put in a tank and pushed up to the application gun, which houses the artificial intelligence that regulates the material flow (Aziz, 2014). Furthermore, AI is employed in the automobile sector to screw, assemble, mark, modify, and control quality. By automating production, industrial AI helps several manufacturing industries increase productivity and improve product quality. Picking, packing, and palletizing are all critical steps in the process. Work that is labor-intensive and time-consuming can be done hand-in-hand. It is unrealistic to expect humanity to function with infinite energy reserves. Furthermore, mistakes will be made throughout tasks (Ferreira et al., 2021).

Aspects of AI Critical For Implementing Industry 4.0

To put the Industry 4.0 theory into effect, many fundamental robotics elements are necessary. There are real-time features, historical trends, up-and-downtime monitoring, PLC and other alarms, critical performances, and so on. These recommended robotics-based features will increase the utility of robots while developing an industry 4.0 culture. These particular tasks also serve as a basis for adopting industry 4.0 cultures smoothly and properly. AI innovations improve the dependability and safety of new operations. Autonomous robots, cobots, interactive autonomous smart robs, humanoids, mobile robots, cloud robots, pick and place robots, and robotic swarms are among the most notable robotic technologies influencing growth. Robotics improves the ability to produce customized robots quickly and with greater precision and durability (Rakshit et al., 2022).

Human capital is frequently encouraged to focus on other high-priority or non-repeatable activities by robots. The networked flow of information in Industry 4.0 raises concerns regarding stability, openness, and privacy. Data processing operations may have a big influence on the brand's appeal both outside and within the shop floor as production processes become more personal and adaptive (Javaid et al., 2022). As a result, in order to prevent assaults on critical industrial infrastructure, the transmission and retrieval of sensitive industrial data must be done safely utilizing robots technology. Software ethics and protection, privacy-enhancing technologies, intelligent encryption, zero-trust security, and end-to-end contact security are some of the most recent advancements in this field. Integrity and secrecy must be coupled with cyber security. Industry 4.0 advancements are critical, and robots can help to accelerate them (Niewiadomski et al., 2019).

Various Industries Artificial Intelligence Solicitations in 4.0 Perspectives

It addresses a wide range of industry 4.0 problems, including agile assembly and production, training and learning processes, energy management, and so on. It demonstrates several Industry 4.0 methods to robotics applications. It then examines techniques to improve touch, including quality control, efficacy, error-free operation, soft grabbing, satisfaction, rapid processing, and downtime reduction. By merging complex real-time sensor and simulation data, digital twin technology generates virtual industrial asset models. Prospective digital doubles uses include model-driven programming, virtual prototyping, and virtual machine validation. The adoption of digital twins results in the industrial sector being hyper-integrated. Digital twins provide useful information at every level of the manufacturing process (Parente et al., 2020).

Industry 4.0, also known as Connected Industry, is a concept in which humans and robots may collaborate securely and share knowledge to improve operations and decision-making. The flexible and automated industrial automation necessary to develop Smart Factories is provided by mobile robots and mobile manipulators (Kalsoom et al., 2021). Knowledge sharing facilitated by integrating cutting-edge intelligent technology with robots, such as the Internet of Things, Artificial Intelligence, and Big Data, is the most valuable commodity. The merging of ICT and intelligent technology in the age of mobile robots has resulted in capabilities that extend their industrial uses; these abilities can analyze data, perform measures, and respond to a variety of working conditions (Mastos et al., 2021). Mobile robotics applications in metrology, quality management, servicing on big components or packing, cleaning, polishing, screwing, or drilling have all improved as a result of this (Eriksson et al., 2022).

ARTIFICIAL INTELLIGENCE'S MAJOR POTENTIAL SKILLS IN MANY FIELDS/INDUSTRIES

Advanced artificial intelligence and its ties to machine vision are critical to the advancement of Industry 4.0 projects and have an impact on every stage of the manufacturing process. Machine vision was crucial to the development of industrial AI, and the two grew more interwoven (Javaid et al., 2021). The main reason for this is that cameras in harsh production settings are now more powerful and dependable than they have ever been. In comparison, AI technologies have grown in popularity, indicating that this is the most profitable and successful method. The addition of a sophisticated camera to a robotic arm opens up a world of possibilities, since the arm will move the investigated component to check a variety of needs. For AI-based inspection, machine vision brings together a variety of techniques that yield useful results from picture gathering and interpretation (Ejsmont, 2021).

Artificial intelligence (AI) technologies provide building blocks for developing successful product testing, quality management, problem detection, and data gathering systems. These characteristics are ideal for manufacturers that want to increase efficiency and complete procedures with the least amount of human intervention (Xu & Lv, 2021). A vehicle painter's job is difficult, and booting is exceedingly dangerous. Due to a shortage of employment, finding qualified, experienced painters might be difficult. Because each coat of paint requires durability, the robotic arms will fill the vacuum. Robots should follow a predetermined path that covers large regions and reliably reduces waste. Adhesive, sealant, and other application machines are also useful. This work is ideal for large-scale production robotics. Smaller cobots also perform machinery tendering and loading/unloading operations for relatively minor production activities (Sterne & Razlogova, 2019).

VARIOUS INDUSTRIAL AI OPTIONS 4.0 ACTIVITIES AT THE BOTTOM LEVEL

AI is used in a range of industries for a variety of applications and purposes. These intended roles improve the overall efficacy of the process being implemented/explored in order to enable the industry 4.0 mindset across the board. It investigates numerous robotic versions that are used in the implementation of industry 4.0 cultures at various levels (Bag et al., 2021). Welding robots, material handling robots, pick and place robots, and other miscellaneous applications robots are only a few of the important fields. Collaborative, dispensing, plasma, and spot-welding robots, vision, press robots, assembly, paint, and routing robots are only a few examples of the different robots that allow industry 4.0 deployments (Felsberger et al., 2022). Montage and inspection are two of the most common industrial AI applications. AI in assembly is projected to expand as a result of the high cost of physical labor in these tasks. Because AI can be trained, one assembly work strategy is to produce multi-style batches and reprogram AI (Mittal et al., 2018). Another option is to group multiple product kinds together in a single mounting cell that each robot need. In computer process control, a digital computer is used to guide the procedures of a manufacturing process. Continuous material processing is widespread in computer control of various automatic device operations. In these activities, products are often created in the gas, liquid, or powder phase to facilitate material transfer between phases of the production cycle employing robots (Felsberger et al., 2022).

AI TECHNOLOGY USAGES IN CONTEXT TO INDUSTRY 4.0

Robotics is a cutting-edge technology that is applied in a variety of developing industries. Many personnel on the floor suffer from weariness, weakness, and other bodily discomforts as a result of their repetitive and boring jobs (Bécue et al., 2021). This technology benefits the floor, helps personnel do their responsibilities more successfully, and minimizes or eliminates any physical stress. Network and connection are two of the most important factors in enabling Industry 4.0 (Zhong et al., 2017). A variety of technological advancements, including edge-to-cloud, gigabit ethernet time-sensitive networks, wide-area low-power networks, 5G machine-to-machine connectivity, real-time determinist ethernet networking, omnipresent radio access and unified IoT platform, and zero-touch networks, are enabling factories to implement IIoT and become Industry 4.0 facilities. Machine-machine and human-machine connection and data transfer are constantly being improved to drive advancements in modern businesses. Network and connection are two of the most important elements in facilitating Industry 4.0. These advances are constantly improving the technologies employed in this revolution to aid in data transport (Sanchez et al., 2020).

As long as they are part of the product selection to be processed by the device, new design types will be put into production using a flexible manufacturing system. As a result, such a system is appropriate for items with low to medium demand, and demand changes are predicted (Bibby & Dehe, 2018). Although the flexible production technique is highly automated, people are still necessary for system administration, component loading and unloading, adjustment tools, and machinery maintenance and repair. Robotics is utilized in the industrial industry to increase flexibility (Keleko et al., 2022). Another area where robots are increasingly being employed in the manufacturing is inspection. When the robot inserts a sensor into a routine inspection task involving the work piece, it determines whether or not the component meets consistency standards. In each industrial robot application, different end-of-arm tools, basic reach, and payloads are required and adjustable. It keeps a large inventory and work cells, allowing us to quickly integrate. For a long time, the industrial sector has excelled in making substantial use of contemporary technology (Lu, 2017).

LIMITATIONS

AI can execute difficult and demanding tasks that humans are incapable of. Humans, on the other hand, can do far more than robots. One way humans outperform machines is through their capacity to make decisions. More significantly, they are stronger problem solvers who can think outside the box and adapt more flexibly (Alcácer & Cruz-Machado, 2019). Furthermore, many businesses rely on imagination, originality, and personality, which a robot-powered organization will lack. As a result, even once Industry 4.0 is fully established, human feedback will be required. Many people have raised concerns and misunderstandings regarding robots that take humans (Oztemel & Gursev, 2020). AI often requires significant investment at an early stage. When determining the business case for purchasing, the industry must consider all costs, including installation and setup. Industrial robotics requires specialized operation, maintenance, and programming (Frank et al., 2019). While the number of people with these skills is growing, it is still modest at the moment. As a result, the sector must consider staffing and obtaining this talent or tool for present personnel to carry out the purpose.

DIRECTIONS FOR FUTURE RESEARCH

The future factory has been taking shape for a long time. People and robots collaborate in everyday life, and product creation is getting more productive. Furthermore, new concepts are necessary for the increasingly automated intralogistics operations. Industry 4.0 will only work well if manufacturing and logistics systems are fully integrated. Artificial intelligence will ultimately make its way into a wide range of industrial applications. Robotics will efficiently do welding, fabrication, distribution, raw material processing, assembly, and packaging. AI will grow in popularity as more vendors provide it for a wider range of applications. This AI technology will be widely used in the automotive industry and will play an important role in many production processes. Industrial AI will enable the integration of a wide range of demanding processes on the manufacturing line while boosting operational flexibility. The employment of these technologies will change the essence of people's jobs. Electronic machines will do physically demanding jobs in order to enhance employee health and safety.

CONCLUSION

Industrial production is always improving to help businesses satisfy growing client demand while remaining competitive in the global market. Robotic robots are making advances into a range of industrial fields with the aid of AR and VR. As robotic devices become more affordable, customers will be able to purchase them in a range of combinations, with the potential to touch our lives in a variety of ways. Robotics applications in manufacturing have improved business security, quality, and sustainability. With the arrival of Industry 4.0, robotics integration benefits the manufacturing industry for a multitude of reasons, including reliability, precision, performance, and resilience to hazardous situations (Chang et al., 2021). It has the potential to enable more logical decision making in Industry 4.0. It may also be used with business processes to improve cooperation across several data platforms. It enables a more efficient and dependable manufacturing process. Many industrial production processes have been simplified by intelligent robots that work with exceptional precision and speed. Because of the current demand, highly flexible technologies that allow for regular product revisions at a cheap cost are required. As a result, factory robots have emerged as the most cost-effective choice for assembly automation. As a result, in the next years, automation will provide significant profit prospects to the industrial industry (Zhou et al., 2015; Müller et al., 2018; Aly et al., 2021).

REFERENCES

Al-Smadi, A. M., Alsmadi, M. K., Baareh, A., Almarashdeh, I., Abouelmagd, H., & Ahmed, O. S. S. (2019). Emergent situations for smart cities: a survey. *International Journal of Electrical & Computer Engineering (2088-8708), 9*(6).

Alaloul, W. S., Liew, M. S., Zawawi, N. A. W. A., & Kennedy, I. B. (2020). Industrial Revolution 4.0 in the construction industry: Challenges and opportunities for stakeholders. *Ain Shams Engineering Journal, 11*(1), 225–230. doi:10.1016/j.asej.2019.08.010

Albantani, A. M., & Madkur, A. (2019). Teaching Arabic in the era of Industrial Revolution 4.0 in Indonesia: Challenges and opportunities. *ASEAN Journal of Community Engagement, 3*(2), 12–31. doi:10.7454/ajce.v3i2.1063

Alcácer, V., & Cruz-Machado, V. (2019). Scanning the industry 4.0: A literature review on technologies for manufacturing systems. *Engineering science and technology, an international journal, 22*(3), 899-919.

Aly, M., Khomh, F., & Yacout, S. (2021). What do practitioners discuss about iot and industry 4.0 related technologies? characterization and identification of iot and industry 4.0 categories in stack overflow discussions. *Internet of Things, 14*, 100364. doi:10.1016/j.iot.2021.100364

Ariffin, K. A. Z., & Ahmad, F. H. (2021). Indicators for maturity and readiness for digital forensic investigation in era of industrial revolution 4.0. *Computers & Security, 105*, 102237. doi:10.1016/j.cose.2021.102237

Aziz, M. Y. (2014). Business intelligence trends and challenges. In *The Fourth International Conference on Business Intelligence and Technology (BUSTECH) 2014 Proceedings,* (pp. 1-7).

Bag, S., Gupta, S., & Kumar, S. (2021). Industry 4.0 adoption and 10R advance manufacturing capabilities for sustainable development. *International Journal of Production Economics, 231*, 107844. doi:10.1016/j.ijpe.2020.107844

Bécue, A., Praça, I., & Gama, J. (2021). Artificial intelligence, cyber-threats and Industry 4.0: Challenges and opportunities. *Artificial Intelligence Review, 54*(5), 3849–3886. doi:10.100710462-020-09942-2

Belfiore, M. P., Urraro, F., Grassi, R., Giacobbe, G., Patelli, G., Cappabianca, S., & Reginelli, A. (2020). Artificial intelligence to codify lung CT in Covid-19 patients. *La Radiologia Medica, 125*(5), 500–504. doi:10.100711547-020-01195-x PMID:32367319

Bibby, L., & Dehe, B. (2018). Defining and assessing industry 4.0 maturity levels–case of the defence sector. *Production Planning and Control, 29*(12), 1030–1043. doi:10.1080/09537287.2018.1503355

Bradu, P., Biswas, A., Nair, C., Sreevalsakumar, S., Patil, M., Kannampuzha, S., Mukherjee, A. G., Wanjari, U. R., Renu, K., Vellingiri, B., & Gopalakrishnan, A. V. (2022). Recent advances in green technology and Industrial Revolution 4.0 for a sustainable future. *Environmental Science and Pollution Research International*, 1–32. doi:10.100711356-022-20024-4 PMID:35397034

Brodny, J., & Tutak, M. (2022). Analyzing the Level of Digitalization among the Enterprises of the European Union Member States and Their Impact on Economic Growth. *Journal of Open Innovation*, *8*(2), 70. doi:10.3390/joitmc8020070

Chahkoutahi, F., & Khashei, M. (2017). A seasonal direct optimal hybrid model of computational intelligence and soft computing techniques for electricity load forecasting. *Energy*, *140*, 988–1004. doi:10.1016/j.energy.2017.09.009

Chang, S. C., Chang, H. H., & Lu, M. T. (2021). Evaluating industry 4.0 technology application in SMES: Using a Hybrid MCDM Approach. *Mathematics*, *9*(4), 414. doi:10.3390/math9040414

Chen, J., Chen, S., Liu, Q., & Shen, M. I. (2021). Applying blockchain technology to reshape the service models of supply chain finance for SMEs in China. *The Singapore Economic Review*, 1–18. doi:10.1142/S0217590821480015

Ching, K. H., Teoh, A. P., & Amran, A. (2020, November). A Conceptual Model of Technology Factors to InsurTech Adoption by Value Chain Activities. In *2020 IEEE Conference on e-Learning, e-Management and e-Services (IC3e),* (pp. 88-92). IEEE. 10.1109/IC3e50159.2020.9288465

Dantas, T. E., De-Souza, E. D., Destro, I. R., Hammes, G., Rodriguez, C. M. T., & Soares, S. R. (2021). How the combination of Circular Economy and Industry 4.0 can contribute towards achieving the Sustainable Development Goals. *Sustainable Production and Consumption*, *26*, 213–227. doi:10.1016/j.spc.2020.10.005

Dewi, M. V. K., & Darma, G. S. (2019). The Role of Marketing & Competitive Intelligence In Industrial Revolution 4.0. *Jurnal Manajemen Bisnis*, *16*(1), 1–12. doi:10.38043/jmb.v16i1.2014

Ejsmont, K. (2021). The Impact of Industry 4.0 on Employees—Insights from Australia. *Sustainability*, *13*(6), 3095. doi:10.3390u13063095

Eriksson, T., Bigi, A., & Bonera, M. (2020). Think with me, or think for me? On the future role of artificial intelligence in marketing strategy formulation. *The TQM Journal*, *32*(4), 795–814. doi:10.1108/TQM-12-2019-0303

Felsberger, A., Qaiser, F. H., Choudhary, A., & Reiner, G. (2022). The impact of Industry 4.0 on the reconciliation of dynamic capabilities: Evidence from the European manufacturing industries. *Production Planning and Control*, *33*(2-3), 277–300. doi:10.1080/09537287.2020.1810765

Feng, Y., Lai, K. H., & Zhu, Q. (2022). Green supply chain innovation: Emergence, adoption, and challenges. *International Journal of Production Economics*, *248*, 108497. doi:10.1016/j.ijpe.2022.108497

Ferreira, R., Pereira, R., Bianchi, I. S., & da Silva, M. M. (2021). Decision factors for remote work adoption: Advantages, disadvantages, driving forces and challenges. *Journal of Open Innovation*, *7*(1), 70. doi:10.3390/joitmc7010070

Frank, A. G., Dalenogare, L. S., & Ayala, N. F. (2019). Industry 4.0 technologies: Implementation patterns in manufacturing companies. *International Journal of Production Economics*, *210*, 15–26. doi:10.1016/j.ijpe.2019.01.004

Georgantzinos, S. K., Giannopoulos, G. I., & Bakalis, P. A. (2021). Additive manufacturing for effective smart structures: The idea of 6D printing. *Journal of Composites Science*, *5*(5), 119. doi:10.3390/jcs5050119

Guergov, S., & Radwan, N. (2021). Blockchain Convergence: Analysis of Issues Affecting IoT, AI and Blockchain. *International Journal of Computations, Information and Manufacturing (IJCIM), 1*(1).

Hamidi, S. R., Aziz, A. A., Shuhidan, S. M., Aziz, A. A., & Mokhsin, M. (2018, March). SMEs maturity model assessment of IR4. 0 digital transformation. In *International Conference on Kansei Engineering & Emotion Research,* (pp. 721-732). Springer, Singapore.

Hassoun, A., Aït-Kaddour, A., Abu-Mahfouz, A. M., Rathod, N. B., Bader, F., Barba, F. J., Biancolillo, A., Cropotova, J., Galanakis, C. M., Jambrak, A. R., Lorenzo, J. M., Måge, I., Ozogul, F., & Regenstein, J. (2022). The fourth industrial revolution in the food industry—Part I: Industry 4.0 technologies. *Critical Reviews in Food Science and Nutrition*, 1–17. doi:10.1080/10408398.2022.2034735 PMID:35114860

Jamwal, A., Agrawal, R., Sharma, M., & Giallanza, A. (2021). Industry 4.0 technologies for manufacturing sustainability: A systematic review and future research directions. *Applied Sciences (Basel, Switzerland)*, *11*(12), 5725. doi:10.3390/app11125725

Javaid, M., Haleem, A., Singh, R. P., & Suman, R. (2021). Substantial capabilities of robotics in enhancing industry 4.0 implementation. *Cognitive Robotics*, *1*, 58–75. doi:10.1016/j.cogr.2021.06.001

Javaid, M., Haleem, A., Singh, R. P., & Suman, R. (2022). Artificial intelligence applications for industry 4.0: A literature-based study. *Journal of Industrial Integration and Management*, *7*(01), 83–111. doi:10.1142/S2424862221300040

Jung, W. K., Kim, D. R., Lee, H., Lee, T. H., Yang, I., Youn, B. D., Zontar, D., Brockmann, M., Brecher, C., & Ahn, S. H. (2021). Appropriate smart factory for SMEs: Concept, application and perspective. *International Journal of Precision Engineering and Manufacturing*, *22*(1), 201–215. doi:10.100712541-020-00445-2

Kahle, J. H., Marcon, É., Ghezzi, A., & Frank, A. G. (2020). Smart Products value creation in SMEs innovation ecosystems. *Technological Forecasting and Social Change*, *156*, 120024. doi:10.1016/j.techfore.2020.120024

Kalsoom, T., Ahmed, S., Rafi-ul-Shan, P. M., Azmat, M., Akhtar, P., Pervez, Z., Imran, M. A., & Ur-Rehman, M. (2021). Impact of IoT on Manufacturing Industry 4.0: A new triangular systematic review. *Sustainability*, *13*(22), 12506. doi:10.3390u132212506

Kalsoom, T., Ramzan, N., Ahmed, S., & Ur-Rehman, M. (2020). Advances in sensor technologies in the era of smart factory and industry 4.0. *Sensors (Basel)*, *20*(23), 6783. doi:10.339020236783 PMID:33261021

Kaushik, M. (2022). Artificial Intelligence (Ai). In Intelligent System Algorithms and Applications in Science and Technology, (pp. 119-133). Apple Academic Press.

Keleko, A. T., Kamsu-Foguem, B., Ngouna, R. H., & Tongne, A. (2022). Artificial intelligence and real-time predictive maintenance in industry 4.0: a bibliometric analysis. *AI and Ethics*, 1-25.

Khairani, N. A., & Rajagukguk, J. (2019, December). Development of Moodle E-Learning Media in Industrial Revolution 4.0 Era. In *4th Annual International Seminar on Transformative Education and Educational Leadership (AISTEEL 2019),* (pp. 559-565). Atlantis Press.

Khan, S. A. R., Godil, D. I., Jabbour, C. J. C., Shujaat, S., Razzaq, A., & Yu, Z. (2021). Green data analytics, blockchain technology for sustainable development, and sustainable supply chain practices: Evidence from small and medium enterprises. *Annals of Operations Research*, 1–25. doi:10.100710479-021-04275-x

Khazaei, H. (2020). Integrating cognitive antecedents to UTAUT model to explain adoption of blockchain technology among Malaysian SMEs. *JOIV: International Journal on Informatics Visualization*, *4*(2), 85–90. doi:10.30630/joiv.4.2.362

Ko, T., Lee, J., & Ryu, D. (2018). Blockchain technology and manufacturing industry: Real-time transparency and cost savings. *Sustainability*, *10*(11), 4274. doi:10.3390u10114274

Kumar, A., Pujari, P., & Gupta, N. (2021). Artificial Intelligence: Technology 4.0 as a solution for healthcare workers during COVID-19 pandemic. *Acta Universitatis Bohemiae Meridionales*, *24*(1), 19–35. doi:10.32725/acta.2021.002

Kumar, S., Lim, W. M., Sivarajah, U., & Kaur, J. (2022). Artificial intelligence and blockchain integration in business: Trends from a bibliometric-content analysis. *Information Systems Frontiers*, 1–26. doi:10.100710796-022-10279-0 PMID:35431617

Lanzini, F., Ubacht, J., & De Greeff, J. (2021). Blockchain adoptioin factors for SMEs in supply chain management. *Journal of Supply Chain Management Science*, *2*(1-2), 47–68.

Lee, H. (2019). Real-time manufacturing modeling and simulation framework using augmented reality and stochastic network analysis. *Virtual Reality (Waltham Cross)*, *23*(1), 85–99. doi:10.100710055-018-0343-6

Leng, J., Ruan, G., Jiang, P., Xu, K., Liu, Q., Zhou, X., & Liu, C. (2020). Blockchain-empowered sustainable manufacturing and product lifecycle management in industry 4.0: A survey. *Renewable & Sustainable Energy Reviews*, *132*, 110112. doi:10.1016/j.rser.2020.110112

Li, Z., Chen, S., & Zhou, B. (2021). Electric vehicle peer-to-peer energy trading model based on smes and blockchain. *IEEE Transactions on Applied Superconductivity*, *31*(8), 1–4. doi:10.1109/TASC.2021.3091074

Lim, C. H., Lim, S., How, B. S., Ng, W. P. Q., Ngan, S. L., Leong, W. D., & Lam, H. L. (2021). A review of industry 4.0 revolution potential in a sustainable and renewable palm oil industry: HAZOP approach. *Renewable & Sustainable Energy Reviews*, *135*, 110223. doi:10.1016/j.rser.2020.110223

Lu, Y. (2017). Industry 4.0: A survey on technologies, applications and open research issues. *Journal of Industrial Information Integration*, *6*, 1–10. doi:10.1016/j.jii.2017.04.005

Luo, S., & Choi, T. M. (2022). Operational research for technology-driven supply chains in the industry 4.0 Era: Recent development and future studies. *Asia-Pacific Journal of Operational Research, 39*(01), 2040021. doi:10.1142/S0217595920400217

Malik, A. A., Masood, T., & Bilberg, A. (2020). Virtual reality in manufacturing: Immersive and collaborative artificial-reality in design of human-robot workspace. *International Journal of Computer Integrated Manufacturing, 33*(1), 22–37. doi:10.1080/0951192X.2019.1690685

Mandic, V. (2020). Model-based manufacturing system supported by virtual technologies in an Industry 4.0 context. In *Proceedings of 5th International Conference on the Industry 4.0 Model for Advanced Manufacturing,* (pp. 215-226). Springer, Cham. 10.1007/978-3-030-46212-3_15

Maryanti, N., Rohana, R., & Kristiawan, M. (2020). The principal's strategy in preparing students ready to face the industrial revolution 4.0. *International Journal of Educational Research, 2*(1), 54–69.

Mastos, T. D., Nizamis, A., Terzi, S., Gkortzis, D., Papadopoulos, A., Tsagkalidis, N., Ioannidis, D., Votis, K., & Tzovaras, D. (2021). Introducing an application of an industry 4.0 solution for circular supply chain management. *Journal of Cleaner Production, 300*, 126886. doi:10.1016/j.jclepro.2021.126886

Merugula, S., Dinesh, G., Kathiravan, M., Das, G., Nandankar, P., & Karanam, S. R. (2021, March). Study of Blockchain Technology in Empowering the SME. In *2021 International Conference on Artificial Intelligence and Smart Systems (ICAIS),* (pp. 758-765). IEEE. 10.1109/ICAIS50930.2021.9395831

Mittal, S., Khan, M. A., Romero, D., & Wuest, T. (2018). A critical review of smart manufacturing & Industry 4.0 maturity models: Implications for small and medium-sized enterprises (SMEs). *Journal of Manufacturing Systems, 49*, 194–214. doi:10.1016/j.jmsy.2018.10.005

Mohanta, B. K., Jena, D., Satapathy, U., & Patnaik, S. (2020). Survey on IoT security: Challenges and solution using machine learning, artificial intelligence and blockchain technology. *Internet of Things, 11*, 100227. doi:10.1016/j.iot.2020.100227

Mohd Salleh, N. H., Selvaduray, M., Jeevan, J., Ngah, A. H., & Zailani, S. (2021). Adaptation of industrial revolution 4.0 in a seaport system. *Sustainability, 13*(19), 10667. doi:10.3390u131910667

Morrar, R., Arman, H., & Mousa, S. (2017). The fourth industrial revolution (Industry 4.0): A social innovation perspective. *Technology Innovation Management Review, 7*(11), 12–20. doi:10.22215/timreview/1117

Müller, J. M., Kiel, D., & Voigt, K. I. (2018). What drives the implementation of Industry 4.0? The role of opportunities and challenges in the context of sustainability. *Sustainability, 10*(1), 247. doi:10.3390u10010247

Nica, E., Stan, C. I., Luțan, A. G., & Oașa, R. Ș. (2021). Internet of things-based real-time production logistics, sustainable industrial value creation, and artificial intelligence-driven big data analytics in cyber-physical smart manufacturing systems. *Economics, Management, and Financial Markets, 16*(1), 52–63. doi:10.22381/emfm16120215

Niewiadomski, P., Stachowiak, A., & Pawlak, N. (2019). Knowledge on IT tools based on AI maturity–Industry 4.0 perspective. *Procedia Manufacturing, 39*, 574–582. doi:10.1016/j.promfg.2020.01.421

Oztemel, E., & Gursev, S. (2020). Literature review of Industry 4.0 and related technologies. *Journal of Intelligent Manufacturing*, *31*(1), 127–182. doi:10.100710845-018-1433-8

Pan, X., Pan, X., Song, M., Ai, B., & Ming, Y. (2020). Blockchain technology and enterprise operational capabilities: An empirical test. *International Journal of Information Management*, *52*, 101946. doi:10.1016/j.ijinfomgt.2019.05.002

Parente, M., Figueira, G., Amorim, P., & Marques, A. (2020). Production scheduling in the context of Industry 4.0: Review and trends. *International Journal of Production Research*, *58*(17), 5401–5431. doi:10.1080/00207543.2020.1718794

Paschek, D., Luminosu, C. T., & Ocakci, E. (2022). Industry 5.0 Challenges and Perspectives for Manufacturing Systems in the Society 5.0. *Sustainability and Innovation in Manufacturing Enterprises*, 17-63.

Pereira, N., Lima, A. C., Lanceros-Mendez, S., & Martins, P. (2020). Magnetoelectrics: Three centuries of research heading towards the 4.0 industrial revolution. *Materials (Basel)*, *13*(18), 4033. doi:10.3390/ma13184033 PMID:32932903

Peres, R. S., Jia, X., Lee, J., Sun, K., Colombo, A. W., & Barata, J. (2020). Industrial artificial intelligence in industry 4.0-systematic review, challenges and outlook. *IEEE Access : Practical Innovations, Open Solutions*, *8*, 220121–220139. doi:10.1109/ACCESS.2020.3042874

Popkova, E. G., Ragulina, Y. V., & Bogoviz, A. V. (Eds.). (2019). *Industry 4.0: Industrial revolution of the 21st century,* (Vol. 169, p. 249). Springer.

Rafiola, R., Setyosari, P., Radjah, C., & Ramli, M. (2020). The effect of learning motivation, self-efficacy, and blended learning on students' achievement in the industrial revolution 4.0. [iJET]. *International Journal of Emerging Technologies in Learning*, *15*(8), 71–82. doi:10.3991/ijet.v15i08.12525

Ragulina, Y. V., Alekseev, A. N., Strizhkina, I. V., & Tumanov, A. I. (2019). Methodology of criterial evaluation of consequences of the industrial revolution of the 21st century. In *Industry 4.0: Industrial Revolution of the 21st Century,* (pp. 235–244). Springer. doi:10.1007/978-3-319-94310-7_24

Rahardja, U., Aini, Q., Graha, Y. I., & Tangkaw, M. R. (2019, December). Gamification framework design of management education and development in industrial revolution 4.0. *Journal of Physics: Conference Series*, *1364*(1), 012035. doi:10.1088/1742-6596/1364/1/012035

Raja, G. B. (2021). Impact of Internet of Things, Artificial Intelligence, and Blockchain Technology in Industry 4.0. In *Internet of Things, Artificial Intelligence and Blockchain Technology,* (pp. 157–178). Springer. doi:10.1007/978-3-030-74150-1_8

Rakshit, S., Islam, N., Mondal, S., & Paul, T. (2022). Influence of blockchain technology in SME internationalization: Evidence from high-tech SMEs in India. *Technovation*, *115*, 102518. doi:10.1016/j.technovation.2022.102518

Rocha, D. M., Brasil, L. M., Lamas, J. M., Luz, G. V., & Bacelar, S. S. (2020). Evidence of the benefits, advantages and potentialities of the structured radiological report: An integrative review. *Artificial Intelligence in Medicine*, *102*, 101770. doi:10.1016/j.artmed.2019.101770 PMID:31980107

Romero, D., Stahre, J., Wuest, T., Noran, O., Bernus, P., Fast-Berglund, Å., & Gorecky, D. (2016, October). Towards an operator 4.0 typology: a human-centric perspective on the fourth industrial revolution technologies. In proceedings of the international conference on computers and industrial engineering (CIE46), Tianjin, China, (pp. 29-31).

Roy, A., Boninsegni, M. F., Kumar, S., Peronard, J. P., & Reimer, T. (2020). Transformative Outcomes of Consumer Well-Being in the Era of IR 4.0: Opportunities and Threats of Physical, Biological and Digital Technologies Across Sectors. *Consumer Interests Annual, 66*.

Rymarczyk, J. (2020). Technologies, opportunities and challenges of the industrial revolution 4.0: Theoretical considerations. *Entrepreneurial Business and Economics Review*, *8*(1), 185–198. doi:10.15678/EBER.2020.080110

Rymarczyk, J. (2021). The impact of industrial revolution 4.0 on international trade. *Entrepreneurial Business and Economics Review*, *9*(1), 105–117. doi:10.15678/EBER.2021.090107

Sanchez, M., Exposito, E., & Aguilar, J. (2020). Industry 4.0: Survey from a system integration perspective. *International Journal of Computer Integrated Manufacturing*, *33*(10-11), 1017–1041. doi:10.1080/0951192X.2020.1775295

Shahroom, A. A., & Hussin, N. (2018). Industrial revolution 4.0 and education. *International Journal of Academic Research in Business & Social Sciences*, *8*(9), 314–319. doi:10.6007/IJARBSS/v8-i9/4593

Sherwani, F., Asad, M. M., & Ibrahim, B. S. K. K. (2020, March). Collaborative robots and industrial revolution 4.0 (ir 4.0). In *2020 International Conference on Emerging Trends in Smart Technologies (ICETST),* (pp. 1-5). IEEE. 10.1109/ICETST49965.2020.9080724

Silva, A. J., Cortez, P., Pereira, C., & Pilastri, A. (2021). Business analytics in industry 4.0: A systematic review. *Expert Systems: International Journal of Knowledge Engineering and Neural Networks*, *38*(7), e12741. doi:10.1111/exsy.12741

Stăncioiu, A. (2017). The Fourth Industrial Revolution 'Industry 4.0'. *Fiabilitate Şi Durabilitate*, *1*(19), 74–78.

Sterne, J., & Razlogova, E. (2019). Machine learning in context, or learning from landr: Artificial intelligence and the platformization of music mastering. *Social Media + Society*, *5*(2), 2056305119847525. doi:10.1177/2056305119847525

Sudiwedani, A., & Darma, G. S. (2020). Analysis of the effect of knowledge, attitude, and skill related to the preparation of doctors in facing industrial revolution 4.0. *Bali Medical Journal*, *9*(2), 524–530. doi:10.15562/bmj.v9i2.1895

Tambare, P., Meshram, C., Lee, C. C., Ramteke, R. J., & Imoize, A. L. (2021). Performance measurement system and quality management in data-driven Industry 4.0: A review. *Sensors (Basel)*, *22*(1), 224. doi:10.339022010224 PMID:35009767

Tang, C. S., & Veelenturf, L. P. (2019). The strategic role of logistics in the industry 4.0 era. *Transportation Research Part E, Logistics and Transportation Review*, *129*, 1–11. doi:10.1016/j.tre.2019.06.004

Treiblmaier, H., & Špan, Ž. (2022). Will blockchain really impact your business model? Empirical evidence from Slovenian SMEs. *Economic and Business Review*, *24*(2), 132–140. doi:10.15458/2335-4216.1302

Tsang, Y. P., & Lee, C. K. M. (2022). Artificial intelligence in industrial design: A semi-automated literature survey. *Engineering Applications of Artificial Intelligence*, *112*, 104884. doi:10.1016/j.engappai.2022.104884

Tseng, M. L., Tran, T. P. T., Ha, H. M., Bui, T. D., & Lim, M. K. (2021). Sustainable industrial and operation engineering trends and challenges Toward Industry 4.0: A data driven analysis. *Journal of Industrial and Production Engineering*, *38*(8), 581–598. doi:10.1080/21681015.2021.1950227

Wang, R., Lin, Z., & Luo, H. (2019). Blockchain, bank credit and SME financing. *Quality & Quantity*, *53*(3), 1127–1140. doi:10.100711135-018-0806-6

Xu, Y., & Lv, L. (2021, August). A New Method of Free Combat Teaching Based on Artificial Intelligence. *Journal of Physics: Conference Series*, *1992*(4), 042010. doi:10.1088/1742-6596/1992/4/042010

Zhong, R. Y., Xu, X., Klotz, E., & Newman, S. T. (2017). Intelligent manufacturing in the context of industry 4.0: A review. *Engineering*, *3*(5), 616–630. doi:10.1016/J.ENG.2017.05.015

Zhou, K., Liu, T., & Zhou, L. (2015, August). Industry 4.0: Towards future industrial opportunities and challenges. In *2015 12th International conference on fuzzy systems and knowledge discovery (FSKD)*, (pp. 2147-2152). IEEE.

ADDITIONAL READING

AlShamsi, M., Salloum, S. A., Alshurideh, M., & Abdallah, S. (2021). Artificial intelligence and blockchain for transparency in governance. In *Artificial intelligence for sustainable development: Theory, practice and future applications,* (pp. 219–230). Springer. doi:10.1007/978-3-030-51920-9_11

Angelis, J., & Da Silva, E. R. (2019). Blockchain adoption: A value driver perspective. *Business Horizons*, *62*(3), 307–314. doi:10.1016/j.bushor.2018.12.001

Baabdullah, A. M., Alalwan, A. A., Slade, E. L., Raman, R., & Khatatneh, K. F. (2021). SMEs and artificial intelligence (AI): Antecedents and consequences of AI-based B2B practices. *Industrial Marketing Management*, *98*, 255–270. doi:10.1016/j.indmarman.2021.09.003

Bhatt, P. C., Kumar, V., Lu, T. C., & Daim, T. (2021). Technology convergence assessment: Case of blockchain within the IR 4.0 platform. *Technology in Society*, *67*, 101709. doi:10.1016/j.techsoc.2021.101709

Wagire, A. A., Joshi, R., Rathore, A. P. S., & Jain, R. (2021). Development of maturity model for assessing the implementation of Industry 4.0: Learning from theory and practice. *Production Planning and Control*, *32*(8), 603–622. doi:10.1080/09537287.2020.1744763

Wanto, A., & Hardinata, J. T. (2020). Estimations of Indonesian poor people as poverty reduction efforts facing industrial revolution 4.0. *IOP Conference Series. Materials Science and Engineering*, *725*(1), 012114. doi:10.1088/1757-899X/725/1/012114

Wogu, I. A. P., Misra, S., Assibong, P. A., Ogiri, S. O., Damasevicius, R., & Maskeliunas, R. (2018). Super-Intelligent Machine Operations in Twenty-First-Century Manufacturing Industries: A Boost or Doom to Political and Human Development? In *Towards Extensible and Adaptable Methods in Computing,* (pp. 209–224). Springer. doi:10.1007/978-981-13-2348-5_16

Žigienė, G., Rybakovas, E., & Alzbutas, R. (2019). Artificial intelligence based commercial risk management framework for SMEs. *Sustainability*, *11*(16), 4501. doi:10.3390u11164501

Chapter 3
Exploring the Incredible Potential and Opportunity of the Metaverse World

Mitali Chugh
University of Petroleum and Energy Studies, India

Sonali Vyas
https://orcid.org/0000-0003-2348-3394
University of Petroleum and Energy Studies, India

ABSTRACT

Technology is playing a very important role in transforming the industry. Currently, the advancement has been to Web3 technology, where we are using mutual interaction online in order to communicate with each other, or to organize the meetings or any social gathering with the help of internet. In the evolution of technology from radio to internet, metaverse is the current and upcoming technology supporting web3. Metaverse is expanding exponentially, the implementation of new technology is necessary for the progression to deliver the best experience using AR (augmented reality), VR (virtual reality), XR (extended reality). This chapter discusses the incredible potentials in which one can achieve outstanding results for implementing Metaverse technology, or to establish connection and connect with people in more live manner (in virtual spaces) and to achieve real-time experience.

INTRODUCTION

In simplest language "Metaverse" can be defined as a space where users(people) can traverse in a virtual environment that is completely similar to the real world using the latest technologies such as augmented reality (AR), virtual reality (VR), artificial intelligence (AI). The word Metaverse is created from two words i.e., The Greek word "meta" (means the post or beyond) and "verse" refers to the universe. It can be termed a multiuser environment consisting integration of physical reality with digital virtual (Mystakidis,2022). Metaverse is a hypothetical synthetic environment linked to the physical world. The

DOI: 10.4018/978-1-6684-5732-0.ch003

Metaverse is a hugely versatile, determined organization of interconnected virtual universes zeroed in on continuous cooperation where individuals can work, socially collaborate, execute, play, and even make. Many techno enthusiasts trust that the ideal cutting-edge adaptation of "The Metaverse" there would be one single stage where you have your persona, your character, and stage administrations associated under which numerous virtual spaces get made where you can get entrance (Lee, L. H. et.al, 2021). Like a world with many sub-universes which you can join, leave, or even make. Significant variables are as yet that there is a definition for a computerized character, advanced proprietorship, advanced monetary standards, and the general adaptability of advanced resources - Thus empowering a completely working economy in a virtual world. This way the Metaverse could supplant a few parts of how the travel industry functions, going on a show, how to find workmanship presentations yet particularly likewise how individuals learn, study, communicate and, surprisingly, meet companions (Nevelsteen, 2018). Metaverse is the new era of technology, while there is no reasonable assent about the meanings of Web3 and the distinction of the Metaverse, there is a lot of research supporting Metaverse.

The technological convergence of the Metaverse: With the fourth industrial revolution integration of integrated microchips, memory units, networks, sensor-based devices, and software applications allows the creation of a new edge technology generally known as the evolution of web-1, web-2, and web-3. The cutting-edge technologies brought a revolution by transforming independent technologies such as IoT (internet of things), big-data, robotics, artificial intelligence, blockchain, and virtual reality i.e., integrating physical, digital, and virtual worlds (Harwood, T., 2011).

Figure 1. Into the Metaverse

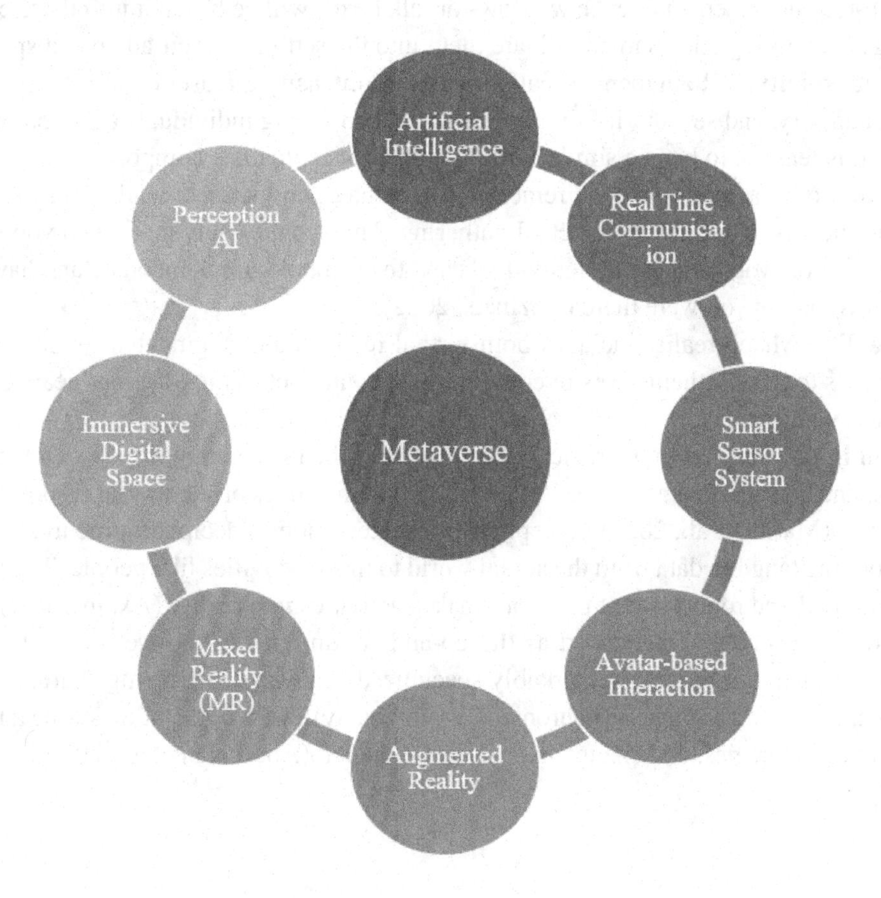

Into the Metaverse, Figure 1 explains the Metaverse as a combination of different technologies to enable social and fiscal activity. Which includes social interaction with avatars, different sensors for creating a real-time environment, and opportunities for earning. Different supporting technologies make Metaverse more live i.e., creating immersive digital space, ai (artificial intelligence), real-time networks and communication, augmented reality, and mixed reality.

Immersive Digital Space: The immersive digital space is an interactive environment that is a mixture of the digital and physical worlds. Like avatars in games or on social media websites. This integration of virtual environments in which creating avatars, help the users to identify and socialize users with each other in the digital world.

Artificial Intelligence: The new version of AI (Generation AI) is developing the technology by using less complex code tools, which will take AI to next level in the Metaverse (Kim, J., 2021). Generative AI will address that by taking inputs from clients and rapidly creating content to give the ideal Metaverse background. Generation AI can make content that is significant in light of regular data sources like language or visual data. For instance, a client could provide voice orders and have a Metaverse environment made or take photographs of a genuine item from various points to have its similarity can be imported digitally.

Real-Time Communication: In the world of superfast internet and network connectivity real-time communication plays a very important role in supporting the Metaverse for creating value in the current generation. On the off chance that the Metaverse will be a decent spot to team up, correspondence should feel as normal as it does in reality. Sound will be a significant piece of the correspondence experience, expanded by expressive symbols and text. Users will likewise have to speak with individuals, not in the Metaverse. More usual videoconference windows or talk boxes will be brought into these conditions as 2D technology, permitting clients to coordinate them into the setting of their advanced space.

Augmented Reality: AR augmented reality is used to intensify the user experience by adding layers of visual, auditory, and sensory information. Rather than seeing individuals on a screen with some video blocks, it is feasible to be in a similar virtual room, conceptualize, compose on a whiteboard and even change the room as per your requirements. Augmented workspaces would join these elements, permitting individuals to partake in an actual gathering. This would imply that when you have 3D images in the room, you would encounter individuals and the symbols simultaneously and have the option to collaborate as though you were there (Kozinets, 2022).

Mixed Reality: Mixed reality includes both virtual realities and augmented reality. Both include permitting clients to drench themselves in computerized content utilizing different gears or with a cell phone for a less vivid impact.

Perception in AI: Perception in artificial intelligence is the most common way of decoding vision, sounds, smell, and contact. Perception assists with building machines or robots that respond like people. (Cheong, 2022), (Marini et al., 2022) Perception is an interaction to decipher, procure, select, and afterward sort out the tangible data from the actual world to make activities like people. The fundamental contrast between AI and robots is that the robot makes activities into reality (Akour, 2022).

The Metaverse can be best perceived as the up-and-coming age of the web: it will expand upon and iteratively change it. It requires remarkably specialized headways (delivering shared, relentless reenactments that a great many clients synchronized continuously), and will include a strong administrative structure, business strategies, and change in customer conduct. (Aburbeian et al., 2022).

RELATED WORK

Metaverse was initially utilized in Neil Stevenson's sci-fi novel Snow Crash in 1992 and alluded to an existence where reality and the virtual collaborate and make esteem via different social exercises (Rillig et al., 2022). Using the mark of the Metaverse is extensive and ceaselessly developing, different descriptions and comparative ideas exist.

Figure 2. Metaverse World

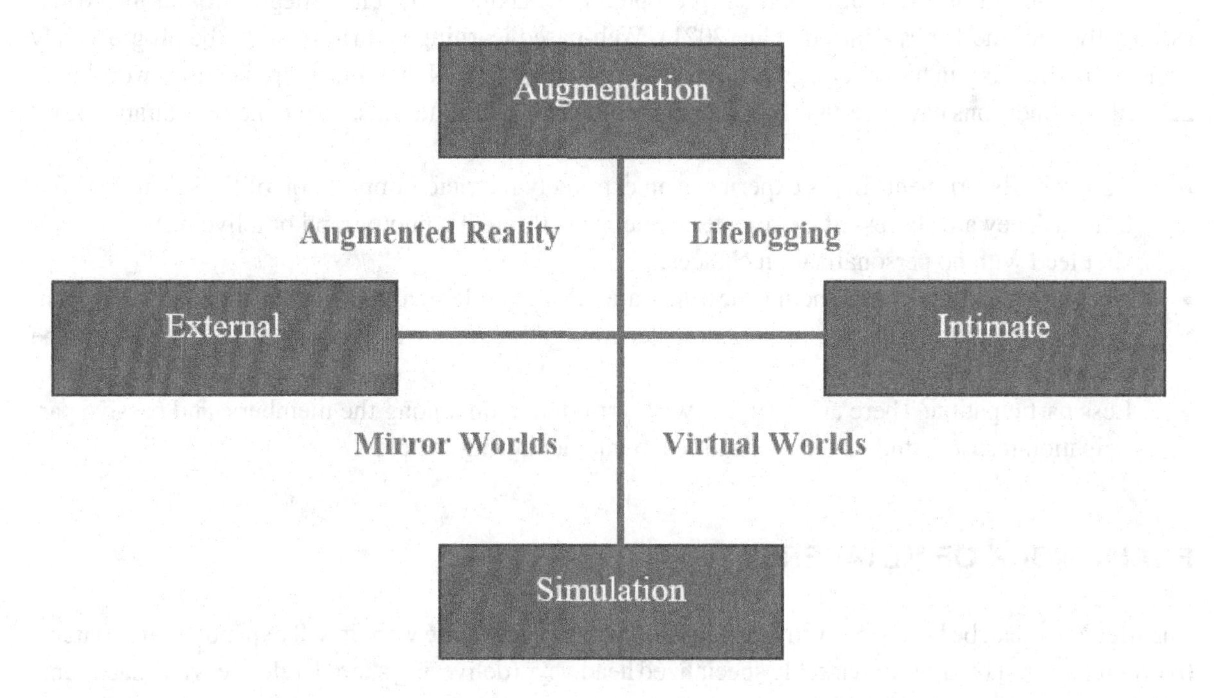

The initial article on Metaverse deliberates on strategic issues and the use of Metaverse (Lee, S. G., et.al, 2011). The Metaverse is divided into virtual, augmented reality, lifelogging, and mirror world, as depicted in Figure 2 depending on whether the site used is virtual or real, or whether the information used is external or internal. Metaverse is motivated on building the virtual environment, instead as per recent studies, it has been portrayed as an exchange of interests and a center for the social media content (Čopič Pucihar, K., & Kljun, M., 2018). As these days' vehicles are equipped with strong computational limits and high-level sensors, associated vehicles with 5G or much further developed organizations might go past the vehicle-to-vehicle associations, and ultimately interface by Metaverse (Riegler, A., Riener, A., & Holzmann, C., 2021). The current applications are countless for example, Computer-Aided Design (CAD) for item planning and constructing models, brilliant metropolitan preparation, AI-helped modern frameworks, and robot-upheld risky activities. To understand the Metaverse, advances other than the Internet, interpersonal organizations, gaming, and virtual conditions, ought to be taken into contemplation. The approach of VR and AR, rapid organizations AI, Edge Computing, and Blockchain act as the main components of the Metaverse. (Kress, B. C., & Peroz, C., 2020). The Metaverse depends on advancements that empower multisensory associations with virtual conditions, computerized items, and individuals.

The illustrative constancy of the XR framework is empowered by 3-D showcases that can express the view of understanding (El Beheiry et al., 2019). We have reviewed the existing literature by taking the example of online learning which can be improved by using technologies that support Metaverse.

LIMITATIONS OF 2-D LEARNING ENVIRONMENT

Virtual distance education is related to the development and theory of Open Education (Bailenson, 2021). The Open Education development prompted the making of Open Colleges around the world, essentially after the 1960s (Barteit et al., 2021). Web-based learning is turning out to be progressively standard particularly in higher and grown-up, ceaseless schooling. Nonetheless, applications working in 2D, online conditions have irrefutably factual impediments and limitations. Three major limitations are:

- Low self-discernment: Users experience an extremely restricted impression of the self in 2D conditions. They are addressed as immaterial elements through a photograph or a live webcam headshot feed with no personalization choices.
- No presence: Web conferencing meetings are seen as video calls to join as opposed to virtual aggregate gathering places. Members in lengthy gatherings will quite often incline out and be occupied.
- Less participation: There are very less ways for interaction among the members, and passive participation restricts students' opportunities to interact during the class.

FRAMEWORK OF METAVERSE

The Metaverse can be best perceived as the up-and-coming age of the web: it will expand upon and iteratively change it. It requires remarkably specialized headways (delivering shared, relentless re-enactments that a huge number of clients synchronized progressively), and will include a strong administrative structure and business strategies. (Chia, 2022) Metaverse is very dynamic and developing in the technology space, with all the advancements the most recent version of the internet is Metaverse. (Duan, H., et.al, 2021) The Metaverse can be utilized effectively by E-Learning (i.e., E-Learning, M-Learning, Blended Learning, Virtual Learning, Distance Learning, and Online Learning) as an answer for the subjects that rely absolutely upon a combination and can't be shown on the web or in distance learning, such as clinical and designing courses. Even though E-Learning conditions have a large number of ways to be delivered. (Dahan et.al, 2022) There are many existing frameworks, that were implemented with many different applications for example Gaming or education. As Metaverse is now taking place in diverse fields there is a requirement for different frameworks according to match the different models (Han et al., 2021).

Figure 3. Metaverse Building Blocks

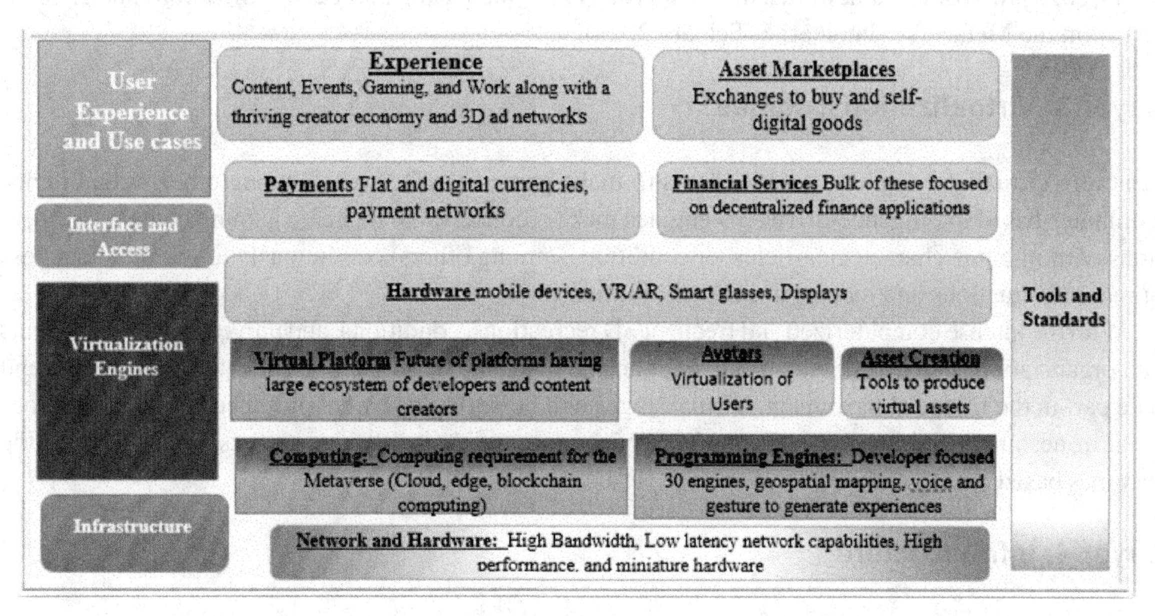

According to (Dionisio et al., 2013) there are five main layers for the basic framework of the Metaverse that adapts all the technical aspects to meet the requirements of the different types of users. The discussed layers are shown in Figure 3:

Layer 1: User Experience and use cases.
Layer 2: Interface and Access.
Layer 3: Virtualization Engines.
Layer 4: Infrastructure.
Layer 5: Tools and Standards

Layer 1: User Experience and Use Cases

The creation, deal, exchange, capacity, instalment, and monetary administration of advanced resources. This contains all business and administrations "based on top of" or potentially which "administration" the Metaverse, and which are not upward coordinated into a virtual stage, including content which is fabricated explicitly for the Metaverse, free of virtual stages. This includes various services like asset marketplaces (for buying and selling goods), Payments (Digital payments and cryptocurrencies), financial services, content, games, movies, and work applications provided by big enterprises (Cai et al., 2022).

Layer 2: Interface and Access

This is the main layer that helps us to use the Metaverse, this layer includes everything which is mobile devices or AR and VR devices. Mobile phones (smartphones) are just getting all the more extraordinary. With additional scaling down (size of chipsets), sensors, inserted AI, and low-inertness admittance to strong edge processing frameworks, they'll ingest an ever-increasing number of utilizations and encounters from the Metaverse (Jungherr & Schlarb 2022).

Layer 3: Virtualization Engines

Centrally virtualized engines, permit clients to make games or 3D spaces, among other virtual items. Computer-based intelligence-driven 3D engines make it conceivable to create gaming content (connections with in-game characters, articles, or conditions, among others) continuously to permit an endless number of situations and make games more intuitive.

With the intense computerized and frequent 3D recreations, conditions, and universes wherein clients and organizations can investigate, make, mingle, and take part in a wide assortment of encounters, and take part in the financial movement. Virtual stages will be portrayed by the presence of a huge environment of designers and content makers who produce most of the content as well as gather most of the incomes based on top of the concealed stage (Balica et al., 2022).

Layer 4: Infrastructure

The Metaverse will be based on the provisioning of steady, constant associations, high transfer speed, and decentralized information transmission by spine suppliers, the organizations, trade focuses, and benefits that course among them, as well as those making due 'last mile' information to shoppers. 5G organizations will decisively further develop transfer speed while decreasing organization disputes and dormancy. 6G will speed up by one more significant degree. Completely new advances, business lines, and administrations are being created to take special care of the developing requirement for ongoing data transmission applications. Utilizing the high configuration hardware for the more performance-oriented tasks (Balica et al., 2022).

Layer 5: Tools and Standards

The tools, conventions, arrangements, administrations, and standards act as genuine norms for interoperability and empower the development, activity, and progress upgrades to the Metaverse. These norms support exercises, for example, material science, AI, resource designs, the similarity of the board, tooling, and data of the executives (Valaskova, Machova, & Lewis, 2022).

All of these layers create a well-defined structure for the Metaverse, Big Giants like Meta (Facebook), Microsoft, and Google are working to make the best of the technology to improve advanced user experience.

METAVERSE-ENABLING TECHNOLOGIES

Metaverse is a universe of computerized items, virtual symbols, and utilitarian economies, where innovation is more than a utility; it's a lifestyle. Metaverse is a non-actual universe where individuals can impart through different types of virtual innovation. The Metaverse of the VR world comprises a variety of virtual stages and conditions connected to individuals from around the globe. There is no question that this innovation is staying put for quite a while. The key Technologies of Metaverse are shown in Figure 4.

Figure 4. Key Technologies of Metaverse

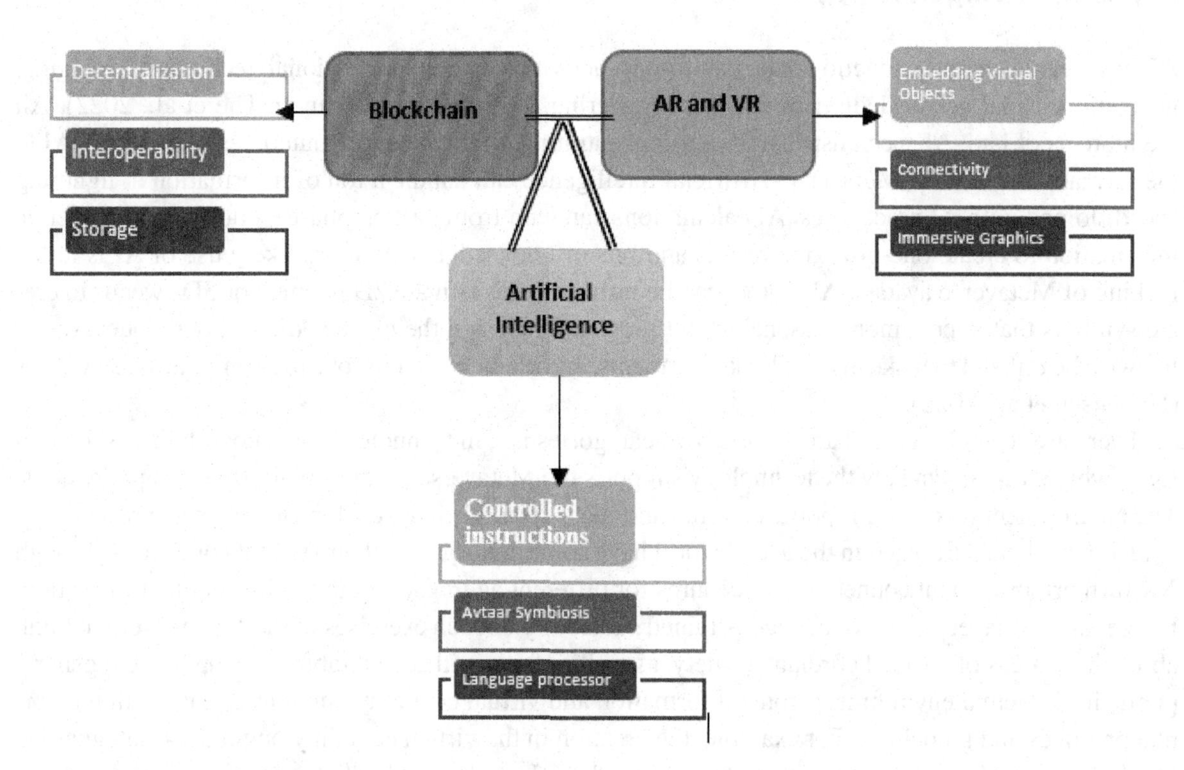

Cryptocurrency and Blockchain

Blockchain innovation gives a distributed and straightforward answer for computerized verification of proprietorship, advanced collectability, move of significant worth, administration, openness, and interoperability (Gadekallu et al., 2022). Cryptographic forms of money empower clients to move esteem although they exert and associate in the 3D world. Afterward, cryptocurrency might boost individuals to make use of the Metaverse. As additional organizations take the workplaces virtually for remote working resulting in the rise of relative Metaverse positions.

Virtual Reality (VR) and Augmented Reality (AR)

AR and VR can give us a vivid connection with 3D familiarity. AR utilizes computerized visual components and characters to transform this present reality. It's further open than VR and may be utilized on practically any cell phone or computerized gadget with a camera. Through AR uses, clients can see their environmental elements with intuitive computerized visuals. VR works unexpectedly. Likewise, Metaverse delivers a PC-produced virtual environment. Clients can then investigate it utilizing VR gloves, headsets, and sensors (MacCallum & Parsons, 2019). How AR and VR work displays an initial prototype of the Metaverse. VR is as of now making a computerized world that integrates fictitious visual substance. As its innovation turns out to be fuller grown, VR can extend the Metaverse understanding to include actual reproductions with VR hardware. Clients will want to sense, receive and collaborate with individuals from different areas of the world.

Artificial Intelligence (AI)

AI has been generally functional in day-to-day activities lately: professional technique arranging, navigation, facial acknowledgment, quicker registering, and then some (Huynh-The et al., 2022). All the more as of late, AI specialists have been concentrating on the potential outcomes of applying AI to the formation of a vivid Metaverse. Artificial intelligence can handle a ton of information at lightning speed. Joined with AI procedures, AI calculations can gain from past emphases, considering authentic information to create one-of-a-kind results and bits of knowledge. One more likely use of AI is in the making of Metaverse avatars. AI motors can be utilized to investigate 2D pictures or 3D sweeps to create symbols that appear more reasonable and precise. To design the cycle additionally unique, AI can likewise be utilized to make diverse looks, garments, and elements to improve the computerized systems (Hollensen et al., 2022).

There are several focussed areas under two categories i.e., in technology and ecosystem as shown in fig.5, where it is shown how the technology supports the Metaverse ecosystem for diverse applications. The environment (ecosystem) portrays a free and meta-sized virtual world, reflecting this present reality. Human clients arranged in the actual world have some control over their symbols or avatars through XR furthermore, client connection procedures for different aggregate exercises like content formation. Hence, the virtual economy is an unconstrained subsidiary of such exercises in the Metaverse. We think about three areas of societal cordiality, safety, as well as faith, and accountability. Simple to the general public in the actual environment, content formation and virtual economy ought to line up with the normal practices and guidelines. For example, the creation in the virtual economy ought to be safeguarded by possession, while such creation results ought to be acknowledged by different symbols (i.e., human clients) in the Metaverse, also there is a huge risk for humans to rely upon the Metaverse as some of the privacy risks and security threats can occur. Figure 5 depicts Metaverse Technologies and Ecosystem.

Figure 5. Ecosystem and Technology of the Metaverse

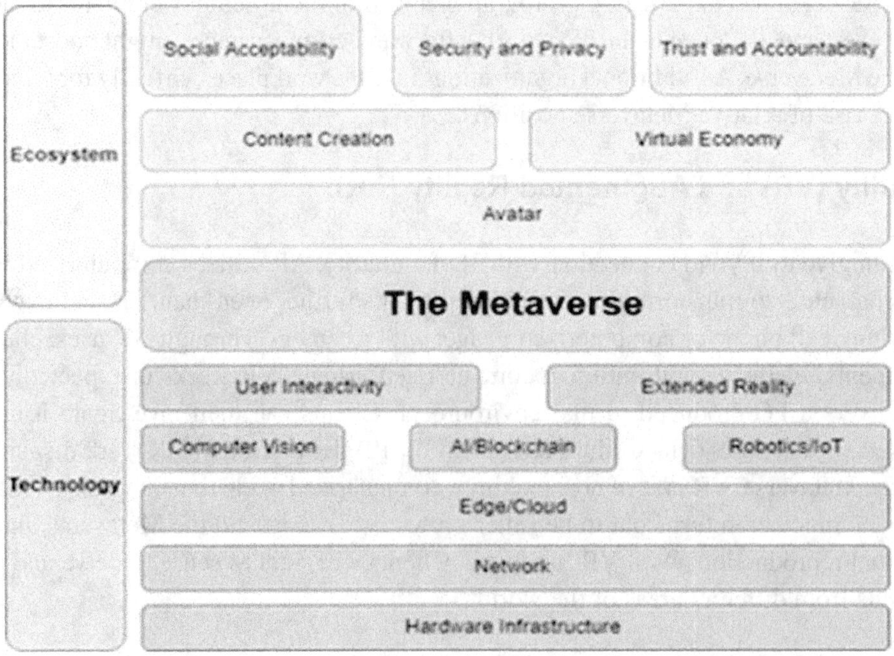

METAVERSE DOMAINS AND OPPORTUNITIES

Immersive multi-dimensional telepresence through AR or VR equipment is seen by a larger number of people as a basic empowering influence to conveying an authentic Metaverse experience. Rather than being on a PC, in the Metaverse, you could utilize a headset to enter a virtual world that interfaces a wide range of computerized conditions. In this situation, an individual would have the option to detect different clients possessing that equivalent common area as opposed to taking a gander at faces on a level screen show. AR and VR could in this way mirror the legitimate step past the cell phone and the ongoing versatile web period (Rillig, M. C., et.al, 2022). Metaverse is having a versatile technological aspect that favors developing new technologies with a different perspective. There is different sort of fields in which Metaverse can be utilized as a tool for advancement and progression to deliver the best results. Today there are numerous singular use cases and items, all making their renditions of a Metaverse. Valuable open doors across different enterprises include, including advanced education, clinical, military, and different sorts of exchanges can convey a more vivid growth opportunity (Zhao et al., 2022). They don't have to make their foundation, as the Metaverse will give the structure. Virtual cases, having acquired prominence throughout recent years, can now introduce more incorporated contributions. Retail can stretch out its scope to a vivid shopping experience that considers more intricate items (Dincelli, E., & Yayla, A., 2022), (Park & Kim 2022).

According to a business viewpoint, innovation administration organizations will require an empowering administrative environment that encourages development. According to a populace viewpoint, legislatures will require procedures to cultivate reception, which will incorporate reasonable customer admittance to the innovation and network that they need to take part in the Metaverse (Ning et al., 2021).

There is no lack of revolutionary thoughts for medical care in the Metaverse, from utilizing Blockchain innovation to appropriately 'claiming' and overseeing and sharing clinical records to the birthing of a 'computerized twin' for the occasion, yet additionally at a future age. Metaverse gives patients a vivid encounter enough to be utilized in psychotherapy. Individuals know that fantasies and books are not practical, however, they are moved. Also, Metaverse isn't this present reality yet can give a substantial feeling, so benefits because of vivid client intelligent stories can give (Wang et al., 2022).

METAVERSE CHALLENGES AND IMMINENT SCOPES

Metaverse is another kind of Internet utilization and social structure that coordinates an assortment of novel advances (Hwang & Chien, 2022). It gives a vivid encounter because of augmented reality innovation, makes a perfect representation of this present reality in light of computerized twin innovation, fabricates a financial framework because of blockchain innovation, and firmly coordinates the virtual world and this present reality into the monetary framework, the social framework, and the personality framework, permitting every client to deliver content (Leenes, 2007). There are several challenges in Metaverse. Some of them are discussed below:

Interface Problem

The connecting devices by which the technology is being used are not that energy efficient or the compatibility between the intermediate devices is not up to the mark, which lead sometimes in improper results. The transparency between the devices and interactivity is the major challenge. While utilizing the technology directly upon users to provide services, can affect the quality of the service, or delivery of service may not be assured.

Computational Problems

Computing power alludes to the capacity to handle data, not set in stone by evaluation, storage, and data transmission. It is essential to Compute power efficiency in the advanced economy period, its offices are a significant help for technological development. Metaverse implies a bigger number of clients, more extravagant organization assets, and computing assets, and computing power is a significant help for Metaverse. Metaverse requirements to constantly further develop handling rate, intricacy, and power utilization (Bauer et al., 2019).

Ethical Problems

The Metaverse has given individuals another character and made a new, exceptionally free space forever and exercises. It contains more convoluted social connections. Right now, the job of ethics becomes vital. The first set of principles has been impacted, and the definition of the encoding of ethics is lingering behind, and can't stay aware of the improvement of the Metaverse. Hence, the management of the Metaverse ought to be fortified, and significant regulations and guidelines ought to be formed and refreshed as quickly as possible (Papagiannidis et al., 2008).

Privacy Problems

The Metaverse is firmly connected to this present reality and compares to the genuine personality. As the development of another age of organizations, the Metaverse should assess data security assurance issues, very much like the past group environment.

Consequently, users need to understand the unintended return compromises of taking part in intense conditions, know about digital dangers, and lead their exploration before going into the Metaverse. (Papagiannidis et al., 2008) The scope of Metaverse is huge, arising advances have carried employees numerous helpful options in contrast to conventional work in the Metaverse. With the utilization of VR, the employees or workforce will be changed. Users or employees will never again need to pick either remote work or going to the workplace because the workplace will be available at any place.

CONCLUSION

In this chapter, we have comprehensively investigated and discussed the incredible potentials and opportunities of Metaverse. Metaverse is still under improvement, and various developers and researchers are researching its potential. We have discussed important models that are used to implement the Metaverse and can be used as a future technology to work upon. We have listed major challenges that can be faced while using Metaverse. The improvements are constant with the help of surrounding technologies such as (AR, VR, and XR) which in turn enhance the performance and practicality of the Metaverse services in the virtual world. The potential of Metaverse has a lot to do from now and it will add value to the life of people by creating a whole different virtual world that is similar to the real.

REFERENCES

Aburbeian, A. M., Owda, A. Y., & Owda, M. (2022). A Technology Acceptance Model Survey of the Metaverse Prospects. *AI, 3*(2), 285-302.

Akour, I. A., Al-Maroof, R. S., Alfaisal, R., & Salloum, S. A. (2022). A conceptual framework for determining metaverse adoption in higher institutions of gulf area: An empirical study using hybrid SEM-ANN approach. *Computers and Education: Artificial Intelligence*, *3*, 100052. doi:10.1016/j.caeai.2022.100052

Bailenson, J. N. (2021). Nonverbal overload: A theoretical argument for the causes of Zoom fatigue.

Balica, R. Ş., Majerová, J., & Cuţitoi, A. C. (2022). Metaverse Applications, Technologies, and Infrastructure: Predictive Algorithms, Real-Time Customer Data Analytics, and Virtual Navigation Tools. *Linguistic and Philosophical Investigations*, 21.

Barteit, S., Lanfermann, L., Bärnighausen, T., Neuhann, F., & Beiersmann, C. (2021). Augmented, mixed, and virtual reality-based head-mounted devices for medical education: Systematic review. *JMIR Serious Games*, *9*(3), e29080. doi:10.2196/29080 PMID:34255668

Bauer, T., Antonino, P. O., & Kuhn, T. (2019, May). Towards architecting digital twin-pervaded systems. In *IEEE/ACM 7th International Workshop on Software Engineering for Systems-of-Systems (SESoS) and 13th Workshop on Distributed Software Development, Software Ecosystems and Systems-of-Systems (WDES),* (pp. 66-69). IEEE. 10.1109/SESoS/WDES.2019.00018

Cai, Y., Llorca, J., Tulino, A. M., & Molisch, A. F. (2022). Compute-and data-intensive networks: The key to the Metaverse.

Cheong, B. C. (2022). Avatars in the metaverse: potential legal issues and remedies. *International Cybersecurity Law Review*, 1-28.

Chia, A. (2022). The metaverse, but not the way you think: Game engines and automation beyond game development. *Critical Studies in Media Communication*, *39*(3), 1–10. doi:10.1080/15295036.2022.2080850

Čopič Pucihar, K., & Kljun, M. (2018). ART for art: augmented reality taxonomy for art and cultural heritage. In *Augmented reality art,* (pp. 73–94). Springer. doi:10.1007/978-3-319-69932-5_3

Dahan, N. A., Al-Razgan, M., Al-Laith, A., Alsoufi, M. A., Al-Asaly, M. S., & Alfakih, T. (2022). Metaverse Framework: A Case Study on E-Learning Environment (ELEM). *Electronics (Basel)*, *11*(10), 1616. doi:10.3390/electronics11101616

Dincelli, E., & Yayla, A. (2022). Immersive virtual reality in the age of the Metaverse: A hybrid-narrative review based on the technology affordance perspective. *The Journal of Strategic Information Systems*, *31*(2), 101717. doi:10.1016/j.jsis.2022.101717

Dionisio, J. D. N., III, W. G. B., & Gilbert, R. (2013). 3D virtual worlds and the metaverse: Current status and future possibilities. *ACM Computing Surveys (CSUR), 45*(3), 1-38.

Duan, H., Li, J., Fan, S., Lin, Z., Wu, X., & Cai, W. (2021, October). Metaverse for social good: A university campus prototype. In *Proceedings of the 29th ACM International Conference on Multimedia*, (pp. 153-161). 10.1145/3474085.3479238

El Beheiry, M., Doutreligne, S., Caporal, C., Ostertag, C., Dahan, M., & Masson, J. B. (2019). Virtual reality: Beyond visualization. *Journal of Molecular Biology*, *431*(7), 1315–1321. doi:10.1016/j.jmb.2019.01.033 PMID:30738026

Gadekallu, T. R., Huynh-The, T., Wang, W., Yenduri, G., Ranaweera, P., Pham, Q. V., & Liyanage, M. (2022). Blockchain for the Metaverse: A Review.

Han, Y., Niyato, D., Leung, C., Miao, C., & Kim, D. I. (2021). A dynamic resource allocation framework for synchronizing metaverse with IoT service and data.

Harwood, T. (2011). Convergence of online gaming and e-commerce. In *Virtual worlds and e-commerce: Technologies and applications for building customer relationships,* (pp. 61–89). IGI Global. doi:10.4018/978-1-61692-808-7.ch004

Hollensen, S., Kotler, P., & Opresnik, M. O. (2022). Metaverse–the new marketing universe. *The Journal of Business Strategy*. doi:10.1108/JBS-01-2022-0014

Huynh-The, T., Pham, Q. V., Pham, X. Q., Nguyen, T. T., Han, Z., & Kim, D. S. (2022). Artificial Intelligence for the Metaverse: A Survey.

Hwang, G. J., & Chien, S. Y. (2022). Definition, roles, and potential research issues of the metaverse in education: An artificial intelligence perspective. *Computers and Education: Artificial Intelligence*, 100082.

Jungherr, A., & Schlarb, D. B. (2022). The Extended Reach of Game Engine Companies: How Companies Like Epic Games and Unity Technologies Provide Platforms for Extended Reality Applications and the Metaverse. *Social Media+ Society, 8*(2), 20563051221107641.

Kim, J. (2021). Advertising in the Metaverse: Research agenda. *Journal of Interactive Advertising*, *21*(3), 141–144. doi:10.1080/15252019.2021.2001273

Kozinets, R. V. (2022). Immersive netnography: A novel method for service experience research in virtual reality, augmented reality and metaverse contexts. *Journal of Service Management*. doi:10.1108/JOSM-12-2021-0481

Kress, B. C., & Peroz, C. (2020, February). Optical architectures for displays and sensing in augmented, virtual, and mixed reality (AR, VR, MR). *Proceedings of the Society for Photo-Instrumentation Engineers*, *11310*, 1131001.

Lee, L. H., Braud, T., Zhou, P., Wang, L., Xu, D., Lin, Z., & Hui, P. (2021). All one needs to know about metaverse: A complete survey on technological singularity, virtual ecosystem, and research agenda.

Lee, S. G., Trimi, S., Byun, W. K., & Kang, M. (2011). Innovation and imitation effects in Metaverse service adoption. *Service Business*, *5*(2), 155–172. doi:10.100711628-011-0108-8

Leenes, R. (2007, August). Privacy in the Metaverse. In *IFIP International Summer School on the Future of Identity in the Information Society,* (pp. 95–112). Springer.

MacCallum, K., & Parsons, D. (2019, September). Teacher perspectives on mobile augmented reality: The potential of metaverse for learning. In *World Conference on Mobile and Contextual Learning*, (pp. 21-28).

Marini, A., Nafisah, S., Sekaringtyas, T., Safitri, D., Lestari, I., Suntari, Y., Umasih, Sudrajat, A., & Iskandar, R. (2022). Mobile Augmented Reality Learning Media with Metaverse to Improve Student Learning Outcomes in Science Class. *International Journal of Interactive Mobile Technologies*, *16*(7), 99–115. doi:10.3991/ijim.v16i07.25727

Mystakidis, S. (2022). Metaverse. *Metaverse. Encyclopedia*, *2*(1), 486–497. doi:10.3390/encyclopedia2010031

Nevelsteen, K. J. (2018). Virtual world, defined from a technological perspective and applied to video games, mixed reality, and the Metaverse. *Computer Animation and Virtual Worlds*, *29*(1), e1752. doi:10.1002/cav.1752

Ning, H., Wang, H., Lin, Y., Wang, W., Dhelim, S., Farha, F., & Daneshmand, M. (2021). A Survey on Metaverse: the State-of-the-art, Technologies, Applications, and Challenges.

Papagiannidis, S., Bourlakis, M., & Li, F. (2008). Making real money in virtual worlds: MMORPGs and emerging business opportunities, challenges and ethical implications in metaverses. *Technological Forecasting and Social Change*, *75*(5), 610–622. doi:10.1016/j.techfore.2007.04.007

Park, S. M., & Kim, Y. G. (2022). A Metaverse: Taxonomy, components, applications, and open challenges. *IEEE Access : Practical Innovations, Open Solutions*, *10*, 4209–4251. doi:10.1109/ACCESS.2021.3140175

Riegler, A., Riener, A., & Holzmann, C. (2021). Augmented reality for future mobility: Insights from a literature review and hci workshop. *i-com, 20*(3), 295-318.

Rillig, M. C., Gould, K. A., Maeder, M., Kim, S. W., Dueñas, J. F., Pinek, L., Lehmann, A., & Bielcik, M. (2022). Opportunities and Risks of the "Metaverse" For Biodiversity and the Environment. *Environmental Science & Technology*, *56*(8), 4721–4723. doi:10.1021/acs.est.2c01562 PMID:35380430

Rillig, M. C., Gould, K. A., Maeder, M., Kim, S. W., Dueñas, J. F., Pinek, L., Lehmann, A., & Bielcik, M. (2022). Opportunities and Risks of the "Metaverse" For Biodiversity and the Environment. *Environmental Science & Technology*, *56*(8), 4721–4723. doi:10.1021/acs.est.2c01562 PMID:35380430

Valaskova, K., Machova, V., & Lewis, E. (2022). Virtual Marketplace Dynamics Data, Spatial Analytics, and Customer Engagement Tools in a Real-Time Interoperable Decentralized Metaverse. Linguistic and Philosophical Investigations, 21.

Wang, Y., Su, Z., Zhang, N., Liu, D., Xing, R., Luan, T. H., & Shen, X. (2022). A survey on metaverse: Fundamentals, security, and privacy.

Zhao, Y., Jiang, J., Chen, Y., Liu, R., Yang, Y., Xue, X., & Chen, S. (2022). *Metaverse: Perspectives from graphics, interactions and visualization*. Visual Informatics.

Chapter 4
The Fourth Illusion:
How a New Economy of Consumption Is Being Created in the Metaverse

Alexandre Ruco
https://orcid.org/0000-0002-8863-6606
Institute of Accounting and Administration, Institute Politecnico de Coimbra, Portugal

ABSTRACT

This chapter aims to analyze the business models' opportunity represented by the Metaverse (Foundation, n.d.), proposing a framework composed of four interdependent blocks; the "illusions" (Slater, 2003). The word Metaverse has been used to describe a sort of thing, mainly related to virtual reality (VR) (Lanier, 1992) and the non-fungible tokens (NFTs). The immersive experience of virtual reality is characterized by three illusions: the illusion of place, the illusion of embodiment, and the illusion of plausibility (Slater et al., 2009). This chapter describes the use of blockchain technology, namely the NFTs, as the origin of a fourth illusion.

INTRODUCTION

This chapter aims to analyze the business model opportunities offered by the Metaverse (Everything You Need to Know about the Metaverse. | Foundation, n.d.) and proposes a framework composed of four interdependent blocks; the "illusions" (Slater, 2003). The word *metaverse* has been used to describe a type of thing mainly related to virtual reality (VR) (Lanier, 1992) and non-fungible tokens (NFTs). The immersive experience of virtual reality is characterized by three illusions: the illusion of place, the illusion of embodiment, and the illusion of plausibility (Slater, 2009) (Slater et al., 2009). In this chapter, the use of blockchain technology, namely NFTs, is described as the origin of a fourth illusion. Together, the four illusions make the metaverse a highly potentially profitable business.

The place illusion (Pi) (Slater, 2009). is the feeling of being in a virtual place even though you know you are not there. The plausibility illusion (Psi) is the belief that the events you see on VR are actually happening. The embodiment illusion (Ei) is the perception that the virtual body, the avatar, is the user's

DOI: 10.4018/978-1-6684-5732-0.ch004

actual body. Once immersed in this digital experience, the user can interact with an unreal environment as if it were real.

An immersive experience where pi and psi occur is sufficient for users to react realistically to the VR (Slater, 2009). The more realistic the user's reactions, the stronger Pi and Psi can be. Strong Pi depends on a number of factors, including the ability of VR to respond interactively to users' actions. It is a sensory rather than an intellectual phenomenon. Stronger Psi, on the other hand, depends on a cognitive factor, probability. Once interrupted, Pi can be resumed. An interruption of psi caused by an unexpected fact that lacks probability usually interrupts the experience.

Immersive experiences have been offered by the entertainment industry for decades with great profit. Books, cinema, theme parks, games and others are among the biggest industries in the world. VR offers this kind of experience in a deeper, much more immersive way. Competition within the entertainment industry has been the main avenue for VR startups and businesses - at least until the advent of blockchain.

The first generation of the blockchain, represented by Bitcoin, is a combination of different technologies and mathematical concepts (Nakamoto, 2009). Cryptography, distributed computing and game theory work together to create a way to shop, exchange and create value in a secure, independent, persistent and decentralized way. This value is represented by cryptographic codes called tokens. Bitcoin was created by a man, woman or group under the pseudonym Satoshi Nakamoto and solved two main problems of digital assets.

The first is what is called the double spending problem. How can I know that a digital currency sent to me as payment for a service is not copied and sent to someone else? Or how can I know that the platform VR, which is selling me this house, is not selling the same house to someone else?

In the Bitcoin White Paper (2009), Nakamoto proposes a solution to the double-spending problem: a peer-to-peer network with blocks of timestamps of each transaction cryptographically hashed into subsequent blocks. Each transaction record consumes a lot of computing power, and every change to a transaction in a past block means a change in all subsequent hashes. The computing power required to maliciously alter the ledger makes fraud mathematically impossible.

As soon as a consumer in the metaverse buys a new and unique car, dress, or glasses for his avatar and this transaction is recorded in the permanent ledger of a blockchain, he can feel like the real and only owner of that good - even knowing that the good is virtual.

The second problem of digital assets that the blockchain has solved is the problem of persistence. As a decentralised network, Bitcoin cannot be shut down by a server error, moved to another URL, destroyed by hackers, discontinued due to a management decision, or blocked by a government act.

The same cannot be said about websites, traditional databases and digital goods hosted on a server or even in the computer cloud. The Bitcoin network provides value to the participants who register the blocks - they are called miners. Once you buy a fraction of a bitcoin (a "satoshi"), that satoshi is always yours because you can never sell it or transfer it to another wallet. The house, car and sunglasses of a customer's avatar on a Metaverse platform are always registered as belonging to their wallet.

The most important principle in economics is scarcity (Robbins, 1932). It was not present in the digital world before the blockchain. An image of a dollar bill could be infinitely reproduced. The value of an infinite good is always the same: zero. Bitcoin, the first blockchain, solved this problem by creating an immutable and transparent digital ledger through which every fraction of Bitcoin belongs to one person - one wallet - and no one else. Bitcoin brought scarcity to the digital world.

The second generation of the blockchain (What is Ethereum? n.d.) also offers the possibility to shop and process not only cryptocurrencies, but any kind of data. This immutable, permanent and secure data storage has given rise to the concept of NFT - Non-Fungible Tokens. Since cryptocurrencies are fungible - they can be merged and separated without losing their properties - NFTs are unique, indivisible, permanent digital assets that belong to someone (a wallet) (Find an Ethereum Wallet, n.d.).

In combination with VR, the second-generation blockchain has given rise to the metaverse. A virtual dimension in which a person can reside, visit places and, thanks to NFTs, also become the owner of things, places and even virtual bodies. This is what this chapter proposal calls THE FOURTH ILLUSION: the illusion of ownership.

The Fourth Illusion enables a brand new, extremely profitable business model - the sale of digital assets in the metaverse. Although marginal costs are close to zero, NFTs can be perceived as real commodities and even sold in secondary markets. The desire to own objects and places in the real world can be transferred to the metaverse in a limited way, creating more than a multitude of marketplaces-- even a new economy. As Negroponte (1996) said, "the change from atoms to bits is irrevocable and unstoppable". First we replaced information stored on paper with databases. Then we abandoned physical audio discs, newspapers, videotapes, photographs, and other media. With the advent of mobile apps, we traded night houses for Tinder, cars for Uber and beach houses for Airbnb. Since the events of the early 2020s, we have traded offices for virtual meetings and conferences for digital events. With the possibility of uniqueness, the Oi makes consumers feel like the sole owner of places, wearables, collectibles, artwork, avatars and other digital goods.

HOW THE CHAPTER IS ORGANIZED

This work starts by proposing a definition for the term Metaverse- a statement is always a good starting point to work over a new concept - next, this definition is analyzed and contextualized to introduce the concept of the fourth illusion; then, the way the digital assets that sustain the fourth illusion is detailed; and finally, the economic impact of this phenomenon is approached, with a focus on the Metaverse economy.

THE METAVERSE

The term metaverse, coined for the novel Snow Crash (Stephenson, 2008), has found new applications since the advances of blockchain and novel virtual reality headsets. Some definitions include "an emerging shared digital space that can be navigated without corporate interference" (Everything You Need to Know about the Metaverse. | Foundation, n.d.) and "a digital reality that combines aspects of social media, online gaming, augmented reality (AR), virtual reality (VR) and cryptocurrencies to allow users to interact virtually" (Folger, 2022).

According to the Foundation, the Metaverse consists of spaces that are, by definition, independent of intermediaries. The Metaverse is a trustless environment, just like Bitcoin. The economy of the Metaverse is governed by open and immutable algorithms. This definition applies to the decentralised Metaverse. Meta (the former Facebook), on the other hand, defines the Metaverse as follows: 'The Metaverse will be a place where we can work, play and connect with others in immersive online experiences' (Meta Platforms and Technologies | Meta, n.d.). In 2022, Meta defines the Metaverse in the future as something that the company may build, a kind of "official" Metaverse.

In the absence of a comprehensive and universally accepted definition of the metaverse, and in the belief that definitions are a good starting point, I will propose one:

The metaverse is a conceptual network of virtual spaces that augments the physical world: where people participate in immersive experiences and engage in social, economic, and hedonistic activities.

This definition will serve as a guide for the introduction of some metaverse-related concepts.

A Network of Virtual Spaces

By our definition, the Metaverse is not owned by any company and is not necessarily open source, permission-free or decentralised: It is a network of virtual spaces of different kinds.

These virtual spaces are not necessarily a copy of the physical world, like the Mirror World (Gelernter, 1991). Gelernter's Mirror World is a metaphor for the digital world, a vast and distributed software that contains a complete mirror image of physical reality as a hole. Reality can be visited, simulated or downloaded to study as an exact copy of the actual world (Sonvilla-Weiss, 2008).

An incalculable number of digital spaces can coexist. Every person, historical fact, place and situation can have multiple copies in these spaces, linked by similarities, portals, shared spaces and assets, and users.

Immersive Experiences

If the perception of reality is a mental construction, physical reality cannot be considered "real" vis-à-vis an "unreal" reality constructed by computers. Immersive realities are also real for those who experience them.

Bostrom goes further and postulates that it is impossible to establish that we are not living in a computer simulation (Bostrom, 2003). If we assume that the computing power of humanity or a post-human civilization will one day be sufficient to create a kind of mirror world whose virtual habitats are sophisticated enough to have a conscience, then we could be living in *the Matrix* (D. J. Chalmers, 2003), (D. J. Chalmers, 2017).

Real or not, virtual reality has a remarkable effect: its users are aware that what is happening in the immersive experience is not happening in the physical world. Despite this awareness, users react to facts as if those facts were really happening; they react to places as if they were really there, even though they know they are not; and they react to avatars as if the avatars were their actual bodies. These reactions to immersion are called the three illusions (Slater, 2009).

The three illusions have to do with attention and coherence. When we pay more attention to the physical or virtual world around us than to our inner thoughts and feelings, presence emerges (Waterworth & Tjostheim, 2021). And the more coherent what we see in a virtual space is to what we expect to be "real", the stronger the presence effect of immersion. In one of its various experiments, Professor Slater's team found that the illusion of plausibility increases when users see a bar fight and are immersed in a scenario of the bar with decorations typical of a bar, which people tend to see as more appropriate for a bar fight, as opposed to a scenario of a bar decorated like a sophisticated restaurant (Skarbez et al., 2017), (Bergstroem et al., 2017).

Another important component for immersion is the background of the characters, namely the user's avatar. Avatars are a representation of the user's personality, more than a personality they think is ideal (Ruedinger, 2022). The more realistic the avatar, the better the illusion of embodiment (Maselli & Slater, 2013); and the ability to customise an avatar, which is more than a static feature, is also a component of the immersive experience (Ruedinger, 2022).

Studies on people's motivations to engage with video games show that the main motivational components are socialisation (interaction with other people, friendships, teamwork), achievement (progress, competition, optimisation) and immersion, characterised by role-playing, discovery, fantasy and escapism, among others (Yee, n.d.). Given the importance of video games as immersive experiences, these findings can be extrapolated to a limited extent to understand the motivations for engagement in other digital spaces of the metaverse.

Social and Economic Activities

As a motivating factor for people to engage in digital experiences, the social component is characterised by meeting people, building meaningful relationships and working in a team (group benefits and collaboration) (Yee, n.d.).

The social aspect, like the hedonic and functional attributes, is considered a driver for purchasing virtual goods. Social aspects are closely linked to rarity attributes through the distinction between the owner's inner circle and the ordinary group consisting of non-owners (Lehdonvirta, 2009).

Futurist Cathy Hackl sees the metaverse as a transformation of our physical and digital selves, where our digital identities can consume content and generate revenue streams. Unlike pre-blockchain immersive experiences, there the capture of value (such as Mario Bros coins) does not only exist during the game, but can be brought into the real world as cryptocurrency and used through physical things or experiences (Hackl et al., 2022).

Users' digital selves can be represented in digital spaces by avatars (Kocur et al., 2021). Changes in these representations can even lead to behavioural changes in users, for example by making them more confident by choosing more attractive avatars (Yee et al., 2009).

Hedonic Activities

Another group of attributes of virtual objects that act as purchase drivers are the hedonic ones. Background fiction, origin, customisability and appearance of avatars, locations and other items are some of these attributes (Lehdonvirta, 2009).

Why do users spend money to enhance their collections with these kinds of attributes? To enhance the hedonistic aspect of digital experiences. I call the use and search for this kind of upgrade hedonic activities.

Hedonic activities are related to a group of motivations for immersive experiences (Yee, n.d.), Achievement. This group of hedonic motivations includes advancement (achieving goals and accumulating digital assets), mechanics (the satisfaction that comes from understanding the logic and quantitative aspects of the system) and competition (users enjoy beating, dominating and challenging others in games or economic battles.

AN EXTENSION OF THE PHYSICAL WORLD

Chalmers' (2017) virtual-digital view of how real virtual reality is states that virtual objects are not unreal, but digital objects that actually exist, even if only in the memory of a computer machine; virtual events are digital events that, like digital objects, actually take place; virtual reality experiences involve a non-illusory perception of a virtual world, and immersive experiences may involve valuable experiences in the material world.

In the physical world, people also engage in social, hedonistic, and economic activities. The metaverse can complement these activities, for example when a brand uses a virtual shop to sell physical goods, or when a company hosts a meeting in a digital space attended by employees from different parts of the world.

The metaverse can also be used to carry out activities that would be too expensive, too difficult, too dangerous or even impossible in the physical environment. In this case, bits act as an extension of the atom's environment and provide a way to satisfy human needs that cannot be met with traditional products or services.

THE FOURTH ILLUSION

While three of Slater's illusions are perceptual rather than cognitive - users know that facts do not take place in the material world - the fourth illusion, the illusion of possession, has a stronger cognitive component. Things, places and avatars owned by users in the metaverse can be exchanged for real money - there are reasons to believe that they own these items. Apart from the fact that the existence of objects is an illusion, any ownership of them is also an illusion.

Ownership of Digital Assets

Ownership of digital assets has been used by video games since the second generation of consoles. What has changed with the advent of blockchains? At least two things: the persistence of items and their inventory control.

In Web 1.0 and Web 2.0, digital objects were inexhaustible. A copy of a picture of a dog could be reproduced infinitely without losing a single pixel. What is not limited cannot be considered a product. One could sell the intellectual property, the rights to reproduce the image; reproducing it without permission is considered piracy. It is also possible to sell a data carrier, e.g. a CD-ROM from the 1990s, which contains an image file of the picture. But it is impossible to sell the image itself.

In Web3.0, selling a digital image is still impossible. Blockchain technology offers a way to prevent the problem of double spending: selling a second copy of the image. A fully decentralised network of computers receives internally generated cryptocurrency to validate transactions using cryptography techniques, and all nodes on this network have copies of the immutable database. The problem of double spending: solved (Nakamoto, 2009).

If selling the image of the dog itself is still impossible, how does solving the problem of double spending enable the fourth illusion? By transferring the public registration of the image from one crypto-wallet to another.

NON-FUNGIBLE TOKENS

Every new invention causes some confusion, but in reality, NFT technology could not be simpler: While Bitcoin, the first-generation blockchain, is used to prevent the problem of double spending by publicly registering ownership of a set of cryptocurrency, Ethereum, the second-generation blockchain, is used to transfer ownership of a computer file that represents what you want.

In our example, a block in Ethereum contains the hash of the owner's wallet (public information); some optional metadata (age, breed, hair colour...); and a hash of a public address where the dog's picture can be found. When this NFT is transferred, the "owner's wallet" field is simply registered in the blockchain and receives a new value.

The sale of an NFT is authorised by a decentralised application such as OpenSea (OpenSea, the Largest NFT Marketplace, n.d.) to change this field and at the same time send some Ether to the seller's wallet as payment for this transfer. (Non-Fungible Tokens (NFT), n.d.), (NFT Basics, n.d.). It brings to digital assets the two game-changing properties that enable the fourth illusion: Inventory Control and Durability.

Inventory Control

Inventory control is possible as soon as the creator of an NFT can determine how many "copies" there may be or how many purses can be considered the owner of that asset. Items can have different levels of rarity and can even be transferred from one digital space to another. A magic sword can be obtained by a character in a 16-bit game and transferred to a 32-bit game, with equivalent properties; a different appearance, the "same" sword.

Persistence

Persistence is also very important for the illusion of ownership. In Web2.0, digital databases and computer files such as HTML pages or jpg images have arbitrary logical addresses based on the Hyper Text Transfer Protocol (the "HTTP://" and "HTTPS://" that precede domain names such as thomasedison.org, for example). How could we sell ownership of a file hosted at any address? Let us say the Edison Innovation Foundation decides for some reason to change its domain to thomasalvaedison.org. What then happens to all the links pointing to the former address? The buyer of an NFT based on an arbitrary address could see the asset disappear from one day to the next.

Web 3.0 solved the problem of persistence in hosting digital files in a decentralised storage service called IPFS, where files are given non-arbitrary addresses that are determined automatically and cryptographically based on the binary code of the computer file itself. This persistence database is managed by nodes that are paid for with cryptocurrency, such as Bitcoin - in this case a token called FileCoin (IPFS Powers the Distributed Web, n.d.).

The fourth illusion, enabled by inventory control and persistence, can make digital goods scarce and even rare. Now they can be treated like products by marketing (Lynn, 1991). Now they are an economic problem (Robbins, 1932). Now the exchange of atoms for bits (Negroponte, 1996) is more unstoppable than ever.

Motivations for Digital Assets Purchasing

Marketing efforts aimed at increasing sales of digital assets can explore two types of purchase motivations, among others: the need to belong and collector behaviour.

Under conditions of limited resources, making interpersonal connections can be vital for survival. Being part of a group that shares scarce resources has clear advantages over being alone. A possible dispute over a resource between a lone wolf and a group brings an even greater advantage to the group. This can partly explain the human need for group membership (Baumeister & Leary, 1995). Ownership of some digital assets may in some cases serve as a symbol of belonging to groups of holders.

Formanek (2003) examined the motivations that determine the behaviour of collectors. These motivations include the need to preserve (preserving so that other people can enjoy them), continuity (e.g. reliving the past), investment (making money by exploring price asymmetries) and addiction, whereby the objects in a collection may even lose meaning and the search for the objects becomes obsessive and compulsive (Belk, 1992).

Incentives for Virtual Goods Purchases

Incentives for purchasing virtual goods mentioned by (Hamari & Lehdonvirta, 2010) included inserting inconveniences into the game experience (including levels that are particularly difficult to overcome), providing free currency as an incentive for consumption behaviour, and artificial scarcity. A scarce virtual item cannot be seen as a traditional commodity that fetches higher prices due to the scarcity x necessities ratio, but as a feature that makes creative use of the blockchain more interesting and, why not, more fun (Chohan, 2021).

The success of NFT Bored Apes was followed by a sea of monkey-related projects. NFT images with profile pictures of monkeys became increasingly rare. But prices for the original Bored Apes continued to rise. This market behaviour illustrates Cohan's perspective of the influence of investment as a form of leisure that follows rules that distinguish the metaverse market from the traditional financial market.

The increased prices of Bored Apes also illustrate another trend: the value of originality. Bored Apes is the original Apes NFT collection. Bonnet & Bloom (2008) created a fake "copying machine" and convinced dozens of 6-year-old children that the artefact produced an exact copy of an object that belonged to Queen Elizabeth II. Most of the little volunteers rated the original object higher, and only one child (2%) chose the copy.

THE FOURTH ILLUSION AND THE METAVERSE ECONOMY

Ownership Illusion and Place Illusion

Place illusion is the perception of being in a virtual space when you know you are not (Slater, 2009). The pleasure of being in a place that is yours can be reproduced in the metaverse. Not only can you feel like you are in a place, but you can also be the owner of the place. To be a "neighbour" of the most famous Metaverse user, Snoop Dog, an investor paid $450,000 for a piece of virtual land. (Day, 2022).

Ownership Illusion and Embodiment Illusion

Embodiment Illusion is the perception of a virtual body as your own body (Slater et al., 2009). Ownership Illusion enables the commercialisation of virtual bodies. Users can perceive a virtual body not only as their physical body, but as an asset, a body that can even be sold to another user and replaced by another.

Virtual bodies as products represent a market opportunity. It is a way to enhance the experience by providing avatars with exciting backgrounds, customised features such as exclusive accessories and clothing, and avatars with physical attributes that are hard to obtain and even harder to maintain, such as sculptured bodies and top model hair.

Ownership Illusion and the Metaversian Dream

Freedom, democracy, opportunity, equality, prosperity and success - the American dream - are no more than a dream for most people. At least in the world of the atom.

The metaverse can be an environment where people "collect" experiences, objects, relationships and status that they would otherwise never have access to. Like Hiro, the protagonist of the novel Snow Crash, who lives in an atomic container and owns an ultra-high-priced spacein the metaverse (Robinson, 2017).

While in the world of the atom, connections, chance and family background can decide a person's future (Yee, n.d.), in the Metaverse any user can become a rich, powerful and admired person, even earning real money.

REFERENCES

Baumeister, R. F., & Leary, M. R. (1995). The need to belong: Desire for interpersonal attachments as a fundamental human motivation. *Psychological Bulletin*, *117*(3), 497–529. doi:10.1037/0033-2909.117.3.497 PMID:7777651

Belk, R. W. (1992). Attachment to Possessions. In I. Altman & S. M. Low (Eds.), *Place Attachment,* (pp. 37–62)., doi:10.1007/978-1-4684-8753-4_3

Bergström, I., Azevedo, S., Papiotis, P., Saldanha, N., & Slater, M. (2017). The Plausibility of a String Quartet Performance in Virtual Reality. *IEEE Transactions on Visualization and Computer Graphics*, *23*(4), 1352–1359. doi:10.1109/TVCG.2017.2657138 PMID:28141523

Bostrom, N. (2003). Are You Living in a Computer Simulation? *The Philosophical Quarterly*, *53*(211), 243–255. doi:10.1111/1467-9213.00309

Chalmers, D. J. (2003). The Matrix as metaphysics. *Science Fiction and Philosophy: From Time Travel to Superintelligence.*

Chalmers, D. J. (2017). The Virtual and the Real. *Disputatio*, *9*(46), 309–352. doi:10.1515/disp-2017-0009

Chohan, U. W. (2021). *Non-Fungible Tokens: Blockchains, Scarcity, and Value.* Social Science Research Network. doi:10.2139/ssrn.3822743

Day, C. D. Andrea. (2022, January 12). *Investors are paying millions for virtual land in the metaverse*. CNBC. https://www.cnbc.com/2022/01/12/investors-are-paying-millions-for-virtual-land-in-the-metaverse.html

Ethereum. (n.d.). *Find an Ethereum Wallet*. Ethereum.Org. https://ethereum.org

Ethereum. (n.d.). *Non-fungible tokens (NFT)*. Ethereum.Org. https://ethereum.org

Ethereum. (n.d.). *What is Ethereum?* Ethereum.Org. https://ethereum.org

Folger, J. (2022, February 15). *Metaverse*. Investopedia. https://www.investopedia.com/metaverse-definition-5206578

Formanek, R. (2003). Why they collect: Collectors reveal their motivations. In *Interpreting objects and collections*. Routledge.

Foundation Team. (2021). *Everything you need to know about the metaverse. Foundation.* https://foundation.app/blog/enter-the-metaverse

Gelernter, D. (1991). *Mirror Worlds: Or: The Day Software Puts the Universe in a Shoebox...How It Will Happen and What It Will Mean*. Oxford University Press., doi:10.1093/oso/9780195068122.001.0001

Hackl, C., Lueth, D., di Bartolo, T., Arkontaky, J., & Siu, Y. (2022). *Navigating the metaverse: A guide to limitless possibilities in a Web 3.0 world*. https://search.ebscohost.com/login.aspx?direct=true&scope=site&db=nlebk&db=nlabk&AN=3273058

Hamari, J., & Lehdonvirta, V. (2010). Game design as marketing: How game mechanics create demand for virtual goods. *Journal of Business Science and Applied Management*, *5*(1), 17.

Hood, B. M., & Bloom, P. (2008). Children prefer certain individuals over perfect duplicates. *Cognition*, *106*(1), 455–462. doi:10.1016/j.cognition.2007.01.012 PMID:17335793

IPFS. (n.d.). How IPFS Works. *IPFS*. https://ipfs.io/#how

Kocur, M., Henze, N., & Schwind, V. (2021). *The Extent of the Proteus Effect as a Behavioral Measure for Assessing User Experience in Virtual Reality*, 4.

Lanier, J. (1992). Virtual reality: The promise of the future. *Interactive Learning International*, *8*(4), 275–279.

Lehdonvirta, V. (2009). Virtual item sales as a revenue model: Identifying attributes that drive purchase decisions. *Electronic Commerce Research*, *9*(1–2), 97–113. doi:10.100710660-009-9028-2

Lynn, M. (1991). Scarcity effects on value: A quantitative review of the commodity theory literature. *Psychology and Marketing*, *8*(1), 43–57. doi:10.1002/mar.4220080105

Maselli, A., & Slater, M. (2013). The building blocks of the full body ownership illusion. *Frontiers in Human Neuroscience*, *7*. https://www.frontiersin.org/article/10.3389/fnhum.2013.00083. doi:10.3389/fnhum.2013.00083 PMID:23519597

Meta Platforms and Technologies | Meta. (n.d.). Meta Platforms and Technologies. https://about.facebook.com/technologies/

Nakamoto, S. (2009). *Bitcoin: A Peer-to-Peer Electronic Cash System, 9.*

Negroponte, N. (1996). *Being digital* (1. Vintage Books ed). *NFT basics*. https://nftschool.dev/concepts/non-fungible-tokens/

OpenSea. (n.d.). *the largest NFT marketplace*. Opensea. https://opensea.io/

Robbins, L. (1932). *An Essay on the Nature and Significance of Economic Science*. doi:10.2307/2342397

Robinson, J. (2017, June 23). The Sci-Fi Guru Who Predicted Google Earth Explains Silicon Valley's Latest Obsession. *Vanity Fair.* https://www.vanityfair.com/news/2017/06/neal-stephenson-metaverse-snow-crash-silicon-valley-virtual-reality

Ruedinger, B. M. (2022). *Social Rejection, Avatar Creation, & Self-esteem*. University Of Oklahoma.

Skarbez, R., Neyret, S., Brooks, F. P., Slater, M., & Whitton, M. C. (2017). A Psychophysical Experiment Regarding Components of the Plausibility Illusion. *IEEE Transactions on Visualization and Computer Graphics*, *23*(4), 1369–1378. doi:10.1109/TVCG.2017.2657158 PMID:28129171

Slater, M. (2003). A note on presence terminology. *Presence Connect*, *3*(3), 1–5.

Slater, M. (2009). Place illusion and plausibility can lead to realistic behaviour in immersive virtual environments. *Philosophical Transactions of the Royal Society of London. Series B, Biological Sciences*, *364*(1535), 3549–3557. doi:10.1098/rstb.2009.0138 PMID:19884149

Slater, M., Pérez Marcos, D., Ehrsson, H., & Sanchez-Vives, M. (2009). Inducing illusory ownership of a virtual body. *Frontiers in Neuroscience*, *3*(2), 214–220. https://www.frontiersin.org/article/10.3389/neuro.01.029.2009. doi:10.3389/neuro.01.029.2009 PMID:20011144

Sonvilla-Weiss, S. (2008). *(In)visible: Learning to act in the metaverse*. Springer.

Stephenson, N. (2008). *Snow crash*. Bantam Books.

Wright, J. (2016, October 21). Nosedive. Black Mirror.

Yee, N. (n.d.). *Motivations of Play in MMORPGs, 46.*

Yee, N., Bailenson, J. N., & Ducheneaut, N. (2009). The Proteus Effect: Implications of Transformed Digital Self-Representation on Online and Offline Behavior. *Communication Research*, *36*(2), 285–312. doi:10.1177/0093650208330254

Zutter, N. (2016, October 24). *Trying Too Hard: Black Mirror, "Nosedive."* Tor.Com. https://www.tor.com/2016/10/24/black-mirror-season-3-nosedive-television-review/

Chapter 5
User Acceptance Towards Non-Fungible Token (NFT) as the FinTech for Investment Management in the Metaverse

Ree Chan Ho
Taylor's University, Malaysia

Bee Lian Song
Asia Pacific University of Technology and Innovation, Malaysia

ABSTRACT

The growth of the metaverse is an imminent phenomenon for the world to tap its economic opportunities. Non-fungible token (NFT) is the choice of a new form of financial technology in metaverse to authenticate digital asset ownership. The objective of this study was to examine the impacts of NFT on the metaverse communities. Based on the unified theory of acceptance and use of technology model, the sample was financial users in the metaverse who are familiar with NFT. The findings validated that UTAUT constructs influence the intention to use NFT payments. Also, self-efficacy was confirmed as the antecedent to explain the need for managing the processes while payments were made. This study provides an understanding of the use of NFT by examining the theoretical aspect of the current usage scenario in the metaverse. Furthermore, the required personal ability of the users in determining the benefits of NFT offers practical consideration for wider adoption. Hence, it contributes to the research on the behavioral finance aspect of NFT usage.

INTRODUCTION

The arrival of the metaverse has witnessed another phase of digital transformation driven by virtual reality for new business challenges and opportunities. Following this trend, the increased financial transactions traded on this shared virtual business platform have changed the way customers deal with

DOI: 10.4018/978-1-6684-5732-0.ch005

their purchases. Non fungible token (NFT) is the choice of a new form of cryptocurrency in the metaverse (Bao & Roubaud, 2022). Businesses and consumers are exploring the advantages of NFT tokens as their main exchange medium to transact digital assets. NFT provides security over the digital asset as it ensures one rightful owner at any given time. The tampering and dispute over asset ownership have been prevented. Hence, this would pave the way for more rapid development of the digital assets market.

The unique features have accelerated the use of NFT in the financial market. Its benefits such as simple to invest in, highly accessible by anyone, and proven security protocol (Schaar & Kampakis, 2022). The ease of digital asset ownership is supported well to avoid duplication. A list of studies hailed NFT as the de facto payment mode for the digital asset in the metaverse environment (Arcenegui Almenara et al., 2021; Maouchi et al., 2022; Rafli, 2022). The security protection was strong and not easily tampered with the backing of blockchain technology. Also, the crypto mechanism provides the transparency of digital ownership records. These benefits directly led to the fast acceptance of the NFT payment mode.

The present literature supports the non-fungible nature of the NFT token as irreplaceable in the bidding of virtual collectibles and assets (Bolton & Cora, 2021; Chandra, 2022; Doan et al., 2021). This promotes digital products such as painting, music, caricature and games (Choi et al., 2021). NFT make asset ownership faster. Guadamuz (2021) showed that the bidding process of assets and obtaining ownership were expedited and convenient. The solid security features of NFT causes the financial transaction in the metaverse safe. This is also supported by Chandra (2022)'s study of the globalization effect in making NFT tokens available to anyone from four corners of the world. The high acceptance level by the global market attracted more financial investors in choosing the NFT method for acquiring digital assets because anyone can purchase and transfer digital ownership with ease.

Aims and Contributions

Although consumers are using NFT for transactions in the metaverse, mixed reactions were on its adoption and satisfaction. Many NFT users were concerned over the reported scams and fraud in the NFT space (Slater-Robins, 2022). The users need to register and use a crypto-based digital wallet for the NFT marketplace. Many incidents of transaction discontinuation where the users were distrust and unfamiliar with the process (Cornelius, 2021). The prerequisite to know sufficient information and knowledge to manage it well. Self-efficacy represents one's confidence in carrying out the task in any given context. It is validated as the precursor to the acceptance and link to the familiarity of the new of new technological tools and applications (Latikka et al., 2019).

Hence, this study adopted the theoretical lens of the Unified Theory of Acceptance and Use of Technology (UTAUT) to unearth consumer reaction and attitude toward the NFT usage (Venkatesh et al., 2016). The use of NFT requires self-efficacy to reduce the learning efforts and increased the users' confidence in investing in the metaverse world. Hence, the need for self-efficacy in exploring the new financial cryptocurrency is worth further exploration. This study examined the effect of technology adoption factors in UTAUT theory on the use of NFT in the metaverse. Furthermore, the need for antecedent for the variables is required in this highly virtual environment. Hence, self-efficacy is required as a precursor to the UTAUT variables warrants investigation. This study offers a conceptual framework for predicting the adoption of the NFT method for digital asset purchase and examined its use based on the technology acceptance perspective.

LITERATURE REVIEW

Non-Fungible Token

Non fungible token (NFT) has revolutionized the way we invest for monetary gain in the virtual financial markets. Its non-fungible nature dictates that it is unique and non-replaceable to safeguard digital items bought over the open metaverse marketplace. It is a kind of virtual financial transaction that is available in a range of different formats and platforms (Valeonti et al., 2021). The ability to authenticate and confirm the ownership is it main feature of its popular adoption. Other benefits include immutability, ease of transfer, and unique identification because NFTs are available and supported by blockchain technology.

The investors have been purchasing NFT collectibles in the form of art, music, in-game items, virtual real property, and other digital artifacts. They are traded and purchased online with the use of cryptocurrency supported by related software applications. This is due to the establishment and the high connectivity of internet and mobile communication and the widespread of NFT-related information. Having said that, some challenges from making NFT a mainstream option for many investors. The diversification of rules and complex infrastructure to streamline the transaction procedure is deterring novice and new users from using it (Park et al., 2022). Furthermore, the need to use cryptocurrencies that exist in different formats and platforms poses another hindrance (Dowling, 2022). New users would find it hard to rely on and learn from the internet in searching for the right and suitable tokens.

Theoretical Development

Unified Theory of Acceptance and Use of Technology (UTAUT)

The growth of financial technology in the investment markets is backed by emerging technologies which include artificial intelligence, machine learning, big data, and other advanced information systems (Ghazali et al., 2018). The advancement of technology has paved way for the use of cryptocurrency for financial investment. It also streamlined the business process for the use of NFT in the metaverse environment. Many empirical studies on business technology applied the Unified Theory of Acceptance and Use of Technology (UTAUT) theory to uncover the consumer behavior on new technology implementation (Venkatesh et al., 2016).

The theory examined the impact of the technology from the functional aspect gained by the users. It provided the theoretical lens to explain technology adoption and acceptance in an business and organizational context, i.e. online shopping (Celik, 2016), to predict use of travel itinerary (Ho & Amin, 2019), telepresence robots (Han & Conti, 2020) and use of enterprise resource planning systems (Soliman et al., 2019). Compared to other technology theories such as the Technology Acceptance Model (TAM), Theory of Panned Behavior (TPB), and the Uses Gratification Theory (UGT), UTAUT offers a comprehensive model based on functional and adoption dimensions of the novice technology (Im et al., 2011). It examines the intention of the users of using the new tools and then explains the user behavior. This is because UTAUT is supported by the underlying user perception to uncover the attitude and its related actions (Zhou et al., 2010). it is applied in this study because it can provide explanatory power based on the functional usage dimension of technology acceptance. Hence, the use of the UTAUT model was appropriate in this study and we include performance expectancy, effort expectancy, social influence and facilitating condition as the main determinant for NFT adoption. Figure 1 depicts the conceptual framework of this study.

Figure 1. The conceptual framework

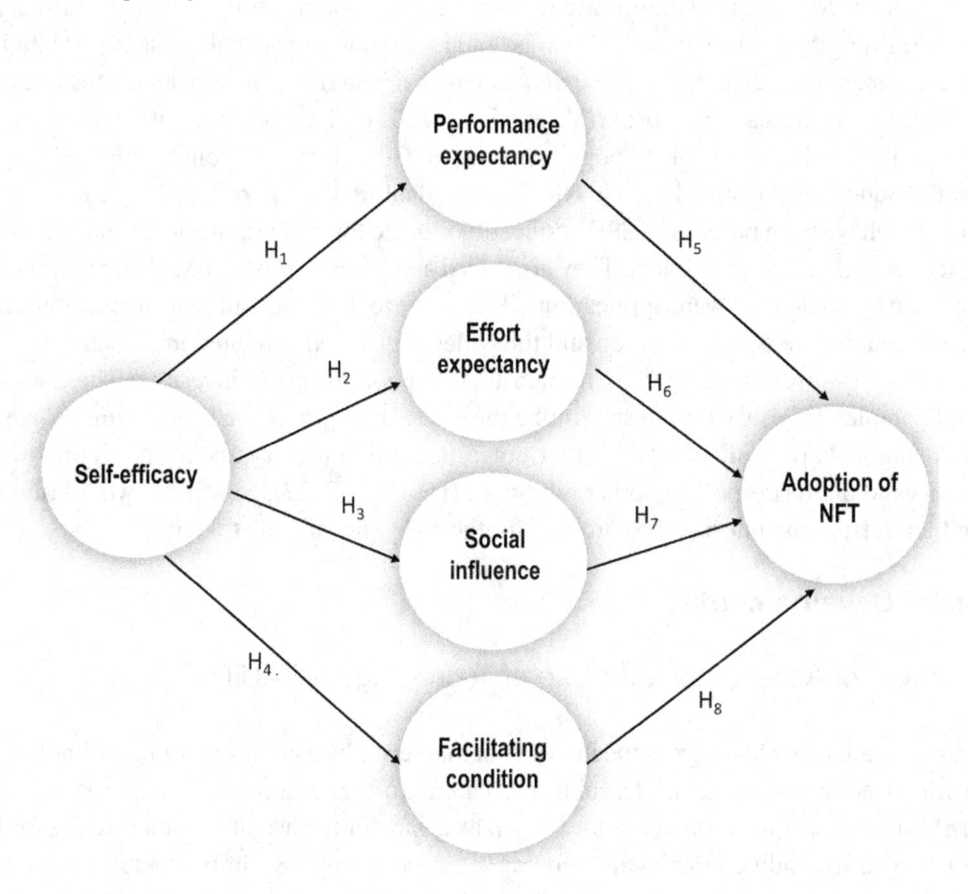

Self-Efficacy

Self-efficacy is the self-belief about own capability to perform a particular task confidently (Williams & Rhodes, 2016). It is regarded as the confidence in performing a specific task without the need for guidance from others. Hence, human action can be productive with a high level of self-efficacy and links to the anticipated performance required for the given tasks (Mao et al., 2019). Self-efficacy is important to online investment because the user needs to be able to complete the business process successfully. Hence, customers who possess more self-efficacy have no difficulty and managed the process well (Ho, 2021). It was evidenced that many who quit the online process were those people with a low level of self-efficacy.

Furthermore, the difficulty of understanding the processes involved in NFT transactions due to its multiple forms and variety of cryptocurrencies could be resolved if the users were able to manage it well. Self-efficacy has proven to be necessary for handling complex processes (Wood et al., 2000). Therefore, it can be applied to handle the NFT transactions and expedite the adoption process. The four main constructs under UTAUT, i.e. performance expectancy, effort expectancy, social influence, and facilitating condition can be strengthened by the use of NFT with the influence of self-efficacy. Regarding the online shopping environment, the confidence of the users in their capability to handle the shopping process increased the performance of the shopping tasks (Ho, 2019; Tims et al., 2014). This is supported by Schwoerer et al., (2005)'s study in combing the need for effort expectancy and self-efficacy for productive shopping activities. In the same vein, Ramadania and Braridwan (2019) validated that self-efficacy has a direct relationship with the perceived ease of use required in completing the tasks effectively. The first and second hypotheses were examined based on the above discussion.

H_1: Self-efficacy is positively influence on the performance expectancy required to use NFT.

H_2: Self-efficacy is positively influence on the effort expectancy required to use NFT.

Li et al. (2018) validated the positive effect of self-efficacy on managing the transaction process for online shopping. Furthermore, self-efficacy was critical to have a high level of controllability in facilitating conditions in obtaining the needed financial information for online stock investment. There are also existing studies relating self-efficacy to self-efficacy and social influence (Bhattarai et al., 2021; Lucas et al., 2006; Sung et al., 2015). Humayun Kabir (2018) confirmed the role of self-efficacy to be critical in relating to the herb behavior of financial investment. After all, users would approach their peers when they faced difficulty in online financial robot transactions (Ge et al., 2021). Hence, this study intended to examine how self-efficacy could provide the inertia to seek facilitating conditions and social influence for NFT investment. Hence, the following hypotheses are tested in this study.

H_3: Self-efficacy is positively influence on the social influence required to use NFT.

H_4: Self-efficacy is positively influence on the facilitating condition required to use NFT.

Performance Expectancy

Performance expectancy refers to the users' expectation of the use of technology tools in enabling them to achieve the desired performance (Diep et al., 2016). The performance expectancy is considered critical in determining the user adoption of the tools they intend to use (Ho et al., 2021). The usefulness of the technology was validated as a key determinant to gaining satisfaction (Wu, 2013). Furthermore, it was also confirmed as the precursor to the subsequent adoption of the crypto currency technology (Miraz et al., 2022). Hence, performance expectancy is regarded as the required expectation of getting the NFT investment transaction process completed successfully. Furthermore, NFT transactions would produce the unique identification for the digital asset ownership under the blockchain network (Karandikar et al., 2021). In this study, the users sought to have the registration of the NFT, and this could inspire the users for the performance they expected. Hence, the following hypothesis was formulated:

H_5: Performance expectancy is positively influence on the intention to use the NFT.

Effort Expectancy

Effort expectancy means the efforts devoted to the use of technological tools when the tasks were achieved (Venkatesh et al., 2016). The efforts are desired to be manageable and the time taken should be shorter to enjoy the satisfaction of job completion. Technological tools and applications should be developed to make it unsophisticated and easy to use. The less effort required to use the systems, the more likely it satisfied the effort expectancy for the users. This would promote self-directed to gain the product knowledge (Ho & Chua, 2015). In general, most studies have proven that ease to learn and ease of use are the main reasons for the adoption of technological tools. The existing literature has abundant studies to support effort expectancy as the key to intention to use (Chong, 2013; Iranmanesh et al., 2022; Zhang et al., 2018). In the realm of financial technology, Miraz et al., (2022) indicated that effort expectancy has a direct influence on the re-purchase of cryptocurrency for investment. In this study, the digital asset ownership would be produced directly and stored in the bitcoin network without any effort from the users. Hence, it is expected that effort expectancy may have a different impact on NFT usage. With that, the following hypothesis was developed to investigate further.

H_6: Effort expectancy is positively influences on the intention to use the NFT

Social Influence

In a business transaction setting, social influence is derived from the social factors imposed on the customers' choice to carry out a certain behavior (Cialdini & Goldstein, 2004). This involved the purchase decision or discontinuing the shopping task which influenced by social impact (Tarhini et al., 2018). The most critical impact stems from the social groups, particularly the peers or people who are closely connected to us i.e. friends and family members. Interestingly, an individual would conduct certain behavior because he or she would perceive what other people want him or her to perform. This scenario applies to financial investment activities. The herd behavior in financial markets is common and abundant in related studies in the financial literature (Choijil et al., 2022; Cipriani & Guarino, 2014; Yang et al., 2018). With the highly pervasive nature of social media, information related to financial markets is shared extensively in popular social media applications. Furthermore, social media users and influencers tended to recommend financial products and services to other users. The existing literature presented that social influence exerted influence on the choice of cryptocurrency (Jalal et al., 2021; Yeong et al., 2022; Younus et al., 2022). Having said that, the complexity of the nature of NFT would be hotly discussed on social media. Therefore, the following hypothesis is worth to be investigated:

H_7: Effort expectancy is positively influence on the intention to use the NFT

Facilitating Condition

Facilitating condition is the business process and technical support offered by the platform or systems (San Martín & Herrero, 2012). it offers the required technological components of NFT, such as the communication channel bandwidth, the storage space of the systems and users' devices, and even the familiarity of the users with the NFT infrastructure operations (Schaar & Kampakis, 2022). A list of extant studies has confirmed the positive relationship between facilitating conditions and the adoption of technology innovation (Bervell et al., 2022; Peñarroja et al., 2019; Ramírez-Correa et al., 2019). However, this study covers more than the cryptocurrencies as it also includes the NFT ownership record under the bitcoin networks. The capability of the ownership management of the NFT is also crucial in facilitating the entire process. Therefore, the following hypothesis was developed to evaluate this new facilitating condition as well:

H_8: Facilitating condition positively influences tourists' behavioral intention to adopt the itinerary from the smart travel app.

RESEARCH METHODS

Sample and Data Collection

The sample of this study consisted of business students from a large Malaysian university. It was chosen based on the premise that young business university students were ardent and familiar with the NFT investment. A qualification question was included and respondents who were unfamiliar and did not explore the NFT investment were excluded. 206 qualified respondents filled in and submitted the online questionnaire. Table 1 demonstrates the demographic distribution and features of the respondents of the study.

Table 1. Descriptive statistics (N=206)

	Characteristics	Frequency	%
Gender	Male	102	49.51
	Female	104	50.49
Age	Between 18 – 20	122	59.23
	Between 21 – 30	78	37.86
	More than 30	6	2.91
No. of years – exposed of cryptocurrency	Less than 1 year	95	46.12
	1 < X £ 2 years	101	49.03
	More than 2 years	10	4.86
Importance of NFT	Strongly agree	85	41.26
	Agree	72	34.95
	Neutral	32	15.53
	Disagree	12	5.83
	Strongly disagree	5	2.43

Questionnaire Development

The questionnaire was developed by referring to previous studies. Item measurements were designed with careful consideration of the established scales. The item measurements of UTAUT constructs were adapted from Okumus et al. (2018), self-efficacy from Akhter (2014), and the behavioral intention scale was derived Pavlou and Fygenson (2006). The detail of the scale items is listed in Table 2.

Table 2. Measurement scale items

Construct	Scale	Source
Behavioral intention, BI	BI1: Given the chance, I intend to use the NFT. BI2. Given the chance, I predict that I should use the NFT in the future. BI3. It is likely that I will transact with NFT in the future.	Pavlou 2003
Effort expectancy, EE	EE1: It is easy to learn how to use NFT. EE2: Interacting with NFT is clear and easy to understand. EE3: NFT is easy to use.	Okumus et al., 2018
Facilitating conditions, FC	FC1: I had no difficulty in finding and using the NFT. FC2: I had no difficulty in customizing the NFT for my use. FC3: Overall, the NFT has good performance.	Okumus et al., 2018
Performance expectancy, PE	PE1: NFT can be useful in managing my investment. PE2: NFT can be valuable to my investment. PE3: NFT can be advantageous in better managing my investment.	Okumus et al., 2018
Self-Efficacy, SE	SE1: I have the necessary skills to fully use the NFT. SE2: I have the necessary ability to fully use the NFT. SE3: I am confident that I can solve any problems in using the NFT.	Akhter, 2014
Social influence, SI	SI1: I want to use the NFT because my friends do so. SI2: Using the NFT reflects my personality to other people. SI3: According to people who are important to me, I should use the NFT.	Okumus et al., 2018

DATA ANALYSIS

A two-stage confirmatory factor statistic, structural equation modeling (SEM) was employed to evaluate the measurement model as well as the structural model. SEM was appropriate in this study to enable the latent variables used to be validated by the measurement model, and subsequently the structural model to test the hypotheses derived from the conceptual framework. This approach permits the validation of the relationship of the item measurements with constructs developed for the study (Valaei et al., 2019). The selected data analysis software was SmartPLS, a partial least square method to test the inter-relationship of the variables based on their path analysis. It was relevant as it investigated the interdependency of the paths of the multiple independent and dependent variables of the study.

Measurement Model

The measurement model was considered good with the reliability and validity of the conceptual model supported. Both the Average Variance Extracted (AVE) and communality scores obtained were more than the cut-off value of 0.5, and further confirmed the achievement of internal consistency. Cronbach's alpha for all the variables was evaluated and achieved the minimum score of 0.7 to dictate the acceptable convergent validity. Composite reliability (CR) and AVE were also higher than the threshold value. validity. Therefore, the reliability and validity of the variables were good, and concluded good quality of measurement model for the study. Table 3 presents the summarized results of the measurement model.

Table 3. Summary results for the measurement model

Construct	Indicators*	Loadings	Average Variance Extracted	Composite Reliability	Cronbach's alpha
Behavioral intention (BI)	BI1	0.877	0.610	0.953	0.854
	BI2	0.8710		0.923	
	BI3	0.790		0.932	
Effort expectancy (EE)	EE1	0.833	0.730	0.946	0.895
	EE2	0.883		0.918	
	EE3	0.810		0.823	
Facilitating condition (FE)	FE1	0.819	0.671	0.879	0.792
	FE2	0.780		0.856	
	FE3	0.829		0.893	
Performance expectancy (PE)	PE1	0.842	0.719	0.930	0.857
	PE2	0.810		0.928	
	PE3	0.822		0.862	
Self-efficacy (SE)	SE1	0.811	0.725	0.898	0.875
	SEL	0.800		0.916	
	SE3	0.840		0.950	
Social influence (SI)	SI1	0.772	0.746	0.924	0.884
	SI2	0.849		0.919	
	SI3	0.836		0.925	

The discriminant validity was assessed with the use of the Heterotrait-Monotrait Ratio of Correlations (HTMT). The results confirmed that discriminant validity was achieved and are listed in Table 4.

Table 4. Discriminant Validity Heterotrait-Monotrait Ratio of Correlations (HTMT)

	Behavioral intention	Effort expectancy	Facilitating condition	Performance expectancy	Self efficacy
Effort expectancy	0.553				
Facilitating condition	0.617	0.698			
Performance expectancy	0.667	0.661	0.612		
Self efficacy	0.530	0.636	0.592	0.692	
Social influence	0.532	0.661	0.543	0.427	0.652

Structural Model

The structural model was evaluated with the use of bootstrapping re-sampling method. The way of blindfolding was applied was based on the instructions provided by Chin et al. (2003). Refer to Table 5 for the R^2 and Q^2 values obtained from the blindfolding approach. The Q^2 values showed that the predictive relevance of the model was achieved. The R^2 value of behavioral intention was 54.3% and supported the adoption of NFT. The R^2 for performance expectancy, effort expectancy, social influence and facilitating conditions were recorded as 48.4%, 46.9%, 39.9% and 28.6% respectively. Hence, the variance obtained for the UTAUT variables were also influential to the use of NFT.

Table 5. Blindfolding indexes for constructs

Construct	R^2	Q^2
Behavioral intention	0.543	0.432
Effort expectancy	0.469	0.334
Facilitating condition	0.286	0.167
Self-efficacy	N/A	0.665
Social influence	0.399	0.299
Performance expectancy	0.484	0.287

Self-efficacy was tested by using the Stone-Geisser's predictive relevance test. Cross-validated redundancy Q^2 value was calculated with the use of blindfolding process based on the guides provided by Eom and Ashill (2016). It was recorded as 0.665 and exceeded the threshold value of the predictive power of the model. The standard error, t-value, and p-value of the study were used to analyze the paths. Structural model measurement was recorded in Table 6.

Table 6. Test results for structural model

	Path coefficient	Standard deviation	t-value	p-value (2-tailed)	Supported
H_1: SE → PE	0.292	0.083	2.567	0.000	Yes
H_2: SE → EE	0.434	0.051	6.161	0.000	Yes
H_3: SE → SI	0.187	0.076	2.974	0.000	Yes
H_4: SE → FC	0.390	0.032	2.232	0.000	Yes
H_5: PE → BI	0.250	0.023	4.325	0.006	Yes
H_6: EE → BI	0.288	0.067	3.729	0.000	Yes
H_7: SI → BI	0.398	0.059	4.143	0.000	Yes
H_8: FC → BI	0.499	0.046	3.265	0.001	Yes

The tests results from the structural model indicated that the inter-relationships paths and the corresponding hypotheses were accepted.

CONCLUSION

Discussion

This study aims to investigate the determinants required to drive the adoption of NFT under the metaverse environment. The four main constructs under UTAUT theory were examined for their effects on driving the behavioral intention to use NFT. Furthermore, the study also extended the UTAUT theory with the inclusion of self-efficacy to act as the precursor that influences the determinants of NFT adoption.

The findings demonstrated the need for performance expectancy, effort expectancy, social influence, and facilitating conditions were critical in influencing the users to invest in NFT. The findings were aligned with the previous studies on technological tool adoption in other contexts (Al-Saedi et al., 2020; Chao, 2019; H. Wang et al., 2020). In addition, self-efficacy was found to be the contributor to enabling the users to experience the NFT features. The need for self-efficacy in understanding and operating the complex nature of NFT investment transactions and processes. After all, proficiency in the use of new financial technology was linked directly to a high level of self-independence.

Theoretical Implications

A theoretical conceptual framework was developed for the adoption of a novel financial investment tool, i.e. NFT. This framework was empirically tested and confirmed the main components for users in appreciating NFT investment are performance expectancy, effort expectancy, social influence, and facilitating conditions. Among these factors, performance expectancy was the most salient factor. The next factor was effort expectancy and explaining the users' satisfaction over the efforts needed would be critical. Following the order of importance, we have social influence for the users to seek and relate their investment activities with their peers. Facilitating conditions were concluded as least important compared to the other three factors. This is not surprising as the new applications are becoming more user-friendly with better interface design (Jang et al., 2019). These findings were consistent with other financial technologies studies (Jin et al., 2019; Shaikh et al., 2020). The other contribution was the confirmation of self-efficacy as the antecedent for the adoption factors. This is because innovative technologies require a high level of self-directed learning in making the users more proficient in handling the technology (Ho & Song, 2021). Self-efficacy serves as the psychological construct needed to manage and monitor task performance (Afzal et al., 2019). This involves the operations and reactions of handling the expected outcomes in business transactions. It is important to carry out the procedures and processes in completing the financial investment activities with ease. Therefore, technological constructs derived from UTAUT are not sufficient to use NFT. This is aligned with the practical imperatives just like other technological innovations for online business services (Latikka et al., 2019; Y.-S. Wang et al., 2013). We thus concluded that have the self-managed capability in self-efficacy without the assistance and learning effort to familiar and use the NFT comfortably. Hence, this study empirically validated the requirement of self-efficacy in enabling the acceptance of NFT as a financial technological tool.

In summary, this research integrated the adoption factors from UTAUT and self-efficacy to provide a theoretical explanation of the use of NFT. Hence, this extended framework would be useful to facilitate the operations of NFT and explicate the behavior of the users in NFT. Hence, it provides clarification on the new development of financial technology for NFT. With the expected inpouring NFT transactions in near future, this theoretical explanation would be useful for better investment in the metaverse environment.

Managerial Implications

The users of the financial technologies are more independent and possess the self-efficacy needed to use the relevant technological platform and applications. These independent customers have their own knowledge, preference, and beliefs dictate their behavioral intention. Furthermore, it is imperative to have the commitment in learning new product knowledge (Ho & Amin, 2022). Hence, the NFT operators should cater to their needs by providing more specialized functions or services. Some suggested services include recent transactions review, purchases comparison, and other product-relevant information. Furthermore, this study also validated the importance of social influence in managing the NFT activities as they are socially connected well in the social media space with their mobile devices. Hence, mobile transactions are likely to be the focus of the NFT. The NFT operators have relied on the established NFT marketplace app such as Opensea, Binance NFT, Rarible, etc. The user interface design of the NFT mobile apps has considered users' needs, particularly from the mobile user perspective.

Limitation and Future Direction

The study has a few limitations and was presented as follows. The study has a small sample size because the NFT marketplace is not the mainstream of financial investment as compared to other more established financial markets. The respondents were young investors with limited investment funds. Therefore, the study was not able to paint the full picture with this low generalizability power. The finding is possibly different if all other age groups were included. Also, it would be better if we restricted to the one NFT marketplace app. However, this would further reduce the number of qualified respondents.

This study focused on the behavioral and functional aspects of NFT users. Therefore, it is suggested that future studies investigate other consumer behavior such as cultural, social, and psychosocial perspectives to offer more explanation and comparison. Future research can delve into the different types of consumers who have different characteristics, such as habitual buying behavior, variety-seeking behavior, dissonance-reducing buying behavior, and complex buying behavior. It could offer other dimensions for investigation and provide more insightful clarification. In addition, artificial intelligence features i.e. machine learning has been introduced to enhance the performance of many financial technologies (Chen & Chang, 2021; Jagtiani & Lemieux, 2019). It would be relevant to investigate how this advanced artificial intelligence tool attracts more people to the NFT marketplace.

REFERENCES

Afzal, S., Arshad, M., Saleem, S., & Farooq, O. (2019). The impact of perceived supervisor support on employees' turnover intention and task performance: Mediation of self-efficacy. *Journal of Management Development*, *38*(5), 369–382. doi:10.1108/JMD-03-2019-0076

Akhter, S. H. (2014). Privacy concern and online transactions: The impact of internet self-efficacy and internet involvement. *Journal of Consumer Marketing*, *31*(2), 118–125. doi:10.1108/JCM-06-2013-0606

Al-Saedi, K., Al-Emran, M., Ramayah, T., & Abusham, E. (2020). Developing a general extended UTAUT model for M-payment adoption. *Technology in Society*, *62*, 101293. doi:10.1016/j.techsoc.2020.101293

Arcenegui Almenara, J., Arjona, R., Román Hajderek, R., & Baturone Castillo, M. I. (2021). Secure combination of iot and blockchain by physically binding iot devices to smart non-fungible tokens using pufs. *Sensors, 21* (9), 3119. doi:10.3390/s21093119

Bao, H., & Roubaud, D. (2022). Non-Fungible Token: A Systematic Review and Research Agenda. *Journal of Risk and Financial Management*, *1*(1), 44–46. doi:10.3390/jrfm15050215

Bervell, B. B., Kumar, J. A., Arkorful, V., Agyapong, E. M., & Osman, S. (2022). Remodelling the role of facilitating conditions for Google Classroom acceptance: A revision of UTAUT2. *Australasian Journal of Educational Technology*, *38*(1), 115–135.

Bhattarai, M., Jin, Y., Smedema, S. M., Cadel, K. R., & Baniya, M. (2021). The relationships among self-efficacy, social support, resilience, and subjective well-being in persons with spinal cord injuries. *Journal of Advanced Nursing*, *77*(1), 221–230. doi:10.1111/jan.14573 PMID:33009842

Bolton, S. J., & Cora, J. R. (2021). Virtual Equivalents of Real Objects (VEROs): A type of non-fungible token (NFT) that can help fund the 3D digitization of natural history collections. *Megataxa*, *6*(2), 93–95. doi:10.11646/megataxa.6.2.2

Celik, H. (2016). Customer online shopping anxiety within the Unified Theory of Acceptance and Use Technology (UTAUT) framework. *Asia Pacific Journal of Marketing and Logistics*, *28*(2). doi:10.1108/APJML-05-2015-0077

Chandra, Y. (2022). Non-fungible token-enabled entrepreneurship: A conceptual framework. *Journal of Business Venturing Insights*, *18*, e00323. doi:10.1016/j.jbvi.2022.e00323

Chao, C.-M. (2019). Factors determining the behavioral intention to use mobile learning: An application and extension of the UTAUT model. *Frontiers in Psychology*, *10*, 1652. doi:10.3389/fpsyg.2019.01652 PMID:31379679

Chen, T.-H., & Chang, R.-C. (2021). Using machine learning to evaluate the influence of FinTech patents: The case of Taiwan's financial industry. *Journal of Computational and Applied Mathematics*, *390*, 113215. doi:10.1016/j.cam.2020.113215

Chin, W. W., Marcolin, B. L., & Newsted, P. R. (2003). A partial least squares latent variable modeling approach for measuring interaction effects: Results from a Monte Carlo simulation study and an electronic-mail emotion/adoption study. *Information Systems Research*, *14*(2), 189–217. doi:10.1287/isre.14.2.189.16018

Choi, S.-W., Lee, S.-M., Koh, J.-E., Kim, H.-J., & Kim, J.-S. (2021). A Study on the elements of business model innovation of non-fungible token blockchain game: Based on'PlayDapp'case, an in-game digital asset distribution platform. *Journal of Korea Game Society*, *21*(2), 123–138. doi:10.7583/JKGS.2021.21.2.123

Choijil, E., Méndez, C. E., Wong, W.-K., Vieito, J. P., & Batmunkh, M.-U. (2022). Thirty years of herd behavior in financial markets: A bibliometric analysis. *Research in International Business and Finance*, *59*, 101506. doi:10.1016/j.ribaf.2021.101506

Chong, A. Y.-L. (2013). Predicting m-commerce adoption determinants: A neural network approach. *Expert Systems with Applications*, *40*(2), 523–530. doi:10.1016/j.eswa.2012.07.068

Cialdini, R. B., & Goldstein, N. J. (2004). Social influence: Compliance and conformity. *Annual Review of Psychology*, *55*(1), 591–621. doi:10.1146/annurev.psych.55.090902.142015 PMID:14744228

Cipriani, M., & Guarino, A. (2014). Estimating a structural model of herd behavior in financial markets. *The American Economic Review*, *104*(1), 224–251. doi:10.1257/aer.104.1.224

Cornelius, K. (2021). Betraying blockchain: Accountability, transparency and document standards for non-fungible tokens (nfts). *Information*, *12*(9), 358. doi:10.3390/info12090358

Diep, N. A., Cocquyt, C., Zhu, C., & Vanwing, T. (2016). Predicting adult learners' online participation: Effects of altruism, performance expectancy, and social capital. *Computers & Education*, *101*, 84–101. doi:10.1016/j.compedu.2016.06.002

Doan, A. P., Johnson, R. J., Rasmussen, M. W., Snyder, C. L., Sterling, J. B., & Yeargin, D. G. (2021). NFTs: Key US Legal Considerations for an Emerging Asset Class. *Journal of Taxation of Investments*, *38*(4), 63–69.

Dowling, M. (2022). Is non-fungible token pricing driven by cryptocurrencies? *Finance Research Letters*, *44*, 102097. doi:10.1016/j.frl.2021.102097

Eom, S. B., & Ashill, N. (2016). The determinants of students' perceived learning outcomes and satisfaction in university online education: An update. *Decision Sciences Journal of Innovative Education*, *14*(2), 185–215. doi:10.1111/dsji.12097

Ge, R., Zheng, Z., Tian, X., & Liao, L. (2021). Human–robot interaction: When investors adjust the usage of robo-advisors in peer-to-peer lending. *Information Systems Research*, *32*(3), 774–785. doi:10.1287/isre.2021.1009

Ghazali, E. M., Mutum, D. S., Chong, J. H., & Nguyen, B. (2018). Do consumers want mobile commerce? A closer look at M-shopping and technology adoption in Malaysia. *Asia Pacific Journal of Marketing and Logistics*, *30*(4), 1064–1086. doi:10.1108/APJML-05-2017-0093

Guadamuz, A. (2021). The treachery of images: Non-fungible tokens and copyright. *Journal Of Intellectual Property Law and Practice*, *16*(12), 1367–1385. doi:10.1093/jiplp/jpab152

Han, J., & Conti, D. (2020). The use of UTAUT and post acceptance models to investigate the attitude towards a telepresence robot in an educational setting. *Robotics*, *9*(2), 34. doi:10.3390/robotics9020034

Ho, R. C. (2019). *The outcome expectations of promocode in mobile shopping apps: An integrative behavioral and social cognitive perspective*, 74–79.

Ho, R. C. (2021). Chatbot for Online Customer Service: Customer Engagement in the Era of Artificial Intelligence. In Impact of Globalization and Advanced Technologies on Online Business Models, (pp. 16–31). IGI Global.

Ho, R. C., & Amin, M. (2019). What drives the adoption of smart travel planning apps? The relationship between experiential consumption and mobile app acceptance. *KnE Social Sciences*, 22–41.

Ho, R. C., & Amin, M. (2022). Exploring the role of commitment in potential absorptive capacity and its impact on new financial product knowledge: A social media banking perspective. *Journal of Financial Services Marketing*, 1–14. doi:10.105741264-022-00168-7

Ho, R. C., Amin, M., Ryu, K., & Ali, F. (2021). Integrative model for the adoption of tour itineraries from smart travel apps. *Journal of Hospitality and Tourism Technology*, *12*(2), 372–388. doi:10.1108/JHTT-09-2019-0112

Ho, R. C., & Chua, H. K. (2015). The influence of mobile learning on learner's absorptive capacity: a case of bring-your-own-device (BYOD) learning environment. In Taylor's 7th Teaching and Learning Conference 2014 Proceedings (pp. 471-479). Springer, Singapore.

Ho, R. C., & Song, B. L. (2021). Immersive live streaming experience in satisfying the learners' need for self-directed learning. *Interactive Technology and Smart Education*, *19*(2), 145–160. doi:10.1108/ITSE-12-2020-0242

Humayun Kabir, M. (2018). Did investors herd during the financial crisis? Evidence from the US financial industry. *International Review of Finance*, *18*(1), 59–90. doi:10.1111/irfi.12140

Im, I., Hong, S., & Kang, M. S. (2011). An international comparison of technology adoption: Testing the UTAUT model. *Information & Management*, *48*(1), 1–8. doi:10.1016/j.im.2010.09.001

Iranmanesh, M., Min, C. L., Senali, M. G., Nikbin, D., & Foroughi, B. (2022). Determinants of switching intention from web-based stores to retail apps: Habit as a moderator. *Journal of Retailing and Consumer Services*, *66*, 102957. doi:10.1016/j.jretconser.2022.102957

Jagtiani, J., & Lemieux, C. (2019). The roles of alternative data and machine learning in fintech lending: Evidence from the LendingClub consumer platform. *Financial Management*, *48*(4), 1009–1029. doi:10.1111/fima.12295

Jalal, R. N.-U.-D., Alon, I., & Paltrinieri, A. (2021). A bibliometric review of cryptocurrencies as a financial asset. *Technology Analysis and Strategic Management*, *1*(1), 1–16. doi:10.1080/09537325.2021.1939001

Jang, J., & Choi, J., Jeon, H., & Kang, J. (. (2019). Understanding US travellers' motives to choose Airbnb: A comparison of business and leisure travellers. *International Journal of Tourism Sciences*, *19*(3), 192–209. doi:10.1080/15980634.2019.1664006

Jin, C. C., Seong, L. C., & Khin, A. A. (2019). Factors affecting the consumer acceptance towards fintech products and services in Malaysia. *International Journal of Asian Social Science*, *9*(1), 59–65. doi:10.18488/journal.1.2019.91.59.65

Karandikar, N., Chakravorty, A., & Rong, C. (2021). Blockchain based transaction system with fungible and non-fungible tokens for a community-based energy infrastructure. *Sensors (Basel), 21*(11), 3822. doi:10.339021113822 PMID:34073110

Latikka, R., Turja, T., & Oksanen, A. (2019). Self-efficacy and acceptance of robots. *Computers in Human Behavior, 93*, 157–163. doi:10.1016/j.chb.2018.12.017

Li, Y., Xu, Z., & Xu, F. (2018). Perceived control and purchase intention in online shopping: The mediating role of self-efficacy. *Social Behavior and Personality, 46*(1), 99–105. doi:10.2224bp.6377

Lucas, T., Alexander, S., Firestone, I. J., & Baltes, B. B. (2006). Self-efficacy and independence from social influence: Discovery of an efficacy–difficulty effect. *Social Influence, 1*(1), 58–80. doi:10.1080/15534510500291662

Mao, J., Chiu, C., Owens, B. P., Brown, J. A., & Liao, J. (2019). Growing followers: Exploring the effects of leader humility on follower self-expansion, self-efficacy, and performance. *Journal of Management Studies, 56*(2), 343–371. doi:10.1111/joms.12395

Maouchi, Y., Charfeddine, L., & El Montasser, G. (2022). Understanding digital bubbles amidst the COVID-19 pandemic: Evidence from DeFi and NFTs. *Finance Research Letters, 47*(1), 102584. doi:10.1016/j.frl.2021.102584

Miraz, M. H., Hasan, M. T., Rekabder, M. S., & Akhter, R. (2022). Trust, transaction transparency, volatility, facilitating condition, performance expectancy towards cryptocurrency adoption through intention to use. *Journal of Management Information and Decision Sciences, 25*, 1–20.

Okumus, B., Ali, F., Bilgihan, A., & Ozturk, A. B. (2018). Psychological factors influencing customers' acceptance of smartphone diet apps when ordering food at restaurants. *International Journal of Hospitality Management, 72*, 67–77. doi:10.1016/j.ijhm.2018.01.001

Park, A., Kietzmann, J., Pitt, L., & Dabirian, A. (2022). The evolution of nonfungible tokens: Complexity and novelty of NFT use-cases. *IT Professional, 24*(1), 9–14. doi:10.1109/MITP.2021.3136055

Pavlou, P. A., & Fygenson, M. (2006). Understanding and predicting electronic commerce adoption: An extension of the theory of planned behavior. *Management Information Systems Quarterly, 30*(1), 115–143. doi:10.2307/25148720

Peñarroja, V., Sánchez, J., Gamero, N., Orengo, V., & Zornoza, A. M. (2019). The influence of organisational facilitating conditions and technology acceptance factors on the effectiveness of virtual communities of practice. *Behaviour & Information Technology, 38*(8), 845–857. doi:10.1080/0144929X.2018.1564070

Rafli, D. P. A. D. (2022). NFT Become a Copyright Solution. *Journal of Digital Law and Policy, 1*(2), 43–52.

Ramadania, S., & Braridwan, Z. (2019). The influence of perceived usefulness, ease of use, attitude, self-efficacy, and subjective norms toward intention to use online shopping. *International Business and Accounting Research Journal, 3*(1), 1–14.

Ramírez-Correa, P., Rondán-Cataluña, F. J., Arenas-Gaitán, J., & Martín-Velicia, F. (2019). Analysing the acceptation of online games in mobile devices: An application of UTAUT2. *Journal of Retailing and Consumer Services*, *50*, 85–93. doi:10.1016/j.jretconser.2019.04.018

San Martín, H., & Herrero, Á. (2012). Influence of the user's psychological factors on the online purchase intention in rural tourism: Integrating innovativeness to the UTAUT framework. *Tourism Management*, *33*(2), 341–350. doi:10.1016/j.tourman.2011.04.003

Schaar, L., & Kampakis, S. (2022). Non-fungible tokens as an alternative investment: Evidence from cryptopunks. *The Journal of The British Blockchain Association*, *5*(2), 2516–3949. doi:10.31585/jbba-5-1-(2)2022

Schwoerer, C. E., May, D. R., Hollensbe, E. C., & Mencl, J. (2005). General and specific self-efficacy in the context of a training intervention to enhance performance expectancy. *Human Resource Development Quarterly*, *16*(1), 111–129. doi:10.1002/hrdq.1126

Shaikh, I. M., Qureshi, M. A., Noordin, K., Shaikh, J. M., Khan, A., & Shahbaz, M. S. (2020). Acceptance of Islamic financial technology (FinTech) banking services by Malaysian users: An extension of technology acceptance model. *Foresight*, *22*(3), 367–383. doi:10.1108/FS-12-2019-0105

Slater-Robins, M. (2022). Salesforce employees are reportedly unhappy about its NFT plans. *Techradar,* (22). https://www.techradar.com/news/salesforce-employees-are-reportedly-unhappy-about-its-nft-plans

Soliman, M. S. M., Karia, N., Moeinzadeh, S., Islam, M. S., & Mahmud, I. (2019). Modelling intention to use ERP systems among higher education institutions in Egypt: UTAUT perspective. *International Journal of Supply Chain Management*, *8*(2), 429–440.

Sung, H.-N., Jeong, D.-Y., Jeong, Y.-S., & Shin, J.-I. (2015). The relationship among self-efficacy, social influence, performance expectancy, effort expectancy, and behavioral intention in mobile learning service. *International Journal of U-and e-Service. Science and Technology*, *8*(9), 197–206.

Tarhini, A., Alalwan, A. A., Al-Qirim, N., Algharabat, R., & Masa'deh, R. (2018). An analysis of the factors influencing the adoption of online shopping. [IJTD]. *International Journal of Technology Diffusion*, *9*(3), 68–87. doi:10.4018/IJTD.2018070105

Tims, M., Bakker, A. B., & Derks, D. (2014). Daily job crafting and the self-efficacy–performance relationship. *Journal of Managerial Psychology*, *29*(5), 490–507. doi:10.1108/JMP-05-2012-0148

Valaei, N., Rezaei, S., Ho, R. C., & Okumus, F. (2019). Beyond structural equation modelling in tourism research: Fuzzy Set/qualitative comparative analysis (fs/QCA) and data envelopment analysis (DEA). In Quantitative Tourism Research in Asia, (pp. 297–309). Springer.

Valeonti, F., Bikakis, A., Terras, M., Speed, C., Hudson-Smith, A., & Chalkias, K. (2021). Crypto collectibles, museum funding and OpenGLAM: Challenges, opportunities and the potential of Non-Fungible Tokens (NFTs). *Applied Sciences (Basel, Switzerland)*, *11*(21), 9931. doi:10.3390/app11219931

Venkatesh, V., Thong, J. Y., & Xu, X. (2016). Unified theory of acceptance and use of technology: A synthesis and the road ahead. *Journal of the Association for Information Systems*, *17*(5), 328–376. doi:10.17705/1jais.00428

Wang, H., Tao, D., Yu, N., & Qu, X. (2020). Understanding consumer acceptance of healthcare wearable devices: An integrated model of UTAUT and TTF. *International Journal of Medical Informatics*, *139*, 104156. doi:10.1016/j.ijmedinf.2020.104156 PMID:32387819

Wang, Y.-S., Yeh, C.-H., & Liao, Y.-W. (2013). What drives purchase intention in the context of online content services? The moderating role of ethical self-efficacy for online piracy. *International Journal of Information Management*, *33*(1), 199–208. doi:10.1016/j.ijinfomgt.2012.09.004

Williams, D. M., & Rhodes, R. E. (2016). The confounded self-efficacy construct: Conceptual analysis and recommendations for future research. *Health Psychology Review*, *10*(2), 113–128. doi:10.1080/17437199.2014.941998 PMID:25117692

Wood, R., Atkins, P., & Tabernero, C. (2000). Self-efficacy and strategy on complex tasks. *Applied Psychology*, *49*(3), 430–446. doi:10.1111/1464-0597.00024

Wu, L. (2013). The antecedents of customer satisfaction and its link to complaint intentions in online shopping: An integration of justice, technology, and trust. *International Journal of Information Management*, *33*(1), 166–176. doi:10.1016/j.ijinfomgt.2012.09.001

Yang, X., Gao, M., Wu, Y., & Jin, X. (2018). Performance evaluation and herd behavior in a laboratory financial market. *Journal of Behavioral and Experimental Economics*, *75*, 45–54. doi:10.1016/j.socec.2018.05.001

Yeong, Y.-C., Kalid, K. S., Savita, K., Ahmad, M., & Zaffar, M. (2022). Sustainable cryptocurrency adoption assessment among IT enthusiasts and cryptocurrency social communities. *Sustainable Energy Technologies and Assessments*, *52*, 102085. doi:10.1016/j.seta.2022.102085

Younus, A. M., Tarazi, R., Younis, H., & Abumandil, M. (2022). The Role of Behavioural Intentions in Implementation of Bitcoin Digital Currency Factors in Terms of Usage and Acceptance in New Zealand: Cyber Security and Social Influence. *ECS Transactions*, *107*(1), 10847–10856. doi:10.1149/10701.10847ecst

Zhang, Y., Weng, Q., & Zhu, N. (2018). The relationships between electronic banking adoption and its antecedents: A meta-analytic study of the role of national culture. *International Journal of Information Management*, *40*, 76–87. doi:10.1016/j.ijinfomgt.2018.01.015

Zhou, T., Lu, Y., & Wang, B. (2010). Integrating TTF and UTAUT to explain mobile banking user adoption. *Computers in Human Behavior*, *26*(4), 760–767. doi:10.1016/j.chb.2010.01.013

KEY TERMS AND DEFINITIONS

Non-Fungible Token (NFT): Unique identifier used to validate and authenticate the ownership of digital assets ownership.

Performance Expectancy: The users' expectation of the technology tools in enabling them to achieve the desired performance.

Effort Expectancy: Effort devoted to the use of technological tools when the tasks were achieved.

Social Influence: Emotional feeling derived from the socially related support from carry out a certain behavior.

Facilitating Conditions: Business process and technical support offered by the platform or systems.

Self-Efficacy: Self-belief about own capability to perform a particular task confidently.

Chapter 6
Metaverse in Investment Using Sentiment Analysis and Machine Learning

Eik Den Yeoh
https://orcid.org/0000-0001-6232-2345
HELP University, Malaysia

Tinfah Chung
https://orcid.org/0000-0002-3993-6637
HELP University, Malaysia

Yuyang Wang
Institute of Automation, Chinese Academy of Sciences, China

ABSTRACT

At the end of 2019, individuals' outdoor activities were restricted due to the emergence of COVID-19. As a result of this phenomenon, interest in online activities and interaction in the metaverse environment has increased. Online games have exploded in popularity with the young generation in Metaverse where they can earn money through the platforms. Thus, it is desirable to investigate emerging technology and analyse how to invest using techniques, such as sentiment analysis and machine learning (ML), to predict crypto trends. This study analysed time series data for crypto price and text, where information like news, articles, and feedback from social media can use the input to generate the sentiment score to understand the crypto trends. FinBERT is a sentiment model that was used for this study to generate the result. The AI investing framework is built to incorporate both sentiment analysis technique and predictive model for this chapter, to address the research questions and enable one to make more informed decisions.

DOI: 10.4018/978-1-6684-5732-0.ch006

STUDY BACKGROUND

Metaverse investments can be made in the typical manner, for example, by buying stock related to metaverse technology companies such as Facebook, Nvidia, AMD (Robertson, 2021). Another approach is to concentrate on the emerging market of cryptocurrency economies, which is the frontier of investment on the metaverse and the forefront of this Web 3.0 internet revolution. Metaverse is an interconnected, immersive, 3D virtual environment in which people from all over the world may interact in real-time to create a lasting, user-owned internet economy that spans both the digital and physical worlds (Lee, 2021). However, several crucial components of the metaverse are already in place and are reshaping the worlds of business, entertainment, and even real estate. An open-world metaverse like Decentraland is a good example of the hype: users can log in to play games, earn MANA (Cryptocurrency – the native token with which users can purchase non-fungible tokens (NFT), including land or collectibles, and vote on economy governance commonly known as decentralised autonomous organisations (DAO)), or create NFTs, giving them real world interoperability for the value of their time spent in-game (Han et al., 2021). New waves of cryptocurrency investment have been accelerated by Facebook's decision to change its name from "Facebook" to "Meta" US (Ritter, 2021) as a sign of the internet's potential to evolve, as well as the spotlight on this emerging market.

The cryptocurrency (crypto) market has thousands of different varieties of crypto, making it challenging to apply predictive algorithms for investments that rely mainly on previous data. Investors who are new to cryptocurrency investment will face difficulties analyzing massive data from internet, especially when looking for Metaverse-related token or coins. With this challenge, investors can explore the innovative technique to make decisions by apply sentiment analysis and artificial intelligence (AI) algorithms to anticipate crypto trends based on social media and financial news data. An important consideration in investing decisions is the use of sentiment analysis, commonly known as "opinion mining," (Pascual-Ezama et al., 2014) which gathers information about what people think and feel about a certain subject of interest.

Making an informed decision on whether to sell or buy crypto requires the ability to predict the direction of the crypto market. In order to maximise profits, the investor buy crypto that is expected to rise in value and sells it at a high price to make money in the crypto market, an investor needs to be able to anticipate market moves. The crypto market has been impacted by a current trend in 2021 where the CEO of Facebook, Mark Zuckerberg, has announced that the company's name will be changed to "Meta" effective Thursday, Oct 28, 2021, in New York City (Ritter, 2021). One of the factors influencing market and investor decisions is news and search frequency for Metaverse. Investors who are well-versed in the news, particularly those who have previously made a significant profit in gaming related to the metaverse concept, may be able to anticipate market emotion and price movements in the cryptocurrency market. In contrast, the price of metaverse-related currencies like Decentraland (MANA), The Sandbox (SAND), and Enjin (ENJ) rise to all-time high from 517.94%, 314.15% and 234.02% (Refer to 1.1, 1.2 and 1.3) respectively after Mark Zuckerberg announced the changes to Facebook's name. By analysing a large amount of data in a short period of time, the compute engine helps investors react and respond more swiftly. Scientific evaluation of this study will take place, but first we must look at the methodology used.

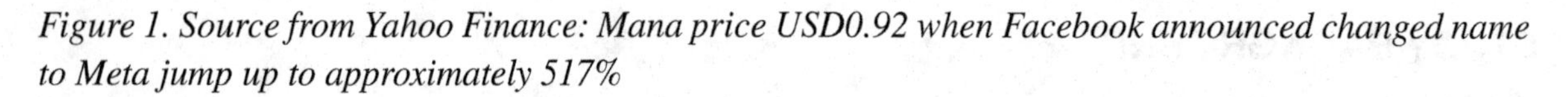

Figure 1. Source from Yahoo Finance: Mana price USD0.92 when Facebook announced changed name to Meta jump up to approximately 517%

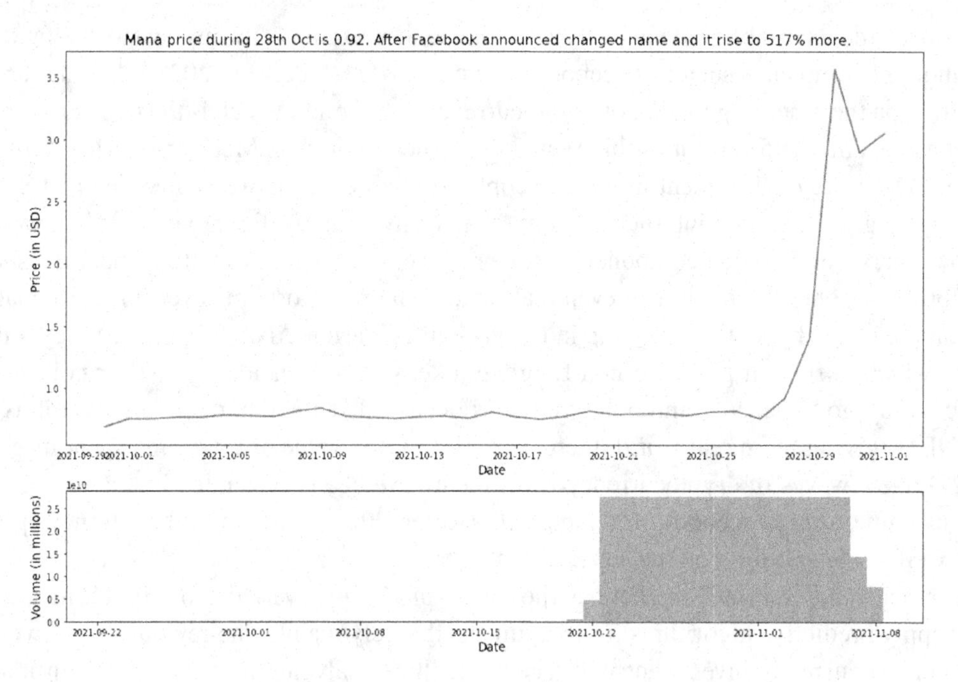

Figure 2. Source from Yahoo Finance: Sandbox price USD0.94 when Facebook announced changed name to Meta jump up to approximately 314%

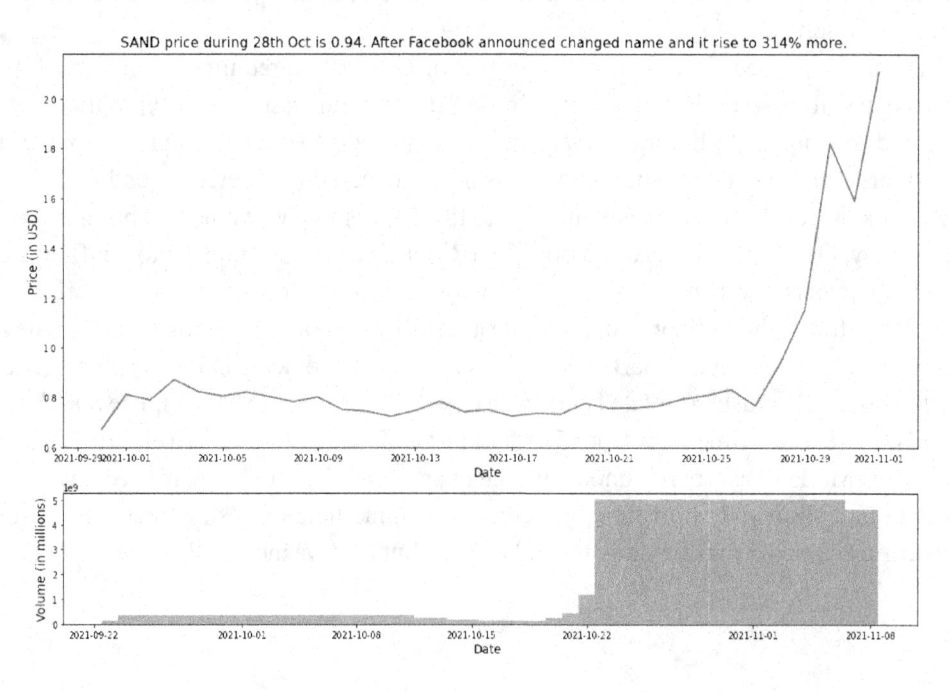

Figure 3. Source from Yahoo Finance: Enjin price USD2.32 when Facebook announced changed name to Meta jump up to approximately 234%

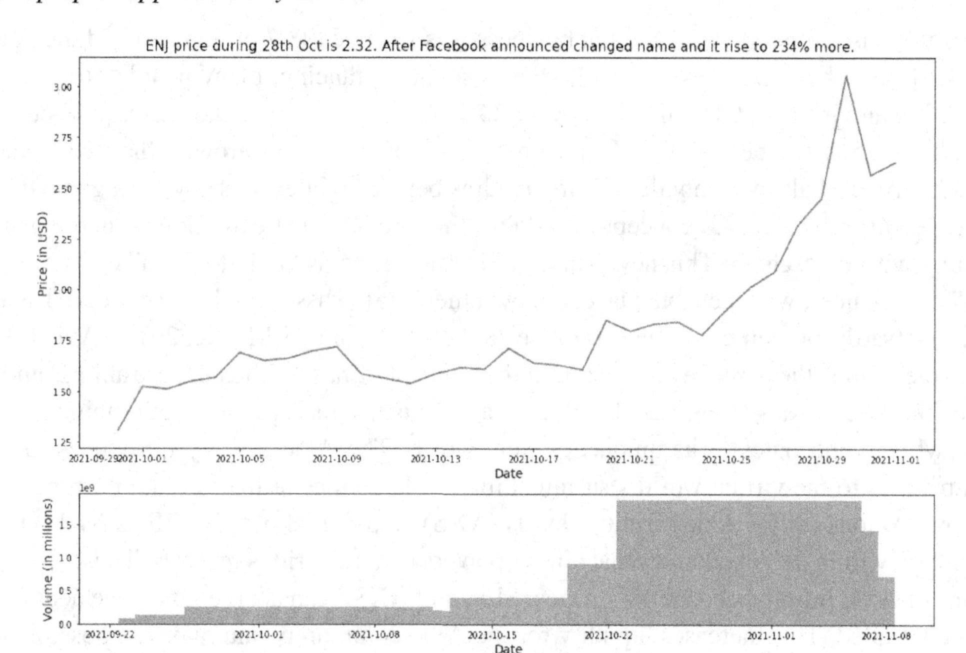

THE EVOLUTION OF WEB TECHNOLOGY

Look back from two decades ago, the Internet from Web 1.0 to Web 2.0 and now Web 3.0. From internet relay chat (IRC) – Web 1.0 to modern social media platforms – Web 2.0, we have been experienced it. From simple digital payments – Web 2.0 to sophisticated online banking services -Web 2.0 and now cutting-edge Internet-based technologies such as cryptocurrency and blockchain – Web 3.0. The Internet has evolved into a critical component of human interaction and connectivity – and it will continue to do so. Even though major corporations still control most of Web 2.0, the trend is shifting to Web 3.0 which is toward decentralized. One of the reasons why Web 2.0 is changing to Web 3.0 is because the biggest corporations have full control over access to the service. When a user agrees to the terms of a service, they agree to give up their private information in exchange for access. For instance, Instagram's photos that users have uploadeddo not, in fact, entirely belong to the users. Every piece of content that Instagram users create is under their control. If they want to block you, they have the authority to do so, and data privacy is a major concern with Web 2.0 as well. However, the Web 3.0 notion of decentralization refers to the future generation of the internet, which will see an evolution in how people can own and own their creations and online information, digital assets, and identities. To make the Metaverse function, numerous new technologies, protocols, enterprises, breakthroughs, and discoveries will be required. Additionally, there will be no "Pre-Metaverse" or "Post-Metaverse." As a result, it will take time until as many goods, services, and capacities as possible emerge together. Especially how metaverse that create a sustainable virtual economy.

THE EVOLUTION OF METAVERSE IN TOKENOMICS

In 2021, a study discovered that the gaming business was worth $173.7 billion and had the potential to reach $314.4 billion by 2026 (Wood, 2021). Due to social distancing, many people turned to gaming as a form of fun and interest. Globally, there were 2.2 billion gamers, with the figure predicted to reach 2.7 billion by 2025 (J Clement, 2021). Unfortunately, despite its rapid growth, the sector has always been lopsided. Almost always, only the gaming studios benefit, while users pay for a game they do not own. Whereas this new web 3.0 concept introduces them to the next evolution of gaming that combines gaming with blockchain. This new paradigm is referred to as "Play-to-Earn" or "Game Finance (GameFi)" experiences, which enable players to own their digital assets in the form of NFT and play in order to earn rewards for their contributions to the gaming community (Merwe, 2021). Axie Infinity is a prime example of this, the game revolves around the concept of raising, breeding, training, and battling digital pets known as Axies (Blaise et al., 2022). Each Axis has its own distinct combination of genes and rarity, which contributes to its uniqueness and worth. The Axie Infinity introduces tokenomics, which contributes to the virtual world's sustainability. Tokenomics utilizes digital currencies such as Smooth Love Potions (SLP), Axie Infinity Shards (AXS), and digital pets (AXIE as NFTs) to create a unique function within the tokenomics that allows players to earn various rewards. These three primary components of Axie Infinity's economy - Axies, SLP, and AXS - can accrue (and have acquired) value and are freely traded. This increases liquidity for the Axie community and makes it possible to earn a sustainable environment in metaverse. Innovation with blockchain and cryptocurrencies has helped turn gaming into a source of revenue rather than just entertainment for thousands of people. The concept of in-game rewards has proven to be successful, allowing users to earn crypto for actions they take while playing the game. However, to identify sustainable metaverse platform is not easy.

Problem Statement

Due to a lack of understanding and experience, prominent investors find it difficult to participate in the crypto market. Even though investing in the cryptocurrency (crypto) market using a systematic manner will result in higher earnings but the risk is huge. Making money is impossible for even the most experienced investors to predict with any degree of certainty. One of the contributing causes to price adjustments is the impact of the news. News affects the cryptocurrency markets, therefore it's important to keep an eye on to monitor it closely. When it comes to buying and selling cryptocurrency, there are several factors to consider, such as the crypto's news, the integration of smart contracts like Ethereum (ETH), Solana (SOL), and Polkadot (DOT) into future projects, and the impact of government legislation on the cryptocurrency market (Khan et al., 2021). News, social media, and forums can all provide insight into how the market react, particularly economic news, is trending and how this may affect crypto values (Abraham et al., 2018). Although most crypto market data is symmetrical or efficient (Tran & Leirvik, 2020), it can be difficult to obtain the information needs due to the difficulty of reliable data available in public. Human effort is required to ensure that collected data is properly interpreted. Time becomes critical for them to examine and analyse the material they have gathered; particularly information that could easily delay their decision and result in monetary loss.

The Metaverse Play-to-Earn Challenge for Investor

According to BscProject.org, there are over 210 games running on the Binance Smart Chain, servicing an average of 890,000 players each day during December 2021 (Alvin, 2021). Imagine that other blockchain networks like Ethereum (ETH), Solana (SOL), Fantom (FTM), Polkadot (DOT), Avalanche (AVA) and etc, added up may have more than thousand over games exist in the market. This is where most games or metaverse platforms struggle to differentiate themselves with maintain a stable economics into their platform. In order to preserve the metaverse's environment, sustainability is a critical component of determining a good investment. Investors who need to understand what information is critical to extract in order to identify a viable sustainable platform, key features required to incorporate into an artificial intelligence model to assist in detecting this critical information. For example, the total number of NFT units created daily, is the platform provides a method for burning the NFT to continue daily unit creation maintain in healthy level, the number of unique addresses that hold the assets, and whether the project's continuous to enhance it for attract players continue stay in their platform (Lee, 2021). The important to sustain the entire ecosystem is to control the supply and demand from the gamers, and investors. According to supply and demand theory, the interaction between sellers and buyers of sources determined the relationship between the price of specific NFTs and people's willingness to buy or sell. Generally, when prices rise, people are willing to supply more and demand less and vice versa when the prices fall. This requires applying of the law of supply and demand in order to generate the equation that will be used in the Artificial Intelligence model to quantify risk. The additional features and sentiment result are additional information to help anticipant the trend of crypto's price.

Objectives

The current crypto market investment options may difficult and unfamiliar to investors. Because they have no idea which cryptocurrency to buy or sell, they would be unable to gain a greater return. Investment in the cryptocurrency market is dependent upon reliable information. The efficiency of their trades is relying on latest news or up-to-date information on crypto market as well as the able to obtain the latest information near to live, particularly from the United States and Europe. Crypto market is 24 hours x 7, they never close the market even holiday. This is where the information is difficult for human to detect strong signal that need to follow from time to time. Identify suitable crypto assets that are connected to the metaverse in order to apply demand and supply laws. Understanding the link between investors, gamers, and creators can assist us in determining the platform's viability. This is capable of defining the strengths and weaknesses of cryptocurrencies prior to applying sentiment analysis and artificial intelligence to predictive trends. Sentiment analysis techniques are crucial for generating the polarity of the score, and the FinBERT model was used to classify "positive", "neutral" and "negative" scores from financial news and social media comment. This classification later applies into machine learning (ML) which is gradient boosting to generate predictive trends that recommend potential direction. This approach provides more comprehensive method for investor to apply better strategy and mitigate risk for the investment portfolio. Hence, the objectives for this chapter are as follows:

1. Identify 5 good crypto linked to metaverse. List the advantages and disadvantages of tokenomics from those crypto.

2. Apply sentiment analysis based on social media news about the selected cryptocurrency and generate recommendation to assist investors in deciding whether to buy or sell. By making more informed planning recommendations and mitigating risk in the face of changing trends.
3. Examine the machine learning model including sentiment analysis result to improve and predict the overall crypto price trends.

Research Questions

1. What are the top 5 potential cryptocurrencies related to Metaverse? What is their weakness and strengths that helps investors to understand before investing on them?
2. What are the trend and which type of textual could impact on Metaverse's top 5 selected cryptocurrencies?
3. Is the price prediction trend for this top 5 selected cryptocurrencies are better performance with sentiment scores?

Significance of the Study

Majority studies are targeted to explore stock market how sentiment analysis could help them make better decision even research papers studies about cryptocurrency. However, there is limit study regarding metaverse linked cryptocurrencies. Therefore, this chapter will provide insights about metaverse-linked crypto by using sentiment analysis and machine learning that create a unique AI investment framework for this chapter.

Investing in crypto if able to foresee the crypto market's direction could help make better decision on whether to sell or buy. The investor buys crypto that is expected from the lower price and sells it for a profit at a higher price could make better profit from the crypto market. Generally, investors who are familiar with the news can predict the crypto market's direction especially they understand well with behaviour finance. In addition, whether metaverse is correlate with breaking news could influence them move to the same direction with stock index or other major crypto. With the theory apply and building the AI investment framework, it could resolve the challenge of analysis a mass data with a short period of time to identify high accurate trend.

Limitation of Study

In today's digitally transformation world, an innovative technology emerges every day that alters our perception of reality. Detecting metaverse-related crypto is a significant difficulty, given it is subject to change on a daily basis. Even machine learning (ML) and sentiment analysis models must be updated on a regular basis to provide a high degree of accuracy in the AI investment model. As a result, the AI investment framework for this research could be refined and continue take over by new researcher continue study and explore the new method to enhance further. Moreover, several types of textual sentiment analysis may be influenced by distinct crypto properties. For instance, game-related crypto in a community's discussion may not have an effect on the price of crypto, as their discussion may be limited on how to play more effectively. There are still many different types of news that use it for textual in sentiment analysis for detect the trends that doesn't conduct in these experiments where researchers could further analyse it especially breakdown into further details such as News content from "English", "Mandarin", or other languages that could impact even better. In this study, we obtained news exclusively from Forbes; nevertheless, additional news sources such as CNN, Blomberg, and others may have an impact on the overall research that can be used for future study.

The results of analysing intervals of minutes, hours, daily, or monthly can vary, and it may be difficult to extract data for intervals of minutes or hours, as few data sources provide information for these intervals. If so, paying for API subscriptions can be extremely expensive. If data is provided at minutely intervals, it may be valuable for the investigation of high frequency trading. However, the concept presented in this chapter could be applied to the analysis of high frequency trading.

Methodology

To construct an AI investing framework that enables investors to make more informed decisions about metaverse-related crypto, it is critical to develop a proper methodology approach. The methodology approach's purpose is to address the 3 major research questions raised in this chapter. This necessitated the use of sentiment analysis techniques on social media data in order to forecast the crypto price trend, which needed the integration of two parts of techniques into one framework (Refer to 1.7.1). The solution is developed primarily in Python and makes use of a variety of libraries for preprocessing, sentiment analysis models such as Finance Bidirectional Encoder Representations Transformers (FinBERT), and predictive models such as machine learning – gradient boosting, a type of regression model.

This involved the use of sentiment analysis techniques on social media, Discord community discussion and Discord announcement from top 5 metaverse crypto in order to anticipant the crypto price movement, which requires the merging of two parts of approaches into one. Sentiment score computed using FinBERT where a set of variables is to be defined so that regression model can be predict the crypto price movement. In this case, the dependent variable will be defined as the price of selected crypto linked to metaverse. While the independent variables will be represented by a sentiment analysis score (news, or social media data). The data collection is primarily based on time-series algorithms that extract data from Oct 2021 to Apr 2022 for crypto price and textual analysis used, a period that includes noteworthy events such as the excitement around the Metaverse and the Russia-Ukraine war. A comprehensive of processes that can found in 1.7.1 of the AI investment frameworks.

AI Investment Framework

Figure 4. AI Investment Framework - Extract data, Preprocessing Data, Feature Engineering, and Classification of NLP for Sentiment Analysis

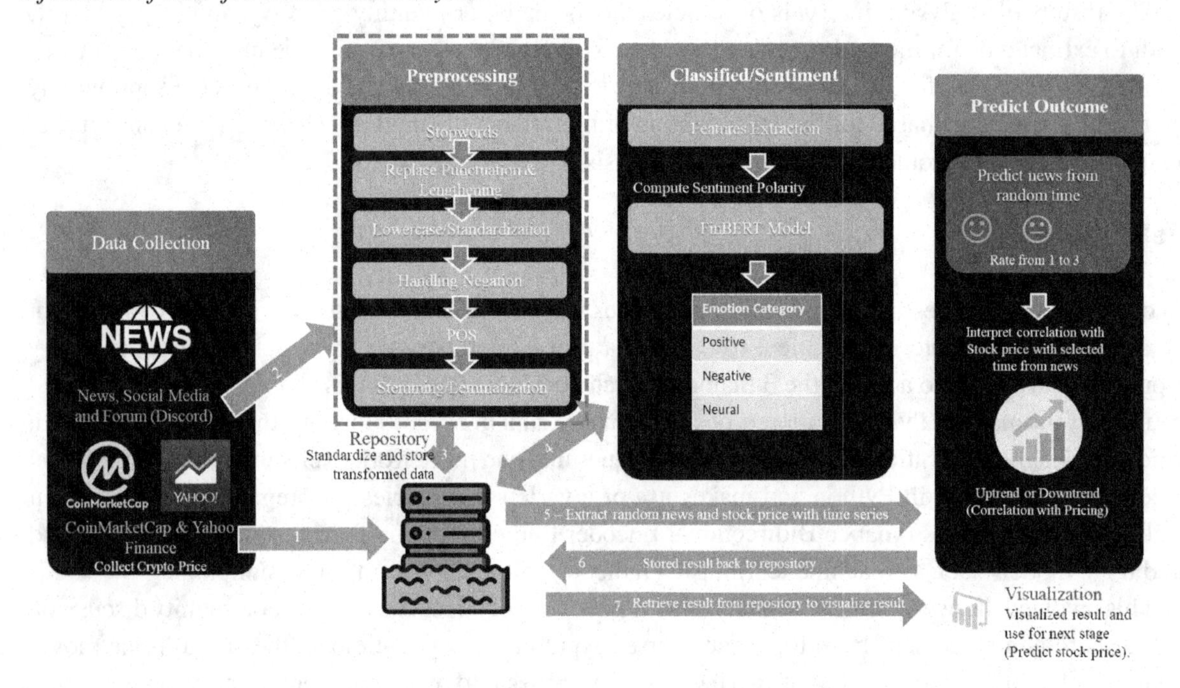

The proposed framework provides sentiment analysis on investor review that predefined investment sentiment data which can be used in FinBERT. FinBERT is state of art which already pre-train on financial term that can be direct use for identifying sentiment polarity. The proposed framework consists of the four components: Data collection, preprocessing, classification, and prediction. We summarise the process flow of how NLP derives from this chapter as below:

1. CryptoMarketCap and Yahoo Finance data extraction to help address the statistical result to understand the fundamental of crypto and use for sentiment and predictive model later. Data are kept in the storage, such as stock closing price, open price, adjusted price, volume etc.
2. News (Forbes), social media or forum (Discord) data extract from different time series to keep in repository to identify the sentiment polarity later. It is required to go through the preprocessing step to remove noise.
3. Keep the preprocessing data into excel (repository) for use to run the sentiment analysis model.
4. Data transfer to ML & DL modelling to classify the sentiment polarity and compare the performance. The result generated will be stored in the storage as a dictionary to retrieve the model later.
5. Retrieve strong sentiment score on news and discord's data against Yahoo Finance data with time to ensure alignment to generate the correlation result for interpretation whether is bullish, bearish, or neutral are closely correlated. Then generate the prediction result to analyze crypto price trend.
6. The result is stored back to the repository where it will be used for visualizing or analysis and later.

7. Visualization tools extract it out to analyze the data further to understand the crypto price trend as well as overall sentiment result.

DATA COLLECTION AND STATISTICAL ANALYSIS

In the following part, we detail the procedure of data collection, which will subsequently be used to provide sentiment results and predictive technique from the top five cryptocurrencies' price trends that have been defined.

Coinmarketcap Data

The approach in which Twitter, Facebook, and other social media firms structure their APIs places restrictions on how researchers may design their studies and what they can study using APIs as a methodological instrument. To assess the utility of API data in empirical research, it is crucial to consider the sorts of user data that may be collected, by extension, the categories of users that APIs enable us to study. CoinMarketCap API (Haslhofer et al., 2021) provided the necessary information for this chapter. The APIs[1] connector must register with CoinMarketCap prior to allowing us to gather data from them. There are several subscriptions that need to pay attention. They provided basic, hobbyist, startup, standard, professional and enterprise model where basic is the free subscription that use for this research. The basic subscription is only use for personal use and not for commercial purpose. Hence, the data collection via this API has to be ethically use for research only. The CoinMarketCap API is a collection of high-performance RESTful JSON endpoints that researchers can use to explore their research. To decide the top five metaverse crypto, the CoinMarketCap API is an indispensable connection for extracting the data needed to analyse the crypto's fundamentals. To understand their market capitalization, we must consider their trading volume, liquidity, circulating supply, total supply, and Fundamental Crypto Asset Score (FCAS) for us to identify the top five metaverse crypto for this study.

Cryptocurrency Market Capitalization

To find top 5 metaverse related cryptocurrencies, we have to analyse the market capitalization in the first step. Market capitalization (van Tonder et al., 2019) of the crypto asset is calculated by multiplying the existing price of crypto asset by the current circulating supply. Let's take the market capitalization of Bitcoin as an example to understand the concept:

Let (C) be the last known reference price of Bitcoin from CoinMarketCap in USD.
Let (S) be the current circulating supply of Bitcoin
Let (D) be the derived market capitalization of Bitcoin.
For this example, let (C) = $10,000/1 BTC and let (S) = 17,000,000 BTC
D = C * S
D = $10,000/1 BTC * 17,000,000 BTC
BTC = $170,000,000,000
Therefore, the derived market capitalization for Bitcoin is USD170,000,000,000.

Trading Volume & Liquidity

A crucial metric, the trade volume statistic reflects how much is bought and traded over time. Liquidity is a number that indicates how easily a crypto asset can be bought or sold, which is crucial since it enables you to buy and sell at any moment (Bianchi et al., 2019). This is especially important when evaluating low-volume projects where you may not be able to sell out at your preferred time. A medium-cap project's low trading volume may also signal that it has been abandoned, lacks a real-world use case, or has a small community, all of which are red signals.

Circulating Supply & Total Supply

It is essential to differentiate between the circulating supply and total supply of a cryptocurrency project. The circulating supply is the number of coins/tokens that are now in circulation, whereas the total supply is the greatest number of coins/tokens that might be produced. Contrary to the total supply and maximum supply, the circulating supply is the number of coins available to the public (Bianchi et al., 2019). The total supply is used to determine the total number of coins in circulation, which is equal to the total number of coins issued minus the total number of coins burnt. The entire supply is equal to the sum of the coins in circulation and those held in escrow. Escrow refers to staking, farming, or restricted-for-a-period-of-time-sale assets held by founders or first investors (locked).

Bitcoin serves as a useful reference point for figuring out the circulating supply and overall supply of the project. There are only 21 million Bitcoins in existence, with 19 million now in circulation. The scarcity of Bitcoin coins is obviously higher value when compared to a project like Ripple (XRP), which has a total quantity of 100 billion and a circulating supply of about 47 million (thus the higher value for one Bitcoin). In situations where the total supply is many orders of magnitude more than the circulating supply, a significant increase in the circulating supply can rapidly depress the price.

Fundamental Crypto Asset Score (FCAS)

FCAS stands for Fundamental Crypto Asset Score, a single consistently comparable value for measuring crypto project health. FCAS measures users activity, developer behavior and market maturity.

Identify Top 5 Metaverse Crypto (Coins)

The CoinMarketCap API can extract the circulating supply, total supply, maximum supply, FCAS score, and grade for this market capitalization. The API generated results for the top five metaverse coins are shown in Table 1 on 25 April 2022.

Table 1. Top Five Metaverse Coins by CoinMarketCap API

	Name	Symbol	Rank	Market Cap Dominance	Market Cap	Price	Circulating Supply	Total Supply	Max Supply	Volume (24h)	Date Added	Score	Grade	Last Updated
1	Decentraland	MANA	35	0.1959	3.52B	1.912301	1.84B	2.19B	None	313.7M	2017-09-17 00:00:00+00:00	899	A	2022-04-25T12:34:00.000Z
2	The Sandbox	SAND	40	0.1695	3.05B	2.631628	1.16B	3B	3B	503.55M	2020-08-05 00:00:00+00:00	0	No Grade	2022-04-25T12:35:00.000Z
3	Theta Network	THETA	41	0.1602	2.88B	2.882165	1B	1B	1B	203.49M	2018-01-17 00:00:00+00:00	486	F	2022-04-25T12:35:00.000Z
4	Axie Infinity	AXS	46	0.1409	2.53B	41.580707	60.91M	270M	270M	275.32M	2020-08-31 00:00:00+00:00	0	No Grade	2022-04-25T12:34:00.000Z
5	Stacks	STX	64	0.0782	1.41B	1.073816	1.31B	1.35B	1.82B	43.33M	2019-10-28 00:00:00+00:00	760	A	2022-04-25T12:34:00.000Z

Decentraland (MANA), The Sandbox (SAND), Theta Network (THETA), Axie Infinity (AXS), and Stacks (STX) are ranked 25, 40, 41, 46, and 64 respectively in the overall market on 25 April 2022. This is quite high compared to other currencies on the market. The FCAS score, MANA, THETA, and STX were assessed further. The scores of 899, 486, and 760 correspond to an A, F, and A, respectively. SAND and AXS do not disclose the score and grade, however this does not imply that a coin with no FCAS score, and grade is worthless. Is still needed to look into the overall market capitalization, total supply and circulating supply. Notice that the maximum supply of MANA coins is unlimited, whereas other coins have a limited supply (O'Leary, 2017). These MANA coins are burned or used to buy virtual land. In order for us to understand the important of maximum supply with or without limit. The benefit of a limited supply is that inflation can gradually destroy the value of crypto price over time. Almost every crypto is released with a limit supply at the time of its launch. Each coin's source code decides its quantity; for example, only 21 million Bitcoins have been issued worldwide. Thus, as demand increases, its value will increase, keeping pace with the market and preventing inflation eventually. Similar to Ethereum, the downside of an endless supply is that there is no defined limit. Thus, Ethereum has the potential to cause inflation. As with Ethereum, there is no hard limit for Bitcoin. This would imply that Ethereum is susceptible to inflation. In order to prevent inflation, the amount "burned" or "lost" is determined.

Circulating supply is importance metric for us to evaluate. Like explain early, circulating supply refers to the total number of an asset's crypto that are publicly tradable via centralized exchanges (CEX) or decentralized exchanges (DEX) at a given time. Generally used to calculate market capitalization given these crypto are those that reflect market demand most directly. In Table 1 shown that Decentraland, The Sandbox and Theta Network provided quite huge amount of circulate supply. Decentraland larger than the other 4. Decentraland have 1.84B follow by Stacks 1.41B, The Sandbox 1.16B, Theta Network 1B and Axie Infinity 60.91M. Low circulate supply might also be easier to influence the price. If the demand increase, the price will go up even higher. From each of their whitepaper mentioned in tokenomics. Decentraland, The Sandbox, and Stacks only have one token to contribute into the tokenomics whether Axie infinity have AXS, they also include Smooth Love Potion (SLP) into tokenomics, Theta Network have other token which is TFuel that use for payment transaction. Obviously, Axie infinity price is higher than the rest as the circulating supply is less than the rest. However, in Axie infinity another coin Smooth Love Potion (SLP) mainly use to reward player from play to earn and the max supply is unlimited. In fact, this broken the tokenomics that cause the price keep fallen since 2021. Example, the game keeps rewarding gamers whatever gamers win the games but the consume of SLP in the supply is less than what been generated. In other hand, if the generated coin is less than consume, then the price will move up. Let's check up the price of SLP with Figure 4. The SLP price over the time, the coin became over supply and cause the pricing drop drastically. When the coin keep generated from the game and doesn't have enough mechanism to burn the coin. The price will go down. Hence, the supply unlimited have this disadvantage which is one of the areas that Decentraland may happened.

Figure 5. Source from Yahoo Finance: SLP price from Jan 2021 to Apr 2022

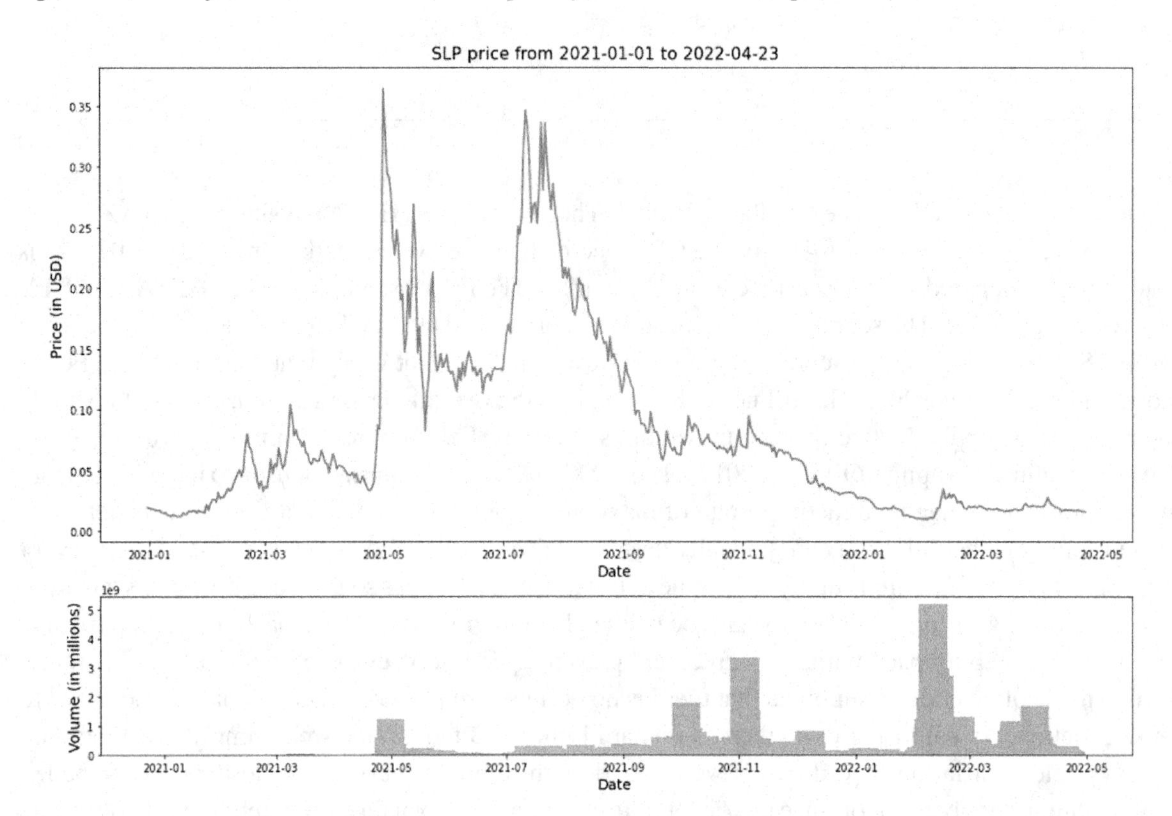

Another good example occurred on November 25, 2011, when the volatility of the coin ETERNAL, which is used in the game CryptoMines, dropped sharply from USD801 to around USD5 in two days before gradually recovering to USD95 (refer figure 5). As a result of the event, gamers were scared by the price drop and withdrew all the rewards, effectively rendering the accumulated coins worthless due to a lack of liquidity to supply the reward. Despite the company's announcement that it would launch a new game on a new platform, the crypto ETERNAL became useless. As a result, liquidity within the circulating supply is critical. If there is no remaining liquidity, like in the case of ETERNAL, your coin will be worthless.

Figure 6. Source from Yahoo Finance: ETERNAL price from Jan 2021 to Apr 2022

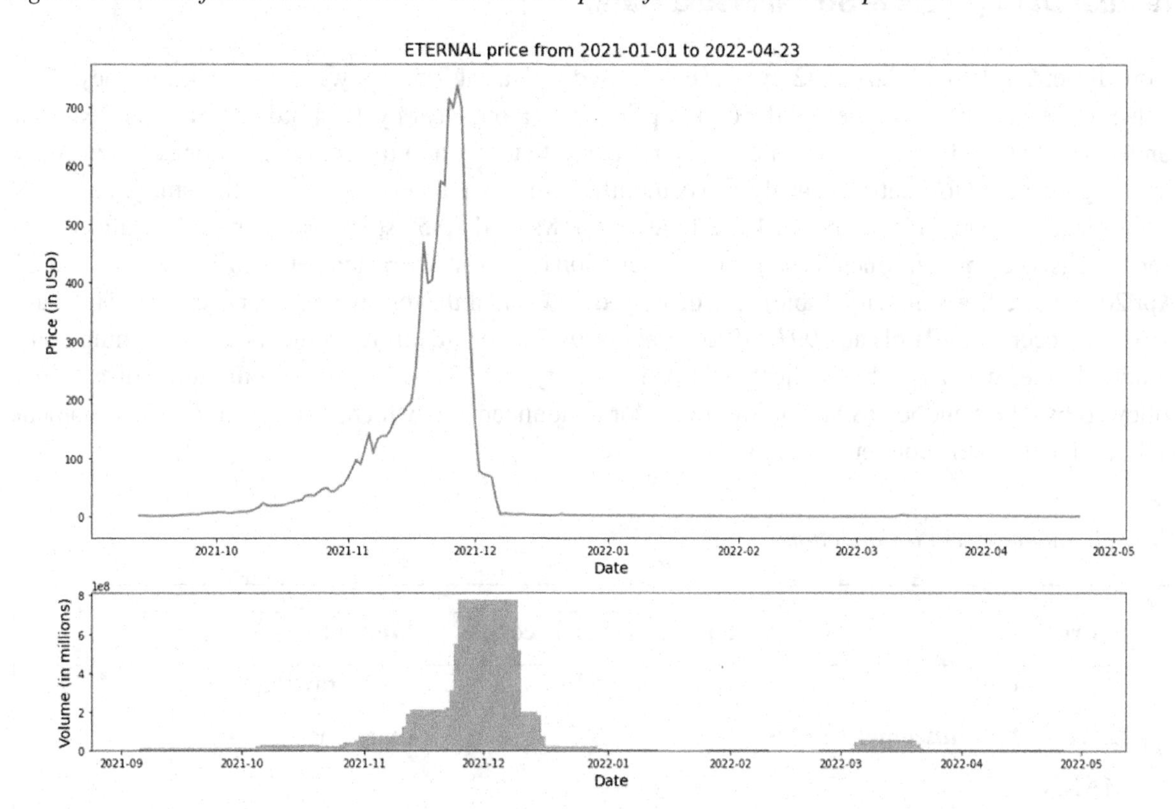

Hence, the top five identify metaverse's coins are Decentraland (MANA), The Sandbox (SAND), Theta Network (THETA), Axie Infinity (AXS) and Stacks (STX) based on the Table 2 listed based on the analysis. We can conclude to use the top five metaverse crypto for this research.

Yahoo Finance Crypto Price Data

We acquired price information for MANA, SAND, THETA, AXS, and STX using API queries to Yahoo Finance. Yahoo Finance's API (Kingaby, 2022) is open source where anyone can access the data without required to register. This study collected data from 1 Jan 2021 to 30 Apr 2022 to understand the historical pattern. The data is break to two parts. First dataset filter from 1 Oct 2021 to 30 Apr 2022 just for use to streamline the study for sentiment analysis with develop a daily price charge to analysis the relationships. Second dataset is collecting the full set which is from 1 Jan 2021 to 30 Apr 2022 for historical data statistical study purpose only. The Yahoo Finance API can be accessed with web data reader[2] from pandas module in Python's library. This library provides functions for extract data from numerous internet sources, including the World Bank, OECD, Eurostat, Quandl, Yahoo Finance, etc. The Yahoo Finance API endpoints are accessible for retrieving daily stock prices, including the price of cryptocurrency (high, open, close, volume, adjusted close).

Textual Data (News & Social Media Data)

From 1 Oct 2021 to 29 Apr 2022, we have collected a total of 7,530 news from Forbes Money. This is due to the fact that we collected the crypto price from a one-year cycle in order to find the seasonal tendencies that could be used for forecasting purposes that mention in section 2.2. For news, we only need to gather data for sentiment analysis; six months' worth of data is sufficient for this study. The news categorized into 960 crypto's news, 4,715 finance's news, and 1,855 stock market's news. Data collection for Discord announcements and general discussion data were extracted between 1 Oct 2021 and 29 Apr 2022 as well. A summary Table 2.1 is displayed for each of the top five metaverse crypto. There are 476 announcement records and 997,179 general discussion records from members of the gaming community. In the summary table, indicate that Axie Infinity (AXS) has the active community discussion, followed by The Sandbox (SAND). Similarity for announcements which Axie Infinity's development activity update to the community is the most active.

Table 2. Summary Data Collections

Source	Total number records	Method
Forbes Money	7,530	Web Crawling
Discord Announcement Total	**476**	Extractor Tool
MANA	74	
SAND	83	
THETA	69	
AXS	208	
STX	42	
Discord General Total	**997,179**	Extractor Tool
MANA	92,502	
SAND	366,716	
THETA	24,696	
AXS	501,921	
STX	11,344	

Technically, web crawling (Najork, 2017) was used to extract data for Forbes news. The major purpose of this Python scrapy-based method is to extract structured data from unstructured web pages. Scrapy features the Selenium library[3], which extracts a website via its URL and aids in crawling the HTML value from the targeted URL, enabling us to extract the metadata in Excel, including title, description, author, and publication date. Following this, we categorised the news into Finance, Stock Markets, and Cryptocurrencies (crypto) in order to evaluate the relationship between crypto pricings and their underlying fundamentals.

Discord data collected via export tools where we can be obtained by filter the date that required and specific group of discussion to extract. Refer to the export tools from GitHub repository that created by Tyrrrz[4] for further details. Figure 7 shows the breakdown of the transactions by week for the top five metaverse crypto.

Figure 7. Discord Data by Week frequency since Oct 2021 to Apr 2022

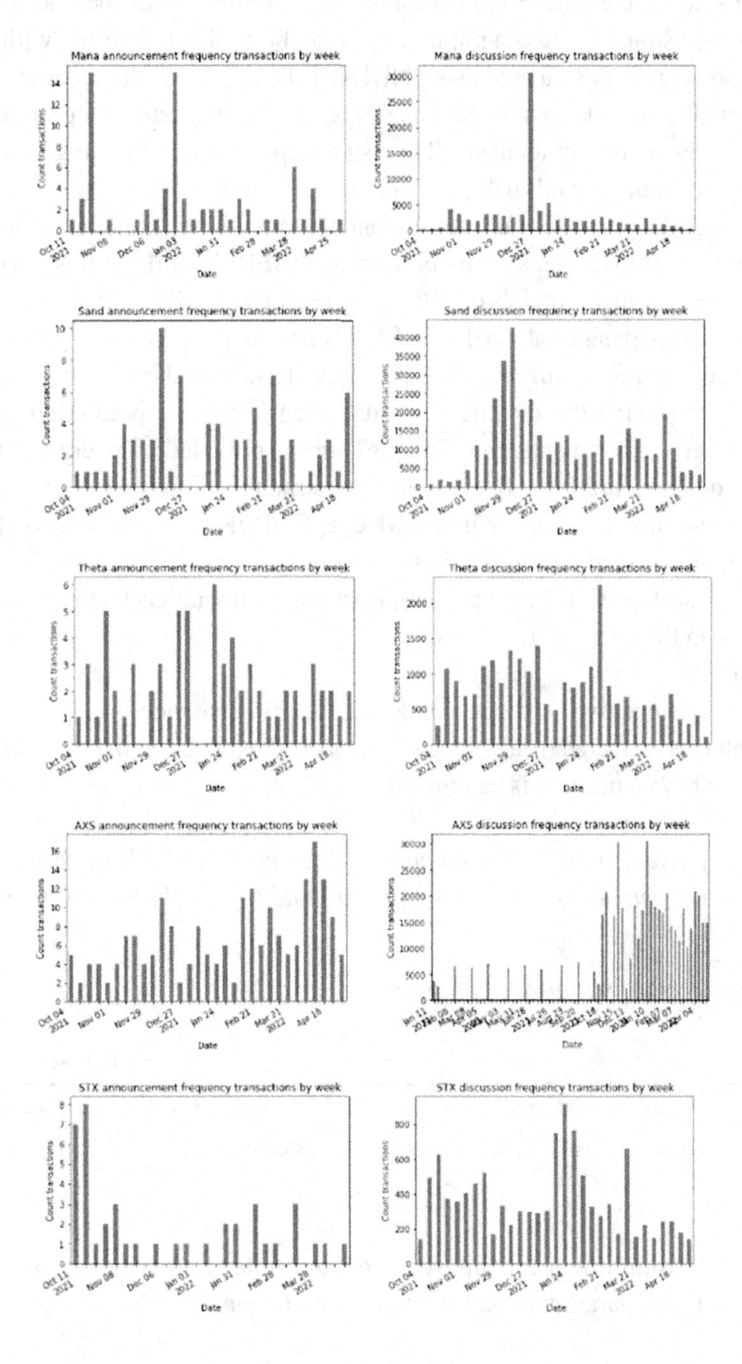

With the observation, Sand and AXS is most active discussion among others. There are 1780 discussions in Sand and 2699 discussions in AXS average per day.

SENTIMENT ANALYSIS

After data collection, the textual data are preprocessed in a variety of ways. Regular expressions and the Natural Language Toolkit (NLTK) (Loper & Bird, 2002) can be utilized to remove stopwords, punctuation, special characters, lower case, negation handling, and stemming. This method will aid in boosting the sentiment score's precision. NLTK is a popular Python library for interacting with human language data. Moreover, we tokenize every sentence using NLTK. Tokenization is the process of dividing a huge volume of text into smaller parts known as tokens. Tokenization is a crucial step in translating text to numeric data, followed by sentiment analysis. This is to helps machine learning (ML) models as ML require numerical data for training and making predictions.

After structuring and cleaning the data, the Finance Bidirectional Encoder Representations from Transformers are used to compute the sentiment score (FinBERT). FinBERT is derived from BERT, one of the most effective sentiment models developed in recent years. BERT was initially developed by Google and Devlin's team (Hoang et al., 2019). BERT is an open-source library that has been trained on millions of Wikipedia terms. Using the BERT model, it guesses the masked words. The strategy consists of attempting to predict the disguised words, often known as predicting the next sentence. Based on the Dogu Araci's experiments, the FinBERT model provided an overall of the result of 86% of accuracy which is quite accurate for finance and investment used (Araci 2019). Due to its unique focus on the financial and investment domain knowledge, FinBERT is a suitable methodology for this research. Each sentence computed by this FinBERT will produce a classification showing whether it is positive, negative, or neutral, as well as the probability associated with each classification (logits). The sentiment is calculated in the following:

$$\text{Sentiment} = \text{Logit}_{\text{positive}} - \text{Logit}_{\text{negative}}$$

Where, the sentiment is positive sentence minus the negative sentence.

The following content is extracted from a crypto-related Forbes article and preprocessed for sentiment analysis to demonstrate how sentiment is calculated.

"stepn cutting services players mainland china move major impact app tokens. move earn lifestyle app announced thursday would cut access users playing mainland China abide local regulations."

Figure 8. First sample result of sentiment analysis

	sentence	logit	prediction	sentiment_score
0	stepn cutting services players mainland china ...	[0.07829892, 0.34680462, 0.57489645]	neutral	-0.268506
1	move earn lifestyle app announced thursday wou...	[0.006916296, 0.84532917, 0.14775454]	negative	-0.838413

The FinBERT model assigns the preceding statement generated to two paragraphs with the result of -0.2685 and -0.8384 and both paragraphs result an average of negative value of -0.55. This result in the form of a represents the prediction of the model. The following example demonstrates the use of numerous sentences within a paragraph. Some of the statements may have alternative sentiment interpretations, which we can average to produce a more positive sentiment result.

"good news markets fed preferred inflation metric may peaked. today pce inflation number relatively good news markets suggesting inflation could trend lower summer. annual price change April 2022 6.3 percent. 6.6 percent march prices goods rose slower pace previously price increases services rose broadly similar rate recent months. stripping food energy annual inflation fell back 4.9 percent rate growth last saw december 2021. early sure inflation may trend lower. course inflation still well federal reserve 2 percent target direction potentially positive."

Figure 9. Second sample result of sentiment analysis

	sentence	logit	prediction	sentiment_score
0	good news markets fed preferred inflation metr...	[0.36811048, 0.61065453, 0.02123492]	negative	-0.242544
1	today pce inflation number relatively good new...	[0.96358657, 0.028451834, 0.007961505]	positive	0.935135
2	annual price change april 2022 6.3 percent .	[0.024601964, 0.014657142, 0.96074086]	neutral	0.009945
3	6.6 percent march prices goods rose slower pac...	[0.9722979, 0.01217896, 0.01552313]	positive	0.960119
4	stripping food energy annual inflation fell ba...	[0.0101604, 0.9754249, 0.01441473]	negative	-0.965264
5	course inflation still well federal reserve 2 ...	[0.95280415, 0.016268332, 0.030927548]	positive	0.936536

The above result indicates that the several paragraphs result with some neutral, negative and positive values with an average sentiment of 0.27. Another example below that show positive value of 0.9631.

"animoca leads million funding round hong kong nft platform amid crypto craze ucollex latest funding round comes sales digital collectibles gaining ground city seen increasing number projects launched past year."

Figure 10. Third sample result of sentiment analysis

	sentence	logit	prediction	sentiment_score
0	animoca leads million funding round hong kong ...	[0.9674181, 0.0042448044, 0.028337164]	positive	0.963173

The indicators for sentiment score are in three category, example 1 is extremely positive, 0 is neutral and -1 is extremely negative. This is known as lexicon-based sentiment analysis. In this research, we filtered 0.3 (0.3 to 1) above positive sentiment and -0.3 (-0.3 to -1) below negative sentiment result for us to step by step explore the analysis.

Run the Sentiment Analysis

Each headline and description are passed to the function (Python script) that generates sentiment scores for each statement. The content contained crypto news, stock market news, finance news, announcements, and general discussion for each of the top five metaverse crypto, including Decentraland (MANA), Sandbox (SAND), Theta Network (THETA), Axie Infinity (AXS) and Stacks (STX). The processed result summarised in Table 3. Crypto, stock market, finance news, Theta announcement, and AXS announcement all meet the criteria based on the observation that defined more than 0.3 and less than -0.3 sentiment scores.

Table 3. Summary of Sentiment Analysis for Crypto, Stock Market, and Finance's News and Top Five Metaverse's Announcement & General Result

	count	mean	std	min	25%	50%	75%	max
Crypto News Sentiment	208.0	0.051392	0.296211	-0.964458	-0.078611	0.065063	0.213769	0.830385
Stock Market News Sentiment	196.0	0.094445	0.275697	-0.860075	-0.060330	0.106294	0.234096	0.972315
Finance News Sentiment	211.0	-0.051767	0.127265	-0.493201	-0.118738	-0.041994	0.024828	0.374840
MANA Price Change	210.0	0.913592	12.909687	-18.863676	-3.862073	-0.377413	3.218068	154.738406
MANA Announcement Sentiment	47.0	0.037853	0.063858	-0.073532	0.008516	0.024366	0.053424	0.252991
MANA General Sentiment	211.0	0.036445	0.013780	-0.026008	0.028464	0.036839	0.045100	0.068979
SAND Price Change	210.0	0.854451	8.988221	-17.644210	-4.089958	-0.480892	3.614917	57.795167
SAND Announcement Sentiment	63.0	0.051709	0.084827	-0.297889	0.011075	0.039414	0.063605	0.289323
SAND General Sentiment	211.0	0.037858	0.006802	0.016068	0.033287	0.038149	0.042335	0.056771
THETA Price Change	210.0	-0.230857	5.925377	-18.709536	-3.617005	-0.568374	3.451576	21.630936
THETA Announcement Sentiment	60.0	0.047114	0.104060	-0.198348	0.000000	0.016442	0.046130	0.458129
THETA General Sentiment	211.0	0.035554	0.024585	-0.060468	0.021380	0.036561	0.048317	0.160607
AXS Price Change	210.0	-0.390995	5.823485	-15.854488	-3.667536	-0.899772	2.592650	28.141489
AXS Announcement Sentiment	108.0	0.014809	0.162097	-0.795129	-0.000354	0.023040	0.062977	0.490823
AXS General Sentiment	169.0	0.018484	0.008721	-0.017480	0.014646	0.018674	0.024764	0.036950
STX Price Change	210.0	0.080339	6.640183	-15.079642	-3.657351	-0.272690	2.936249	32.699087
STX Announcement Sentiment	32.0	0.037699	0.042884	-0.086523	0.008570	0.029814	0.066320	0.132280
STX General Sentiment	211.0	0.045833	0.032471	-0.054379	0.024460	0.046090	0.062974	0.135645

Follow the criteria more than 0.3 sentiment and less than -0.3 sentiment. The sentiment was consolidated the daily average result from original records. We further clarified again the consolidate result with explore the total count of sentiment for positive, neutral and negative of each category. As shown in figure 11 below, crypto news obtained of total 946 data points that aggregate the daily average sentiment, with 563 neutral, 228 positive, and 155 negative sentiments. There is a total of 196 of data in the stock market news, with 13 negative, 148 neutral, and 35 positive. There are a total of 211 sentiment results for finance, with 6 negative, 203 neutral, and 2 positive. There are 60 data into total for the Theta announcement and break down to 57 are neutral and 3 are positive. The overall number of AXS reviews is 108, with 7 negative, 96 neutral, and 5 positive.

Figure 11. Total number of sentiment types for Crypto, Stock Market, Finance News, Theta announcement, and AXS announcement

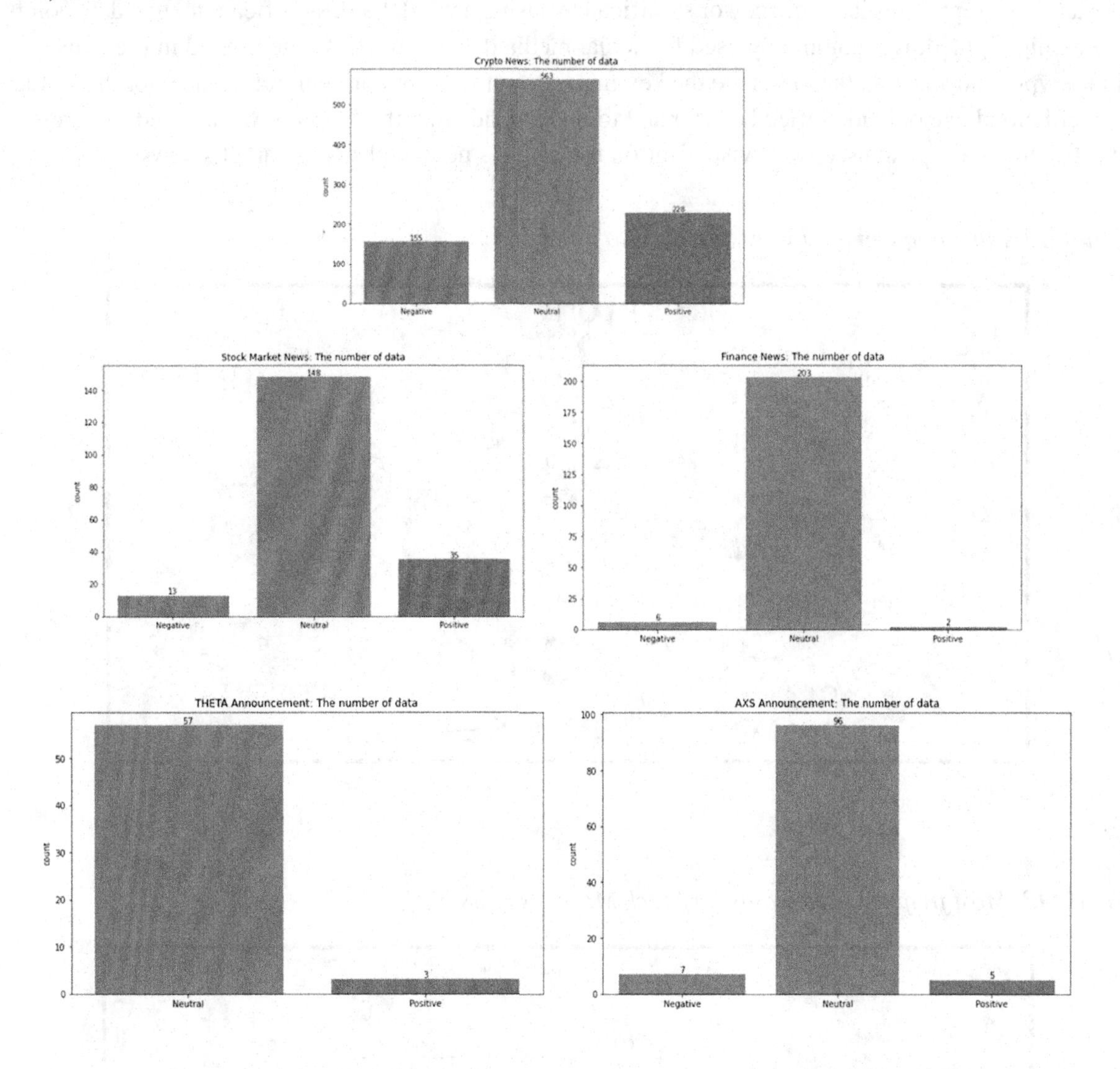

We conclude that the data generated for finance, Theta announcement, and AXS announcement are insufficient as data that generated is too less and it will influence the predictive result later. Hence, for the remaining of the analysis, only crypto and stock market news will be considered as important impacts on the crypto price trend.

Discover Most Frequent Words Used

Next, we further analyse the most frequent words to justify the Section 3.1 hypothesis. To extract the most commonly used words, we install the wordcloud[5] library, which can quickly extract textual data for visualisation (Heimerl et al., 2014). Wordcloud is a text data representation technique in which the size of each word represents its frequency or significance. Using a wordcloud, significant textual data points can be highlighted. It is commonly used for social media data analysis. As mentioned in previous section, crypto and stock market news are the key impacts on the crypto price trend. Hence, for the extract of word cloud, we only identified both textual to understand what the frequent words used. Figures 12 and 13 are the frequent used word visualization for crypto's news and stock market's news.

Figure 12. Most frequent used words for Crypto's news

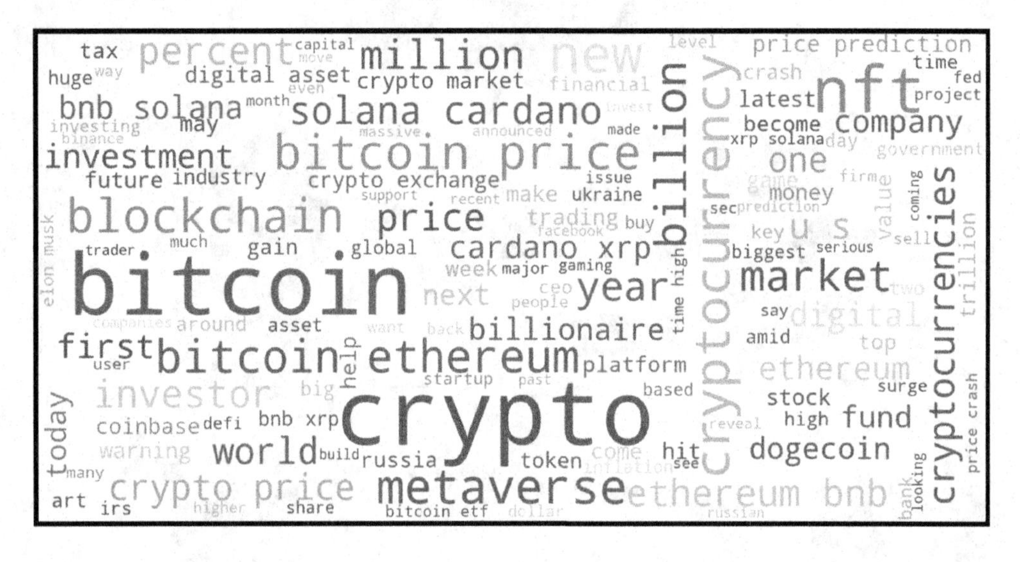

Figure 13. Most frequent used words for Stock Market's news

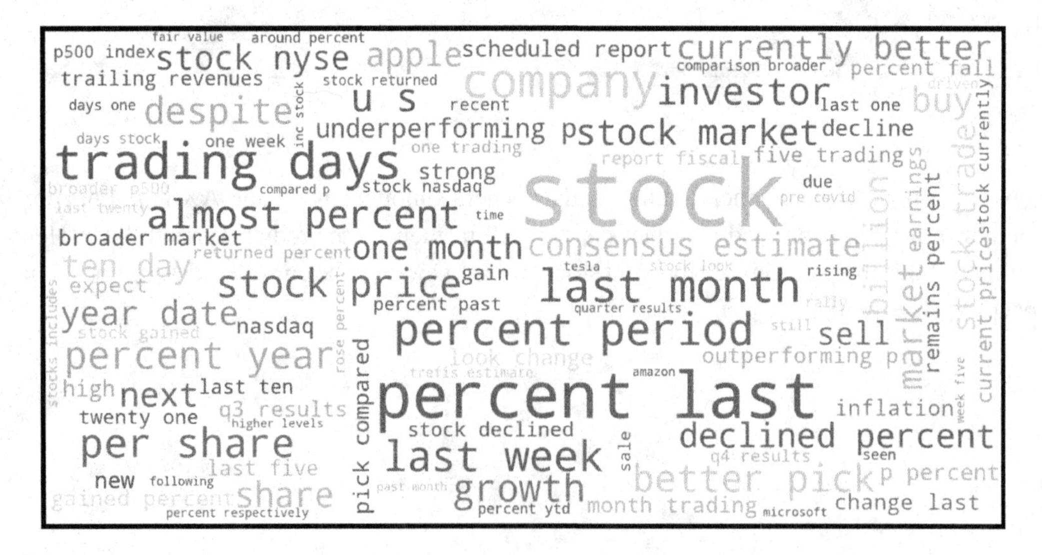

Bitcoin, Ethereum, Metaverse, and NFT are a few of the words that stand out as very attractive in Figure 12. The extracted terms are relevant to crypto category. The percentage, stocks, and trade appear repeatedly in Figure 13. The classification that has been established as stock market news is also relevant.

Correlation Sentiment Score Against Prices Change

After confirmed the textual classification for crypto's news and stock market's news, we identify the correlations between crypto and stock market news to further justify for the findings. This includes price change of crypto with the sentiment score for the study. In this situation, we do not cover all sentiment scores for further analysis, as the analysis revealed that not all sentiment scores are effective in predicting price trend. Observations revealed that the Discord data for announcements may still have an effect, but the sentiment score does not indicate significance data. Other than the AXS and THETA announcements, the average positive sentiment is less than 0.3 and has a substantial impact. Literally, the overall number of impact results from AXS and THETA is insufficient for this study. As depicted in Figure 11, Theta has just three positive sentiments, while AXS has seven negative and five positive sentiments. The Discord data for general discussion are considerably worse, since the average generated result is primarily neutral and close to 0 sentiment score. Analysis from the general data, this is due to most discussion is asking how to play game in overall prospective (Refer most words in Figure 14 as examples for general discussion). Table 3 displays the results for the whole sentiment score that has been calculated.

Figure 14. Most frequent words for MANA and AXS's general discussion

The subsequent section will associate a lagging sentiment score with crypto price movement. It is worth to understand the sentiment generated today and the impact of next day price changed findings as well. Which we can utilize to comprehend the relationship pattern between today's results and tomorrow's outcomes. However, different interpretations may result in different circumstances. The resulting disparity between a five-minute delay and a one-day delay is significant. This research was unable to apply close to near-time analysis due to the constraints of extraction methodologies, which necessitated costly subscriptions to extract data like API's services and costly infrastructure deployment. This could be a limited study that may be able to continue in future. To identify the correlation between today and one day delay sentiment, time series technique is a key to associate between the data. Hence, the date that use in the dataframe must standardize across before able to use it to analyze the correlation. The one-day sentiment score is calculated as Figure 15: if the sentiment score on 2021-10-01 is 0.5549, then this value is shifted one day to 2021-10-02. A result with a one-day delay (Fang et al., 2018; Linhao, 2013) is displayed as sentiment score 1 at Figure 11 as example.

Figure 15. Sentiment score lagged one day result

date	sentiment_score	sentiment_score_1
2021-10-01	0.554997	NaN
2021-10-02	0.629249	0.554997
2021-10-03	-0.834325	0.629249
2021-10-04	0.514501	-0.834325
2021-10-07	0.402989	0.198556
...	...	...
2022-03-27	0.634299	0.133464
2022-04-10	0.405320	0.076061

The correlation results generated based on each crypto with crypto and stock market's news. Figure 16 shown comparison between sentiment result with lagged sentiment result. The correlation for MANA demonstrated a weak association when compared to the lagging result. It is also not significant.

Figure 16. MANA price change correlation with sentiment score

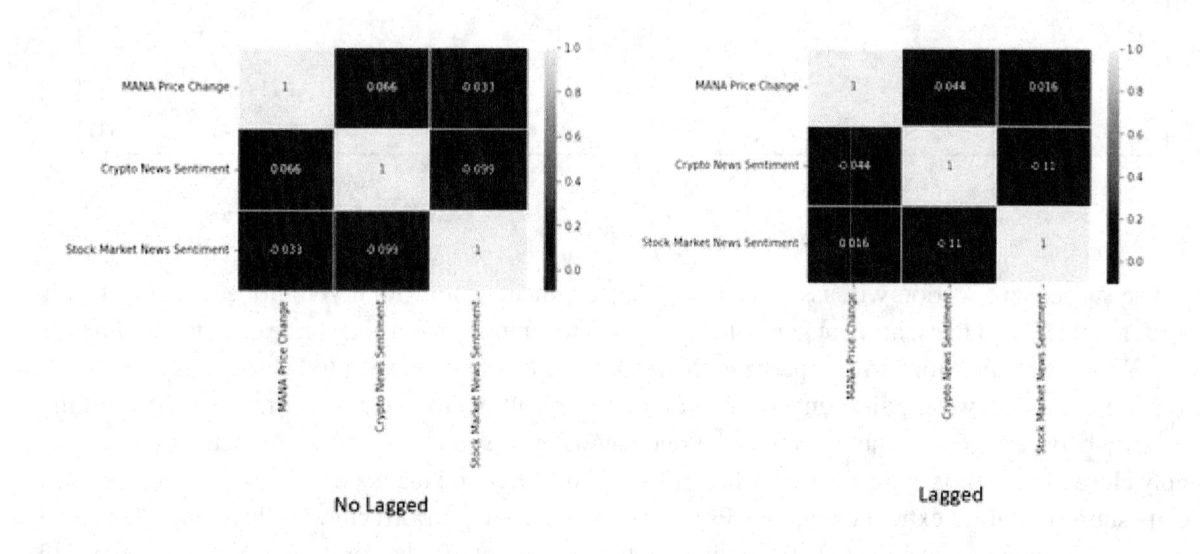

Figure 17 generated the rest of metaverse crypto's price correlation with sentiment and lagged sentiment result.

Figure 17. Crypto's price change correlation with sentiment score

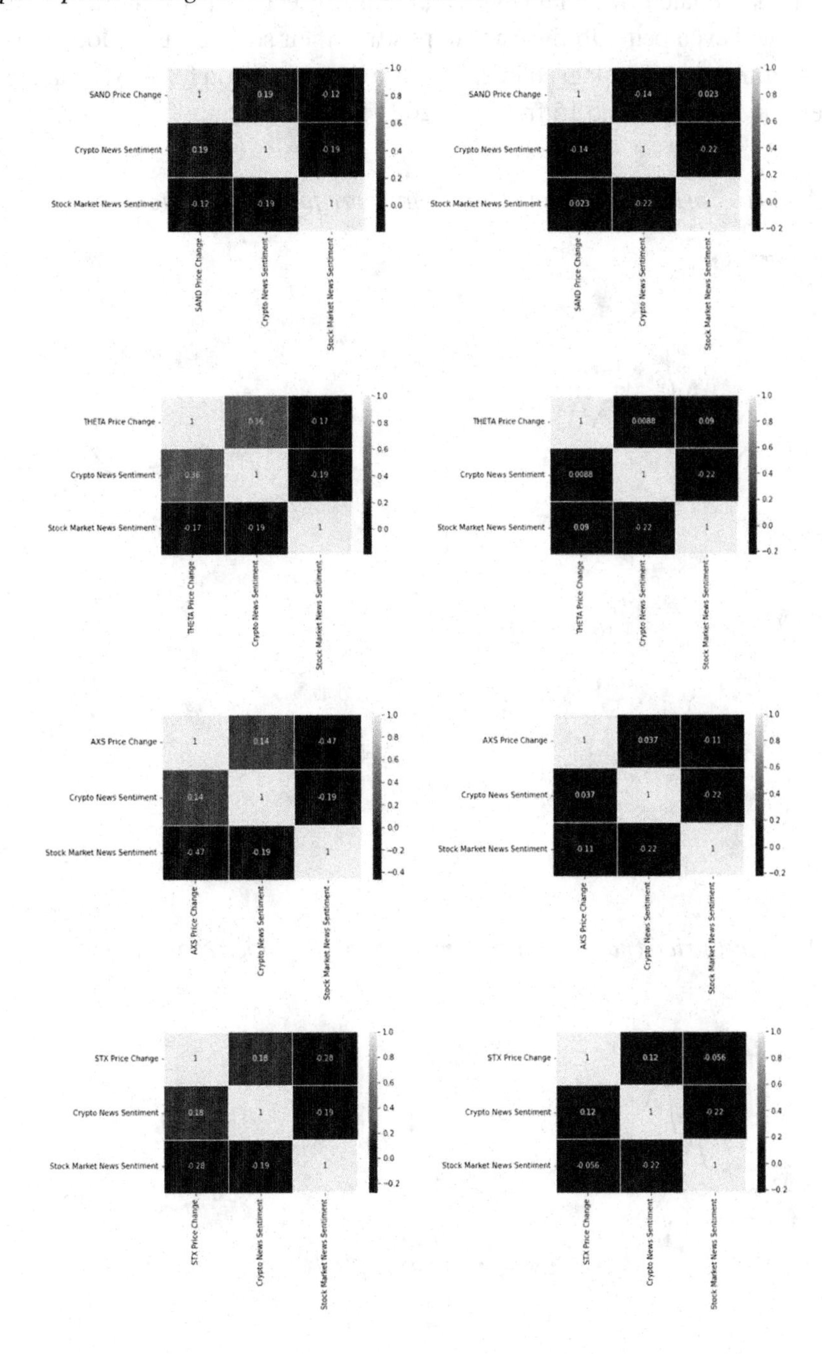

The only association demonstrated in Figure 17 to be greater than 0.30 is the THETA sentiment result. There were 0.36 results generated for crypto news. However, the delayed result is 0.0088, which is not clearly distinguishable. In view of the remaining results, it is clear that sentiment data that is one day behind is not statistically significant. With this finding, sentiment analysis of the crypto market is suited for high frequency trading. If the sentiment score is strongly positive or strongly negative, every bit of crypto and stock market-related financial news can readily affect the price trend. The data that can be process in real time will even better for investor to position their strategy. Let's look closely with price, price change and sentiment without lagged in daily basic as comparison for MANA against crypto's and stock market's news in Figure 18 and 15 from Oct 2021 to Apr 2022.

Figure 18. MANA - price, price change and sentiment score for Crypto's news

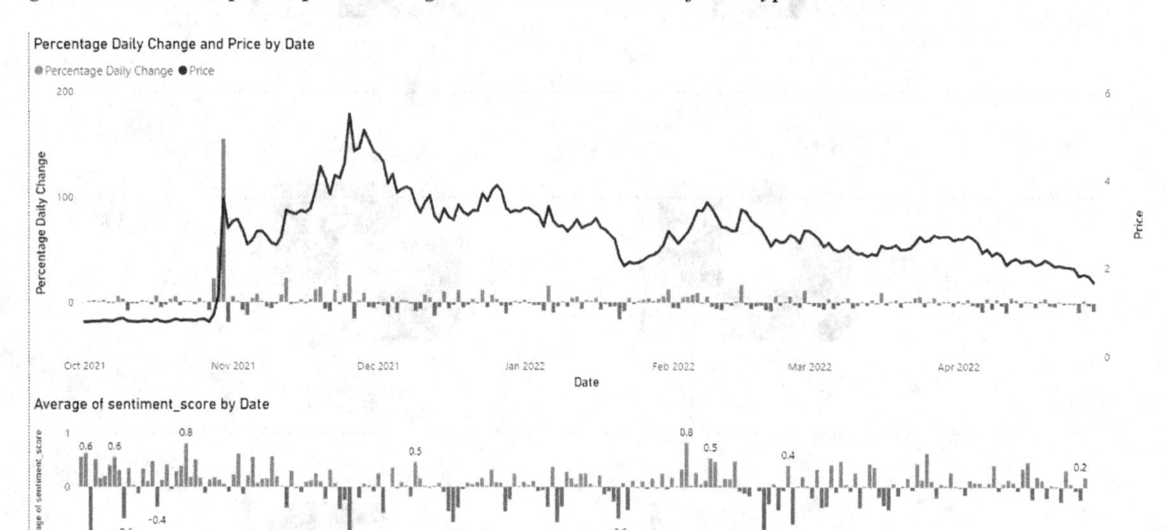

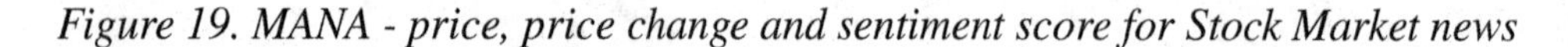

Figure 19. MANA - price, price change and sentiment score for Stock Market news

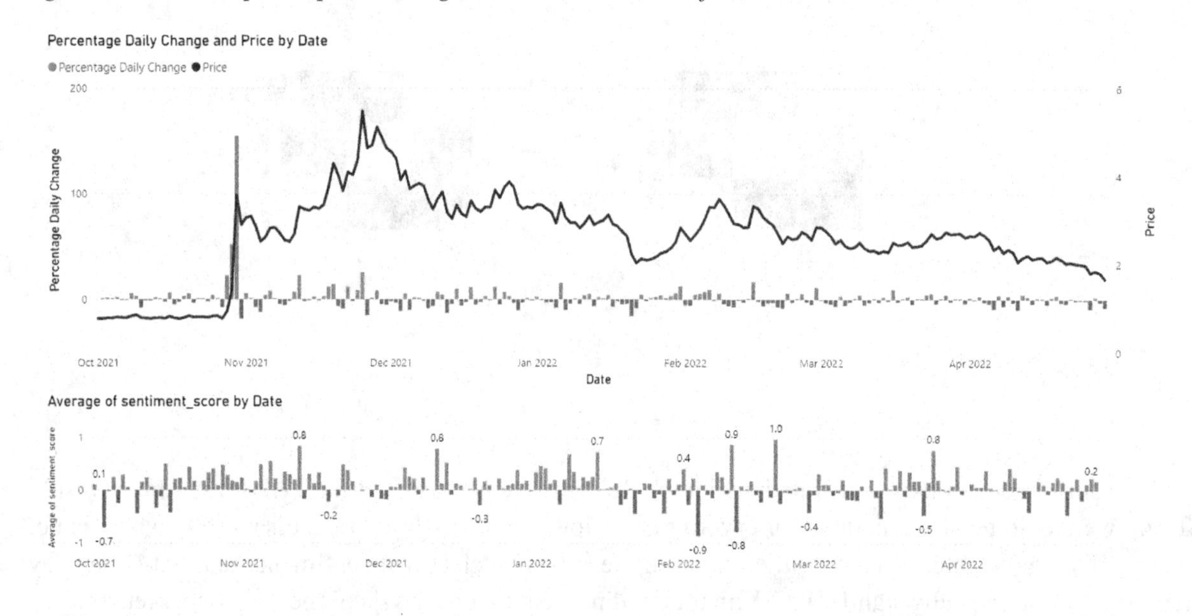

The overall result for both crypto & stocks market news's shown parallel move on the pattern. When the sentiment score is strong positive, the price moves up. When strong negative, the price drop. We take some random period to further analyse the result as sample below especially Facebook changed name to Meta before and after:

MANA Price, Price Change and Sentiment from 15 Oct 2021 to 14 Dec 2021

Figure 20. Crypto's News

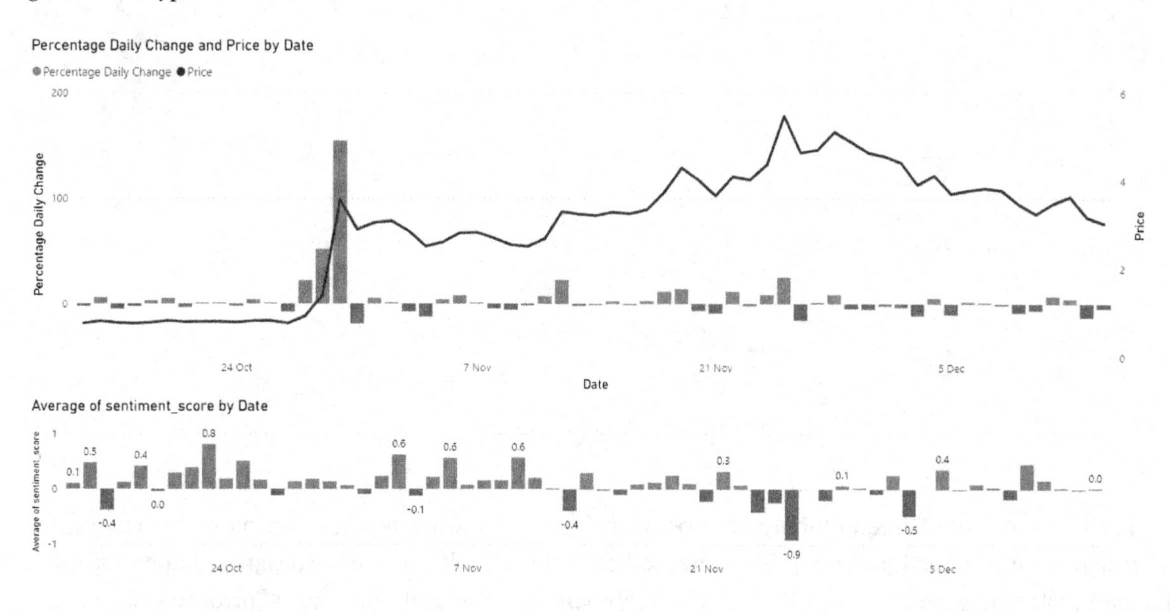

Figure 21. Stock Market's News

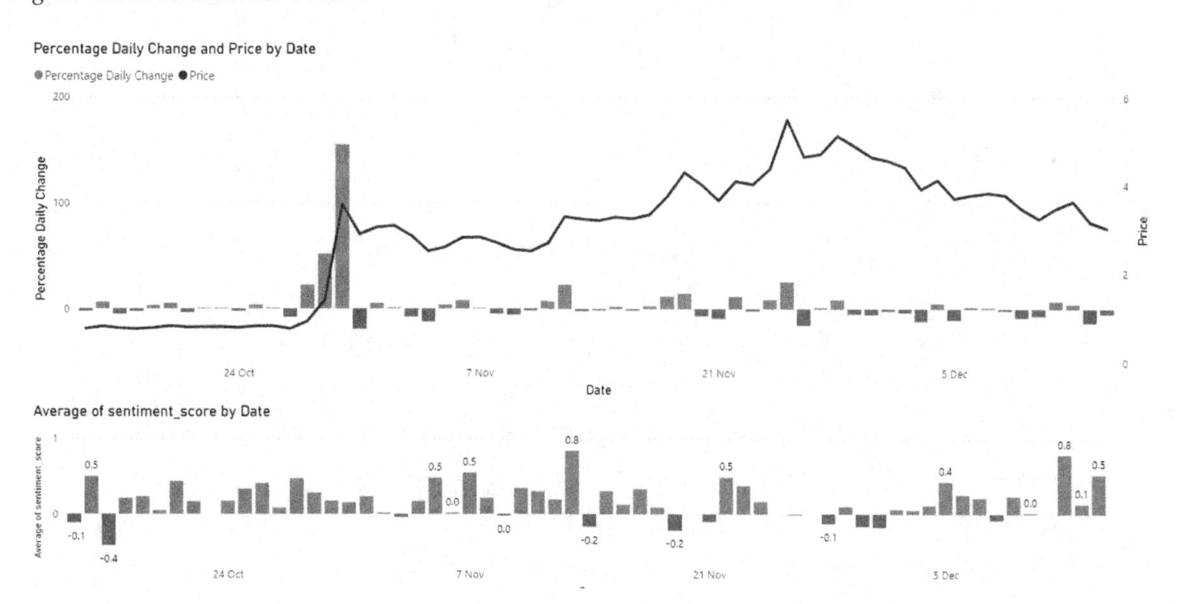

There are few interesting observations shown during the 2 months periods. Take closely below indication for explanation.

Figure 22. Facebook changed name period

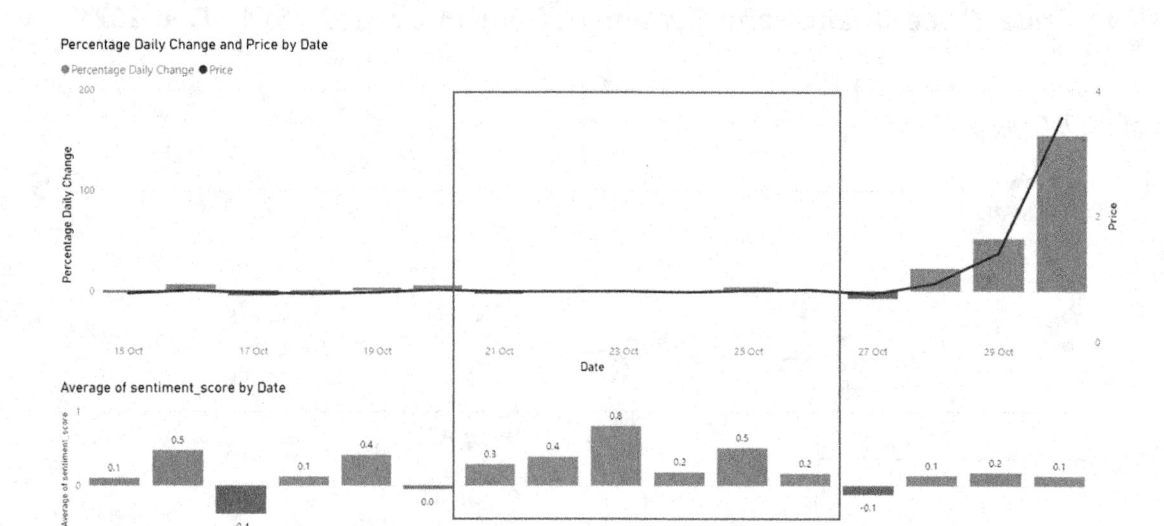

The news indicated in red for both cryptocurrencies and the stock market continue to indicate strong sentiment result that led to a big price increase before Facebook's announced name changed to Meta. The red rectangle serves as an explanation of sentiment score polarity that reacts prior to price change. Where this type of signal has the potential to prompt investors to execute their buy or sell strategy prior to the next market movement. Closely look into 15 Nov 2021 to 15 Dec 2021 below comparison:

Figure 23. Sentiment result with price change trend analysis

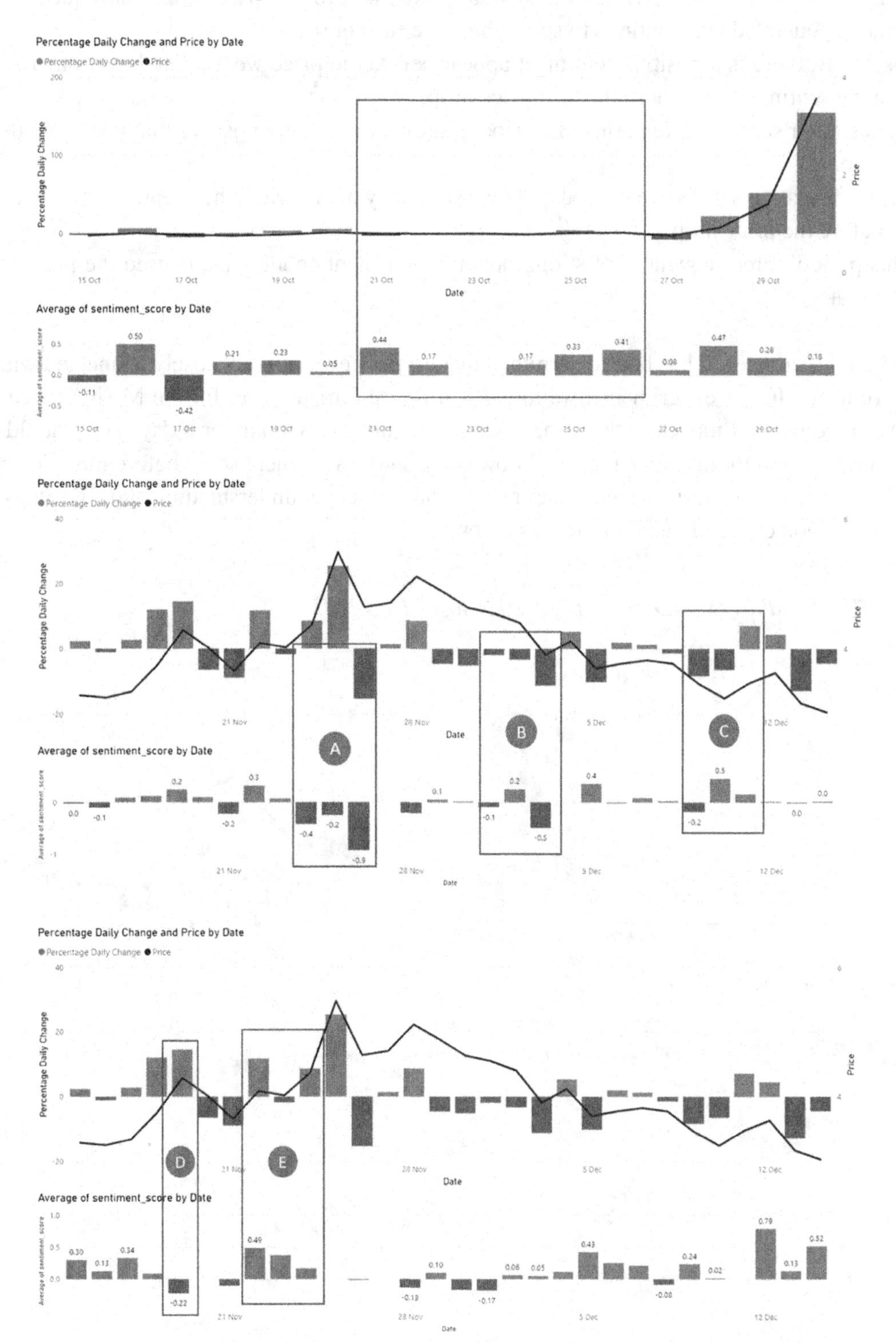

(A) During the period A, negative sentiment was spotted before the price drop. Subsequently, strong negative generated and continue to stress the price further drop.
(B) In period B, there is a positive sentiment appear before the price went up and next day the strong negative sentiment that cause the next price drop.
(C) This is similar scenario like period B, period C shown that strong positive that influence the price went up.
(D) In the stock market's news events, it also shown similarity of A period, the negative sentiment caused the before the price drop.
(E) In this period, three sequences of strong positive sentiment spotted and caused the price went up after that.

Based on the samples, it has been determined that crypto news is more closely connected with price trend movements. In our experiments, we found that the sentiment score for the MANA price comes before the price parallel moves in the same direction within a few hours or a day. This should be the preferred information for investors that can know early on the sentiment score before the price change. We took a detailed look at various metaverse crypto to have a deeper understanding of the findings before and after Facebook changed name to Meta as below:

Figure 24. The Sandbox (SAND) from 1 Oct 2021 to 15 Dec 2021

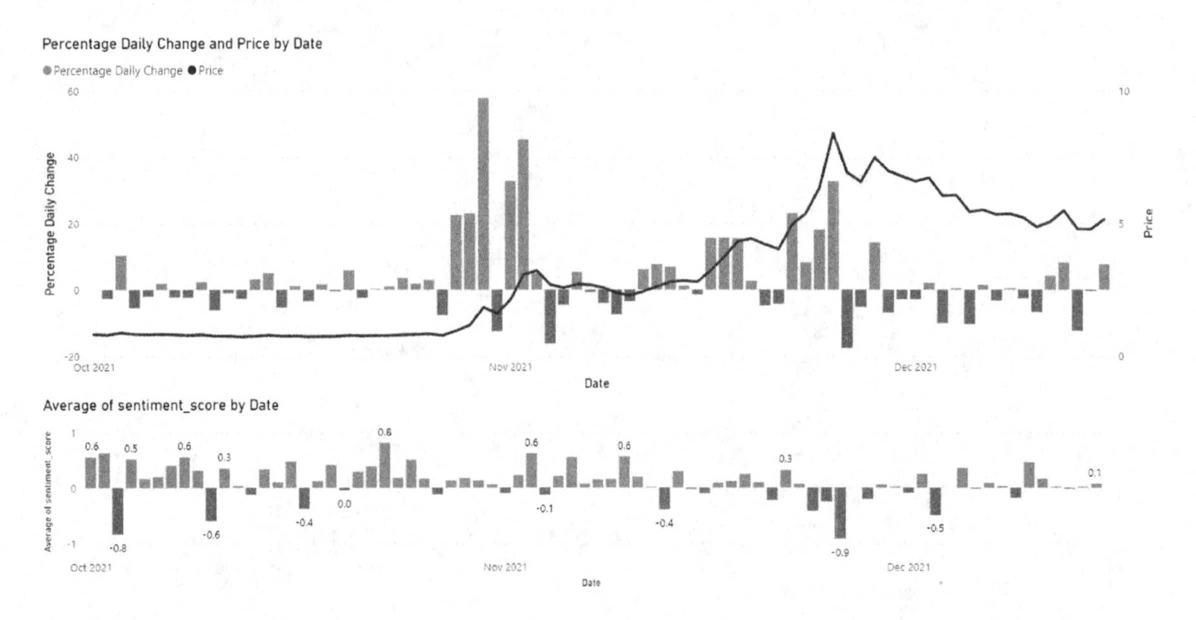

Figure 25. Theta Network (THETA) from 1 Oct 2021 to 15 Dec 2021

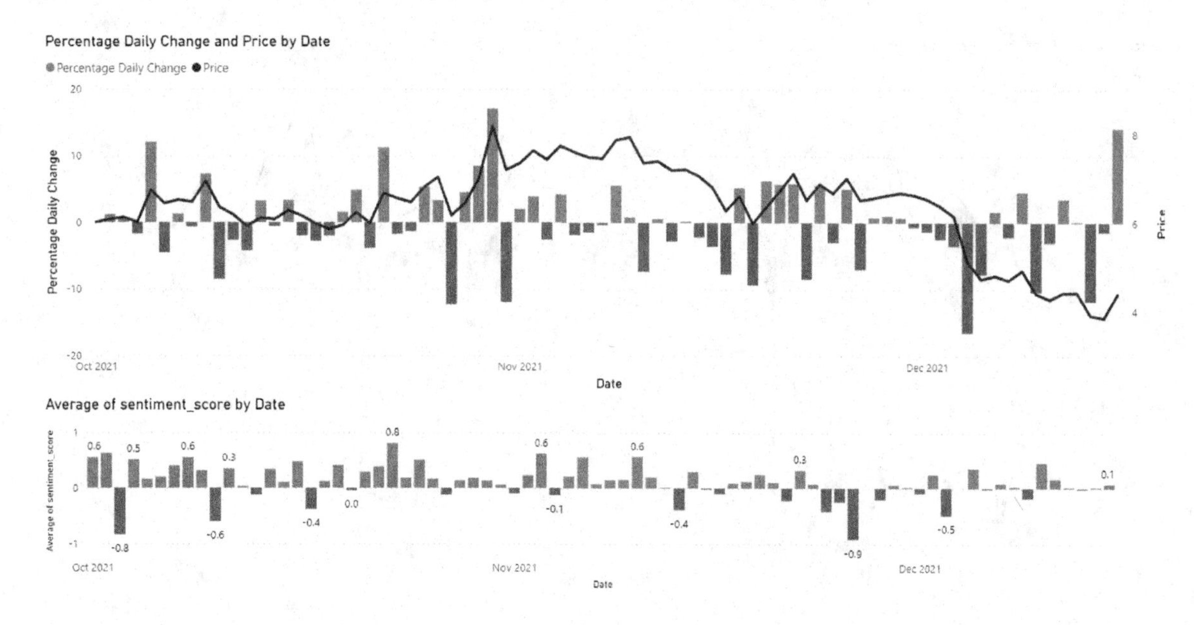

Figure 26. Axie Infinity (AXS) from 1 Oct 2021 to 15 Dec 2021

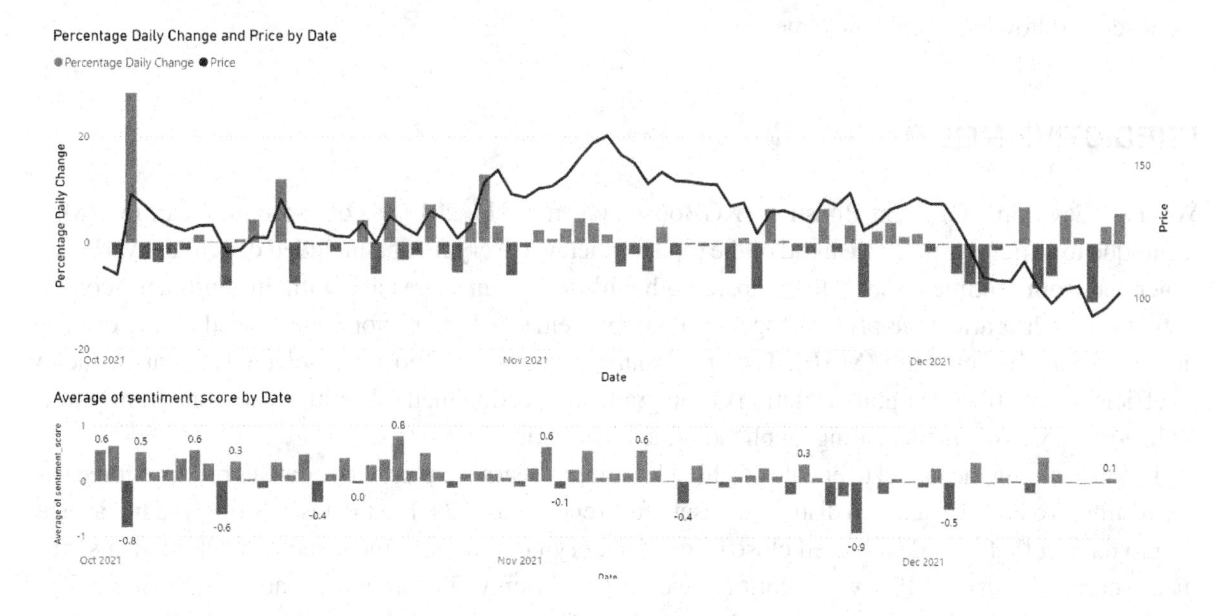

Figure 27. Stacks (STX) from 1 Oct 2021 to 15 Dec 2021

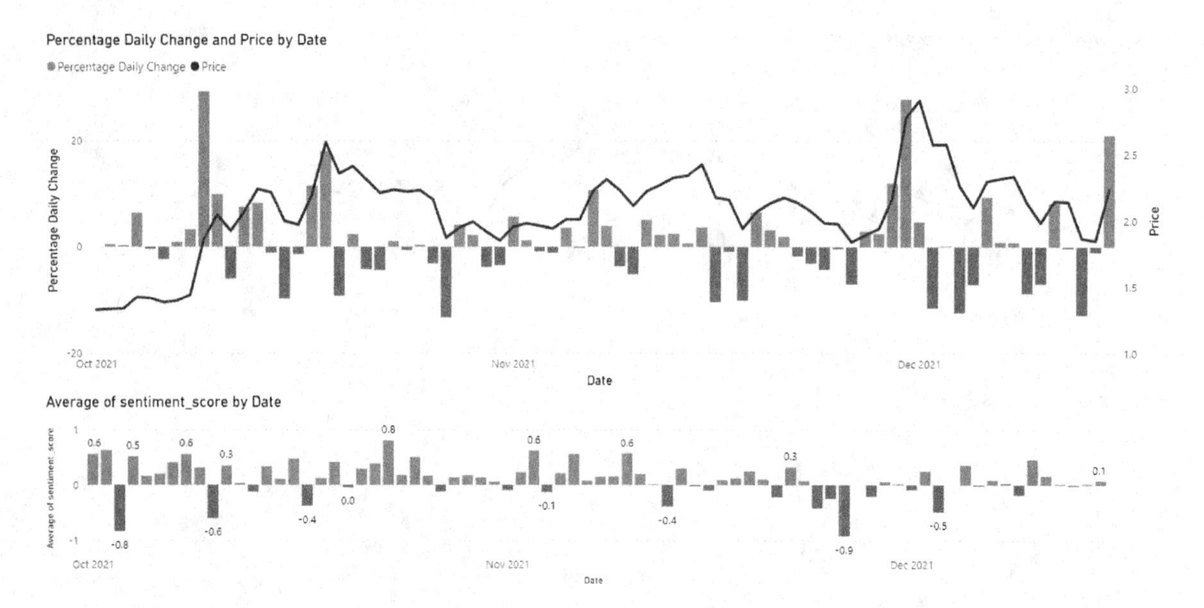

Surprisingly, STX sentiment behavior is distinguished from the norm. The sentiment result does not match the explanation for MANA. The pattern is not parallel, which seems to be the case with MANA, SAND, THETA, and AXS that move parallel within hours or day that similarly in response to the sentiment score following a price movement.

PREDICTIVE MODEL

We used E(x)treme Gradient Boosting (XGBoost) (Chen & He, 2017) as our machine learning (ML) technique to anticipate the price trend of the top five metaverse crypto and included dependent variables generated from sentiment scores to compare both with sentiment scores and without sentiment score to review whether it affect the price change on the improvement of the performance based on the evaluation via mean absolute error (MAE). The open-source package XGBoost was selected primarily due to its efficient and effective implementation of the gradient boosting method. Within the same framework, XGBoost may assist model fitting, application, and validation.

Before training the model to enable machine learning to learn from previous data to produce forecasting results, we constructed two distinct datasets for trend prediction. First dataset is top five metaverse crypto dataset that include adjusted close price, date, crypto news sentiment score, stock market sentiment score. The dataset that with sentiment score should only filter those 0.5 above and -0.5 below sentiment for the experiments. Second dataset is top five metaverse dataset that only adjusted close price. Both datasets, we split to 70% train data and 30% test data for the preparation of predictive model (Vabalas et al., 2019). After preparing the dataset, we run the predictive model (XGBoost) to assess the mean absolute error (MAE) to compare one without sentiment score and another one with sentiment score. Hypothetically, the one with sentiment score MAE should lower than without sentiment score. This could prove that sentiment score is key influence on the price change. Like mentioned, the historical data from Oct 2021 to Apr 2022 is used for simulating the whole concept that compare against the actual “adj close” price value from crypto and use XGBoost to predict the result side by side to identify the trend. However, when come to future/forecast result, this may not be accurate as future price of crypto likely to be alter as forecast result can be influenced by many factors. Specifically, it is possible for a forecast to be objective yet wholly incorrect. The bias relates exclusively to the tendency of the forecasting model to overestimate or underestimate the future. The data intended to assist investors in making informed decisions. Humans are still necessary to review and evaluate the situation, using AI as a resource to make better decisions.

Figure 28. MANA – Line chart compare with expected values and predicted values for last 200 days

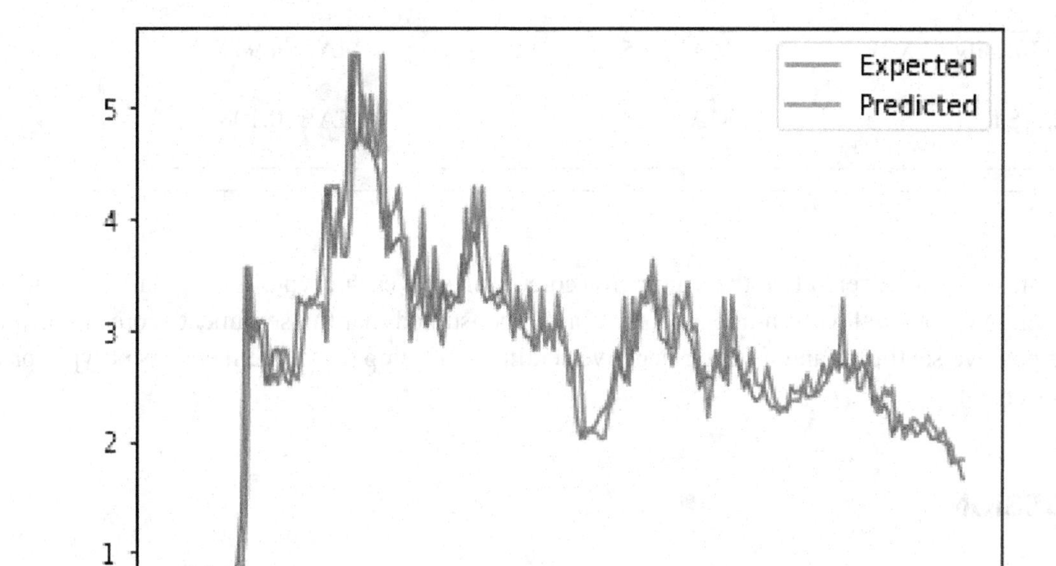

In Figure 28 shown the line chart that created the result from XGBoost to compare against the expected and predicted “adj price” of MANA for the last 200 days of the historical data that without sentiment score. This provides a geometric representation of the model’s performance on the test set. We repeat the predictive model for MANA, SAND, THETA, AXS, and STX in order to get the MAE results with sentiment score (crypto news and stocks market news) and without sentiment score, as shown in Table 4.

Table 4. Top five metaverse mean square error comparison between sentiment and without sentiment score

Crypto Quote	Without Sentiment Score (Adjusted Price)	With Sentiment Score (Adjusted Price, Crypto News Sentiment and Stocks Market News Sentiment)
Decentraland (MANA)	MAE: 0.234	MAE: 0.218
The Sandbox (SAND)	MAE: 0.362	MAE: 0.328
Theta Network (THETA)	MAE: 0.300	MAE: 0.289
Axie Infinity (AXS)	MAE: 5.519	MAE: 4.942
Stack (STX)	MAE: 0.124	MAE: 0.108

In Table 4, we discovered that the sentiment score results for each crypto is less than the result of the prediction model without sentiment score. This demonstrated that the sentiment scores indicating a strong positive sentiment and a strong negative sentiment are impact on the metaverse crypto price trend movement.

DISCUSSION

There are many noises in the market, many factors affect crypto's price change. Headlines like crypto news or stock markets news are some of the factors that could influence the price change. With sentiment analysis techniques, we can alleviate this problem by hedging it and strategy your trading method that include sentiment scores as predictor in risk management strategy. If you investigate the textual like gaming discussion or announcement, then this type of sentiment probably not going to have much impact. Based on the findings, crypto news and stocks market news could provide those insights that helps investors to have some time to plan their trading strategy. In this research, the news that we collected from Forbes may not be provide a safer sentiment result. We should also investigate different sources that from bigger news channels. Mixture with different sources especially reliable news will increase the objective of reliable data that may prevent biased. To succeed in trading, you must discover ways to outsmart others. Since everyone is constantly attempting to outsmart one another, you must be innovative and inventive to stay in the game.

CONCLUSION

In this chapter, we have observed the pattern of news influences the crypto prices of top five metaverse crypto. To achieve that, the crypto is efficient market where crypto's news, and stocks market's news is important key to identify the trend of crypto price trend. Announcement and general discussion from Discord that extract it from the target group of each metaverse's crypto doesn't provide strong signal. In summary, the sentiment score needs to be higher positive or negative sentiment especially break 0.5/-0.5 sentiment score will provide stronger influence on the price move. In general, announcement and general discussion that provided from Discord target group shown in the result doesn't break more than 0.15 which doesn't show much on overreaction in the market. This is considering a neutral environment that doesn't give good indication and most of the discussion or announcement is discussed about games related topics. Hence, Discord data that extract from the target group doesn't provide good indication but News like Finance, Stock Market, Crypto news provide certain degree of strong impact on the crypto pricing.

With the sentiment score that proven from News that have strong positive or negative sentiment score, this research could prove with the sentiment scores results and without sentiment score result to experiment with the predictive model which is Gradient Boosting model to evaluate the comparison. The comparison result that applied predictive model that generated the predictive result which includes sentiment score is provided a good indication as it helps to improve the prediction model in overall for all top five metaverse crypto. The MAE result is lower than the one without sentiment score. Thus, the experiments for this research proof that sentiment score with relevant news able to give good indication for trend prediction.

In future, we will improve the predictive part to include more data for better tuning the performance. We even could include back test and the technical analysis to generate the momentum that potential could predict the forecast better. Such as relative strength index (RSI) and other technical methods. If there were a method to collect minutes interval data, it would be ideal for high-frequency trading experiments using sentiment analysis to discover trends. The source code for this paper we keep at an open source repository[6].

REFERENCES

Abraham, J., Higdon, D., & Nelson, J. (2018). Cryptocurrency price prediction using tweet volumes and sentiment analysis. *SMU Data Science Review*, *1*(3), 22.

Alvin, K. (2021). *Sustainable GameFi: To Play or To Earn?* https://www.binance.org/en/blog/sustainable-gamefi-to-play-o r-to-earn/

Bianchi, D., Dickerson, A., & Babiak, M. (2019). *Trading volume and liquidity provision in cryptocurrency markets. SSRN.* Electronic Journal., doi:10.2139/SSRN.3239670

Blaise, S., de Jesus, D., Marcelo, D. R., Ocampo, C., Is, J., Colleges, L., & Tibudan, A. J. (2022). *Play-to-Earn: A Qualitative Analysis of the Experiences and Challenges Faced By Axie Infinity Online Gamers Amidst the COVID-19 Pandemic.* doi:10.6084/m9.figshare.18856454.v1

Fang, L., Yu, H., & Huang, Y. (2018). The role of investor sentiment in the long-term correlation between U.S. stock and bond markets. *International Review of Economics & Finance*, *58*, 127–139. doi:10.1016/j.iref.2018.03.005

Han, J., Heo, J., & You, E. (2021). Analysis of Metaverse Platform as a New Play Culture: Focusing on Roblox and ZEPETO. *2nd International Conference on Human-Centered Artificial Intelligence, Computing4Human* .

Haslhofer, B., Stütz, R., Romiti, M., & King, R. (2021). *GraphSense: A General-Purpose Cryptoasset Analytics Platform*. doi:10.48550/arxiv.2102.13613

Heimerl, F., Lohmann, S., Lange, S., & Ertl, T. (2014). Word cloud explorer: Text analytics based on word clouds. *Proceedings of the Annual Hawaii International Conference on System Sciences*, 1833–1842. 10.1109/HICSS.2014.231

Hoang, M., Alija Bihorac, O., & Rouces, J. (2019). Aspect-Based Sentiment Analysis Using BERT. Linköping University Electronic Press, 187–196.

Clement, J. (2021, November 23). *Video gaming market size worldwide 2020-2025*. Statista. https://www.statista.com/statistics/292056/video-game-market-value-worldwide/

Khan, S. N., Loukil, F., Ghedira-Guegan, C., Benkhelifa, E., & Bani-Hani, A. (2021). Blockchain smart contracts: Applications, challenges, and future trends. *Peer-to-Peer Networking and Applications*, *14*(5), 2901–2925. doi:10.100712083-021-01127-0 PMID:33897937

Kingaby, S. A. (2022). The Stock Market API. *Data-Driven Alexa Skills*, 387–404. doi:10.1007/978-1-4842-7449-1_16

Lee, J. Y. (2021). *A Study on Metaverse Hype for Sustainable Growth. 10*(3), 72–80.

Linhao, Z. (2013). *Sentiment analysis on Twitter with stock price and significant keyword correlation*. Texas ScholarWorks. https://repositories.lib.utexas.edu/handle/2152/20057

Loper, E., & Bird, S. (2002). *NLTK: The Natural Language Toolkit*. http://nltk.sf.net/

Van Der Merwe, D. (2021). The Metaverse as Virtual Heterotopia. *World Conference on Research in Social Sciences*. 10.33422/3rd.socialsciencesconf.2021.10.61

Najork, M. (2017). *Web Crawler Architecture*. doi:10.1007/978-1-4899-7993-3457-3

O'Leary, D. E. (2017). Configuring blockchain architectures for transaction information in blockchain consortiums: The case of accounting and supply chain systems. *Intelligent Systems in Accounting, Finance & Management*, *24*(4), 138–147. doi:10.1002/isaf.1417

Pascual-Ezama, D., Scandroglio, B., & de Lian, B. G. G. (2014). Can we predict individual investors' behavior in stock markets? A psychological approach. *Universitas Psychologica*, *13*(1), 25–36. doi:10.11144/Javeriana.UPSY13-1.cwpi

Ritter, M. (2021). Cryptocurrencies rally on Facebook's "Meta" rebrand - CNN. *CNN Business*.

Robertson, H. (2021). *Jefferies says the metaverse is as big an investment opportunity as the early internet*. Insider.

Tran, V. Le, & Leirvik, T. (2020). Efficiency in the markets of crypto-currencies. *Finance Research Letters, 35*(November), 101382. doi:10.1016/j.frl.2019.101382

Vabalas, A., Gowen, E., Poliakoff, E., & Casson, A. J. (2019). Machine learning algorithm validation with a limited sample size. *PLoS One, 14*(11), e0224365. doi:10.1371/journal.pone.0224365 PMID:31697686

van Tonder, R., Trockman, A., & le Goues, C. (2019). A panel data set of cryptocurrency development activity on GitHub. *IEEE International Working Conference on Mining Software Repositories,* 186–190. 10.1109/MSR.2019.00037

Wood, L. (2021). Global Gaming Industry to Cross $314 Billion by 2026 - Microsoft, Nintendo, Twitch, and Activision Have All Reached New Heights in Player Investment, Helped by COVID-19. *Research and Markets.* https://finance.yahoo.com/news/global-gaming-industry-cross-314-183000310.html

Chapter 7
Wonders of the World:
Metaverse for Education Delivery

Siti Azreena Binti Mubin
https://orcid.org/0000-0003-0121-5049
Asia Pacific University of Technology and Innovation, Malaysia

Vinesh Thiruchelvam
Asia Pacific University of Technology and Innovation, Malaysia

ABSTRACT

This chapter overviewed several existing works being conducted on the metaverse in immersive education. Most of the reviews have clearly stated that the metaverse would be a promising asset and tool in education besides the conventional method. The metaverse gives space where real-world specifications can be replicated via digital twin technology. This enables smarter problem-solving in industries like construction, architecture, healthcare, life sciences, and more. Reviewed studies on the effectiveness of those technologies for education found moderate evidence for the metaverse to assist students in their daily learning. The application of the metaverse to the education sector has been well studied and widely accepted as an engaging and effective method of delivering learning experiences. Hence, the metaverse might be a promising tool to provide immersive learning experiences to users nowadays. An application named wonders of the world (WoW) has been developed using XR technology to provide immersive learning experiences to the users.

INTRODUCTION

Ever since Mark Zuckerberg renamed his company Meta, a large talk in social media has built up around the idea of the metaverse, a venue that Zuckerberg claims we'll all be using in the very near future. He describes artificial intelligence (AI) as "the key to unlocking the metaverse," with AI assistants integral to creating a seamless virtual experience. But is the metaverse just another hype, or a genuine new phase in online interaction? In fact, many of the 3D images, animation, speech and even metaverse artwork

DOI: 10.4018/978-1-6684-5732-0.ch007

will be generated by AI. Also, AI will be called upon to automate smart contracts, decentralized ledgers and other blockchain technologies to enable virtual transactions.

Thus, education and training are evolving at an accelerated rate along with an increased level of integration with ICT systems. Growth in this area will only continue. Education and training integrated with ICT systems is progressing in two ways: one is in the use of virtual digital information without the use of devices, and the other is in the use of both virtual digital information and sensor devices. Both types of education and training will expand based on the advance of ICT including virtual reality (VR) and augmented reality (AR). Note that mixed reality (MR) is a mix of real and virtual reality and is therefore presumed to be included in the continuum of VR and AR in this paper. Virtual education and training systems are typical applications for systems integration. They are based on the technologies of VR and AR with sensors. Although education and training areas are diverse, virtual education and training systems can have unique information modeling, as well as common functionalities. Common teaching and learning technologies for virtual education and training systems can be determined by categorizing use cases for education and training based on VR and AR.

The concept of the metaverse was popularized in the science-fiction novel *Snow Crash* by Neal Stephenson to refer to a digital universe that can be accessed through virtual reality. This bold, exciting space has been part of many recent works of science fiction, such as *the Matrix* and *Ready Player One* movies primarily.

Metaverse maximizes human engagement with digital content and computer-related tasks while it hides the intricacies of the hardware and software from the user. In terms of learning applications, the user is fully immersed in the experience and learning content, and through multimodal or nature interactions, access to the digital content is intuitive, with little to no learning curves for the user in interacting with the metaverse world itself (Nichols, 2018).

Metaverse is defines as a network of 3D virtual worlds focusing on social connectivity. In simple words, metaverse is a second universe instead of our real world. Metaverse utilizes Extended Reality (XR) technology, and the term has been interchangeable over the current period in the industry. The XR term denotes to the new theory of human-computer interaction that is currently predominantly realized with Augmented Reality (AR) and Virtual Reality (VR) as the human-computer interface. The 'X' in XR is a variable that can stand for any letter, and it covers all the computer-altered reality forms comprising of AR, Mixed Reality (MR) and VR. In recent years, there has been an increasing interest in XR technology for various industry such as healthcare, education, and entertainment. The impact and benefits outweigh the cost and provide industry tremendous return while minimize risk at work (Mubin, Thiruchelvam, & Andrew, 2020).

The convergence of three key factors indicate that it could possibly be at a genuine inflection point in the following areas;

- Software: Seeing the creation of meta platforms and standards which enable interoperability and creation of multiple applications in a functioning value chain well beyond just gaming.
- Hardware: Accelerating advances in hardware, at both the infrastructure and consumer levels, are reducing barriers to widespread adoption of VR such as limited quality of the user experience and its relatively high cost.
- Consumers: The user base for VR is exploding, not only among younger consumers, but also across all demographics, thanks to the shared familiarity with virtual interaction brought about by the pandemic.

Individuals can create lifelike worlds by suing digital avatars, where they can immerse themselves in digital environments through VR/AR headsets. Seems like a sci-fi movie, the metaverse is already here and is defined as the next iteration of the internet via Web3.0 which also include early-stage mixed reality with face and eye tracking technology. Several innovations have also been developed in recent times and it is spreading on to more users with awareness moving rapidly.

It is believed that by exploring the combination of using AR, haptics, and computer vision technologies, it is possible to develop a fully immersive learning system that can automate mentoring while detecting and measuring gross and fine motor skills. This can be achieved by enabling the training platform to capture rigorous, objective measures and human performance metrics, which can measure acquisition of clinical procedural skills. In example, medical simulation for teaching and learning. XR environments allow educators to handle learning activities which are challenging to implement during regular laboratory lessons (Kamińska et al., 2019). The experiments or simulations can be done virtually by stimulate the user's physical presence in artificially generated world that allows interacting with the environment.

This chapter overviewed several existing works being conducted on metaverse in immersive education. There is a large volume of published studies describing the metaverse would be a promising asset and tool in education besides the conventional method (Mystakidis, 2022). The metaverse gives a space where users can 3D-model virtually anything, and real-world specifications can be simulated via digital twin technology. This enables smarter problem-solving in industries like construction, architecture, healthcare, life sciences, and more. Reviewed findings on the effectiveness of those technologies for education found moderate evidence for metaverse to support students in their daily learning. The metaverse application in education sector has been well studied and widely accepted as an engaging and efficient method of delivering learning practices.

To realize the concept of metaverse, an experience called Wonders of the World (WoW) has been built on the base of Unity using a propriety software named VolCap and transported to Hololens 2. The VolCap software must be used with the Volumetric Capture (VolCap) station, equipped with six depth cameras. VolCap station is a small area, built for educational purpose in capturing real object and transforming it to 3D object. After the transformation, the technology is being piloted with an XR experience that tells the story unlike any other that includes the Great Wall of China, Pyramids and Colosseum. Each of the Wonder of the World provides a different nature of experience. It is an immersive experience designed to showcase the potency of metaverse that highlights how it can be applied in multiple disciplines of study; such as history, gamification and problem solving skills, shaping the thought processes and skill set required to create effective metaverse experience. Thus, metaverse might be one of the promising tools to provide immersive learning experiences to users nowadays.

This write-up has been divided into six parts (excluded introduction) and has been organized in the following way. The first part deals with the state of the art of XR technology and metaverse. It will then go on to the metaverse prototype, Wonders of the World. The third section presents the immersive experiences and fourth section on the challenges in creating the WoW. The fifth section presents the future research directions and finally, the conclusion gives a summary of the chapter.

STATE OF THE ART

Network settings with several computers, people and data sets make up the immersive metaverse. However, the entirety of the Internet-connected cyberspace is not included in immersive metaverse. In a three-dimensional environment where social connection and communication are of the utmost importance, the immersive metaverse is visually depicted. Through personalized avatars, users of the virtual worlds present their virtual selves. Users of an immersive metaverse can also showcase their talent by creating 3D animated items and environments (Han, 2020).

Nevertheless, Stylianois (2022) defined technologies in metaverse could be virtual reality (VR), augmented reality (AR) and mixed reality (MR). The metaverse is based on technologies that enable multisensory interactions with virtual environments, digital objects, and people. Stereoscopic displays that can transmit the sense of depth enable the representational quality of the XR system. This is feasible because to independent displays for each eye that are somewhat different from one another. High resolution XR displays enable a large user field of vision that can range from 90 to 180 degrees. Compared to 2D systems, XR systems also provide greater aural experiences. The creation soundscapes that notable improve immersion in AR and VR is made possible by 3D, spatial or binaural audio. Users can locate themselves and recognize the direction of sound cues because to the spatial dispersion of sound, making it a potent tool for navigation and user attention attraction (Mystakidis, 2022).

Online education is growing more and more popular, particularly in higher education and adult continuing education. This trend was accelerated by the COVID-19 pandemic, which interfered with attendance-based activities at all educational levels. Due to health-related physical separation requirements, remote emergency instruction has been mandated globally. Since its inception, asynchronous and synchronous e-learning have been the two main system types used in online education. Both types operate in two-dimensional digital environment that span in-plane digital windows with width and height but no depth and rely on software or online applications. Standard asynchronous online learning resources include learning management system like Moodle and Blackboard, as well as social networks and occasionally collaborate web apps. Asynchronous tools facilitate flexible, anytime, anywhere engagement and communication between teachers, students, and material. Synchronous online learning tools are implemented through web conferencing platforms such as Zoom, WebEx, Microsoft Teams, and Skype.

However, the constraints and inefficiencies of programmes running in 2D online contexts are widely known. Zoom fatigue is a phenomenon brought on by the prolonged everyday use of synchronous Internet platforms. Emotional isolation, a bad emotion for participation motivation, is a common problem on asynchronous platforms (Hassan, 2020).

To solve the mentioned issues and challenges of online education delivery, immersive technologies like XR have shown promises in enhancing human cognitive processing capabilities due to their unique cognitive functionalities in learning support. XR can improve representational connection in working memory through active learning. By integrating with effective cognitive strategies, XR can help enhance deep thinking and promote deep learning (Ryan et al., 2022).

There are many educational opportunities in the immersive metaverse, in example XR. It encourages an interactive learning approach, chances for collaboration, and meaningful interaction across time and space. Visual stimulation is another potent result of the immersive metaverse. Learners who are already digital natives are captivated by the visually animated immersive metaverse and are willing to spend more time there. Learning by doing or learning by seeing encourages self-directed learning and everything learners do in the immersive metaverse can be an educational experience (Han, 2020).

High Fidelity Open-Source VR, Engage VR, Minecraft, and Cloud Party are just a few examples of immersive metaverse that have been transformed into learning environments. Second Life, however, is the most well-known immersive metaverse for education (Mystakidis, 2022). Users can buy virtual land from Linden Lab (LL), the company that owns Second Life. Users must pay for the right to add textures, sounds, and animations because all data are LL's property and under its exclusive control. Institutions from the real world and institutions that are exclusive to the virtual world both exist in Second Life.

Pertaining metaverse's potential for educational radical innovation, laboratory simulation such as safety training and procedural skill development are among the first application areas with spectacular results in terms of training speed, performance and retention with AR and VR-supported instruction. The metaverse can enable immersive journalism to inform mass audiences reliably and objectively on unfamiliar circumstances and happening in remote areas by enabling the capture of 360-degree panoramic photographs and volumetric spherical video. Additionally, new forms of distant learning driven by the metaverse may develop to overcome the limits of 2D platforms. When learners are co-owners of the virtual places and co-creators of fluid learning experiences, meta-education can provide rich, hybrid formal and informal active learning experiences in permanent, alternative online 3D virtual campuses (Duan et al., 2021; Kye et al., 2021).

Multistakeholder collaboration and emerging technologies such as AI will play an increasingly important role in enabling the Metaverse to achieve its vast potential. This webinar will explore how governments, industry and academia can work together to ensure that the Metaverse, supported by AI, can be used as a tool for achieving UN Sustainable Development Goals (SDGs) related to education, healthcare, and inclusiveness. The webinar will also look at how collaboration on the global level can address the expected structural challenges such as interoperability, identity management, privacy, security, and the integration of various services.

AI is expected to be a key enabler of the metaverse, and panelists will discuss how the technology can potentially augment the experience of users in areas such as accurate avatar creation, digital humans, multilingual accessibility, VR world expansion at scale, and intuitive interfacing. The panel will also discuss the key challenges in this domain such as ownership for AI-created content, deepfakes and user transparency, fair use of AI and ML, the right to use data for AI model training and crucially how to achieve accountability for AI bias.

WONDERS OF THE WORLD (WoW)

Since the first algorithm animation interactive computer system called BALSA appeared in 1984, many algorithms animation and visualization applications have been developed for educational purposes. Modern game engines (like Unity 3D and Unreal) possess a multitude of features which are vital for algorithm animation applications. Until now, these features have not been systematically exploited for educational applications. The authors circumvent conventional programming languages (such as Java, C++, etc.) and their corresponding toolkits and instead describe our design of an algorithm animation framework built around Unity 3D and demonstrate its application to the educational realm of graph algorithm animation. This project serves as a proof of concept that modern game engines such as Unity 3D can be used to create modern and effective tools for learning concepts contained in the realm of computer science and engineering.

Wonders of the World (WoW) adapted metaverse environment and has been fully created in Unity and 3Ds Max. Unity is being used for the interaction and animation while assets modeling was done in 3Ds Max. This prototype took 6 months to be completed starting from the idea generation and storyboard phase. Users can experience the immersive environment in any MR devices and this project has been developed for HoloLens 2. Before user able to experience the whole journey, the scene will be activated with simple tutorial on navigating the application such as sliding, object rotation, transformation, and scaling. The purpose is to let the users familiarize with navigation and interaction by pinching the objects. The journey begins with the main scene named 'Globe' application. From the main scene, a user can choose their preferred wonder of the world such as the Great Wall of China, Pyramids of Giza, or Rome's Colosseum. Each of these wonders of the world provide different immersive scenarios to the users. Great Wall of China represents games mode, Pyramids of Giza signifies storytelling and Rome's Colosseum on history of development. Figure 1 depicts the first environment of WoW which is the Globe scene.

Figure 1. Globe scene

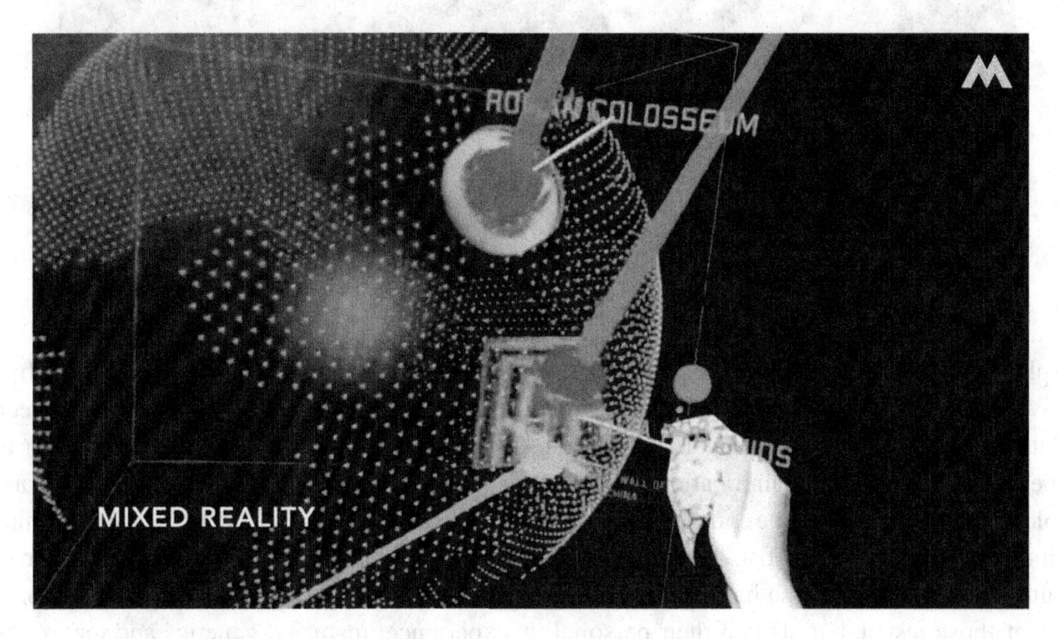

The Great Wall of China

The Great Wall of China is the longest stretched building on earth with a length of 6350KM, built to withstand incursions from the North from as early as 5 B.C. It is estimated by some authors that more than a million lives were lost in building this ancient monument. Users play the role of Chinese soldiers at that period in time taking out the Mongol hordes to defend the wall for the motherland. Basically, in this metaverse experience, The Great Wall of China depict games-based application and provide games experience to the users under the concept of gamification. In the beginning of the scene, users will be presented with short brief of Great Wall of China and the arrow will be shown at the right side. Once user pick up the arrow, they can start play the game and score will be given based on the successful shooting of Mongol hordes. The current version of Great Wall of China games is not good enough and still lack of game elements. It needs to be improved further with more game mechanical and correct score allocation. Figure 2 shows Great Wall of China prototype scene in the WoW.

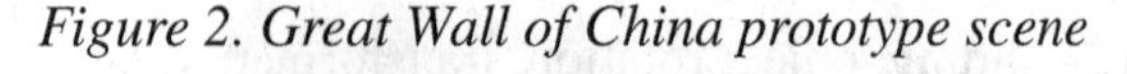

Figure 2. Great Wall of China prototype scene

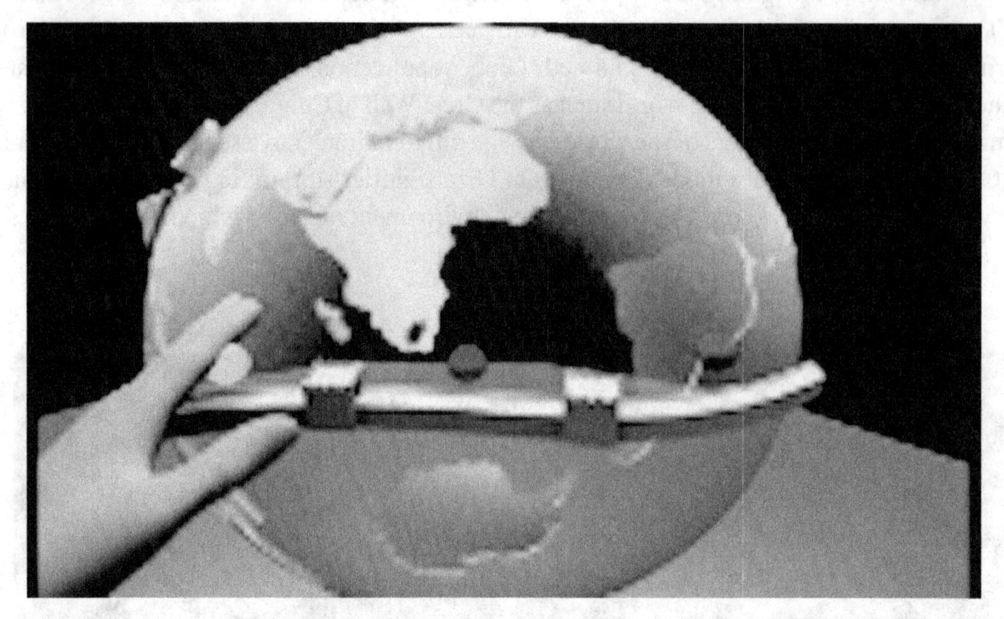

Explaining gamification is relatively simple: according to Investopedia, 'Gamification describes the incentivization of people's engagement in non-game contexts and activities by using game-style mechanics'. Put simply, it is all about using gaming elements as a reward system, which usually motivates users to do better and learn more. Gamification started in the 1990s with the advent of online learning and, as technology developed, it became commonplace in the 2010s. Since then, it has become of serious interest to various companies and scientists, who have studied the effects of gamification on the learning process. The outcomes of their research have shown that the effectiveness of gamification depends not only on the age of those tested, but also on their personality, experience, maturity, genetics and environment. This is clear confirmation that to educate trainees effectively, each course has to be carefully thought out down to the smallest detail, otherwise the outcomes may not be as successful as one might expect.

The Pyramids of Giza

The Pyramids of Giza focuses on designing immersive storytelling and problem-solving skills within a spatial computing environment. Users play the role of investigators on a mission to unearthed ancient technology hidden under the Pyramids of Giza that has been covered up by the Illuminati. Controlling the operation from a high-tech command center, users will have to make snap decisions during the adventure that affects the outcome of the story. Time is of the essence here. The platform allows for a holographic display and adventure searching informative knowledge finding experience. The scene begins with the investigation story and start display for user to proceed with the investigation journey. This scene is relying on user's input and different input will lead to a different output. The aim of this immersive scene is to enrich users problem-solving skills and how they wisely choose input based on the given scenarios. Figure 3 depicts one of the scenes in Pyramids of Giza.

Figure 3. Pyramids of Giza

Holograms, virtual and augmented realities have gradually become mainstream in business and commerce. Providing users with extra valuable information, or creating a brand-new digital world, these technologies allow companies to win big over their competition and lead the race.

In the interaction between humans, the digital and physical worlds, virtual and augmented realities take up opposite positions of the virtuality continuum. The range of virtuality states creates the so-called Mixed Reality spectrum, featuring these three main points:

- Near-digital experience
- Near-physical experience
- Mixed reality

As a part of mixed reality, the holographic display technology can be used in different spheres, the same as the virtual and augmented reality technologies.

Both, conventional holograms, and 3D holographic images add new dimensions to flat pictures and allow learners to get a better idea of any displayed subject. The parallax effect, we can call an "all-round view", inherent to the images created with the authentic holographic technology, is stunning but available only in laboratory conditions. Commonly 3D holographic images cannot boast of having this "wide" angle, but due to the 3D depth and floating effect, they also look highly impressive, which is valuable for many industry areas.

Rome's Colosseum

Rome's Colosseum is designed to showcase spatial computing as an engaging, effective, and intuitive tool through the incorporation of Human Computer Interaction (HCI), Gestural Congruence values. A presentation about the Colosseum's engineering and constructions, allows one to go through the stages of constructions, from beginning until completion, via gesture controls. The Colosseum is the largest amphitheater ever built at that period in time which held 80,000 spectators. It was used for gladiator contests and public spectacles such as animal hunts, executions, dramas, and reenactments of famous battles. Although substantially ruined because of earthquakes and stone-robbers, the Colosseum is still an iconic symbol of Imperial Rome and is listed as one of the New Seven Wonders of the World. In this Rome's Colosseum, the scene is relatively simple without any constructions done to the Colosseum in the beginning. The user must slide the sliders to see the development from one era to the other era. The best part of the scene is the user can clearly see tiny people working on the Colosseum at the center and 360 degrees walk around the object. Rome's Colosseum is focusing on learner's gesture control and showcase spatial computing as an intuitive tool for interaction design. Figure 4 represents the prototype of Rome's Colosseum scene.

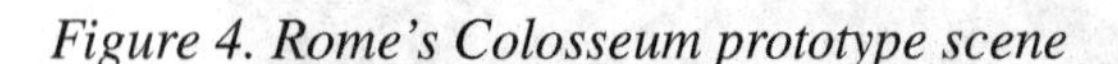

Figure 4. Rome's Colosseum prototype scene

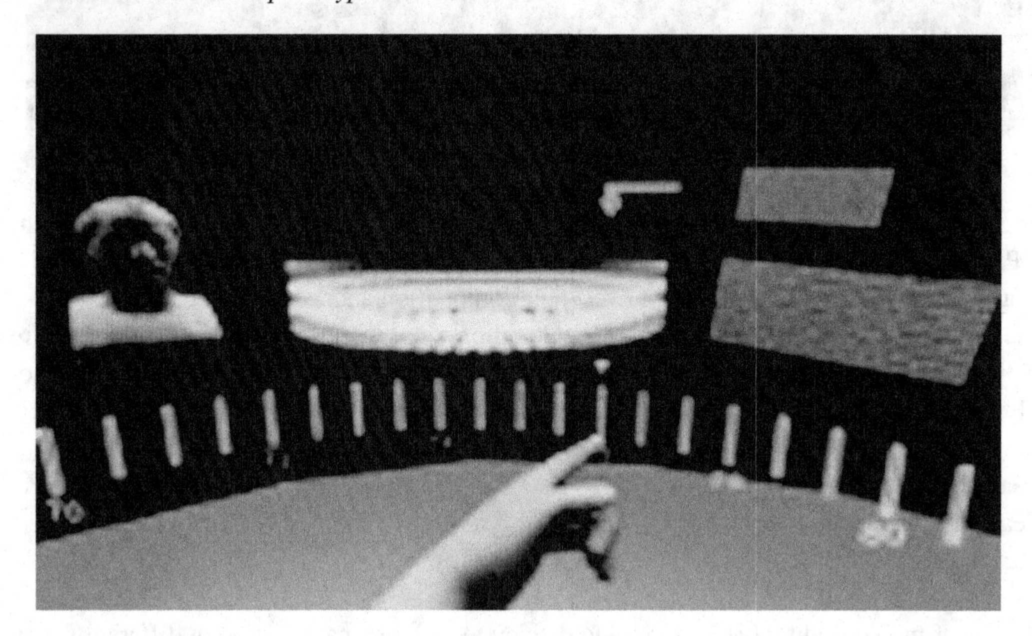

Using metaverse in construction management can bring significant improvements and advantages. Metaverse technology allows for the building of 3D models that will contain information about the future building and provide an immersive walking around that building.

While mainstream virtual reality is currently perceived as a tool for video games and entertainment purposes, developers and construction experts are finding practical applications for the technology in their field. Implementing virtual reality in construction has opened several avenues for improving design, pitching projects, and enhancing training and safety. These aren't just ideas for the distant future – companies are implementing these ideas today to much success.

IMMERSIVE EXPERIENCES

To allow audiences to experience the WoW of metaverse as well, the XR Studio is equipped with a wide screen extended reality holographic display viewing gallery and a mobile application for the audience to have a view at the outcomes. XR Studio is a studio for XR education and research well equipped with various XR devices and facilities including MR device. On top of that, the MR device can also be connected and shared on a big screen of HD TV. The experience via the mobile app is not as immersive as to compared with the end user who is wearing the MR device, the HoloLens 2. Nevertheless, the experience still can be viewed by other audiences in the surrounding area.

Most of the end users who have experienced this wonderfully designed XR Studio have given significant and positive feedback. They are amazed, blown away and enjoyed the experience and the HoloLens 2 which enabled him to connect them to different worlds which includes the metaverse. The MR technology successfully bring the users to the three wonders of the world provide fully immersive experience to them.

This has been designed on to a metaverse, Wonders of the World (WoW) in Metaverse. The metaverse here was to "build the connective tissue" between devices like HoloLens 2 headsets "so one can remove the limitations of physics and move between them with the same ease as moving from one room in your home to the next". A data structure called a spatial anchor, which encapsulates the physical structure of the spatial computing workspace is built and saved by the MR Glasses into the cloud using the Microsoft Azure backend. The Audience Holographic Display and the Spectator XR App both use this same spatial anchor to recognize and track the spatial computing workspace.
Learning outcomes & pedagogy gained by the immersive experiences is as follows;

1. Instructional Methods - the blended approach.
2. Deployment Methods - Case Studies - Time Based Construction, Gamification, Holographic Knowledge Transfer.
3. Content across the board - UNITY 3D, 3D Printing Simulation, Animated Film Making, 3D Stereoscopic, Storyboards, Optical Displays.

The EDEX Station that was designed and built serves as an XR/Spatial Computing laboratory, designed for shared and collaborative education experiences that Excite, Inspire and Stimulate. This immersive space enables technological experiments, research, teachings, and developments to complement available advance courses.

The assignment of tasks to be carried out at The EDEX Station further provide inquisitive minds the advantage of distinctive exposure, practical application, and knowledge. With the integration of multiple technologies, the skills of learners are honed – technically and creatively, to ultimately generate high value creative technologists for the new digital paradigm.

The student or learners projects and assignments carried out at The EDEX Station to develop skills in Spatial Computing are set within the categories of XR Technical Expertise, Integration of Other Courses and XR Creative Expertise. Through an unprecedented collaboration of academia and industry, learners are inspired to design compelling stories via the creation of interactive spatial computing systems that are supportive to the increasing industry demands.

Progressively over the years of studies, learners are encouraged to receive relevant experience by joining organizations on campus; apply or volunteer for available summer jobs and internship programs; get connected with various community organizations to network with like-minded people of similar career and interests; and plan for the next course to sign up for or explore possible career paths and further prepare to join the workforce in an industry of choice.

CHALLENGES

In any prototype development, challenges are around, and those challenges listed in this section are common in any prototype development phase. In WoW development as well, the authors facing several challenges to ensure the completion of the prototype. These challenges are crucial in the beginning as the authors does not have prior experience in handling metaverse project. There are the education content, talent, or skills in developing WoW and suitable devices to be used for the development. For future work, the challenges can be optimized as the past experiences did assist the authors in completing the prototype successfully.

Content

In education delivery, content play an important role and act as the main element. Content creation in immersive environment relatively different from conventional method of education delivery. Several factors such as learner's interactivity, navigational capabilities and fully immersive environment must be considered while maintaining on the engagement. Before development phase, requirements phase or user analysis have been done to gather input from educators on how this content can be implemented in metaverse and give impact to the learners.

For a start, WoW begin with three environments of wonder of the worlds named Pyramid of Giza, Great Wall of China and Rome's Colosseum. The idea comes from the uniqueness of the place itself and it able to provides fully immersive experience to the learners. The navigation and interactions for this version of WoW is quite easy and simple to work around. The objectives are to give metaverse exposure to the learners and educators as well. For future works, more environments will be developed from existing prototype and more user interactions will be added on to enrich the interactivity features. Contents come with a storyboard hence technical or critical writing is also vital. To add on, the UI/UX elements also being designed in this phase. To ensure the content is aligned and suitable with those elements in development phase, developers must keep track for each story flow and design it accordingly.

Talent

It's a challenging part to find experienced XR talent or skills in developing WoW. Most of the developer have experience working with VR and AR, but not with MR. Professional and experience XR developer required high pay and even though they are available, they are not capable to complete it within time frame stipulated due to their other projects and commitment. Though those found talents are inexperience, however they have good skills in game development and 3D modeling. To overcome this issue, they have been sent for extensive training as the talents did not have any experience working with XR technology and metaverse specifically.

Those talent need to have not just knowledge in XR programming, modeling, and animation, but with User Interface (UI)/User Experience (UX) matters. The project spent several days to look for reputable skills and spend few weeks to give training and exposure with MR development to the talent. For future metaverse/XR project, the authors have ready with group of developers and get them prepared with next education delivery project. The current talent pool is small but what makes it more challenging is the hopping or acquiring of talents from one organization to another.

Devices

Realistic XR environments demand gear with strong computing capabilities for rendering, which is also expensive. High expenses associated with creating or obtaining a XR system provide a substantial obstacle. High-end XR experiences are currently provided by tools like the Oculus Rift or HoloLens, which cost $400-$6000, respectively. These tools require a computationally strong PC, which is still a costly alternative to traditional teaching techniques. However, compared to earlier generations of XR gear, employing HMDs brings bright, immersive XR experiences into homes and classrooms with significantly reduced costs and space constraints (Coburn, Freeman, & Salmon, 2017).

Development of WoW is on HoloLens 2 and it is a high cost MR glasses. Thus, the authors future work is on VR development of WoW to make it practical and useful in classroom learning. Additionally, not just to be used in VR HMD, but applicable in non-immersive VR such as web VR environment. Web VR is a non-immersive platform for VR experience. However, it is the most affordable, practical and can be experienced by everyone regardless of the learner's situation. With a desktop or laptop, learners able to view and interact with the VR application. Moreover, to have an immersive experience or three degree of freedoms experience, those web VR can be viewed in mobile phone and slot in any VR box. Though it is not as immersive as using six degree of freedom VR headset or MR glasses, the experience still can be captured.

A key issue with the metaverse is its reliance on hardware to experience the platform. The technology evolution on the hardware is rapid. The metaverse is centered on external digital devices such as the VR headsets that can easily fall prey to hackers if left unprotected. For example, data captured through these headsets, or any other wearable devices that will be introduced in the future, can be very sensitive in nature. This in turn become an opportunity for bad actors.

FUTURE RESEARCH DIRECTIONS

Metaverse environments require computationally powerful hardware for rendering, same goes with the price. Educational simulations might not require the highest available quality but are rather based on the content of experience and possibility to provide many headsets for a class of learners at significantly lower cost. As Wonders of the World (WoW) is represented in Mixed Reality (MR), the prototype is using HoloLens 2 as the headsets which is very costly and precious. In the near future, WoW is going to be developed in VR world and being used in Oculus as well. The VR headsets are more affordable and easier to get compare with MR headset/glasses. The learners can experience in six degrees of freedom of WoW in VR too. On top of that, the next version of WoW might be developed in three degree of freedoms VR platform and AR as well. Example of three degree of freedoms VR is web-based or desktop VR that can be played in smart phone and slot in in any VR box. As for AR nowadays, it is applicable in most of current smartphones. These two platforms have been chosen due to the user's affordability and practical to be used in class setting environment, or even at own comfort at home. Looking at the evolution of XR devices, it might be no surprise if MR glasses can be owned by everyone in the future. For that particular of time, metaverse has become necessity and one of the main tools for education delivery.

For future work, other wonders of the world such as Aurora, Grand Canyon, Everest, and Taj Mahal would be developed as well in the metaverse experience. These prototypes can provide another immersive experience to the learners such as games based, storytelling and virtual tour. Each of this place able to provide exclusive experience to the learners based on the story flow and various interactions involved in the application. For example, Taj Mahal, the uniqueness of the building can capture learner's attention. The architecture and design have symbolic meaning and purpose. For Everest, the experience can be represented as storytelling and enhance learner's problem-solving skills. The prototype can depict basecamp environment and give learners a challenge to be solved while they are navigating around the scene. More interactions and navigations will be explored using the metaverse technology. Prior to the development, an extensive study, reviews, and prototypes will be conducted to get more solid background research work in developing the application models and components involvement.

Metaverse is not restricted to education delivery only. It can be applied to other industry as well such as healthcare, business, manufacturing, and entertainment. Social network is the closest definition and one of main attribute of metaverse. Thus, WoW will be extended to VR social network platform for educators and learners being together in one environment for entertainment purpose. The scenario would be a fun time to hang out together virtually, to know each other more closely (educators and learners) and sharing opinions.

Metaverse is not going to stop here or anywhere soon. Though recent article suggested that the use of Head Mounted Devices (HMD) may lead to unwanted physical side effects such as dizziness and nausea, due to technical advancement in HMD technology, these weaknesses can be overcome (Kamińska et al., 2019). Additionally, every man may have different perception of cyber sickness. Thus, scenarios preparing for education should be accurately evaluated consulted and monitored with educators. Besides disadvantages and other issues arising on metaverse, advantages have been on top of that. It is a promising technology, and many industries are already venturing into it. There are no limitations in terms of growth of the metaverse as it's here to stay.

CONCLUSION

The future of metaverse business, which is expected to be worth trillions of dollars, holds great promise for changing the ways that many of us work, learn, and communicate. But "many of us" does not mean that "all of us". Many of us here means those who have the time and money to purchase headsets, attend school and those who come from families with the ability to introduce them to the newest technology. This group of people will naturally gain the benefits from this historic opportunity in metaverse. They are the ones who will eventually master virtual world of navigation and programming, reaping its biggest benefits. Thus, building a system to connect these worlds, with interoperable assets and avatars, is a key step in enabling a shared metaverse. This will lead to the creation of a new level of reality which will be fully digital, with its own values and economy, and including an expression of ourselves, as in the real world. Many sub-challenges come with this mission: privacy, identity theft and security concerns which should not be dismissed.

The use of metaverse in teaching has several confirmed benefits. First and foremost, metaverse as XR technology offers exceptional visualization that is not possible in a typical classroom. It depicts the setting in which the younger generations feel comfortable with it. It is inclusive, allowing anybody and everyone to take part in the educational process, regardless of status, financial situation, or disability. It provides access to information, books, or articles that is virtually endless. Modern technology such as metaverse increases engagement, stimulates cooperation and involvement. Additionally, it promotes independent learning and highly effective blended learning by allowing self-study. Since the metaverse sector is still in its development, there is still time to learn from the experience and ensure that in the future, the advantage of the metaverse is really distributed across the socioeconomic spectrum.

Thus, for centuries, the education system has been evolved to cater the needs of learners and educators. It has always adapted to the available technology and digital world has become immersive as the real one. Wonders of the World (WoW) is example of applied metaverse prototype for education delivery. Learning in metaverse is as solid as in the physical world. Thus, learners are receiving similar advantages as learning in conventional way.

Great Wall of China, Pyramid of Giza and Rome's Colosseum in WoW provide a different unique immersive experience to the users such as games, storytelling, and spatial interaction. The scenarios depict the physical world and users able to interact, engage and learning with the visual metaverse. Consideration of learning in immersive metaverse should not only focus on the visual aspect, but should also tackle the interactivity, learning content and immersiveness of the environment. Immersiveness plays a vital role in metaverse or XR due to the nature of the technology, spatial computing.

The future of metaverse through online channels will be possible from a conceptual point of view due to introduction of genuine experiences with these XR technologies through the educational delivery. By utilizing the flexibility of collaborative social systems, learners will be able to attend classes and feel as though they are seated in the classroom with their peers and seeing the educators in front of them. The objective of education delivery and with its future, will be to combine the best aspects of both worlds; the advantages of having open access without set hours, adding the convenience of doing it from home, along with the advantages of having experiences, tools, friends, and educators who provide learners with the same value as in-person instruction. The learning opportunity and later for earning is massive. Metaverse can enable learners to travel incredible, virtual learning environments. The learners can discover from the bottom of the ocean about marine biology or art history while inside a virtual Louvre.

Although using modern technology in education delivery is clearly beneficial, still the weaknesses are there. One of the main issues is the lack of flexibility. The users (learners) are restricting to certain conditions and rules. Examples, take part in discussion, ask questions, and receive answers. This might lead to lack of educator-learner interaction. There must be a balance between conventional approach and technology although the authors are tempted to replace all old-fashioned solutions with metaverse applications. However, metaverse have proven numerous advantages as a platform for education delivery. The outstanding visualization and engagement which cannot be obtained in traditional classroom. The key to the engagement is the immersive element. It gives virtually unlimited involvement and stimulates cooperation for individual pursuit of knowledge and self-study encouragement.

ACKNOWLEDGMENT

This research was supported by the Asia Pacific University of Technology & Innovation (APU). The APU XR (Meta) Studio is a first-of-its-kind facility that comprises technologies capable of developing Augmented Reality (AR), Virtual Reality (VR) and Mixed Reality (MR) applications. Developed in partnership with Ministry XR Malaysia, the studio is equipped with state-of-the-art equipment and facilities which include a 360-degree volumetric video capture zone, EDEX (extended education experience) station, VolCap propriety software, mixed reality smart glasses in the form of Microsoft HoloLens and high specification gaming standard workstations.

REFERENCES

Coburn, J. Q., Freeman, I., & Salmon, J. L. (2017). A Review of the Capabilities of Current Low-Cost Virtual Reality Technology and Its Potential to Enhance the Design Process. *Journal of Computing and Information Science in Engineering*, *17*(3), 031013. doi:10.1115/1.4036921

Duan, H., Li, J., Fan, S., Lin, Z., Wu, X., & Cai, W. (2021). Metaverse for Social Good: A University Campus Prototype. In *MM Proceedings of the 29th ACM International Conference on Multimedia*. 10.1145/3474085.3479238

Han, H.-C. (2020). *Sandrine*. From Visual Culture in the Immersive Metaverse to Visual Cognition in Education. doi:10.4018/978-1-7998-3250-8.ch004

Hassan, M. (2020). Online teaching challenges during COVID-19 pandemic. *International Journal of Information and Education Technology (IJIET)*. doi:10.18178/ijiet.2021.11.1.1487

Kamińska, D., Sapiński, T., Wiak, S., Tikk, T., Haamer, R. E., Avots, E., & Anbarjafari, G. (2019). Virtual reality and its applications in education: Survey. *Information (Switzerland)*. doi:10.3390/info10100318

Kye, B., Han, N., Kim, E., Park, Y., & Jo, S. (2021). Educational applications of metaverse: Possibilities and limitations. *Journal of Educational Evaluation for Health Professions*, *18*, 32. doi:10.3352/jeehp.2021.18.32 PMID:34897242

Mubin, S. A., Thiruchelvam, V., & Andrew, Y. W. (2020). Extended Reality: How They Incorporated for ASD Intervention. In *2020 8th International Conference on Information Technology and Multimedia, ICIMU 2020*. doi:10.1109/ICIMU49871.2020.9243332

Mystakidis, S. (2022). Metaverse. *Encyclopedia*, *2*(1), 486–497. doi:10.3390/encyclopedia2010031

Nichols, G. (2018). *Augmented and virtual reality mean business: Everything you need to know An executive guide to the technology and market drivers behind the hype in AR*. VR, and MR.

Ryan, G. V., Callaghan, S., Rafferty, A., Higgins, M. F., Mangina, E., & McAuliffe, F. (2022). Learning Outcomes of Immersive Technologies in Health Care Student Education: Systematic Review of the Literature. *Journal of Medical Internet Research*, *24*(2), e30082. doi:10.2196/30082 PMID:35103607

KEY TERMS AND DEFINITIONS

Augmented Reality: An environment that is enhanced by computer-generated perceptual information during an interactive encounter, sometimes involving many sensory modalities such as visual, aural, haptic, somatosensory, and olfactory.

Education: An intentional practice with a specific goal in mind, such as spreading knowledge or developing talent and character. These objectives might include growth of comprehension, reason, kindness, and honesty.

Extended Reality: A word used to describe all real-and-virtual mixed setting, as well as human-machine interactions produced by wearables and computer technologies. This comprises the regions interpolated between representative forms like augmented reality (AR), mixed reality (MR) and virtual reality (VR).

HoloLens: A set of smart glasses for mixed reality that Microsoft designed and produced. The first head-mounted display to use the Windows Mixed Reality technology and the Window 10 operating system.

Immersive: Giving the player, audience the impression that they are all in on something.

Metaverse: Fictional version of the Internet that is a single, all-encompassing, and immersive virtual environment that is made possible by the usage of virtual reality and augmented reality headsets.

Mixed Reality: The blending of the physical and digital worlds to create new landscapes and visualizations where items may coexist and communicate in real time.

Virtual Reality: A virtual experience that might resemble the real world or be wholly unrelated to it. Virtual reality has applications in business, education, and entertainment.

Chapter 8
Research on Improving the Quality of Talent Training in Higher Vocational Colleges in China

Fengmei Liu
Guangdong Polytechnic of Science and Technology, China & Universiti Tun Abdul Razak, Malaysia

P. C. Lai
Universiti Tun Abdul Razak, Malaysia

ABSTRACT

According to the action plan for improving the quality of vocational education (2020-2023), the vocational school evaluation system includes professional ethics, professional quality, technical skills, employment quality, and entrepreneurial ability as important content areas for assessing the quality of talent training that should be strengthened. This plan was released by the Ministry of Education and nine other government departments. An example higher vocational college in Guangdong is used to illustrate the importance of quality talent training in the new era, and a method for improving talent training quality in vocational colleges is proposed. This chapter is intended to serve as a model for other higher vocational colleges seeking to improve the quality of talent training while also promoting the modernization and internationalization of high-quality vocational education.

INTRODUCTION

In country like China's social industries, socialism with Chinese characteristics is entering a new period, which has also emerged as a critical phase in the development of strong market competition in the country's social industries. General Secretary Xi Jinping issued directives aimed at accelerating the development of vocational education: to accelerate the development of vocational education in order to ensure that everyone has the opportunity to shine in life; to accelerate the development of vocational education in

DOI: 10.4018/978-1-6684-5732-0.ch008

order to ensure that everyone has the opportunity to shine in life. To provide exceptional professionals for the country's social development, higher vocational colleges should be guided by national education policies, further, define the value orientation of talent training in the new era, and constantly study various effective implementation methods according to the ministry of education (Hu, & Liu, 2019). Students should be placed at the center of the learning process, and students should be encouraged to become socialist builders and successors through a comprehensive development of morality, intelligence, body, beauty, and labor. Vocational colleges should also assist students in effectively representing themselves in the workplace and in everyday life, and to gain an advantage while also appreciating its value. Therefore, incorporating the concept of metaverse can bring innovation and enhance education learning. Metaverse technologies must be integrated into education and more efforts are needed from companies to promote the integration of metaverse technologies into the education industry to improve teaching and learning efficiency. Yue Xiwei, co-founder, chairman and CEO of vocational education leader Huike, said as the metaverse continues to heat up, it is important for education companies to really apply such technologies to change education itself, as well as nurture related talents in the field. He made the remarks when the company cooperated with online education firm Kaikeba and a software industrial park of Top East in establishing an education experience center for the metaverse.

A CONCEPTUAL FRAMEWORK FOR VOCATIONAL TRAINING

The Concept of a New Era

Chinese socialism has entered a new era, according to the report of the 19th National Congress, which was held in October 2017 and culminated with the declaration of a new era by the country's 19th National Congress (also known as the "19th National Congress"). In response to the widening gap between social production and people's ever-increasing expectations of a better way of life, the paradox has been turned into a dilemma between unbalanced and inadequate growth. There were approximately 35,100 copies produced between January 2010 and December 2021 (Table 1), as well as 303 foreign language publications that contained the key phrase "New Era" in their titles, according to the CNKI. In the year 2019, the field of "new age" research reached its zenith (Figure 1).

Table 1. 2010-2021 "New Era" related article distribution category statistics table

Academic journals	Dissertations	Conferences	Newspapers	Achievements	Academic series	Featured journals	Total
26800	2441	560	172	1	254	4877	35105

Note: The data comes from the Chinese National Knowledge Infrastructure (CNKI).

Figure 1. From 2010 to 2021, the number of times stories about the "New Era" were read in the media
Note: The data comes from the CNKI.

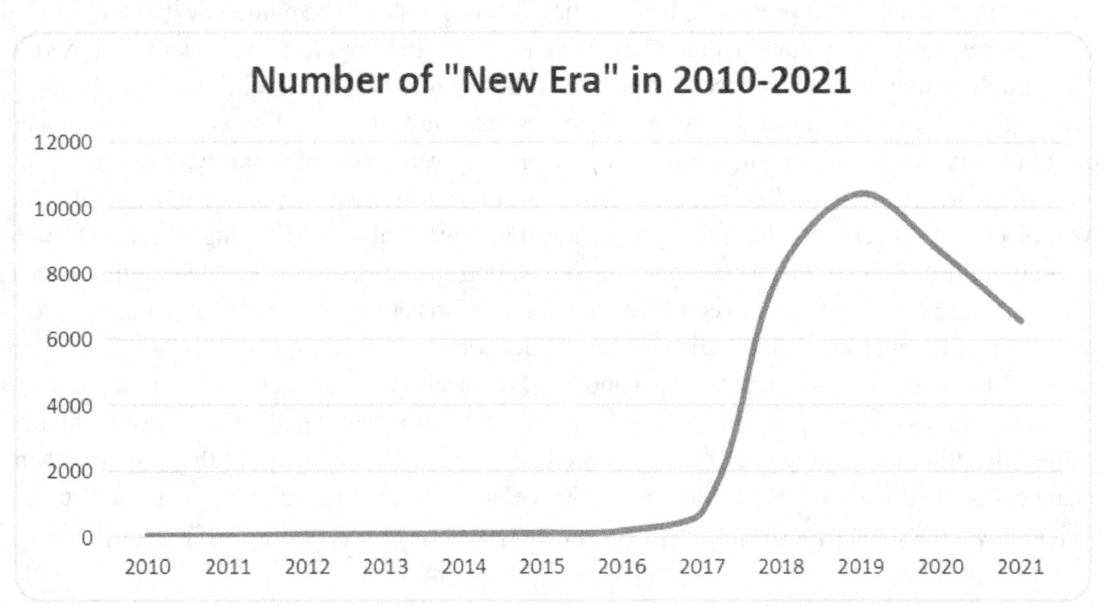

As China's higher education enters a new era, the country finds itself at a critical point in its history. Higher education institutions are taking on new missions and confronting new possibilities and challenges, and there is still a long way to go in terms of establishing a strong country's higher education system, according to the World Bank (Zhong, 2018). According to Yao Wenming (2019), socialism with Chinese characteristics, as well as education, has entered a new era. Updated educational conceptions, enhanced knowledge structure, the transformation of attitude, the transformation of instructional conduct, and the enhancement of educational capacity to cope with educational reform are all priorities for teachers in the new millennium. Universities' connotation construction in the new era should be guided by the unification of educational models and educational laws, the unification of professional settings and social needs, the unification of curriculum systems and educational objectives, and the unification of educational measures and student needs. School administration positioning and stressing the qualities of school administration; developing the educational model and improving the training system; maximizing the curriculum structure and strengthening the development of the discipline (Li, & Jiang, 2018). According to Zhang Wenli and Fan Mingming (2019) as well as Wortley and Lai (2017), the expansion of higher vocational education in the new era is being hampered by a new internal and external environment that is changing rapidly. Higher vocational education should place emphasis on four areas: high-quality talent development, high-level professional development, high-quality production-education integration, and high-level internationalized school management, according to the World Economic Forum.

Talent Identification and Development

In the corporate world, the term "talent training" refers to the process of instructing and training potential employees and employees-to-be. Higher education institutions in the People's Republic of China are mandated to place a high priority on talent cultivation, teaching, scientific research, and social service while also ensuring that the overall quality of education and teaching meets state standards and criteria, according to the Higher Education Law of the PRC. The inner logic of "cultivating who, for whom, and how to cultivate people," which is the inner logic of cultivating newcomers in our country during the era of national rejuvenation, was scientifically explained by General Secretary Xi Jinping in a keynote address delivered on September 10, 2018, at the National Education Conference. If we are talking about educational reform and development, there are a plethora of important practical aspects to take into account (Wortley & Lai, 2017; Sun, 2020; Lai, & Liew, 2021)

Improving the quality of talent training is critical to attaining high-quality development in higher vocational colleges. Establishing moral principles and nurturing persons in higher vocational colleges are critical components of ensuring high-quality growth (Liu, & Wen, 2022). Higher education development of high quality is a strategic goal for the growth of higher education in my country throughout the current "14th Five-Year Plan" period, which is now in effect. In order to construct a high-quality higher education system, which includes a structural framework for higher education, a talent development system, a governance system, as well as involvement in international education governance, among other things, is necessary (Anthony, Rosliza, & Lai, 2019; Lai, 2020; Zhong, 2021).

THE CRITICAL ROLE OF HIGHER VOCATIONAL COLLEGES IN ENHANCING THE QUALITY OF TALENT TRAINING IN THE NEW ERA

Increase the Speed with which China's Social and Industrial Transformation is Carried Out

According to the report of the Communist Party of China's "19th National Congress," the Chinese economy has transitioned from a high-speed growth stage to a high-quality development stage. The country's economy is currently in a critical period of transforming its development mode while also optimizing its economic structure and transforming the driving force of growth. The importance of quality and efficiency in the construction of a current economic system cannot be overstated. Supply-side structural reform is the major purpose of this initiative, and it is intended to assist economic development improvements that improve the overall quality, efficiency, and power of the economy as a whole (Lai, 2015; 2018; 2019; Lai and Lim, 2019; Lai, & Tong, 2022). The primary objective is supply-side structural transformation (referred to as the "three major reforms"). Li Zhihong (2018) believes that the "three essential shifts" have resulted in the development of unique conceptions, novel models, and novel driving factors for my country's industrial development, all of which have evolved into the industrial development methodology. Structure reforms on the industrial supply side should be furthered. Other specific tactics include the establishment of a modern industrial system with coordinated development, as well as the implementation of an industrial innovation-driven development plan (IIDP). These individuals believe that vocational education and industrial development are inextricably linked. This is a belief that both Liu Xiao and Qian Jiannan hold (2020). When it comes to connecting vocational education, specialized construction, and industrial development, the industrial structure, market demand, and industrial technology should be the logical starting points, starting with the industrial catalog, industrial space layout, and labor category demands, and progressing from there. It is possible to experience docking when the five dimensions of labor level and labor skill demand converge and form a docking pattern.

The country's social and industrial structures are currently being upgraded and transformed continuously at this point in history. The level of instruction received by students at top vocational colleges is becoming increasingly important, at the same time. According to an analysis of current industrial structure changes, my country's industries are increasingly becoming dominated by technology-intensive industries, which are gradually displacing old labor-intensive sectors in the process. The raw material industry is changing, and several emergent industries are gradually establishing themselves as the country's present industrial pillars, displacing traditional industries as they grow in importance and influence.

Chinese economic and social development has accelerated dramatically since the reform and opening up, with the country rising to become the world's second-largest economy in terms of nominal GDP since the reform and opening up. However, as our economy continues to grow at a rapid pace, we are increasingly confronted with resource and environmental limits that are becoming more severe. As a result of the Communist Party of China's 19th National Congress, the necessity of harmonious ties between people, as well as between people and society, has been highlighted even more in the process of achieving high-quality economic development (Wu, 2020). This indicator is beginning to demonstrate aspects of the real economy in my country, while also demonstrating the development tendency toward a "low carbon economy," which is a positive development trend. In order to ensure long-term and healthy development, a country development needs the establishment of a significant pool of highly skilled professionals and technical individuals. As a result, boosting the quality of people training is a critical prerequisite for a country's modernization to be successful. In order to achieve rapid adjustment and upgrade of social industries in the new era, it is necessary to pursue multiple disciplines at the same time. It is also necessary for experts who advocate for industrial adjustment and upgrading that they have a thorough understanding of contemporary development as well as the technical competence to innovate within traditional industries. Vocational colleges, which are important educational institutions in this regard, are responsible for the development of exceptional professional talents for the benefit of society. In order to provide more high-quality talent to the country and society, it is necessary to conduct a thorough analysis of the requirements of contemporary industrial development, to further clarify the value orientation centered on talent cultivation, and investigate the connotation of talent cultivation. professional.

Meet People's Spiritual Needs

People's spiritual demands are growing in importance in today's world, and there is a significant increase in the demand for a sense of accomplishment. Because of this, it serves as an important symbol of the great number of employees who make major contributions to the status and dignity of our country, and it also serves as a message of fairness and justice. Higher-quality talent training is essential for meeting people's spiritual demands, and the growing number of students enrolled in higher-level vocational institutions in my country demonstrates the amount of future development that may be predicted in the sector. Higher vocational institutions provide a significant contribution to the improvement of people's lives in the modern-day since they are an important location for directly exporting skills to the broader community in which they operate.

According to the Chinese government's work report for 2021, a high-quality education system, deepening education reform, and implementing measures to improve the quality and capacity of education are all priorities, with the primary goal of increasing the average number of years of education for the working-age population to 11.3 years being the primary focus. According to Xu Juan (2021), there should be a greater emphasis placed on "people satisfaction" as well as the demands of individuals pursuing a higher level of vocational training. She has developed a framework for measuring and evaluating student and family happiness, industry and enterprise satisfaction, school satisfaction, government satisfaction, and societal satisfaction, among other things, based on the perspectives of stakeholders in higher vocational education. System of Evaluation and Assessment (SEA) The new era incorporates a variety of concepts such as knowledge economy, scientific economy, and digital economy, among others. These notions necessitate the development of competencies taught through higher vocational education in order to foster a lifelong learning mindset, which is also the only way to lead a more fulfilling life. Higher vocational education is primarily concerned with developing students' abilities and skills, as well as their social flexibility and adaptability, in order to enable them to make use of their advantages and recognize their significance in the development of society as a whole. Greater attention should be paid by higher vocational education to the spiritual needs of people in terms of their value orientation when it comes to carrying out the work of talent training and developing future leaders.

THE CURRENT SITUATION OF TALENT DEVELOPMENT IN HIGHER VOCATIONAL COLLEGE

A General Overview of a Higher Vocational College

A higher vocational college is a high-level professional group project construction unit with over 28000 full-time pupils that is part of a larger educational system. Guangdong Provincial Higher Vocational College is the largest provincial higher vocational college in the province, and it was established more than 30 years ago on the foundation of and growth of regional advantages. A national high-level university, located in the Pearl River Delta, actively participates in and promotes the "Belt and Road" initiative, as well as the "large bay area" national strategy. It has amassed a number of accomplishments and will be inducted into the national high-level university system on January 1, 2019. (This information comes from the website of a higher vocational college.)

An Examination of the Educational Quality of a Higher Vocational Institution is Carried Out

To ensure that talent development is of high quality, we will strengthen the vocational school evaluation system and evaluate it in terms of professional ethics, professional quality, technical skills, employment quality, and entrepreneurial ability, according to the Action Plan for Improving the Quality and Training in Vocational Education (2020-2023). The survey data from the Annual Report on the Quality of Higher Vocational Education (2022) and the Annual Report on the Employment Quality of Graduates of a Higher Vocational College (2021) are summarized in the following Table 2.

Table 2. List of the whereabouts of graduates from a university from 2019 to 2021

Graduation year	Total number of graduates	Agreement (flexible) employment	Go to a school of a higher grade	start an undertaking	Conscripts	National and local grassroots projects	Unemployed	Employment rate	The employment rate of junior college graduates in Guangdong Province
2021	8022	5631	1748	301	63	3	276	98.22%	96.69%
2020	7594	5996	1074	260	26	5	233	96.93%	82.71%
2019	7805	7103	312	278	1	9	102	98.69%	96.12%

Unit: person

According to Table 2, between 2019 and 2021, the employment rate of graduates from a higher vocational college will be greater than the average employment rate of college graduates in Guangdong Province. Tech knowledge, the standard of living in general, and entrepreneurial aptitude are all improving in the United States of America. While a bigger number of persons complete their education and enlist in the army, the fact that overall student quality has improved suggests that the general quality of students has improved as well. This has occurred as a result of the higher vocational college's longstanding commitment to the school-running philosophy of "establishing quality, developing characteristics, and strengthening innovation" as well as its active development of a "double precision" education model, both of which have resulted in significant growth in recent years for the institution. Consider deepening the reform of teachers, teaching materials, teaching methods, and scientific research, delegating regulation and service, and establishing a "signature achievement library" as an outcome-oriented scientific research mechanism, in addition to measures such as the social service mechanism, the improvement of the scientific research platform, and the establishment of a "signature achievement library" as an outcome-oriented scientific research mechanism, among other things.

An Examination of a Higher Vocational College's Capacity for Social Service

Sun Yongchun (2021) used the principal component factor analysis approach to analyze the social service performance of 43 higher vocational colleges in Guangdong Province in 2021, with the results published in the journal Higher Education. The findings of the study also revealed that the expansion of higher vocational colleges is mostly determined by the level of economic development in the city in which they are located and that the contribution to the city is also more direct as a result of this development. According to a quality report issued by a specific higher vocational college in the last three years, the amount of horizontal technical services, patent accomplishments, and non-academic training will all increase significantly in 2020 and 2021, even if the college is affected by the new crown pneumonia epidemic in 2020 and 2021. (As illustrated in the table below.) The formation of a regional vocational education group, which was chaired by a higher vocational college in August 2015, brought together representatives from the government, schools, businesses, and enterprises, as well as the government's leadership, to discuss issues pertaining to vocational education in their respective regions. This work resulted in the creation of a service platform for information interchange and resource sharing among the region's government, enterprises, and universities, as well as technical and technical staff assistance, as well as intellectual support for the development of the region's economy. Since 2016, a vocational college has been chosen to meet the needs of its surrounding community each year from 2016 to 2018. Higher Vocational Colleges in the United States that have made major contributions to the community are highlighted below.

Table 3. Contribution table for social services for the years 2019-2021

Metric		Unit	2021	2020	2019
1	Horizontal technical service payment amount	Wan Yuan	2420.22	1050.07	601.02
	Economic benefits generated by horizontal technical services	Wan Yuan	5474.70	2801.70	5419.2
2	The amount of longitudinal scientific research funds are received	Wan Yuan	648.27	648.27	513.4
3	Technical transaction amount	Wan Yuan	648.27	226.83	601.02
4	Conversion of patent result into the payment amount	Wan Yuan	30.485	12.00	12.00
5	Amount of non-academic training received	Wan Yuan	879448	699880	1929.73
6	Public welfare training services	class hour	130024	18100	
	The main source of school funds (radio selection): provincial finance (Ö) prefecture-level finance () Industry or enterprise () other ()				

Although a higher vocational college was affected by the new crown pneumonia epidemic between 2019 and 2021, the number of social services received increased during this period, and the number of patent achievements transformed increased by 2.5 times in 2021 compared to the previous year, as illustrated by Table 3. That the school is becoming more capable of providing social help is demonstrated here. Additional activities included participation in national poverty reduction, rural revitalization, and implementation of the East-West Cooperation Vocational Education Action Plan. It has frequently dispatched key teachers and personnel to rural areas, supporting a poverty-stricken town in Meizhou City, Guangdong Province, in overcoming poverty and assisting the entire province of Guangdong in overcoming poverty, among other initiatives. The regeneration and development of a village in Maoming City have been carried out in partnership with a higher vocational institution in Heilongjiang Province, according to a press release. The two schools educated a total of 364 students over the course of two sessions. After being voted as the chairman unit of Guangdong Vocational Education Collaborative Development Alliance (which includes nine pairs (18) paired vocational colleges in Guangdong Province and Heilongjiang Province), the two schools were named as the chairman unit of the alliance in December 2021. By taking this step, a vocational college's social service competence and impact are increased.

INQUIRING INTO THE BEST WAYS TO IMPROVE THE QUALITY OF TALENT TRAINING IN HIGH VOCATIONAL COLLEGES IN THE NEW ERA

Pay Great Attention to the Level of Talent Training that is Provided

The government must do an amazing job of creating top-notch talent training programs in colleges and other educational institutions. The first is to give priority to high-level and highly talented professionals in order to improve relevant treatment; the second is to encourage businesses to actively improve promotion criteria for their original positions in order to protect the economic interests of professional talents. Recognize and promote one's worth; third, publish pertinent policies and make them widely known in order to increase the social recognition and influence of higher vocational colleges; and fourth, government departments at all levels must fully exploit the characteristics of new media communication and conduct a variety of theme activities to bring craftsmen into the spirit of the society. To ensure the quality of talent training, vocational colleges should be guided by the "four services," which should deepen reform, increase the operational quality of schools by a large margin (Li Yuying, 2003; Hamdan, Kassim and Lai, 2021), and work actively with industries and businesses.

Maintain an Awareness of Talent Development Criteria

In recent years, my country has made considerable progress toward its aim of becoming a manufacturing superpower, thanks to the implementation of the "One Belt, One Road" plan. To survive in the face of strong market rivalry, higher vocational schools must build on global development, closely integrate theory and practice, implement the concept of quality education, and collaborate with other organizations. High-level departments are primarily concerned with talent cultivation and reconstruction. These departments collaborate to develop unique higher vocational majors based on an advanced curriculum structure, continuously optimize professional courses based on this structure, organically integrate domestic and international resources, and make use of advanced education technologies such as cloud courses and MOOCs. Students contribute to the development of future professionals all across the world by laying the groundwork for their future careers.

Create a Framework for Collaborative Cultivation that includes Representatives from Government, Education, and Business

China has marked the commencement of the sharing economy in the new millennium by being the first country to do so. Educators at universities and higher vocational colleges should be cognizant of the nuances of the current situation, maintain close ties with government and business alike, and provide a platform for students to educate themselves through collaboration with the government, educational institutions, and businesses alike. Higher vocational institutions must be inspired by the government, and enterprise resources must be used as a springboard for the development of a three-in-one educational system for students if they are to be successful in establishing a collaborative education platform. In the first instance, the extension of the collaborative education platform should be guided by the goal of enhancing local economic development in the community. Both schools and businesses should investigate collaboration options after acquiring a thorough understanding of one another's unique missions. They should also work together to build talents and achieve a tight integration of abilities and requirements.

For practical teaching to take its proper place in higher vocational colleges, higher vocational colleges should establish a "dual-teacher" team, effectively strengthen teachers' teaching abilities, dispatch teachers to train in firms, and recruit high-skilled talents from businesses to enter the classroom. In addition, higher vocational institutions should be open to experimenting with novel teaching formats in order to pique students' interest in both learning and employment while also acknowledging the critical role that collaborative nurturing platforms play in the advancement of knowledge. The fourth point is that both schools and businesses should work closely together to develop curriculum resources, develop teaching content and evaluation standards that are current and relevant to current industry and job requirements, and organically integrate teaching content and evaluation standards into enterprise production and operation (see point three).

Improve the Efficiency of Management Systems in Higher Vocational Institutes by 20 Percent

A growing number of demands are being made on post-secondary vocational education as a result of the upgrading and restructuring of the social economy. Universities and other higher vocational institutions must constantly improve and diversify their management methods in order to remain competitive in today's society. Student management systems in higher vocational colleges, for starters, should be cutting-edge in their conception and implementation. It is currently under progress for major higher vocational institutions in my country to put in place their ongoing enrollment expansion initiatives. Students' entry requirements are being lowered as a result of these programs, allowing a greater number of students to be admitted. The task at hand is fraught with numerous complications. Higher vocational colleges must teach and manage students in accordance with their aptitudes in order to meet the quality requirements of talents in the new era of information technology and automation. They must also construct a learning account for each student using an information technology platform, monitor the situations of students in real-time, and choose scientific and reasonable courses for them to take in order to graduate. As a result of the use of assessment and evaluation standards, students are evaluated in a variety of ways and guided in the right direction for learning, so improving the overall quality of education delivered to students. A further improvement should be made to the faculty administration structure at community and technical colleges. To further implement the school-enterprise collaborative cultivation plan, it is necessary to recruit and train excellent full-time and part-time teachers; implement a performance appraisal system; strengthen the reward and punishment system; establish a high-quality and high-capacity double-teacher team; to define distinct job functions for each managerial staff member in order to further implement the school-enterprise collaborative cultivation plan. Higher vocational colleges must take the third stage, which is to seek government help and submit funding applications for linked activities in order to provide an adequate financial guarantee for talent training in higher vocational colleges.

Promote the Use of Metaverse Technology in Vocational Education

In recent years, the word "metaverse" has been sought after by people from all walks of life and has been selected as the 2021 Collins English Dictionary's Hot Word. The "Metaverse Development Research Report 2020-2021" was published by the New Media Research Center of Tsinghua University in October 2021. The purpose of the report was to clarify the concept, theoretical construction, and industry analysis of the "Metaverse," as well as its development trend and potential risks. In China, the

Metaverse is mostly utilized in the gaming and social industries, and the market is expanding rapidly. With the advancement of science and technology, Metaverse's underlying core supporting technologies such as AI, VR, and AR have been incorporated into the education process, and an increasing number of schools are constructing virtual reality teaching laboratories for experiments and experience-based instruction. In China, vocational education as a form of education cultivates not only the cultural qualities of students but also their vocational abilities. Teachers, teaching scenarios, and training equipment, among other subjective and objective criteria, limit the ability of higher vocational institutions to successfully implement teaching to increase the quality of talent training. The applicability of Metaverse in vocational education extends far beyond the experiential teaching of classroom VR and AR as well as software design and development. Future higher vocational colleges must therefore continue to deepen the in-depth creation of teaching theory and curricular systems and innovate teaching. Develop and investigate additional educational resources and applications. Dai Ruoli, co-founder and chief technology officer of Beijing Noitom Technology Ltd, a virtual reality solutions provider, said: "Digitalization is an irreversible trend. When faced with the metaverse, it is expected to embrace and popularize it to more people through education." Thus, vocational education should investigate incorporating the metaverse into its curriculum.

CONCLUSION

As new technologies like virtual reality. augmented reality, artificial intelligence, and other technologies are present in our daily life so education will need to progress with these technologies, especially vocational education at the forefront of producing these types of skill sets of graduates for the industry. Therefore, vocational education should play an important role in promoting metaverse education and be the pioneer of this field.

REFERENCES

Anthony, N., Rosliza, A. M., & Lai, P. C. (2019). The Literature Review of the Governance Frameworks in Health System. *Journal of Public Administration and Governance*, *3*(9), 252–260.

Cai, S., Jiao, X. Y., & Song, B. J. (2022). Open Another Gate to Education—Application, Challenge and Prospect of Educational Metaverse, *Modern. Educational Technology*, *1*, 16–26.

China Government Network. (2021). *Report on the Work of the Chinese Government in 2021.* http://www.gov.cn/premier/2021-03/12/content_5592671.htm

Duan, C.(2019). Research on the Quality Evaluation System for Talent Training in Higher Vocational Education in China . *Huainan Vocational and Technical College Journal*, (3), 46-47.

Hamdan, N. H., Kassim, S. H., & Lai, P. C. (2021). The COVID-19 Pandemic crisis on micro-entrepreneurs in Malaysia: Impact and mitigation approaches. *Journal of Global Business and Social Entrepreneurship*, *7*(20), 52–64.

Hu, M. B., & Liu, B. Q. (2019). The New Era's Value Guidance and Realization Strategy for the Quality of Higher Vocational Education Talents Training. *Vocational Education Forum, 6*(66), 17-22.

Huang, N. X. (2019). Thoughts on Improving the Quality of Talents in Post-secondary Vocational Education. *Shanxi Youth*, *11*, 188.

Jia, Y. L., & Yang, Y. L. (2020). The Development of a System for Assessing the Quality of Talent Training in Higher Vocational Education. *Shandong Chemical Industry*, *49*, 174–175.

Lai, P. C. (2016). *Smart Healthcare @ the palm of our hand*. ResearchAsia.

Lai, P. C. (2018). Research, Innovation and Development Strategic Planning for Intellectual Property Management. *Economic Alternatives*, *3*(12), 303–310.

Lai, P. C. (2019). *Factors That Influence the tourists' or Potential Tourists' Intention to Visit and the Contribution to the Corporate Social Responsibility Strategy for Eco-Tourism. International Journal of Tourism and Hospitality Management in the Digital Age*. IGI-Global.

Lai, P. C. (2020). Intention to use a drug reminder app: a case study of diabetics and high blood pressure patients *SAGE Research Methods Cases.* doi:10.4135/9781529744767

Lai, P. C., & Liew, E. J. (2021). Towards a Cashless Society: The Effects of Perceived Convenience and Security on Gamified Mobile Payment Platform Adoption. *AJIS. Australasian Journal of Information Systems*, *25*. doi:10.3127/ajis.v25i0.2809

Lai, P. C., & Lim, C. S. (2019). *The Effects of Efficiency, Design and Enjoyment on Single Platform E-payment Research in Business and Management*, *2*(6), 19–34.

Lai, P. C., & Tong, D. L. (2022). An Artificial Intelligence-Based Approach to Model User Behavior on the Adoption of E-Payment. Handbook of Research on Social Impacts of E-Payment and Blockchain, 1-15.

Li, D. Y., Li, J. F., & Yan, D. (2019). The Talent Training Model of Higher Vocational Education in China: Issues and Reforms. *Hebei Vocational Education*, *1*, 30–34.

Li, S. X., & Jiang, L. J. (2019). How to enhance colleges' connotation construction in the New Era. *Chinese University Technology,* 70–72.

Li, Y. Y. (2003). The "four services" should govern higher vocational education reform. *Shandong Youth Political College Journal*, *4*, 76–77.

Li, Z. H. (2018). Theoretical foundations and methods for my country's industrial development's "three major changes". *Reform*, *9*, 91–101.

Liu, X., & Liu,W. K. (2019). The Development of Higher Vocational Education in the Context of Recruitment Expansion: Challenges and Countermeasures. *Education and occupation, 14*, 5-11.

Liu, X., & Qian, J. N. (2020). Major Construction and Industrial Development in Vocational Education: Aligning Logic and Theoretical Framework. *Higher Engineering Education Research*, *2*, 142–147.

Liu, X. J., & Wen, Y. F. (2022). The New Era's Implications for Talent Development in Higher Vocational Colleges. *Education and Career*, *1*, 58–63.

Liu, Z. B. (2018). Strategy for Improving the Quality of Higher Vocational Education Talent Training . *Chinese and international entrepreneurs, 31,* 165.

Lu, S. S. (2019). The Development of a System for Assessing the Quality of Higher Vocational Talent Training. *Henan Education Education, 1*, 53–54.

Meng, J. Z. (2021). Chinese Vocational Education has a distinct value and aim . *Latitude theoretical, 6,* 23-28

Pan, J. S. (2019). Thoughts on the Expansion of One Million Students in Higher Vocational Colleges . *Education and occupation, 14,* 12-16.

Sun, Y. C. (2021). Evaluation of regional higher vocational colleges' social service performance——Based on an empirical analysis of higher vocational colleges in Guangdong Province. *Heilongjiang Journal of Higher Education, 3*(93), 115–120.

Sun, Y. S. (2020). Internal Logic Analysis of Xi Jinping's Talent Development. *Human Resource Development, 19*, 10–11.

The Ministry of Education, et al. (2020). *The Action Plan for Improving the Quality of Vocational Education (2020-2023), 7.*

Tsinghua University New Media Research Center. (2021). *2020-2021 Metaverse Development Research Report.* https://xw.qq.com/cmsid/20210920A0095N00

Wang, Z. B., & Bi, Y. R. (2019). Consider the software technology major at Huai'an Vocational Information Technology College as an example. *Think Tank Times, 45*, 176–177.

Wortley, D. J., & Lai, P. C. (2017) The Impact of Disruptive Enabling Technologies on Creative Education. *3rd International Conference on Creative Education.*

Wu, H. Q. (2020). Conducting research on the social and environmental impacts of my country's industrial growth. *Journal of Zhongnan University of Economics and Law, 5*, 52–63.

Xinhuanet. (2017). *Report of the 19th National Congress of the Communist Party of China.* http://www.news.cn/politics/19cpcnc

Xu, J. (2021). Research on the Development of a System for Evaluating the Quality of Higher Vocational Education Talent Training Using "People's Satisfaction". *Journal of Hunan Industrial Vocational and Technical College, 2*(21), 13–19.

Yang, J. (2015). On China's Vocational Education Modernization Vision. *Higher Vocational Education Exploration, 5*(15), 1–6.

Yao, W. M. (2019). The new era's building of high-quality teachers. *Journal of Chinese Education,* 103-104.

Zhang, F. Q., & Xu, F. K. (2018) . On the Establishment of a Quality Assurance System Throughout the Process of Developing Talent in Higher Vocational Education . *Education and occupation, 2,* 46-51.

Zhang, W. L., & Fan, M. M. (2019). The meaning, fundamental principles, and promotion path for the development of high-quality higher vocational education in the modern era . *Education and occupation, 221,* 25-32.

Zhong, Bi. L. (2022). Preserve a century-old legacy and advance the quality of higher education. Chongqing University of Technology, Chongqing, 1(10), 3-5.

Zhong, B. L. (2018). Ascend to a new era and embrace new possibilities to overcome new obstacles. *Higher Education in China, 1*, 1.

Chapter 9
A Delphi Study on Metaverse Campus Social Perspectives

Glaret Shirley Sinnappan
Tungku Abdul Rahman University College, Malaysia

Liang Han Tay
Tungku Abdul Rahman University College, Malaysia

ABSTRACT

Metaverse is a new learning spectrum for Malaysians where it proffers a parallel virtual universe to academics and students. The metaverse campus (MC) is an idea of converting learning from physical classes into virtual worlds. The initiative of MC allows students worldwide to partake in classes and events in the same virtual world. A study was conducted among 25 Malaysian academicians from local and private tertiary instructions using the fuzzy Delphi method (FDM). This study is aimed to get consensus from the academicians to discover the social perspectives towards MC from accessibility, diversity, equality, and humanity perspective.

INTRODUCTION

The Metaverse is known with many definitions reflecting various acceptances. Many governmental and non-governmental organizations and universities around the world have invested billions in building Metaverse (Laeeq, 2022, Lee et al., 2021). The term Metaverse was coined by Neal Stephenson in 1992. Originally, the concept of Metaverse was defined as an immersive virtual world for leisure (Laeeq, 2022, Lee et al., 2021). Initially, this environment aims to combine Augmented Reality (AR), Virtual Reality (VR), blockchain and rich social medias and others (Dincelli & Yayla, 2022, Warner, 2022, Damar, 2021). According to Mark Zuckerberg, Facebook CEO Metaverse will be the next generation of the internet. He also mentioned that Metaverse can present the user digitally in virtual worlds (Kraus et al., 2022, Laeeq, 2022, Damar, 2021, Lee et al., 2021, Davis et al., 2009, Collins, 2008). Many researchers have suggested that Metaverse can shape the future, avatars can be used to represent any individual in metaworlds to socialize (Damar, 2021). Mixed Reality (MR) combines VR and AR. These technologies

DOI: 10.4018/978-1-6684-5732-0.ch009

that offer a more immersive experience than the internet itself will be able to help universities revolutionize the remote learning experience (Bernard et al., 2015, Benford, 2011, Pan, 2002). Instructors can use MR technology to provide students with engaging and immersive experiences not always achievable in a standard classroom setting (Damar, 2021, Laeeq, 2022, Lee et al., 2021). More students will be able to benefit as MR technology and content becomes more widely available. The main goal of this study is to examine the various social perspectives in the new environment for teachers to teach and students to explore the Metaverse. Students can fully be immersed in a digital classroom environment when universities use this Metaverse to teach. In an interactive environment, students interact with the presenter and learn the topic. Essentially, the Metaverse is an alternate world environment that exists outside of the current world (Kye et al., 2021). Each of these different areas has its own rights and ways of working. Educational experiences are possible, as is the increasing number of students around the world (Lee, 2022). It embraces digital experiences and creates the framework for students to collaborate, communicate, connect, and exist in a virtual environment (Lee et al., 2022, Lee et al., 2021, Pandrangi et al., 2019, Parsons, 2010). Blockchain is another aspect of the Metaverse, where the participants can be exchanged for intangible items like digital photos, audio files, or even ideas like virtual property.

LITERATURE REVIEW

Metaverse

The Metaverse allows people to interact in multimodal ways with virtual environments, digital objects, and with various participants (Dwivedi, et.al., 2022, Han et al.,2022). In Metaverse virtual combines the physical and virtual worlds, as well as provides virtual content material to the human perceptual system by integrating it into our physical environment at the same time (Allam, 2022, Hirsh et al., 2022, Davis et al., 2009). In the same way, the Metaverse indicates that AI is a beneficial technology capable of interacting with millions of people at once (Ning et al., 2021). A virtual aspect that enhances the physical world is referred to as "Metaverse AR." It spatially links the physical and virtual worlds, allowing virtual content resources to be seamlessly integrated into the actual environment for application in human perception systems (Park & Kim 2022, Wang 2022, Louro et al., 2009). Metaverse refers AI as a helpful technology that allows users to engage with millions of people at the same time. For example, a lecturer and students can interact together in class without leaving their homes, users can visit the campus by using Metaverse, a virtual reality environment (Duan et al., 2021).

In the Metaverse environment, the in-built interfaces will connect to the various Metaverse worlds (Louro et al., 2009). The basic equipment can be a computer or a smartphone, which may change to support Metaverse applications as technology evolves. In order to benefit from all of the Metaverse layers and functionalities, servers must be able to replicate, mirror, and visualize augmented and virtual reality (Park & Kim 2022). The receptor devices may provide a more user-friendly experience if the 3D glasses are attached (Kim et al., 2021). The application interface is a component of a user-installable program on the user's device or a link that only functions on a distant server, generally the cloud. The physical connection is the primary means of communication between the user's equipment and the Metaverse (Ning et al., 2021). The virtual learning environment's major files should be kept in the cloud to aid in communication and computation activities on the infrastructure. Furthermore, the cloud will oversee delivering the resources required for programs to run systems, even if the user's devices are incompatible or perform poorly (Fernández et al., 2014). The Metaverse can use security mechanisms such as permitted access and user authentication. A Storage Area Network (SAN), data center, or any other technique of data security can be used to manage the storage structure. Lecturers, students, educational institutions, and central educational agencies should have access to data. Overall to make learning environments accessible the avatar and the virtual campus ecosystem wholly should be kept on cloud servers (Parsons & Stockdale, 2010). The avatar should imitate physical processes and provide it in virtual reality modules. The avatar will first request permission from the virtual learning service provider to connect to the Metaverse world, which will deliver responses based on senses such as virtual worlds, augmentation, blockchain, simulators, and haptics (Dahan et al., 2022). When a user creates new material, both in UGC and in the life tracking record, these modifications are saved for future use. If the procedure requires payment to be accessed, the finance, NFT, and other supporting apps will be launched (Duan et al., 2021). The Metaverse connects the virtual and real worlds by utilizing necessary infrastructure, supporting software, and avatars. The Metaverse can serve as a physical world mirror, reflecting everything back to the digital twin. This framework should be operated on the digital twin and the process is uploaded to the cloud (Lee et al., 2021).

Metaverse Campus

The Metaverse opens great opportunities for educational institutions to extend their virtual copy of their campus. The Metaverse campus is a concept of converting learning into virtual worlds where the differences between the physical and virtual are reduced (Duan et al., 2021, Zallio & Clarkson, 2022). The idea of Metaverse campuses is to allow students from all over the world to participate in classes and events in the same place (Park & Kim, 2022). Metaverse -based classrooms and campuses are likely to offer better educational support to students than physical classrooms. It is expected that Metaverse environments will engage students more effectively than a physical classroom in the real world. Metaverse campuses are projected to reduce the limitations imposed by geographical distance, and it is also predicted that students will be able to work on a unified virtual campus in the future (Basil, 2022, Rospigliosi, 2022).

Social Perspectives

The Metaverse is a virtual environment that has a substantial positive influence on the actual world. According to Duan et al., 2021), Metaverse can show a positive impact on social benefits in terms of accessibility, diversity, equality, humanity. The perspective focused on the accessibility is focused to server different social requirements with great availability. Many events have been converted to virtual form by the Metaverse (Duan, et.al., 2021, Preston, 2021). Metaverse is expected to improve education, connects several educational institutions, online enrolment and other activities that occurs in physical campuses. It is expected to promote globalization and break the geographic distance barriers. The Metaverse may offer excellent accessibility to meet many societal needs for instance, courses may be rendered in virtual form, greater security, reduced cost for the users. In real world, it is impossible to meet and cater everyone's needs in physical campuses. The Metaverse, however, may attain diversity because of its limitless extension space. In the Metaverse, there are many intriguing scenarios that can be played out (Dwivedi et al., 2022, Alexander, 2020). The diversity of the real world ought to be reflected in Metaverse representations. In the future, persons with impairments may use wheelchairs and assistive technologies like cochlear implants, over-the-ear hearing aids, and unique facial shapes (Allam et al.,

2022, Almarzouqi, 2022). Many researchers have indicated that the Metaverse can promote equality. The perspective of equality is focused on race, gender, spiritual goals for humans and etc. (Duan et al., 2021). Customizable avatars are expected to provide a representative feature that enables participants to uphold law, order, and regularity in the Metaverse which is an autonomous ecosystem (Wang et al., 2022).

Accessibility Perspective

The "accessibility" perspective is emphasized by providing students with unrestricted access and digital certification to meet the unique regional curriculum needs of the most modern schools around the world (Hollensen et al., 2022). The Metaverse can enable global collaboration despite the geographic distances of individual universities. In addition, this digital world can provide accessibility for social events such as awareness, hosting concerts, motivation talk which can be attended by thousands of students. These kinds of events can increase students' motivation to learn. Added to this, users can mimic more real events in the Metaverse without worrying about the event place capacity and mass gathering (Bibri, 2022, Dincelli, 2022). Students can also benefit from virtual lessons from academicians and professionals around the world. Additionally, an instructor can talk to the students in virtual classrooms to provide guidance and feedback. It will undoubtedly help prepare students better for what to expect in today's world. For example, students are allowed to access the SEGI University has transformed into a virtual space and operated on the Metaverse from 2022. Multiple 3D worlds are included in this MC that include classrooms, retail stores, student hangouts, conference rooms, libraries, and event spaces. Other educational institutions in Malaysia are expected to slowly convert their campus to MCs in upcoming future.

Diversity Perspective

Diversity perspective is to ensure that the representations in the digital relationships allow wide range of voices, viewpoints, and experiences (Henslin et.al., 2015). In the Metaverse, lecturers can now teach online courses more successfully and create a richer experience than the one-way communication in normal online classrooms. MC also can help save time where students and teachers no longer need to worry about delays or traffic jams. SEGI University for instance can host numerous events or activities in this Metaverse environment such as campaigns, interactions between faculties and students with lower costs, reduction of high-power consumption, and provided good protection from COVID-19 spread. This shows that means that the diversity requirements of physical society are largely met.

Equality Perspective

Equality Perspective make everyone in the Metaverse equal without having to consider their appearance, skin colour, or other characteristics. In MC, it would be possible to hide a person's gender, age, race, assumptions, and biases. MC is expected to provide equality where participants will be able to interact with people in the Metaverse from various geographical locations (Cheng et al., 2022, Seigneur,2022). Regardless of identity, anyone enrolled in the class can use the Metaverse and learn equally. For example, Phoenix Asia Academy Technology College students in Malaysia can created their MC where participants can take advantage of the beautiful view of their main building, numerous classrooms, discussion room etc. This avatar term helps students overcome their introversion and communicate freely.

Humanity Perspective

Humanity perspective in developing new online identities is exhausting after spending our entire lives cultivating our identities in real life (Macionis, 2018). While the Metaverse is a fantastic place to reinvent yourself, it's easier to start with who we are and go from there. Metaverse lowers the barriers to entry for average people who want to enjoy real experiences in virtual worlds (Kalpan et al., 2021, Kaminska et al., 2019). Second, virtual environments and immersive content with photorealistic avatars offer more emotional and authentic encounters for users. This will enable virtual meetings with classes, faculty, and workspaces, expanding the Metaverse beyond games and entertainment. According to the Phoenix Asia Academy of Technology College, the Metaverse will transform all social barriers and break through all boundaries to advance even further. People can express themselves even better, interact with each other without being limited by colour and live the life they have always wanted in this online environment where they act as avatars or their virtual selves. The participants also can host events, meetings, workshop, voice chat, play games, and even host concerts.

METHODOLOGY

The methodology of this study was based on fuzzy Delphi method (FDM) (Chu et al., 2008, Okoli, 2004). The development of the questionnaire items is based on the literature research, pilot study and experiences (Okoli & Pawlowski, 2004). The construction of the items and content elements should be based on literature research in the framework of the study (Okoli & Pawlowski, 2004) According to Ridhuan et al, 2014, it was pointed out that the FDM technique has been used in studies of expert opinions where a group of experts is invited to gather ideas for agreement formation (Mohamed Yusoff et al., 2021, Mohd Jamil et al., 2017, Ridhuan et al., 2014). The FDM approach is preferred over the Delphi technique because it processes questionnaires more efficiently and cost-effectively, and allows specialists to regularly express their opinions, thus achieving great expert consistency. The FDM studies require a minimum sample size of 10 to 50 experts (Ridhuan et al., 2014, Mohd. Jamil et al., 2013).

Experts' Demographic

The items for the FDM were expended based on four social perspectives, the perspectives being Accessibility, Diversity, Equality and Humanity. This study is used to collect the expert consensus of the sampled experts, who are randomly selected tertiary academics from governmental and private institutions in Malaysia. Thus, 25 experts who are academics agreed to participate in this study. In phase 1, initial results were conducted on the use of structural to obtain their consensus to conduct an analysis of their knowledge of the MC and their opinions on social perspectives in the context of Malaysia. The demographics of the experts for this study are collected based on the educational qualifications, work experience and field of study of the experts. The educational qualifications of the experts are shown in Table 1. From the 25 sample of experts (n= 25), 12 experts have a doctorate and 13 have a master's degree.

Table 1. Educational Qualification

Educational Qualification	
Qualification	**Frequency**
PhD	12
Master Degree	13
Total	25

The experts' working experience are categorized into four ranges which are less than 10 years, 11-15 years, 16-20 years, and more than 20 years, as shown in Table 2. From the selected sample, 5 experts have less than 10 years, 10 experts have 11-15 years, 5 experts have 16-20 years and 5 of them have more than 20 years.

Table 2. Working Experience of expert

Working Experience	
Year	**Frequency**
Less than 10 years	5
11-15 years	10
16-20 years	5
More than 20 years	5
Total	25

The experts included in the sample are shown based on their area of expertise, as shown in Table 3. From the selected sample there are 10 experts were from the field of Computing, which consists of experts from Computer Science, Information Technology, Information Systems, and Information Security. The Business field consists of Finance, Marketing and Accounting, in which 9 experts gave their consent to participate. From the field of Engineering, the experts are from the field of Civil Engineering and Mechanical Engineering, in which 6 experts have been agreed to participate.

Table 3. Field of Expertise

Field of Expertise	
Field	**Frequency**
Computing	10
Business	9
Engineering	6
Total	25

DATA COLLECTION

The data collection method used for this study is based on the FDM. The items in this study are based on the four social perspectives which consist of Accessibility, Diversity, Equality and Humanity. Three external reviewers are asked to provide feedback on the items created for the questionnaire. The results of the review are used to modify the items to ensure that the reliability of the items is appropriate to the scope of the study. Each item has achieved an outstanding Cronbach Alpha score of 0.841 for Accessibility, 0.877 for Diversity, 0.922 for Equality and 0.891 for Humanity. The first step in implementing the FDM in this study is based on the four social perspectives from in Phase 1 and the items were able to be mastered by the expert panel (Mohd Ridhuan et al., 2014). During the FD study, the experts interacted with the questionnaire items by indicating their agreement with each item. Items are designed for their validity in terms of content validity, which involved the process of forming sentences, using the correct terms, and maintaining clarity of each item's purpose through a seven-point Likert scale to measure the degree of to obtain consent (Mohd Ridhuan et al., 2014).
For this study, the seven-point fuzzy scale as shown in Table 4 was used.

Table 4. Fuzzy Scale 7 scales (Source: Mohd Jamil and Mat Noh (2020)

Scale	Amount of Consensus	Fuzzy Scale
1	Extremely Strongly Agree	(0.9,1.0,1.0)
2	Strongly Agree	(0.7,0.9,1.0)
3	Agree	(0.5,0.7,0.9)
4	Moderately Agree	(0.3,0.5,0.7)
5	Disagree	(0.1,0.3,0.5)
6	Strongly Disagree	(0.0,0.1,0.3)
7	Extremely Strongly Disagree	(0.0,0.0,0.1)

The data collected based on the Likert scale were translated into fuzzy number data and the Excel spreadsheet was used to reflect the level of agreement. In Phase 2 of the FD technique, two processes are used to analyze data, namely the triangular fuzzy number and the defuzzification process (Mohd Ridhuan et al., 2014). The triangular fuzzy numbers are the values of m1, m2 and m3. Here, m1 means the minimum value, m2 the reasonable value and m3 the maximum value. The degree of agreement for the fuzzy scale is given in odd numbers. The higher the fuzzy scale, the more accurate the data obtained (Mohd Ridhuan et al., 2014)., as indicated in Table 4.

DATA ANALYSIS

During the data analysis, the expert consensus was analyzed using Microsoft Excel (Mohd Jamil & Mat Noh, 2020), Mohd Jamil et al. (2017), Ramlie et al. (2014). The two main assumptions followed in FD technique are the Triangular Fuzzy Number (TZN) and the Defuzzifications Process (DP). The TZN has two conditions where the value of threshold d value is 0.2, where the result value of the expert agreements is less than or equal to 0.2 (Cheng & Lin, 2002; Chen, 2000) The formula below is used to obtain the d value.

$$d\left(\bar{m},\bar{n}\right)=\sqrt{\frac{1}{3}\left[\left(m_1-n_1\right)^2+\left(m_2-n_2\right)^2+\left(m_3-n_3\right)^2\right]}$$

The second condition for the TFN is the percentage of expert agreement, with expert panel agreement greater than 75% being accepted (Chu & Hwang, 2008; Murray & Hammons, 1995). According to Tang & Wu, 2010, the DP is used to determine the fuzzy value (A) based on the cut value of 0.5, with the measured item considered accepted. If the value is less than 0.5, the item is considered rejected. The A value was determined using the formula below:

$$A=\frac{1}{3}*\left(m1+m2+m3\right)$$

The conclusions confirm the experts' agreement and the priority between the dimensions by the value of the dimensions (d), the percentage of experts' agreement on each dimension. Whereas the expert consensus on each item is seen by the value of (d) of the item, the percentage of expert consensus on the item. While considering the rank of the item in each dimension, the item in each dimension is considered by the score of the defuzzification items.

Analysis of the Experts Consensus on the "Accessibility" perspective

Experts reached a consensus based on the items shown in Table 5 for the "Accessibility" perspective.

Table 5. Items for the "Accessibility" perspective

Item Code	Item Details
A1	Provide an organizational dimension on the Metaverse framework where the first- and endpoints is the user who will be a lecturer or a student who interacts with a device.
A2	Provide an organizational dimension that promotes globalization among lecturers and students to use a Metaverse-based application and another form of technology in the future.
A3	Provide an organizational dimension by increasing cooperation among lecturers and students by using receptor devices, such as 3D glasses and any other sensors, which should be connected to offer the best user experience. The user can install the application on their device, use the built-in application in the platform operating the user's device, or use a link that operates only on a remote server, usually the cloud.
A4	Provide a pedagogical dimension by reducing the overall cost by including the files of the applications or systems, such as the files of Numina, InTeRaCt, VoRtex, educational games, or virtual campus ecosystems that are stored on the cloud servers to make it easy to use learning environments for both the real user and the avatar.
A5	Provide a pedagogical dimension by ensuring security practices such as authorized access and user authentication where the data should be accessed by the students, lecturers, educational institutions etc.
A6	Provide a pedagogical dimension by providing a combination of interfaces and a cross-world ecosystem when using the Metaverse special technologies which simulate the physical and real process and provide them in virtual reality modules.
A7	Provides a Technology dimension where Metaverse is an extension of daily campus lives where the system will cause some reactions that users can sense by touch, sight, sound, and possibly also smell in the future.
A8	Provide support on the Technology Dimension on Metaverse that needs an infrastructure to help with communication, processing, computation, rendering, simulation, storage, resource management and operating systems.
A9	Provide support on the Technology Dimension where the infrastructure and implementation of the Metaverse environment is responsible for providing access to the interaction applications or technologies that make it easy to use the Metaverse special technologies.
A10	Provide the support on the Technology Dimension where the Metaverse special technologies are various and helpful to use all the aspects of the Metaverse based on Twin learning environment, Virtual learning world, world Simulator, learning service provider, blockchain, learning Augmentation, Haptic or 3D Touch, Supportive Apps, Personal Life recording, Financing, Sensation Devices, Non-Fungible Token (NFT), User-Generated Content (UGC) etc.

Table 6 shows the threshold d-value, expert consensus rate, defuzzification, and item position for the accessibility perspective.

Table 6. Experts consensus on the "Accessibility" perspective

Item	Triangular Fuzzy Numbers' Condition		Process for Defuzzification Present State				Position	Experts Consensus
	Threshold Value, (d)	(%)of Expert Group Consensus Percentages	M1	M2	M3	Fuzzy Score, (A)		
A1	0.108	88.0%	0.836	0.952	0.980	0.923	6	*Accepted*
A2	0.000	100.0%	0.900	1.000	1.000	0.967	1	*Accepted*
A3	0.000	100.0%	0.900	1.000	1.000	0.967	1	*Accepted*
A4	0.331	36.00%	0.436	0.632	0.788	0.619	-	*Rejected*
A5	0.000	100.00%	0.900	1.000	1.000	0.967	1	*Accepted*
A6	0.076	100.00%	0.804	0.952	1.000	0.919	7	*Accepted*
A7	0.056	100.00%	0.852	0.976	1.000	0.943	4	*Accepted*
A8	0.075	100.00%	0.788	0.944	1.000	0.911	8	*Accepted*
A9	0.066	100.00%	0.836	0.968	1.000	0.935	5	*Accepted*
A10	0.151	100.00%	0.388	0.588	0.788	0.588	9	*Accepted*
Condition: *Triangular Fuzzy Numbers* *1) Threshold Value (d)* ≤ 0.2 *2) Percentage of Experts Consensus* $> 75\%$				*Defuzzification Process* *3) Fuzzy Score (A)* $\geq \alpha -$ *cut value* $= 0.5$				

Analysis of the Experts Consensus on the "Diversity" perspective

Experts have indicated their consensus based on the items shown in Table 7 for the "Diversity" perspective.

Table 8 shows the threshold d-value, expert consensus rate, defuzzification, and item position for the diversity perspective.

Table 8 shows that the perspective of diversity is that 100% of the proposals are accepted. Threshold d-value is ≤ 0.2 was the value that each item recorded. This finding indicates that there is now consensus among experts on each of these points. All items are 100% complete and all item defuzzification values pass the value of α -cut = 0.5.

Table 7. Items for the "Diversity "perspective

Item Code	Item Details
B1	Provide different Metaverse platforms for lecturers and students that have features and functionality that support virtual campuses with the technologies of XR (Extended Reality), VR (Virtual Reality), and AR (Augmented Reality) provide students with the tools to enjoy a fully immersive and meaningful learning experience.
B2	Provides various teaching and learning experiences by connecting the Metaverse and education where lecturers can provide experiences beyond the classroom to students such as students to travel the world, see historical events up close, and take a cosmic tour without ever leaving the classroom and better prepare students for their higher education, and for the requirements of their future workplace.
B3	Provide various ways where the parents would monitor their kids' progress in their studies by getting academic and personal performance by simply logging into the XR campus account.
B4	Provide various ways to lecturers and students to restore and enhance image the parents would monitor their kids' progress e/video quality for achieving better Metaverse where the computer vision allows eXtended Reality (XR) devices to recognize and understand visual information of users' activities and their physical surroundings, helping build more reliable and accurate virtual and augmented environments.
B5	Provide numerous human activities to lecturers and students by using algorithms such as simultaneous Localization and Mapping (SLAM) that support digital twin representatives in the Metaverse which connects human users at the intersection between the physical and digital worlds where the accuracy of object registration and the interaction with the physical world are emphasized.
B6	Provides various Avatar control to lecturers and students that can be achieved through the human body and eye location and orientation using joint positions or key points for each human body part which can reflect the characteristics of human posture, which depict the body parts, such as elbows, legs, shoulders, hands, feet, etc.
B7	Provide several holistic scenes understanding attributes to lecturers and students using stereo matching and depth estimation techniques to understand their role/s, the availability of the content, attributes of the registered objects such as distance, an object characteristic, action recognition, and recognizing of the virtual contents across different virtual worlds etc.
B8	Provide various interactions with lecturers and students where students can obtain access to assignments, announcements, events, timetables, lessons, and feedback from teachers.
B9	Provide numerous benefits for the lecturers and students like obtaining full for all mobility needs, increased attendance, focused on self-paced practices, content review, science experiments in a virtual lab etc.
B10	Provide various career directions and paths by utilizing new Metaverse educational technology in a way that maximizes the advantages for tertiary students both now and in the future.

Table 8. Experts Consensus of the "Diversity" perspective

Item	Condition of Triangular Fuzzy Numbers		Condition of Defuzzification Process				Position	Experts Consensus
	Threshold Value, (d)	Percentages of Expert Group Consensus, (%)	M1	M2	M3	Fuzzy Score, (A)		
B1	0.082	100.0%	0.772	0.932	0.996	0.900	3	*Accepted*
B2	0.056	100.0%	0.852	0.976	1.000	0.943	1	*Accepted*
B3	0.124	100.0%	0.588	0.788	0.944	0.773	9	*Accepted*
B4	0.151	100.00%	0.388	0.588	0.788	0.588	10	*Accepted*
B5	0.134	100.00%	0.596	0.788	0.940	0.775	8	*Accepted*
B6	0.073	100.00%	0.756	0.924	0.996	0.892	6	*Accepted*
B7	0.076	100.00%	0.804	0.952	1.000	0.919	2	*Accepted*
B8	0.089	100.00%	0.740	0.908	0.988	0.879	7	*Accepted*
B9	0.090	100.00%	0.764	0.924	0.992	0.893	4	*Accepted*
B10	0.090	100.00%	0.764	0.924	0.992	0.893	4	*Accepted*
Condition: *Triangular Fuzzy Numbers* *1) Threshold Value (d) ≤ 0.2* *2) Percentage of Experts Consensus > 75%*			*Defuzzification Process* *3) Fuzzy Score (A) ≥ α – cut value = 0.5*					

Analysis of the Experts Consensus on the "Equality" perspective

Experts reached a consensus based on the items shown in Table 9 for the "Equality" perspective.

Table 10 shows the threshold d-value, expert consensus rate, defuzzification, and item position for the diversity perspective.

Table 10 shows that all experts are accepted. Threshold d-value ≤ 0.2 was the value that each item recorded. This finding indicates that there is now consensus among experts on each of these points. All items are above 90% and all defuzzification values for items exceed the value of α -cut = 0.5.

Table 9. Items for the "Equality "perspective

Item Code	Item Details
C1	Provide a platform that allows lecturers and students to access their avatar in specific Metaverse disregarding elements such as language, race, various ethnicities, and other diversity and engage in cross Metaverse scenarios via multiple channels.
C2	Provide the lecturers and the student's Metaverse channels that can discard any kind of discrimination in the real world and it is seen as an equalizer where gender, race, disability, and social status are eliminated.
C3	Provide acceptance of lecturers and students from various backgrounds from campuses all over the world with a unique individual identity like the physical world accessed via one single sim in various domains such as education, entertainment, gaming and financial.
C4	Provide unlimited extension of spaces in Metaverse channels to lecturers and students where their avatars are used as a digital representation where this representation can be various in different applications or games such as avatars like a human shape, imaginary creatures, or animals.
C5	Provide openness in spiritual pursuit for lecturers and students in various Metaverse channels.
C6	Provide Metaverse channels to build fairness that promotes unity, and freedom, limit discrimination and sustainable society for lecturers and students on Metaverse campus.
C7	Provide a Metaverse campus that is based on an autonomous ecosystem that creates an effective equality, opportunities and treatments for the lecturers and students.
C8	Provide a digital twin world that is aimed to enhance the digital space interactions among lecturers and students that allow full participants in all campus activities, provide equal opportunity and equally rewarded based on each individual's contribution.

Table 10. Experts Consensus of the "Diversity" perspective

Item	Condition of Triangular Fuzzy Numbers		Condition of Defuzzification Process				Position	Experts Consensus
	Threshold Value, (d)	Percentages of Expert Group Consensus, (%)	M1	M2	M3	Fuzzy Score, (A)		
C1	0.076	100.0%	0.796	0.948	1.000	0.915	2	*Accepted*
C2	0.062	100.0%	0.844	0.972	1.000	0.939	1	*Accepted*
C3	0.066	100.0%	0.764	0.932	1.000	0.899	3	*Accepted*
C4	0.138	100.00%	0.668	0.848	0.964	0.827	8	*Accepted*
C5	0.095	100.00%	0.660	0.856	0.976	0.831	7	*Accepted*
C6	0.061	100.00%	0.724	0.904	0.992	0.873	5	*Accepted*
C7	0.104	100.00%	0.740	0.904	0.984	0.876	4	*Accepted*
C8	0.084	96.00%	0.700	0.884	0.980	0.855	6	*Accepted*
Condition: *Triangular Fuzzy Numbers* *1) Threshold Value (d) ≤ 0.2* *2) Percentage of Experts Consensus > 75%*			*Defuzzification Process* *3) Fuzzy Score (A) $\geq \alpha$ − cut value = 0.5*					

Analysis of the Experts Consensus on the "Humanity" perspective

Experts reached a consensus based on the items shown in Table 11 for the "Humanity" perspective.

Table 11. Items for the "Humanity" perspective

Item Code	Item Details
D1	Provide a Metaverse campus that is focused on the humanistic spirit that maintenance, pursues, and concerns human dignity, value, and destiny among lecturers and students.
D2	Provide a Metaverse campus platform that will preserve different cultural and artistic forms focused on society, humanity, spirituality and culture.
D3	Provide unlimited community mixtures of lecturers and students within the Metaverse campus with different beliefs and thoughts and allow everyone to be part of the Metaverse.
D4	Provide a Metaverse platform that ensures significant ethical issues such as violating people's privacy, data protection, bullying, cheating, and educational inequality within Metaverse campuses.
D5	Provide a Metaverse campus platform that welcomes all that should be accessible, diverse, and inclusive.
D6	Provide a Metaverse campus platform that emphasizes privacy that is focused on the fundamental right of users to use an avatar that has been created to each individual's choice.
D7	Provide a Metaverse platform that is focused on the reputation that allows users to report malicious users' misbehavior and malpractice using Decentralized Autonomous Organization (DAO).
D8	Provide a Metaverse campus that provides evidence for artefact restoration on the digital reconstruction of cultural relics.
D9	Provide a platform for Metaverse that can be a door to cultural communications and protections.
D10	Provide options to create Metahumans that allow us to perceive our bodies and movements in real-time, represent lecturers' and students' feelings, facial expressions, personal characteristics and are much more realistic than avatars.

Table 12 shows the threshold d-value, expert consensus rate, defuzzification, and item position for the perspective humanity. It shows that 70% of the expert suggestions are accepted and 30% are rejected. The threshold d-value ≤ 0.2 was the value that each item recorded. Only the threshold for D9 is higher than 0.2. Items are rejected at 30% and most defuzzification scores for items are lower.

Table 12. Experts Consensus of the "Humanity" perspective

Item	Condition of Triangular Fuzzy Numbers		Condition of Defuzzification Process				Position	Experts Consensus
	Threshold Value, (d)	Percentages of Expert Group Consensus, (%)	M1	M2	M3	Fuzzy Score, (A)		
D1	0.073	100.0%	0.780	0.940	1.000	0.907	1	*Accepted*
D2	0.293	68.0%	0.256	0.436	0.616	0.436	6	*Rejected*
D3	0.253	56.0%	0.348	0.548	0.732	0.543	4	*Accepted*
D4	0.141	100.00%	0.172	0.372	0.572	0.372	8	*Rejected*
D5	0.082	100.00%	0.772	0.932	0.996	0.900	2	*Accepted*
D6	0.199	80.00%	0.284	0.484	0.676	0.481	5	*Rejected*
D7	0.090	100.00%	0.764	0.924	0.992	0.893	3	*Accepted*
D8	0.184	76.00%	0.092	0.248	0.436	0.259	10	*Rejected*
D9	0.303	52.00%	0.200	0.368	0.560	0.376	7	*Rejected*
D10	0.140	96.00%	0.160	0.356	0.556	0.357	9	*Rejected*
Condition: *Triangular Fuzzy Numbers* *1) Threshold Value (d) ≤ 0.2* *2) Percentage of Experts Consensus > 75%*				*Defuzzification Process* *3) Fuzzy Score (A) ≥ α – cut value = 0.5*				

DISCUSSION

The result obtained on "Accessibility" perspective shows the expert agrees that there should be an organizational component that promotes globalization among lecturers and students so that they can use a Metaverse-based applications with other types of technologies in the future. The receptor devices such as 3D glasses and other sensors, which should be connected to provide the best user experience, the organizational dimension can be increased through greater collaboration between faculties and students. The user has three options to be on the MC which are by install the program on his device, use build-in application or link to a remote server using cloud. The MC is expected to provide pedagogical dimension by maintaining security procedures such as access permission and user authentication in situations where students, lecturers, educational institutions should have access to the MC. Going forward, the technological dimension, where the Metaverse is an extension of student life on campus, can evoke specific responses that users can sense through touch, sight, sound, and maybe even smell. The experts have indicated that the Metaverse technologies should imitate physical and real processes and transmit them in virtual reality modules, adds an educational dimension by offering a mixture of interfaces and world-spanning ecology to support communication, processing, computation, playback, simulation, storage, resource management, and operating systems, the technology dimension of the Metaverse requires infrastructure. The MC also are expected to provide support on other special technologies that can be merged with MC such as twin learning environment, virtual learning world, world simulator, learning service provider, blockchain, learning augmentation, haptic or 3d touch, supportive apps, personal life recording, funding, sensation devices, non-fungible token (NFT), user-generated content (UGC) etc. However, the item code A4 was rejected that MC will reduce the overall operation cost because the experts have agreed that cost to include application on the cloud server that supports students' interaction that matches both real and avatar in twin world will be costly.

From the "Diversity" perspective the experts have indicated that the MC using XR, MR, VR, and AR technologies can enhance students' virtual immersive and engage them for meaningful learning experience. MC is believed can link education beyond classroom by providing experiences such as travelling the world, seeing historical events up close, and taking a cosmic tour without leaving the classroom. This will prepare students to pursue their higher education and they also can meet the current industrial demands. On the other hand, MC can be a platform for parents to check on their children's academic progress by entering their XR campus accounts and viewing their academic and personal performance. XR devices can be used to capture visual data on users' activity and their physical environment to support the creation of more accurate and reliable virtual and augmented environments. The experts have agreed that the simultaneous localization and mapping (SLAM) can be used to connect digital twin representatives in the Metaverse that connects the physical and digital worlds. The joint positions or key points for each human body part, which reflect characteristics of human posture, and which can represent body parts such as elbows, legs, shoulders, hands, feet, etc. It is possible to achieve avatar controls for faculty and students. This will help the faculty and students understand their roles, content availability, registered object perspectives such as distance and an object characteristic, action detection, and virtual content recognition across multiple virtual worlds using stereo matching and depth estimation techniques. Students can access assignments, announcements, class schedules, events, classes, and teacher feedback through a variety of interactions with instructors. There are several advantages for lecturers and students, including full coverage of all mobility requirements, higher attendance rates, an emphasis on self-paced exercises, and expert experiments in a virtual laboratory. To take advantage of the new future, students will benefit now and in the future by using the new Metaverse technology in education.

From the perspective of "Equality," focuses on the gender, race, disability, and socioeconomic status are eliminated in the faculty's and student's in Metaverse campus. MC is expected to reject any form of discrimination in today's world to be eliminated. The platform allows lecturers and students to interact with cross-Metaverse scenarios through different channels while using their avatars in a specific Metaverse, ignoring factors such as language, race, different ethnicities, and other diversity. Many lecturers and students from diverse backgrounds from universities around the world with a distinct individual identity resembling the physical world accessed by a single sim in a variety of fields such as education, entertainment, gaming, and finance. The Metaverse campus is built on an independent ecosystem that provides effective equal opportunity, opportunity, and treatment for faculty and students. At Metaverse Universities, fairness is built using channels that support community, freedom, and sustainability for both faculty and students. It aims to improve digital connections between lecturers and students, allow full participation in all university activities, provide equal opportunities and reward everyone equally based on their contributions and accessibility to spirituality for both faculty and students via numerous Metaverse channels. The infinite expanse of the Metaverse opens the possibility for lecturers and students to use their avatars as digital representations in a variety of games and applications. These avatars can take the form of humans, animals, or other imaginary beings.

On the "Humanity" perspective, experts agree that the campus is aligned with a humanistic ethos that upholds, pursues, and reflects on the human dignity, value and intention of students and the educational institution. It is a platform for a Metaverse campus that welcomes all that should be open, diverse, and inclusive. The Metaverse platform is reputation-focused, and users can use a Decentralized Autonomous Organization (DAO) to report the malicious user crimes and wrongdoing. Allow everyone to participate in the Metaverse and create an infinite community of lecturers and students at MC with different ideologies. The experts believed that the MC should prioritizes privacy and focuses on the fundamental right of users to use an avatar of their choice. They all should be allowed to establish a Metaverse campus platform to preserve many artistic and cultural genres dealing with society, humanity, spirituality, and culture. Metaverse platform also should serve as a gateway for cross-cultural dialogue and security. Build a framework for the Metaverse that safeguards important ethical issues, including violations of people's privacy, data protection, bullying, fraud, and educational inequality at Metaverse universities. MC should provide ways to build metahumans that are far more realistic than avatars, allow us to sense our bodies and movements in real time, and represent the emotions, expressions, and personal traits of academicians and students. The experts believed that the MC should houses cultural artifacts that have been digitally recreated as evidence of artifact restoration. There are the 3 code elements that the experts disagree on, namely D2, D3 and D9. The reason for the rejection of D2 is the creation of a Metaverse campus platform to preserve various artistic and cultural genres dealing with society, humanity, spirituality, and culture. The experts indicate that the MC should be a free and fair ecosystem. Next, the experts rejected D3 because they do not believe that the Metaverse would allow anyone to participate in the Metaverse and create an infinite community of academicians and students on the Metaverse campus with diverse ideologies. Eventually, the expert rejected D9. Because the expert does not believe that the Metaverse platform can build a trust that can open doors to cultural protection and communication.

CONCLUSION

Metaverse is a new learning spectrum for Malaysians that offers academicians and students in virtual parallel universe. The Metaverse Campus (MC) is an idea to transform learning from physical classes into virtual worlds. The MCs initiative enables students worldwide to take courses and events in the same world. A study using the Fuzzy Delphi Method (FDM) was conducted among 25 Malaysian academics from government and private educational institutions. This study aims to create a consensus among scholars to discover the social perspectives regarding MC from the perspective of accessibility, diversity, equality, and humanity. Findings show that Malaysian academics see MC's accessibility perspective as supporting globalization, improving academic-student collaboration, providing the best user experience, protecting security, and providing interfaces for audiences drawn from a worldwide ecosystem. The experts believe that MC will be an enrichment of everyday campus life in the future, using senses such as touch, sight, hearing, and smell. Most of the experts agreed that MC infrastructure ex essential to maintain excellent communication, processing, computation, rendering, simulation, storage, resource management and operating systems. From the expert's perspective, they expect MC to support such as various models of devices, Non-Fungible Tokens (NFT), User-Generated Content (UGC). However, the experts do not anticipate that the MC will reduce operational costs compared to the operational costs of a physical campus.

From a diversity perspective, experts felt that MC will offer features and functionality that support XR, VR, and AR technologies. They advocated that the MC will be a learning environment that provides the right tools to be immersive and experiential for meaningful learning. They supported that MC offers advantages for different parties by providing teaching and learning experiences outside of the classroom, parents would monitor their children's progress and encourage educational institutions to build reliable and accurate virtual and augmented environments. MC is intended to support representatives of digital twins between the physical and digital world. The experts also supported that MC will be able to provide different avatars that will allow faculty and students to control their body and eye alignment by using knots for all human body parts like elbows, legs, shoulders, hands, feet etc.MC should also be able to demonstrate to faculty and students the holistic scenes using stereo matching and depth estimation techniques. This will support faculty and students to understand their role/s, content, objects and their perspectives like distance, characteristic, action/s in different Metaverse worlds. MC will also be able to provide various lectures-student interactions such as access to assignments, announcements, events, lessons, timetables, and feedback. Lecturers and students will be able to take advantage of mobility needs, increase attendance, focus on self-paced practice, content review and science experiments in a virtual lab, etc. Academics also believe that MC will also be able to offer diverse career paths and orientations using new Metaverse education technology in a way that maximizes the benefit to current tertiary students and those in the future.

From the equality perspective, MC is expected to provide a platform that allows teachers and students to use their avatar regardless of differences based on users' language, race, gender, disability, social status, ethnicity, etc. MC is also expected to eliminating discrimination and promoting equality for Malaysians. Academics point out that the MC accepts faculty and students from diverse backgrounds from international universities to participate and join the program offered through MC. The teachers and students can also use different types of avatars like human figures, imaginary creatures, or animals to promote unity and freedom. Academics trust that the MC will be an autonomous ecosystem that creates equality, opportunity and treatment for Malaysian academician and students. Students are expected to fully participate in all on-campus activities, receive equal opportunity, treatment, and opportunities to contribute and show talent.

Finally, from humanity perspective, the MC is expected to encourage humanistic spirit, human dignity, value and destiny among local and international academicians and students. MC will create an unlimited community mix of faculty and students within the MC with different beliefs and thoughts. Academics believe that the MC should be an accessible, diverse, and inclusive platform. MC should also focus on reputation, which allows faculty and students to report malicious misconduct and user misconduct to the Decentralized Autonomous Organization (DAO). However, experts disagree that MC will be a platform that preserves diverse cultural and artistic forms focused on society, humanity, spirituality, and culture. However, the experts do not believe that the MC ensures significant ethical issues such as invasions of individuals' privacy, privacy issues, bullying and fraud. Academics find it difficult to uphold the fundamental right of users to use an avatar, created through the choice of the individual, that shows the importance of culture, cultural activities, norms, and communication values in the Metaverse worlds.

REFERENCE

Alexander, B. (2020). *Academia next: The futures of higher education*. Johns Hopkins University Press.

Allam, Z., Sharifi, A., Bibri, S. E., Jones, D. S., & Krogstie, J. (2022). The Metaverse as a Virtual Form of Smart Cities: Opportunities and Challenges for Environmental, Economic, and Social Sustainability in Urban Futures. *Smart Cities*, *5*(3), 771–801. doi:10.3390martcities5030040

Almarzouqi, A., Aburayya, A., & Salloum, S. A. (2022). Prediction of User's Intention to Use Metaverse System in Medical Education: A Hybrid SEM-ML Learning Approach. *IEEE Access : Practical Innovations, Open Solutions*, *10*, 43421–43434. doi:10.1109/ACCESS.2022.3169285

Basil, W. (2022). *The" Metaverse." The arrival of the future of 3D research, learning, life & commerce.*

Benford, S., & Giannachi, G. (2011). *Performing mixed reality*. MIT Press.

Bibri, S. E. (2022). The Social Shaping of the Metaverse as an Alternative to the Imaginaries of Data-Driven Smart Cities: A Study in Science, Technology, and Society. *Smart Cities*, *5*(3), 832–874. doi:10.3390martcities5030043

Cheng, R., Wu, N., Chen, S., & Han, B. (2022, March). Reality Check of Metaverse: A First Look at Commercial Social Virtual Reality Platforms. In *IEEE Conference on Virtual Reality and 3D User Interfaces Abstracts and Workshops (VRW),* (pp. 141-148). IEEE. 10.1109/VRW55335.2022.00040

Chu, H. C., & Hwang, G. J. (2008). A Delphi-based approach to developing expert systems with the cooperation of multiple experts. *Expert Systems with Applications*, *34*(4), 2826–2840. doi:10.1016/j.eswa.2007.05.034 PMID:32288332

Collins, C. (2008). Looking to the future: Higher education in the Metaverse. *EDUCAUSE Review*, *43*(5), 51–63.

Dahan, N. A., Al-Razgan, M., Al-Laith, A., Alsoufi, M. A., Al-Asaly, M. S., & Alfakih, T. (2022). Metaverse Framework: A Case Study on E-Learning Environment (ELEM). *Electronics (Basel)*, *11*(10), 1616. doi:10.3390/electronics11101616

Damar, M. (2021). Metaverse Shape of Your Life for Future: A bibliometric snapshot. *Journal of Metaverse*, *1*(1), 1–8.

Davis, A., Murphy, J., Owens, D., Khazanchi, D., & Zigurs, I. (2009). Avatars, people, and virtual worlds: Foundations for research in Metaverse s. *Journal of the Association for Information Systems*, *10*(2), 1. doi:10.17705/1jais.00183

Dincelli, E., & Yayla, A. (2022). Immersive virtual reality in the age of the Metaverse: A hybrid-narrative review based on the technology affordance perspective. *The Journal of Strategic Information Systems*, *31*(2), 101717. doi:10.1016/j.jsis.2022.101717

Duan, H., Li, J., Fan, S., Lin, Z., Wu, X., & Cai, W. (2021). Metaverse for social good: A university campus prototype. In *Proceedings of the 29th ACM International Conference on Multimedia,* (pp. 153-161). 10.1145/3474085.3479238

Dwivedi, Y. K., Hughes, L., Baabdullah, A. M., Ribeiro-Navarrete, S., Giannakis, M., Al-Debei, M. M., Dennehy, D., Metri, B., Buhalis, D., Cheung, C. M. K., Conboy, K., Doyle, R., Dubey, R., Dutot, V., Felix, R., Goyal, D. P., Gustafsson, A., Hinsch, C., Jebabli, I., & Wamba, S. F. (2022). Metaverse beyond the hype: Multidisciplinary perspectives on emerging challenges, opportunities, and agenda for research, practice and policy. *International Journal of Information Management*, *66*, 102542. doi:10.1016/j.ijinfomgt.2022.102542

Fernández, A., Peralta, D., Benítez, J. M., & Herrera, F. (2014). E-learning and educational data mining in cloud computing: An overview. *International Journal of Learning Technology*, *9*(1), 25–52. doi:10.1504/IJLT.2014.062447

Han, D. I. D., Bergs, Y., & Moorhouse, N. (2022). Virtual reality consumer experience escapes: Preparing for the Metaverse. *Virtual Reality (Waltham Cross)*, *26*(4), 1–16. doi:10.100710055-022-00641-7

Henslin, J. M., Possamai, A. M., Possamai-Inesedy, A. L., Marjoribanks, T., & Elder, K. (2015). *Sociology: A down to earth approach*. Pearson Higher Education AU.

Hirsh Pasek, K., Zosh, J., Hadani, H. S., Golinkoff, R. M., Clark, K., Donohue, C., & Wartella, E. (2022). *A whole new world: Education meets the Metaverse*. Policy.

Hollensen, S., Kotler, P., & Opresnik, M. O. (2022). Metaverse –the new marketing universe. *The Journal of Business Strategy*. doi:10.1108/JBS-01-2022-0014

Horan, B., Gardner, M., & Scott, J. (2015), MiRTLE: a mixed reality teaching & learning environment, Hybrid Learning and Education, In *International Conference on Hybrid Learning and Education,* (pp. 54-65). Springer, Berlin, Heidelberg.

Hughes, C. E., Stapleton, C. B., Hughes, D. E., & Smith, E. M. (2005). Mixed reality in education, entertainment, and training. *IEEE Computer Graphics and Applications*, *25*(6), 24–30. doi:10.1109/MCG.2005.139 PMID:16315474

Kaminska, D., Sapinski, T., Wiak, S., Tikk, T., Haamer, R. E., Avots, E., & Anbarjafari, G. (2019). Virtual reality and its applications in education: Survey. *Information (Basel)*, *10*(10), 318. doi:10.3390/info10100318

Kaplan, A. D., Cruit, J., Endsley, M., Beers, S. M., Sawyer, B. D., & Hancock, P. A. (2021). The effects of virtual reality, augmented reality, and mixed reality as training enhancement methods: A meta-analysis. *Human Factors*, *63*(4), 706–726. doi:10.1177/0018720820904229 PMID:32091937

Kim, H., Kwon, Y. T., Lim, H. R., Kim, J. H., Kim, Y. S., & Yeo, W. H. (2021). Recent advances in wearable sensors and integrated functional devices for virtual and augmented reality applications. *Advanced Functional Materials*, *31*(39), 2005692. doi:10.1002/adfm.202005692

Kim, J. (2021). Advertising in the Metaverse: Research agenda. *Journal of Interactive Advertising*, *21*(3), 141–144. doi:10.1080/15252019.2021.2001273

Kraus, S., Kanbach, D. K., Krysta, P. M., Steinhoff, M. M., & Tomini, N. (2022). Facebook and the creation of the Metaverse: Radical business model innovation or incremental transformation? *International Journal of Entrepreneurial Behaviour & Research*, *28*(9), 52–77. doi:10.1108/IJEBR-12-2021-0984

Kye, B., Han, N., Kim, E., Park, Y., & Jo, S. (2021). Educational applications of Metaverse: Possibilities and limitations. *Journal of Educational Evaluation for Health Professions*, *18*, 18. doi:10.3352/jeehp.2021.18.32 PMID:34897242

Laeeq, K. (2022). *Metaverse: Why, How and What.* https://www.researchgate.net/publication/358505001_Metaverse _Why_How_and_

Lee, H., Woo, D., & Yu, S. (2022). Virtual Reality Metaverse System Supplementing Remote Education Methods: Based on Aircraft Maintenance Simulation. *Applied Sciences (Basel, Switzerland)*, *12*(5), 2667. doi:10.3390/app12052667

Lee, L. H., Braud, T., Zhou, P., Wang, L., Xu, D., Lin, Z., & Hui, P. (2021). All one needs to know about Metaverse: A complete survey on technological singularity, virtual ecosystem, and research agenda.

Lee, O. (2022). When Worlds Collide: Challenges and Opportunities in Virtual Reality. *Embodied: The Stanford Undergraduate Journal of Feminist, Gender, and Sexuality Studies, 1*(1).

Louro, D., Fraga, T., & Pontuschka, M. (2009). Metaverse: Building Affective Systems and Its Digital Morphologies in Virtual Environments. *Journal of Virtual Worlds Research*, *2*(5). doi:10.4101/jvwr.v2i5.950

Macionis, J. J. (2006). *Society: The Basics* (8th ed.). Pearson/Prentice Hall.

Martín, G. F. (2018). Social and psychological impact of musical collective creative processes in virtual environments; Te Avatar Orchestra Metaverse in Second Life. Musica/Tecnologia Music. *Technology*, 75.

Ming, T. R., Norowi, N. M., Wirza, R., & Kamaruddin, A. (2021). Designing a Collaborative Virtual Conference Application: Challenges. *Requirements and Guidelines. Future Internet*, *13*(10), 253. doi:10.3390/fi13100253

Mohamed Yusoff, A. F., Hashim, A., Muhamad, N., & Wan Hamat, W. N. (2021). Application of fuzzy delphi technique to identify the elements for designing and developing the e-PBM PI-Poli module. [AJUE]. *Asian Journal of University Education*, *7*(1), 292–304. doi:10.24191/ajue.v17i1.12625

Mohd. Ridhuan, M. J., Zaharah, H., Nurul Rabihah, M. N., Ahmad Arifin, S., & Norlidah, A. (2014). *Application of Fuzzy Delphi Method in Educational Research. Design and Developmental Research*. Pearson Malaysia Sdn. Bhd.

Ning, H., Wang, H., Lin, Y., Wang, W., Dhelim, S., Farha, F., & Daneshmand, M. (2021). *A Survey on Metaverse: the State-of-the-art*. Technologies, Applications, and Challenges.

Okoli, C., & Pawlowski, S. D. (2004). The Delphi method as a research tool: An example, design considerations and applications. *Information & Management*, *42*(1), 15–29. doi:10.1016/j.im.2003.11.002

Pan, Z., Cheok, A. D., Yang, H., Zhu, J., & Shi, J. (2006). Virtual reality and mixed reality for virtual learning environments. *Computers & Graphics*, *30*(1), 20–28. doi:10.1016/j.cag.2005.10.004

Pandrangi, V. C., Gaston, B., Appelbaum, N. P., Albuquerque, F. C. Jr, Levy, M. M., & Larson, R. A. (2019). The application of virtual reality in patient education. *Annals of Vascular Surgery*, *59*, 184–189. doi:10.1016/j.avsg.2019.01.015 PMID:31009725

Park, S., & Kim, S. (2022). Identifying World Types to Deliver Gameful Experiences for Sustainable Learning in the Metaverse. *Sustainability*, *14*(3), 1361. doi:10.3390u14031361

Park, S. M., & Kim, Y. G. (2022). A Metaverse: Taxonomy, components, applications, and open challenges. *IEEE Access : Practical Innovations, Open Solutions*, *10*, 4209–4251. doi:10.1109/ACCESS.2021.3140175

Parsons, D., & Stockdale, R. (2010). Cloud as context: Virtual world learning with open wonderland. In *Proceedings of the 9th World Conference on Mobile and Contextual Learning,* Malta, (pp. 123-130).

Preston, J. (2021). Facebook, the Metaverse and the monetisation of higher education. *Impact of Social Sciences Blog*.

Rospigliosi, P. A. (2022). Metaverse or Simulacra? Roblox, Minecraft, Meta and the turn to virtual reality for education, socialisation and work. *Interactive Learning Environments*, *30*(1), 1–3. doi:10.1080/10494820.2022.2022899

Seigneur, J. M., & Choukou, M. A. (2022). How should Metaverse augment humans with disabilities? In *13th Augmented Human International Conference Proceedings*. ACM. 10.1145/3532525.3532534

Wang, Y., Su, Z., Zhang, N., Liu, D., Xing, R., Luan, T. H., & Shen, X. (2022). A survey on Metaverse: Fundamentals, security, and privacy. doi:10.36227/techrxiv.19255058.v2

Warner, D. S. (2022). *The Metaverse with Chinese Characteristics: A Discussion of the Metaverse through the Lens of Confucianism and Daoism* [Doctoral dissertation]. University of Pittsburgh.

Yusoff, A. F. M., Hashim, A., Muhamad, N., & Hamat, W. N. W. (2021). Application of Fuzzy Delphi technique towards designing and developing the elements for the e-PBM PI-Poli Module. *Asian Journal of University Education*, *17*(1), 292–304. doi:10.24191/ajue.v17i1.12625

Zallio, M., & Clarkson, P. (2022). Inclusive Metaverse. How businesses can maximize opportunities to deliver an accessible, inclusive, safe Metaverse that guarantees equity and diversity.

Chapter 10
Research on College Students' Purchasing Using Webcast Platform

WeiXiang Pan
Universiti Tun Abdul Razak, Malaysia

P. C. Lai
Universiti Tun Abdul Razak, Malaysia

ABSTRACT

This paper studies the purchasing behavior of college students using the webcast platform; understanding the factors that college students use the webcast platform and understanding the main factors that promote their consumption so that school managers can effectively manage and guide students to consume rationally. At the national economic level, we need to develop live e-commerce, and at the national education level, we need to ensure that students are not affected by external transition interference. The research uses the technology acceptance model (TAM). Several key factors include webcast platform design, webcast marketing, webcast celebrity webcast, perceived usefulness, and perceived ease of use. The survey data are collected by using the five-point scale, the goodness of fit of the measurement model is used for statistics, and the data are analyzed by using the standardized regression weight of the structural model.

INTRODUCTION

Research Background

According to the 47th statistical report on China's internet development released by CNNIC, as of December 2020, China's live broadcast users had reached 617 million, including 388 million e-commerce live broadcast users, accounting for 39.2% of the total number of Internet users. Due to the impact of the epidemic, many colleges and universities implement closed management on campus, and students

DOI: 10.4018/978-1-6684-5732-0.ch010

are limited to go out (Peishi, 2022). Therefore, online shopping has become the main way of daily consumption (Xingnan, 2020). With the growth of demand, simple online shopping can no longer meet the demand (Yuexi, Mingfeng, & Tongcui, 2020). Mengmeng (2020) stated that more and more college students began to use the emerging live streaming platforms for shopping. In order to meet the needs of college students, businesses have used more marketing means suitable for college students to attract college students to use the live shopping platform(Dengfeng & Zhenpeng, 2021).

According to the data provided by CNNIC, China Internet Information Center, about 16 million teenagers in China may be addicted to webcasts and watch them every day. According to the survey and research of Wen (2019), it is found that most students use webcasts for more than one hour. Beibei (2021) agrees that college students are an important group of webcast audiences. Junjun said that (2020) college students are the main force in using webcasts in the era of new media. Gao Fangyuan believes that (2020) webcast is deeply loved by college students because it is instant and interactive, with a sense of common participation and intuition. Lan (2019) mentioned the issue of consumption. She believes that the rapid development of the Internet provides a favorable platform for webcast marketing. As an active consumer group, college students have a strong ability to accept new things and are easy to be attracted by the new marketing model and produce consumption behavior(Mengnan, 2018). According to the research of Ruonan (2021), e-commerce consumption is addictive. Tai believes that sometimes when college students consume e-commerce, the object of attention will change from the use value of the commodity itself to the sense of achievement and stimulation when paying and checking out, which has become the driving source of College Students' e-commerce consumption. Therefore, in order to balance the needs of economic development and education, this paper analyzes the reasons why college students use the webcast platform and the behavioral factors of shopping through webcast, so as to facilitate education managers to effectively manage and guide college students and avoid affecting learning and excessive consumption. It also provides a reference for businesses to reasonably develop the e-commerce economy as well as the metaverse world.

Research Questions

Li (2021) believes that existing laws and regulations do not supervise online live broadcasts in place, colleges and families lack guidance for college students to watch online live broadcasts, and the content of live broadcasters' performances may be more beneficial, and college students may face negative values when watching online live broadcasts Impact. Facing similar problems, Peijun (2021) believes that people need to maintain the healthy growth of college students in the process of watching webcasts. Liang (2021)'s research believes that college students have virtualized social methods and emotional consumption concepts when watching online live broadcasts. Yuting (2021) believes that webcasting has an impact on college students' impulsive consumption. Therefore, combined with the actual situation of colleges and universities, this paper puts forward the following questions: what are the main factors that promote college students to use the webcast platform and purchase? This paper mainly addresses these problems to provide solutions for college education and management.

LITERATURE REVIEW AND HYPOTHESIS

TAM Model

Davis (1989) proposed the technology acceptance model (TAM), which uses the rational behavior theory to study the user's acceptance of the information system. As we all know, the original purpose of the technology acceptance model is to explain the decisive factors of computer-wide acceptance. The technology acceptance model proposes two main determinants: perceived usefulness and perceived ease of use. In TAM, external variables rather than original subjective normative variables (normative beliefs and compliance motivation) can affect users' internal beliefs (perceived usefulness and perceived ease of use), then affect users' attitudes and behavioral intentions, and finally, affect their actual use of information technology. In TAM, external variables rather than original subjective normative variables (normative beliefs and compliance motivation) can affect users' internal beliefs (perceived usefulness and perceived ease of use), then affect users' attitudes and behavioral intentions, and finally affect their actual use of information technology (Davis, 1989; Lai & Zainal, 2015; Lai & Liew, 2021).

Hui (2017) believes that entertainment, interaction and perceived usefulness will affect consumers' attitudes towards product use. Xiong, Mengmeng, & Wan (2019) studied a webcast with TAM. The conclusion is that users think that the webcast platform is very useful to their life, so they will continue to use the webcast platform. At the same time, in the process of using the webcast platform, the interface is simple and easy to operate. Users have a positive experience and are willing to continue to use the webcast platform. Hausman and Siekpe (2009) confirmed that perceived usefulness can stimulate immersion when studying the impact of website features on consumers' purchase intention. Hui and Sahelan (2021), as well as Lai and Zainal (2015), found that external variables determine individual perceived usefulness, perceived usefulness affects an individual's attitude toward wanting to use, both attitude and perceived usefulness affect an individual's behavior intention, behavior intention ultimately determines an individual's use of the system.

Design of Webcast Platform

Guangyan (2020) believes that the reason why college students use the webcast platform is mainly because the design of the webcast app is simple and easy to use, which can meet the psychological needs of obtaining information, learning skills, and entertainment, psychological identity and personal growth. Fei and Chaoyan (2020) studied interface design and believed that it is a direct medium for the transmission and exchange of information between people and products. It is a cross-research field of computer science and psychology, design art, cognitive science and ergonomics. Baihui (2018) believes that vision is the most intuitive part of human perception of the external world and an important function in processing external graphic information, color, text and other elements. Therefore, when designing the interface, we should fully consider the customer's visual acceptance ability and let the customer accept the whole interface interaction process from many aspects(Jieqiong, 2017). Fengqi, Yutong and Juan (2021) studied users' cognition and emotional response, and believed that the presentation effect of visual design affected users' perception to a certain extent, such as purchase desire, pleasure and presence, and then affected users' purchase behavior and intention. Qin (2020) believes that commonly used function buttons, such as shopping links, interactive buttons and interactive boxes, should be set in easy-to-see positions, such as the four corners of the live broadcast interface. The function of each button should be simple and easy to understand, which can improve the user experience. Shuqin (2021) said in the research: the live broadcast e-commerce takes the network platform as the carrier, and the host transmits information through the body language interaction with the audience in the live broadcast room to facilitate the transaction. In the online broadcast, the interaction between users and anchor, between users and users, and the satisfaction of user functions and entertainment needs will improve their willingness to use the live broadcast (Weisheng, 2022). Figure 1 shows the live broadcast interface of Taobao.

Figure 1. Live broadcast interface seen by mobile phone

Yu Guoming and Yang Jiayi (2020) think that live broadcasting is a full channel media to reproduce the scene of face-to-face communication. Chen Yao (2021) studied the barrage information of the webcast platform and believed that its interactivity, visibility, entertainment and usefulness stimulated users' participation in the barrage information. AI Xiaoya (2018) pointed out that the real social scene and immersive social experience can accelerate the large-scale speed of the webcast platform. The biggest feature that distinguishes webcast from online video is the real-time interaction of webcast, which gives the natural social attribute of the webcast (Xu Ying, 2019). Zhang Bingjie (2021) feels that webcast gives a super long period of time, generous capacity and freestyle, which is not cut and cut due to the scarcity of media resources.

Live Broadcast Platform Marketing

Yongan and Xuetao (2017) said that webcast is a new business model, which transforms the dissemination of information from simple text, pictures, audio and video to a more advanced real-time image form, improves the value and dissemination effect of information, and has attracted high attention from the business and academic circles. Fuqiang and Penghui (2018) believe that most content in a webcast is produced in real-time during the interaction between the anchor and the audience. The interaction between the anchor is actually a kind of marketing(Fan Hongzhao, 2018). Yong and Tianping (2018) felt that in the webcast of entertainment and business marketing, the interpersonal emotional flow maintained by personal performances reconstructed people's relationship scenes. Xin and Lu (2021) believe that the marketing model of the webcast platform is an emerging business model in recent years. Webcast platforms and network anchors cooperate with e-commerce to make profits by selling products, and webcast procurement has become a new path for the development of e-commerce(Guozhang & Yao, 2021). Sihao et al. (2021) found that the webcast platform is a good commodity sales channel, with the rich marketing experience of the anchor, improves the viewing and purchases conversion rate of the live broadcast audience, highlighting the great advantages of the combination of the online celebrity economy and Internet technology. Shue and Yuyi (2014) have predicted that in the webcast platform, the network marketing effect will continue to expand like a snowball, and there is a law of increasing marginal return in network economics. Jing (2019) as well as Lai, Toh and Alkhrabsheh (2020) support the user-oriented network marketing effect. After research, they believes that doing a good job in online marketing content is the key to maintaining the high user stickiness of the webcast platform. Through the analysis of customers, Ning and Yuqi (2020) think that in order to better marketing, the webcast platform provides users with the content they need on the basis of customer segmentation and by analyzing user needs.

Online Celebrity Live Broadcast

Dewey (2021) believes that an online celebrity economy is a form of using celebrities on the Internet to drive economic development. Online celebrities generally refer to those who are concerned and continuously understood by the public because of an event or a behavior point of view on the Internet, or who meet the public through online social media for a long time (Lidan & Zhiwei, 2021). Wen and Kaiyan (2021) think that online celebrities have excellent appearances and unique personal charm, which is easy to attract college students to watch the online live broadcast and feel happy and satisfied, so as to attract college students to use the live broadcast platform. Lili et al. (2021) as well as Lai and Tong (2022) studied the anchor and believed that the e-commerce live broadcast takes the anchor as the carrier and the content as the medium, and the goods are communicated to users through celebrities, so as to form purchasing power. Based on people's needs, the product introduction and promotion interaction in the live broadcast room and the hot atmosphere make consumers have psychology of conformity consumption (Kai, Luchuan, Chenglin, 2022). Hoffman and Novak (1996) as well as Lai and Liew (2021) believe that when people are immersed, they will get the "best experience" and people will naturally participate in it. Nen et al. (2018) studied the factors affecting the viewing intention of live viewers and found that the interaction between online celebrities and viewers, the personality charm of celebrities, and the interest and innovation of live content all have a positive impact on the viewing intention of viewers.

Starting from the motivation of network anchor live broadcasting, Tang et al. (2016) through the interview with the anchor on the periscope live broadcasting platform, concluded that some anchor lives broadcasting for the purpose of shaping their personal brand, while others are for the purpose of obtaining the social support of the audience. Zhao et al. Found in their research in 2017 that online celebrities' perception of the attractiveness of the live broadcasting platform and their personal expected performance level has a positive impact on their willingness to continue live broadcasting. Jingmei (2020) compared the previous web procurement with the current live procurement and believed that online celebrities use the products to describe the advantages and feelings of the products, break through the limitation that the previous products only have a single picture introduction, strengthen the two-way dissemination of product information and enhance consumers' desire to buy. Jinyi (2019) believes that the core asset of a webcast platform is the anchor, and the marketing of the webcast platform is actually the personalized marketing of the anchor. As a core asset, the anchor has a high degree of exposure, and the anchors can attract more users through their personal charm. Kwang and Ju (2017) believe that the live broadcast platform should make full use of the celebrity effect of the anchor, constantly update the functions to meet the social needs of users, and be able to respond to user comments in time. Rose (2013) once pointed out that the effect of content marketing depends on whether the content it delivers is valuable and entertaining. Based on this point of view, Lan and Huicui (2021) thought that online celebrity live broadcasting can make consumers feel happy, curious and interesting content, which is very attractive to College students' use intention.

Celebrity Incentives Drive Purchases

Sihao et al. (2021) studied the purchasing psychology of the audience and believed that when the audience watched the favorite anchor live broadcast, their emotional needs and purchase information acquisition needs were met, so they were more likely to be infected by the atmosphere in the live broadcast room and make impulsive consumption. Lei and Can (2020) give another analysis of consumer psychology. They believe that under the background of collectivism, Chinese consumers tend to follow the choice of most people, so they are more vulnerable to the spread of online celebrity word-of-mouth. Hu and Leonce (2019) studied Confucian culture. In this context, face awareness has become one of the important factors affecting consumers' behavior. They like to follow and imitate celebrities and purchase goods recommended by celebrities. Therefore, Wahab and Tao (2019) believe that online celebrities play an important role in the followers' purchase decision-making and materialism through identity and quasi-social relations, and stimulate the audience's purchase desire. The research results of Lingrong and Lan (2021) believe that each online celebrity's mainstream purchase of goods is related to their personal experience, and has a high degree of matching with their fan base. They believe that consumers trust anchors with similar social attributes or the same experience, which is more likely to cause empathy, quickly enhance the value of trust, and trust further catalyzes consumption. Jiang et al. (2021) think that online celebrities can have close contact with the audience through their communication content to stimulate consumers' purchase desires.

Hypothesis

Based on the above literature, for the purpose of research, this paper puts forward the following assumptions:

H1a: The design of the webcast platform is positively correlated with perceived usefulness;
H1b: The design of the webcast platform is positively related to perceived ease of use;
H2a: Webcast marketing is positively correlated with perceived usefulness;
H2b: Webcast marketing is positively correlated with perceived ease of use;
H3a: Online celebrity webcast is positively correlated with perceived usefulness;
H3b: Online celebrity webcast is positively correlated with perceived ease of use;
H4: Perceived ease of use is positively correlated with perceived usefulness;
H5: Perceived usefulness is positively correlated with college students' shopping through the webcast platform;
H6: Perceived ease of use is positively related to college students' shopping through the webcast platform.

METHODS

As mentioned earlier in this paper, about 16 million teenagers in China are watching the webcast. Taking this as the research base, according to the raosoft sample size requirements, the minimum sample size is 385 (error range = 5%, confidence level = 95%). In order to meet the timeliness of the study, the target population of this study only includes college students who have used the webcast platform in the past 6 months. These respondents come from the data of the webcast platform. In order to meet the sample size, a questionnaire survey was conducted among about 2000 college students, and network access was conducted by sending questionnaire links and QR codes to the respondents, which was used as the method of data collection. A total of 430 questionnaires were received. Finally, 400 data meeting the requirements of the sample were screened, which provided strong support for this study. In this study, the Likert type of five point scale was used to measure the willingness of college students to use the webcast platform to watch the webcast of online celebrities, and then purchase goods. From "strongly agree" to "strongly disagree", they were recorded as 5, 4, 3, 2 and 1 respectively. Cronbach's alpha analysis was carried out by SPSS. The results were the design of webcast platform (0.873) and webcast marketing (0.876), Online celebrity live broadcast (0.824), perceived usefulness (0.869), perceived ease of use (0.864), which shows that the reliability of the model is high.

Figure 2. Theoretical model of college students using webcast platform

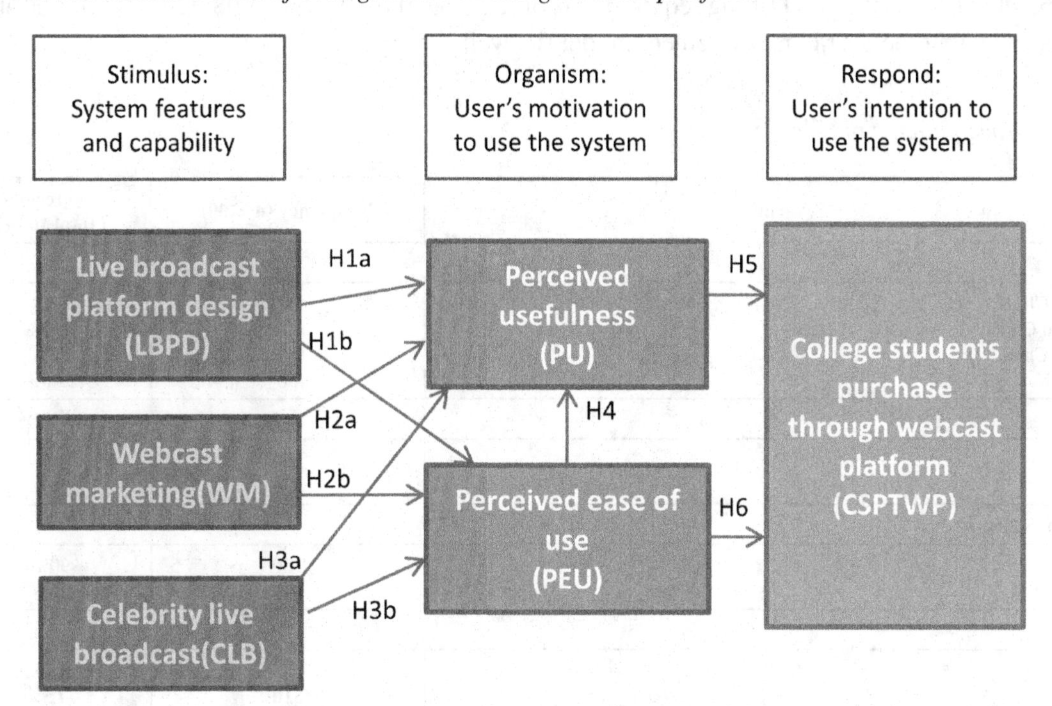

RESULTS (RESPONDENT PROFILE)

In this study, college students were taken as respondents to collect data and analyze each variable respectively, so as to collect the main demographic information of respondents and obtain preliminary data. Table 1 shows the demographic profile of the respondents. 35% of the respondents are in South China, 42% are male. Because they are college students, 90% are single, 77.5% are aged between 18 and 22, 39.5% have college degrees, 76.5% have liberal arts students and 64% have rich families.

Measurement Model

Firstly, the estimated value of standardized factor load is used to check the convergence validity, and then determine the structural validity. For the indicators of perceived usefulness and perceived ease of use, the factor load of Confirmatory Factor Analysis (CFA) ranges from 0.79 to 0.87. By analyzing the estimated value of factor load, the data results of each factor are greater than 0.50. Then the structural validity was analyzed. The Compound Reliability (CR) was 0.87, which met the needs of more than 0.70. The Average Value (AVE) result is 0.82, which meets the demand greater than 0.50 (bagozzi and Yi, 2012; Lai, 2020). The data analysis results show that the goodness of fit indexes meet the requirements of the effectiveness of CFA model. When p = 0.00, the result of chi square test is 289.57 and DF (degree of freedom) is 80. According to tabachnick and Fidell (2007), the relative chi square value is 3.88 (x2 / DF), which meets the requirements of good fitting less than 5.0. In the absolute fitting index, the goodness-of-fit index (GFI) is 0.91, which meets the good fitting demand higher than 0.90 (hair et al, 2010; Lai, 2020). The Comparative Fit Index (CFI) is 0.93, which meets the requirements of good fit higher than 0.90 (Hu and bentler, 1999; Lai, 2020). The approximate root mean square error (RMSEA) is 0.06, which meets the good fitting requirements of less than 0.08 (Byrne 1998; Lai, 2020). The above results show that the overall measurement model fits well.

Table 1. Respondents profile

Variable	Frequency (n=400)	Percent (Total 100%)
Region		
South China	140	35%
Central China	122	30.5%
North China	88	22%
Other	50	12.5%
Gender		
Male	168	42%
Female	232	58%
Marital Status		
Single	360	90%
Married	40	10%
Age		
<18	30	7.5%
18-22	310	77.5%
>22	60	15%
Education		
Junior college	158	39.5%
Undergraduate	220	55%
Graduate student	22	5.5%
Subject Classification		
Liberal arts	306	76.5%
science	74	18.5%
Other	20	5%
Family economic situation		
Family affluence	256	64%
The family economy is medium	112	28%
Family poverty	32	8%

Structural Model

According to the results of the analysis and measurement model, the structural model is effectively tested. The theoretical data are listed in Table 2. All data values meet the requirements of the reference value, indicating that the goodness of fit index is good and the model is acceptable.

Table 2. Goodness-of-fit statistics for measurement model

Goodness-of-fit Statistics		Level of Acceptance	Index Value
Absolute fit Measures			
Chi-square	X^2	p>0.05	5.647(p=0.06)
Degree of freedom	df	$^{3}0$	2
Root mean square error of approximation	RMSEA	<0.08	0.062
Goodness of fit index	GFI	>0.90	0.922
Incremental fit measures			
Comparative fit index	CFI	>0.90	0.93
Parsimonious fit measures			
Relative Chi-Squar	X^2/df	<5	2.971

Table 3 shows the overall structure model data. All standardized regression weight paths shown in the data are statistically significant at the significance levels P £ 0.001 and P £ 0.01.

Table 3. Overal structure model data

Hypothesis	Standardized Regression Weights			S.E.	C.R.	P	Results
Hypothesis 1a	PU	¬	LBPD	0.03	43.81	***	Significant p£0.001
Hypothesis 1b	PEU	¬	LBPD	0.02	7.32	***	Significant p£0.001
Hypothesis 2a	PU	¬	WM	0.03	3.11	0.002	Significant p£0.01
Hypothesis 2b	PEU	¬	WM	0.02	3.69	***	Significant p£0.001
Hypothesis 3a	PU	¬	CLB	0.06	17.36	***	Significant p£0.001
Hypothesis 3b	PEU	¬	CLB	0.05	15.50	***	Significant p£0.001
Hypothesis 4	PU	¬	PEU	0.06	2.91	0.004	Significant p£0.01
Hypothesis 5	CSPTWP	¬	PU	0.06	2.88	***	Significant p£0.001
Hypothesis 6	CSPTWP	¬	PEU	0.05	3.10	***	Significant p£0.001

Note:***p£0.001,***p£0.01,***p£0.05

Hypothesis 1A: the design of the webcast platform is positively correlated with perceived usefulness.

According to the analysis data, it can be seen that there is a strong direct relationship between the design of the webcast platform and perceived usefulness, and the critical ratio is C.R = 43.81, P = 0.00 (P < 0.001). In this case, the higher the support of webcast platform design, the higher its usefulness.

Hypothesis 1b: the design of the webcast platform is positively related to perceived ease of use.

The data analysis results show that this hypothesis is true. This shows that the webcast sales platform with simple operation, complete functions, and convenient selection and payment are directly related to the user experience of students. College students subconsciously like to communicate and shop conveniently and quickly in the process of watching webcasts. Therefore, if the hypothesis is true, there is a significant relationship between the design of the webcast platform and perceived ease of use.

Hypothesis 2A: webcast marketing is positively correlated with perceived usefulness.

The results of the data analysis support this hypothesis. This shows that webcast marketing affects college students to feel the usefulness of webcast platforms.

Hypothesis 2B: webcast marketing is positively correlated with perceived ease of use.

Survey data support this hypothesis. This hypothesis shows that the marketing of webcasts is also related to the ease of use of the platform.

Hypothesis 3A: Online celebrity live broadcast is positively correlated with perceived usefulness.

The analysis of survey data shows that it supports this hypothesis. This shows that college students have a sense of identity with online celebrities, and even some students come to the live broadcast platform for these online celebrities.

Hypothesis 3B: Online celebrity live broadcast is positively correlated with perceived ease of use.

The data survey shows that there is a positive relationship between celebrity live broadcasts and platform ease of use.

Hypothesis 4: perceived ease of use is positively correlated with perceived usefulness.

The data analysis results support this hypothesis, and ease of use has a positive impact on usefulness.

Hypothesis 5: perceived usefulness is positively correlated with college students' shopping through the live broadcast platform.

The results of data analysis show that perceived usefulness has a direct impact on College Students' use of live broadcast platforms and purchases.

Hypothesis 6: perceived ease of use is positively related to college students' shopping through the live broadcast platform.

The data analysis results have high support for this hypothesis, which shows that the ease of use of the platform design has a great relationship with college students' use and acceptance of the platform, and users like to be simple and direct.

DISCUSSIONS

There is a significant positive correlation between the design of webcasting platforms and perceived usefulness and perceived ease of use, which is similar to the studies by Lai and Zainal (2015) and Lai and Liew (2021). The experience of using the webcast platform has met the expectations of college students. Users find this media tool useful. Users think that the platform-tools are easy to use and can be used proficiently. This is similar to Ma Zhihao's research view on youth groups in 2018.

Webcast marketing has a positive impact on perceived usefulness and perceived ease of use. As a way of publicity, webcast marketing can be accepted and used by college students for the first time, which is inseparable from the role of marketing. The better the marketing method, the more it can show its usefulness. Siyue and Dongju (2021) believed that the first step of webcast marketing is to attract college students to use the webcast platform, the second step is to effectively stimulate college students' purchase desire. Whether webcast marketing is good or bad can largely determine whether college students continue to use the webcast platform and buy goods(Jun, Tingting, & Qing, 2021).

Online celebrity live broadcasting has a significant relationship between perceived usefulness and perceived ease of use. Under the influence of online celebrities, some college students use the live broadcast platform to chase celebrities and interact with celebrities, which makes college students a great viscosity with the platform(Shuo, Juan, Jiaxin, & Yu, 2021). The higher the time and frequency of celebrity live broadcasts, the higher the time and frequency of college students using the live broadcast platform (Siting, 2016). The higher the popularity and influence of online celebrities' live broadcasts, the higher the willingness of consumers to shop through live broadcasting (Xiaohan, Ruiliang, & Baolu, 2021). This paper holds that college students have potential psychological recognition for online celebrity live broadcasting, which is manifested in perceived usefulness, which may improve college students' purchase intention and guide consumers to purchase. Online celebrity live broadcasting is an important factor for college students to shop through the live broadcasting platform.

Online celebrities have always been the focus of college students and an important factor to stimulate college students to shop through the live broadcast platform(Sun Ruimeng, 2022). Improving the popularity of online celebrities can significantly increase the sales and sales of live broadcasts(Liu Lingrong, Wang Lan, 2021). In China, some TikTok celebrities are able to sell hundreds of millions of dollars in live broadcasts.

CONCLUSION AND RECOMMENDATIONS

Suggestions for Colleges and Students

Webcast has developed rapidly in recent years and has great attraction to college students (Weiguo, Jinrong, & Huijing, 2020). Colleges and universities should not ignore or avoid this situation, but should correctly guide and make use of it. It is suggested that college administrators should do a good job in ideological education and avoid the blind pursuit of celebrities and blind worship. The celebrity worship of teenagers is more reflected in the warm infatuation and pursuit of the people they admire (Xingli, 2007). They often show considerable enthusiasm, blind obedience, and lack of independence in chasing them (He & Jianzhi, 2011). Some children never ask why they worship their own idols, but blindly flock to worship them when they think they are beautiful (Chengju, 2021). Viewpoint of this article, to worship celebrities, we should grasp their spiritual attributes and social values, have good ideological and moral concepts and the spirit of the times and have a healthy personality.

Individual students blindly indulge in celebrity live broadcasting, resulting in excessive consumption, overdraft of credit cards, and even borrowing usury on the internet (Ling, 2017). They also enter a vicious circle because they can't repay usury, resulting in serious consequences. Yingyan (2019), Wei (2019), Linjing (2021), and others have studied college students' usury, According to their research, some students are overwhelmed and choose to end their lives. Therefore, this paper warns contemporary college students that they can properly participate in watching webcasts and purchasing, but don't be too addicted and don't consume too much. Learning is the key task of college students. Learn socialist values, learn scientific knowledge, be diligent and thrifty, do more outdoor sports, and be positive and healthy.

Suggestions for e-Commerce and Online Celebrities

The current webcast platform is well-designed and successful, with a large number of users and wide influence (Chenghong & Shibin, 2018). The number of college students using webcast platforms is also large, so it has also been greatly affected (Peijun, 2021). Viewpoint of this article, some of the effects are negative and unfavorable to students' growth. In addition to making profits, e-commerce should also bear the responsibility of social education, at least avoiding misleading students.

The view of this article is that the webcast platform should reduce exaggerated marketing advertising. For the sake of sales volume, some online celebrities use vulgar terms, make false propaganda, mislead or even cheat consumers in the live broadcast, and live broadcast marketing chaos is frequent (Yi, 2021). E-commerce live broadcasting platforms, media, advertising, and other channels should actively promote the concept of "moderate consumption" and "rational consumption" and cultivate the style of diligence, thrift, and hard work (Maohua, 2021).

The view of this article is that the live broadcast of online celebrities should reduce grandiose performances. Some online celebrities use low-grade and stimulating performances to attract attention and fans, which has a serious negative impact on college students (Xiaoyin, 2021). We should actively cultivate positive online celebrities to become powerful communicators of socialist core values and bring a positive impact on the values of college students (Xiqun, 2021). Further research is encouraged to be conducted for the metaverse world with the Webcast solution and factors discussed above.

REFERENCES

Bagozzi, R. P., & Yi, Y. (2012). Specifications, evaluation, and interpretation of structural equation models. *Journal of the Academy of Marketing Science*, *40*(1), 8–34. doi:10.100711747-011-0278-x

Baihui, Z. (2019). Research on interactive design of social sharing in Pan entertainment mobile live broadcast [Doctorat dissertation]. Jiangnan University.

Beibei, D. (2021). Research on the influence and Countermeasures of webcast on College Students' Socialist Core Values Education. *Science and technology communication,13*(16), 142-145. doi:.1674-6708.2021.16.050. doi:10.16607/j.cnki

Bingjie, Z. (2021). Webcast: a relational media. *Southeast propagation*, (09), 52-55. doi:.dncb. cn35-1274/j.2021.09.016. doi:10.13556/j.cnki

Byrne, B. M. (1998). *Structural Equation Modeling with LISREL, PRELIS, and SIMPLIS: Basic Concepts, Applications, and Programming*. Lawrence Erlbaum Associates.

Cheng, C., Guohua, W., & Zhihong, Z. (2020). Research on the influence of price discount level on consumers' purchase intention. *Business economics research*, (23), 76-79.

Chengju, J. (2021). New characteristics and problems of teenagers' entertainment idol worship. *Modern youth*, (04), 51-53.

CNNIC. (2021). The 47th Statistical Report on China's Internet Development was released. *China Broadcasting*, (04), 38

Davis, F. D. (1989). Perceived usefulness, perceived ease of use, and user acceptance of information technology. *Management Information Systems Quarterly*, *13*(3), 319–340. doi:10.2307/249008

Dengfeng, C. & Zhenpeng, Y. (2021). Can I make a reservation after a good call? Research on the impact of platform and anchor reputation on consumers' willingness to participate in online live shopping. *Xinjiang Agricultural Reclamation Economy*, (06), 82-92.

Dewey. (2021). Research on the impact of Daren webcast e-commerce on consumers' purchase intention. *Internet Weekly*, (18), 62-64

Fangyuan, G. (2020). Research on the current situation, causes and correction of moral anomie in college students' webcast. *Popular literature and art*, (05): 179-180

Fei, Z. & Chaoyan, H. (2020). Design of home intelligent remote controller based on interactive design. *Industrial design*, (06), 116-117

Fengqi, Z., Yutong, X., & Juan, L. (2021). Research on interface usability based on live broadcast marketing platform. *Science and technology information, 19*(19), 29-31 doi:.1672-3791.2108-5042-1191. doi:10.16661/j.cnki

Fuqiang, Y. & Penghui, H. (2020). Simulacrum, body and emotion: an analysis of webcast in consumer society. *China Youth Research*, (7), 5-12 + 32

Guangyan, H., Xiaofang, Z., & Wanzhen, X. (2020). Investigation and Research on the current situation of Qinghai college students using webcast platform. *Journal of Qinghai Normal University normal college for nationalities*, *31*(01): 83-88. doi:. 63-1060/g4. 2020.01.018. doi:10.13780/j.cnki

Guo J. & Fuli, S. (2019). Explore the impact of webcast platform on traditional culture brand communication. *Think tank era,* (43), 259-260

Guoming, Y. & Jiayi, Y. (2020). Understanding live broadcasting: social reconstruction according to communication logic -- an analysis of the value and influence of live broadcasting from the perspective of media. *Journalist*, (08), 12-19

Harrison, R., Scheela, W., Lai, P. C., & Vivekarajah, S. (2017). Beyond institutional voids and the middle income trap? The emerging business angel market in Malaysia. *Asia Pacific Journal of Management*, 1–27.

He, W., Chen, J. (2011). Brief Discussion on the Implementation of Positive Psychological Education for Contemporary College Students. *Educational Exploration*, (03), 131-133.

Hoffman, D. L., & Novak, T. P. (1996). Marketing in Hypermedia Computer-Mediated Environments: Conceptual Foundations *Journal of Marketing*, *60*(3), 50–68. doi:10.1177/002224299606000304

Hongzhao, F. (2018). Research on Online Live Broadcasting Marketing Strategy and Mode in the Era of Fans Economy. *Modern Economic Information*, (11), 336-337+339.

Hu, L., & Bentler, P. M. (1999). Cutoff criteria for fit indexes in covariance structure analysis: Conventional criteria versus new alternatives. *Structural Equation Modeling*, *6*(1), 1–55. doi:10.1080/10705519909540118

Huang, S., Deng, F., & Xiao, J. (2021). Research on impulse purchase decision of audience on webcast platform -- from the perspective of dual path influence. *Financial science,* (05): 119-132

Hui, L. (2017). Empirical Study on Influencing Factors of webcast use attitude from the perspective of TAM theory. *South China University of technology.*

Hui, W. (2021). sahelan An empirical analysis of the impact of additional comments on e-commerce shopping platforms on consumers' Purchase Intention — Based on the survey data of colleges and universities in Xinjiang. *Journal of Xinjiang Radio and Television University*, *25*(02), 45–51.

James, M. X., Hu, Z., & Leonce, T. E. (2019). Predictors of organic tea purchase intentions by Chinese consumers: Attitudes, subjective norms and demographic factors. *Journal of Agribusiness in Developing and Emerging Economies*, *9*(3), 202–219. doi:10.1108/JADEE-03-2018-0038

Jiang, C., Ying, W., & Lilong, Z. (2021). Research on the influence of e-commerce online Red self presentation on consumers' purchase behavior and its mechanism. *Journal of Chongqing Industrial and Commercial University,* 1-14. https://kns.cnki.net/kcms/detail/50.1154.c.20210506.1637.004 .html

Jie, R. & Peng, Z. (2019). College Students' irrational herd consumption -- from the perspective of social psychology. *Knowledge economy*, (18), 60-61 doi:. zsjj. 2019.18.031. doi:10.15880/j.cnki

Jie, L. (2021). Influence of online live broadcasting on Fujian tea marketing. *Fujian tea, 43*(10), 42-43

Jing, H. (2019). Research on differentiated innovation of webcast platform. *China newspaper industry,* (06), 12-13. doi:. cni. 2019.06.005. doi:10.13854/j.cnki

Jingmei, W. (2020). Analysis on business development model of webcast platform. *News research guide, 11*(09), 214-215

Ju, P. (2021). Research on the Impact of Live Broadcast on College Students' Values and Guiding Countermeasures.. *News Research Guide, 12*(09), 58–59.

Ju, P. (2021). Research on the Influence of Webcasting on College Students' Values and Guiding Countermeasures.. *Journal of News Research, 12*(09), 58–59.

Jun, F., Chen, T., & Qing, Z. (2021). The impact of the host's interaction strategy on the audience's willingness to reward in different types of live broadcast scenes *Nankai Management Review*, *24*(06), 195–204.

Junjun, C., Huiwen, L., Ruijing, S., & Xiaolu, Z. J. (2020). Investigation and Research on the use of College Students' webcast platform -- Taking an agricultural university as an example. *Rural economy and science and technology, 31*(1), 361-362

Kai, S., Liu, L., & Liu, C. (2022). Impulsive purchase intention of live broadcast e-commerce consumers from an emotional perspective.. *China's Circulation Economy*, *36*(01), 33–42. doi:10.14089/j.cnki.cn11-3664/f.2022.01.004

Kline, R. B. (2011). *Principles and practice of structural equation modelling,* (3rd ed.). Guilford Press.

Lai, P. C. (2016). *SMART HEALTHCARE @ the palm of your hand.* ResearchAsia.

Lai, P. C. (2018). Research, Innovation and Development Strategic Planning for Intellectual Property Management. *Economic Alternatives.*, *12*(3), 303–310.

Lai, P. C. (2019). *Factors That Influence the tourists' or Potential Tourists' Intention to Visit and the Contribution to the Corporate Social Responsibility Strategy for Eco-Tourism. International Journal of Tourism and Hospitality Management in the Digital Age*. IGI-Global.

Lai, P. C. (2020). Intention to use a drug reminder app: a case study of diabetics and high blood pressure patients. *SAGE Research Methods Cases*, 1-10. doi:10.4135/9781529744767

Lai, P. C., & Liew, E. J. (2021). Towards a Cashless Society: The Effects of Perceived Convenience and Security on Gamified Mobile Payment Platform Adoption. *AJIS. Australasian Journal of Information Systems*, *25*. doi:10.3127/ajis.v25i0.2809

Lai, P. C., & Lim, C. S. (2019). *The Effects of Efficiency, Design and Enjoyment on Single Platform E-payment Research in Business and Management*, *6*(2), 19–34.

Lai, P. C., Toh, E. B. H. & Alkhrabsheh, A. A. (2020). Empirical Study of Single Platform E-Payment in South East Asia. *Strategies and Tools for Managing Connected Consumers,* 252-278

Lai, P. C., & Tong, D. L. (2022). *An Artificial Intelligence-Based Approach to Model User Behavior on the Adoption of E-Payment. Handbook of Research on Social Impacts of E-Payment and Blockchain.* IGI Global.

Lai, P. C., & Zainal, A. A. (2015). Perceived Risk as an Extension to TAM Model: Users' Intention To Use A Single Platform E-Payment. *Australian Journal of Basic and Applied Sciences*, *9*(2), 323–330.

Lan, L., Jingwen, L., & Yinuo, W. (2019). Investigation on the impact of webcast marketing on College Students' consumption. *China business theory,* (13), 87-90 doi:. issn2096-0298.2019.13.087. doi:10.19699/j.cnki

Lan, Z. & Huicui, Q. (2021). Research on the impact of webcast based on content marketing on consumers' purchase intention. *Mall modernization*, (14), 121-123 doi:. scxdh. 2021.14.042. doi:10.14013/j.cnki

Lei, C., & Feng, L. (2017). Research on consumers' willingness to continue to use in the socialized e-commerce environment -- from the perspective of continuous trust. *Science, technology and economy, 30*(06), 61-65 doi:. cn32-1276n. 2017.06.013. doi:10.14059/j.cnki

Li, J. (2017). On the Optimization Control of Visual Elements in UI Design.. *Art Science and Technology*, *30*(11), 120–121.

Li, Z. (2021). *Guide online live broadcast of college students with socialist core values*. Guizhou Normal University. doi:10.27048/d.cnki.ggzsu.2021.000042

Lian, X. (2021). The influence of "Internet popularity" on college students and its countermeasures.. *Journal of Zunyi Normal University*, *23*(05), 122–125.

Liang, B., Jianchao, Z., & Zhijie, D. (2011). Three-dimensional reflection on the ideological and political education of college students in the era of webcasting. *Jiangsu Higher Education,* (11), 95-98. doi: . doi:10.13236/j.cnki.jshe.2021.11.016

Lidan, C. & Zhiwei, X. (2015). Zhou Liang's Commentary from the perspective of online collective action. *Journalist*, (01), 74-78. doi:10.16057/j.cnki.31-1171/g2.2015.01.011

Lili, L., Xinyue, C., & Shi, C. (2021). Research on e-commerce live broadcasting mode. *China's collective economy*, (s33), 65-68

Lin, Y., Wang, M., & Ma, T. (2020). The impact of online shopping on physical retail — a survey based on the business center of Shanghai Southern Mall.. *World Geographic Research*, *29*(03), 568–578.

Ling, H. (2017). Saving Campus Loans - In depth Scanning of Campus Network Consumption Loans. *China Credit*, (06): 16-27

Ling, L., Meng, Q., Hong, W., Yanzhao, W., Hao, H., Darong, L. (2020). Analysis on the influence of "comparison psychology" on College Students' consumption psychology and behavior -- Taking the research of Shengli University as an example. *Value engineering, 39*(15), 38-40 doi:. cn13-1085/n.2020.15.018. doi:10.14018/j.cnki

Lingrong, L., & Lan, W. (2021). Research on the impact of consumer purchase decisions from the perspective of online celebrity live broadcast with goods. *Modern Commerce,* (28), 48-50. doi:10.14097/j.cnki.5392/202128.015

Lingrong, L., & Lan, W. (2021). Research on the impact of consumers' purchase decision from the perspective of online live broadcasting. *Modern business*, (28), 48-50 doi:. 5392/2021.28.015. doi:10.14097/j.cnki

Liu, L., & Fang, C. C. (2020). Accelerating the Social Media Process: The Impact of Internet Celebrity Word-of-Mouth Communication and Relationship Quality on Consumer Information Sharing.. *International Journal of Human Resource Studies, 10*(1), 201222–201222. doi:10.5296/ijhrs.v10i1.16043

Lu, S. (2016). The Impact of Social Utility Software on College Students' Interpersonal Relations — Taking Live Online Software as an Example.. *Digital Media Research, 33*(10), 42–46.

Maohua, C. (2021). Research on the induction and strategy of capital logic on Contemporary College Students' consumption behavior. *Journal of Mianyang Normal University,40*(12): 15-20. doi:.cn51-1670/g.2021.12.003. doi:10.16276/j.cnki

Mengmeng, D. (2020). A survey on college students' online live broadcast purchase behavior. *Marketing (Theory and Practice).*

Ning, S. & Beibei, L. (2021). Research on the negative impact of Pan entertainment on College Students' values and countermeasures. *Science and technology communication, 13*(16), 66-68. doi:.1674-6708.2021.16.026. doi:10.16607/j.cnki

Ning, Y. & Yuqi, G. (2020). E-commerce's marketing strategy based on webcast platform. *Farm staff,* (22), 172 + 212

Ping, L. (2021). Research on behavior problems of higher vocational students from the perspective of positive psychology-- consumer psychological comparison, etc. *The age of wealth*, (08), 229-230

Qin, Y. & Renjun, L. (2020). Research on the design of Taobao live broadcast interface based on user characteristics. *Packaging engineering, 41*(08), 219-222 doi:.1001-3563.2020.08.032. doi:10.19554/j.cnki

Rong, W. & Juan, L. (2021). Research on the impact of webcast on the mental health of media art college students. *Western radio and television, 42*(05), 54-56

Ruimeng, S. (2022). Research on the Impact of Online Red Economy on College Students' Consumption Behavior. *Time honored Brand Marketing*, (19), 64-66.

Ruonan, T. (2021). The influence of e-commerce consumption on college students and its countermeasures. *China management informatization, 24* (23), 77-80

Shuang, Y. (2021). The influence of Pan entertainment on College Students' Ideological and political education and its countermeasures. *Foreign trade and economic cooperation*, (12), 96-99

Shue, M. & Yuyi, W. (2014). Research on business model of technology trading platform from the perspective of value network. *Scientific and technological progress and countermeasures, 31*(06), 1-5

Shuo, G., Juan, L., Jiaxin, Y., & Yu, H. (2019). An Analysis of the Current Situation of the Development of College Students' Live Broadcasting and the Countermeasures -- Taking Henan University as an Example. *Science and Technology Communication, 11* (07), 131-134+191. doi:10.16607/j.cnki.1674-6708.2019.07.066

Shuqin, B. (2021). An empirical study on College Students' webcast behavior and psychology. *News knowledge*, (10), 65-70

Siyue, Q. & Dongju, W. (2021). Research on the Current Situation and Development Countermeasures of Direct Broadcasting with Goods to Assist Agriculture -- Taking Taobao Direct Broadcasting as an Example.. *China Agricultural Accounting*, (10), 88-89. doi:10.13575/j.cnki.319.2021.10.034

Tabachnick, B. G., & Fidell, L. S. (2007). *Using Multivariate Statistics*. Pearson Education Inc.

Tang, J. C., Venolia, G., & Inkpen, K. (2016). *Meerkat and Periscope:I Stream,You Stream,Apps Stream for LiveStreams*. Human Factors In Computing Systems. doi:10.1145/2858036.2858374

Tian, L. & Yongqi, L. (2021). Analysis of Taobao online marketing strategy. *Market weekly, 34*(08), 72-74

Tian, X., Guo, R., & Wang, B. (2021). Research on the purchase intention of clothing consumers in Taobao live broadcast based on perceived risk theory. *Journal of Beijing Institute of Clothing Technology, 41*(01), 61–66. doi:10.16454/j.cnki.issn.1001-0564.2021.01.010

Törhönen, M., Sjöblom, M., & Hamari J. (2018). Likes and Views:Investigating Internet Video Content Creators Perceptionsof Popularity. New York:The2nd International GamiFIN Conference.

Wahab, H. K. A., & Tao, M. (2019). The Influence of Internet Celebrity on Purchase Decision and Materialism: The Mediating Role of Para-social Relationships and Identification.. *European Journal of Business and Management, 11*, 15.

Wang, P. (2022). Research on the Management of Epidemic Prevention and Control in Colleges and Universities.. *China Higher Education*, (09), 47-49

Wang, M. (2018). On Consumer Knowledge and Consumer Behavior — Taking College Students as an Example.. *Value Engineering, 37*(17), 233–235. doi:10.14018/j.cnki.cn13-1085/n.2018.17.103

Wang, W., Zhang, J., & Huang, H. (2020). Research on Optimization of New Media Platform for Ideological and Political Education of College Students.. *Higher Education Forum*, (12), 91-93

Weisheng, Z. (2022). Key Problems and Solutions of Music Works in Online Live Broadcasting. *Friends of Editors*, (05), 83-87. doi:10.13786/j.cnki.cn14-1066/g2.2022.5.013

Wen, C. & Kaiyan, Z. (2021). Research on the impact of online live broadcasting on consumers' purchase intention. *Communication and copyright*, (07), 55-58. doi:. 45-1390/g2. 2021.07.019. doi:10.16852/j.cnki

Wen, Z. & Xuguang, C. (2019). Research on College Students' motivation and behavior of watching webcast. *Youth exploration*, (02), 78-86. doi:. issn1004-3780.2019.02.008. doi:10.13583/j.cnki

Wortley, D. J., & Lai, P. C. (2017) The Impact of Disruptive Enabling Technologies on Creative Education, *3rd International Conference on Creative Education.*

Wu, J., Yang, C., & Wenbing, X. (2020). Research on College Students' daily reasonable consumption. *Guangxi quality supervision guide*, (01), 220-221

Xiaoya, A. I. (2018). Development status and prospective research of webcast platform in China. *New media research*, 4,(4), 7-10 doi:. issn2096-0360.2018.04.003. doi:10.16604/j.cnki

Xin, L. V., & Lu, D. (2021). Analysis on business development model of webcast platform. *Marketing*, (37), 29-30

Xingli, D. (2007). Discussion on the Morbid Star Chasing of Teenagers. *Contemporary Youth Research*, (08), 70-75

Xingnan, J. (2020). Data Research and Evaluation on the Status Quo of College Students' E-commerce Platform Shopping. *North Economic and Trade Journal*, (06), 48-50

Xiong, G., Mengmeng, L., & Wan, L. (2019). Study on Influencing Factors of College Students' willingness to continue using webcast platform. *Science and technology communication,11*(07):173-176.

Xu, H., Zhao, M., Zhuang, J., Chen, W., & Yang, J. (2021). The influence of "special price economy" on College Students' rational consumption behavior *Collection of award-winning papers of the national statistical modeling competition for college students in 2021*, China Statistical Education Society, (p. 46). doi:. 2021.041766.10.26914/c.cnkihy

Yao, G., & Yao, C. (2021). Research on the Development of Rural E-Commerce under the Network Live Broadcast Mode.. *Tropical Agricultural Engineering*, *45*(05), 89–91.

Yi, F. (2021). Research on online live broadcasting e-commerce model under the background of new media marketing. *China media science and technology*, (11): 109-111 doi:.11-4653/n.2021.11.033. doi:10.19483/j.cnki

Ying, X. (2019). Reflections on the "loss" of social networking in the live webcast from the perspective of strangers.. *Journal of Chongqing University of Science and Technology (Social Science Edition)*, (04), 30-33+40. doi:10.19406/j.cnki.cqkjxyxbskb.2019.04.008

Yong, Z. (2018). The Tianping "Autonomous" situation: Contemporary representation of the construction of the relationship between live broadcasting and social interaction -- a re examination of merowitz's situation theory. International Press.

Yong'an, Z. & Tao, W. X. (2017). Profit model, profit change and driving factors of webcast platform -- an exploratory case study based on the era of happy gathering. *China Science and Technology Forum*, (12), 182-192 doi:.fstc. 2017.12.022. doi:10.13580/j.cnki

Yuan, Z. (2021). From "618" to "double 11": e-commerce festival marketing experience and Its Enlightenment. *Business economics research*, (14), 102-105

Zhang, Y. & Song, B. (2021). The influence of webcasting on the impulse buying behavior of teenagers: An empirical study based on three provinces and cities. *China Business Theory,* (24), 67-73. doi: .issn2096-0298.2021.24.067. doi:10.19699/j.cnki

Zhao, Q., Chen, C., Cheng, H., & Wang, J.-L. (2017). Determinants of Live Streamers'Continuance Broadcasting Intentions on Twitch:A Self-determination Theory Perspective.. *Telematics and Informatics*, *35*(2), 406–420. doi:10.1016/j.tele.2017.12.018

Zhao, X. & Jingli, F. (2021). The negative influence of online celebrity culture on College Students' values and its countermeasures. *Contemporary educational theory and practice*, 13 (06), 68-74. doi:.1674-5884.2021.06.012. doi:10.13582/j.cnki

ADDITIONAL READING

Al-Maweri, N. A. S., Sabri, A. Q. M., Mansoor, A. M., Obaidellah, U. H., Faizal, E. R. M., & C, J. L. P. (2017). Lai, P.C. Metadata hiding for UAV video based on digital watermarking in DWT transform. *Multimedia Tools and Applications*, *76*(15), 16239–16261. doi:10.100711042-016-3906-0

Anthony N. T. R., Rosliza A. M & Lai P. C., (2019). The Literature Review of the Governance Chen Juhong, Shu Shibin. Research on Online Live Broadcasting Marketing Mode of Publishing Industry in the "Internet plus" Era.. *Science and Technology Publishing,* (03), 85-89. doi:10.16510/j.cnki.kjycb.2018.03.012

Cohen, P. N. (1998). Black concentration effects on black-white and gender inequality: Multilevel analysis for US metropolitan areas. *Social Forces*, *77*(1), 207–229. doi:10.2307/3006015

Cooper, D. R., & Schindler, P. S. (2008). *Business research methods,s* (10th ed.). McGraw-Hill.

Hamdan, N. H., Kassim, S. H., & Lai, P. C. (2021). The COVID-19 Pandemic crisis on micro-entrepreneurs in Malaysia: Impact and mitigation approaches. *Journal of Global Business and Social Entrepreneurship*, *7*(20), 52–64.

Chapter 11
Classroom Interaction and Second Language Acquisition in the Metaverse World

Han Wang
Universiti Tun Abdul Razak, Malaysia

P. C. Lai
Universiti Tun Abdul Razak, Malaysia

ABSTRACT

Classroom interaction is one of the most commonly used teaching methodologies that can be applied to develop linguistic competencies in second language instruction. Interaction in the classroom refers to the conversation between teachers and students, as well as interactions between the students. Active participation and interaction are both critical themes and vastly invaluable concerning the learning of students and second language acquisition. This paper examines how classroom interaction helps students to learn and acquire a second language. Using the findings of previous studies as a baseline, this study reveals some factors that affect how instructors should use this teaching methodology in different situations. In addition, this paper mentions applying innovative ideas for second language acquisition with the support of new scenarios and developed applications (e.g., role-play activities in personalized settings, etc.), aimed at exploring these possibilities for language teaching.

INTRODUCTION

Interaction plays an important and constructive role in Second Language Acquisition (SLA). (Zhao, 2013) The role of interaction has long been central to the theory and study of language acquisition. For example, this is clarified by Long in his Interaction Hypothesis. Long (1985) suggests that "negotiation" is indirectly connected with acquisition: since linguistic/conversational adjustments promote the comprehension of input and comprehensible input promotes the acquisition, it can be deduced that linguistic/conversational adjustments promote acquisition. There is a large number of studies dealing with input and

DOI: 10.4018/978-1-6684-5732-0.ch011

interaction in SLA. Although there are promising results of the early research, the effect of interaction on acquisition has remained a complex issue. Many research findings focus on the technical problems about the use of classroom interaction methodology, such as how to classify students' participation turns and how to organize different types of classroom interactions. It can be seen that the relationship between interaction and SLA is hard to quantify because the interaction is a complex and dynamic process and embodies many variables. (Zhao, 2013; Anthony, Rosliza and Lai, 2019).

Traditional general foreign language classroom teaching is difficult to provide its specific teaching practice conditions, which cannot meet the needs of learners to use specialized English to deal with specific areas of things. It has become the main bottleneck for specialized English learners to learn. The development of virtual reality technology in recent years has provided a good way to break through this bottleneck

This paper will synthesize the role of classroom interaction in ESL teaching in general. However, in addition to the methodologies employed in the classroom, classroom interactional patterns also depend on some contextual, cultural and local factors. To be more specific, this paper tries to analyze the following questions: How different types of classroom interaction can be implemented in various ESL classrooms? What factors do teachers need to be concerned with to design a communicative lesson while using the classroom interaction method? For the pedagogical implications, this paper analyzes the following questions: How do ESL teachers use classroom interaction methodology? This includes many types of interactions to design and implement activities that develop language skills that help ESL learners improve their English proficiency. How do teachers remain flexible to use classroom interaction methodology?

HOW MUCH INTERACTION SHOULD BE IMPLEMENTED?

The main objective of second/ foreign language teaching is to facilitate learning by keeping in mind that learners are using language for social communication so it should be used both correctly and appropriately. This indicates that meeting the educational needs of the growing number of ESL students has become an increasingly important and complex concern for educators. (Williams, 2001) Classroom interaction has long been talked about for decades in western countries, it is still a new teaching model in many other areas, especially in developing countries. In traditional English language teaching classes, the teachers mainly focus on reading, writing, and grammar, which results in students who have a hard time successfully becoming competent and effective communicators. (Naheed, 2005) In recent years, more and more instructors are encouraged to change their traditional literature teaching and implement classroom interaction into their ESL teaching. It is a good change, however, when instructors start to use this teaching form, they may also feel confused about how to design a non-traditional lecture class using different types of classroom interaction appropriately. First of all, ESL instructors need to know what classroom interaction is in the ESL classroom setting. Classroom interaction can be the interaction between two students, or teacher and students, or among several students. As Allwright (1984) proposed, "everything that happens in the classroom happens through a process of live person-to-person interaction". Many researchers conclude that since language output and classroom interactions have been shown to have a facilitative impact on language learning, ESL/EFL teachers should stimulate learners' interests and provide as many opportunities as possible for language learners to produce the target language by implementing various classroom interaction tasks. Admittedly, using classroom interaction appropriately has an influence on students' language learning but should teachers do it as much as possible or not? Teachers need to think about many factors before implementing classroom interaction, such as the identity of the students, the personality of the students, the content that they want to teach, and the classroom environment, etc. How many interaction activities should they apply in the class? Is it that the more classroom interaction the better? There have been two case studies about the use of classroom interaction in ESL teaching classes in Iran and Japan. The results of these studies showed that implementing classroom interaction is context-based and teachers cannot use one design, syllabus, method, or curriculum for different situations. In other words, creating classroom interaction in some ways is more flexible than designing a traditional lecturing classroom. Instructors can choose different types of classroom interaction based on various teaching contexts, such as teacher-student interaction, students' pair work, group work, etc.

CLASSROOM INTERACTION AND LEARNING OUTCOME

Researchers are interested in finding out whether more participation in interaction causes greater achievement and has a positive influence on students' learning. They are attempting to find empirical support for the encouragement of more interactions in class. Ellis (2009) argued that "the interaction provides learners with opportunities to encounter input or to practice the target language. It also creates within the learners a 'state of receptivity', defined as "an active openness, a willingness to encounter the language and the culture". For instance, when the students asked the teacher questions, the interaction between the teacher and learner occurs and the resulting teacher talk can attract the learner's attention and maybe more facilitative of acquisition than students just listening to the lecture teaching class. Wang and Castro (2010) did a comparable case study with Chinese ESL students. He designed two tasks: in task A, students in the non-treatment group were asked to complete the task totally by themselves without any discussions among students or asking questions to the teachers; in task B, students in the treatment group were engaged in the classroom interaction, which involved the teachers' answering questions about the target form, and teaching the target form if the students seemed confused about its use. He found that the treatment group involved in classroom interaction and output production performed much better than the output-only group, which suggests that classroom interactions may prompt learners to notice the target form. It is good to use "may" to show the result of this case study because there is a situation that many teachers are worried about happening in class: if they ask students to work through interaction, how can teachers know if all the students have attained the knowledge in a certain exercise? Because even if there is one student in the interaction group who knows how to do the task correctly, the whole group can produce good output. This situation especially happens when the teacher asks students to finish the task, for example, choose the correct phrasal verbs to fill in the sentences; correct the grammar errors in the context, etc. Unlike the discussion of the opening questions, asking students to work in a group to finish this kind of task is a challenge for teachers to know whether all the students have understood the knowledge and can write down the right answers individually.

ACTIVE STUDENTS VS. QUIET STUDENTS

One point that researchers also care about is students' participation. Do the teachers care more about the students who participate in classroom interaction actively? Do the students who participate actively in classroom interaction have better language achievement than those who are relatively quieter?

Researchers set out to investigate and define the relationship between interaction and learning outcomes in the classroom. (Zhao, 2013) There are some researchers such as Seliger think that "learners who initiate interaction are better able to turn input into intake" through the analysis of student participation patterns. There are also researchers like Doughty and Pica who deem that it cannot be drawn that "learners who actively negotiate for meaning actually achieve more, linguistically speaking, than those who do not". Besides that, as Zhao mentioned, from teachers' common observations, active students no matter in whole class work or in group work are not necessarily better achievers in the interaction process. It is difficult to judge whether a student is learning or not, simply by their explicit performance in the interaction. There is no evidence to prove that quiet students are not learning, they can learn through listening or through acquiring different ideas. There is also no evidence to prove that active students can get all the knowledge that was taught by teachers. As ESL instructors, it is unprofessional to easily judge students through one aspect.

SLA IN THE VIRTUAL ENVIRONMENT

Today, more and more facts have proved that artificial intelligence technology will be widely used in language learning and will produce unexpected results. Relevant experts pointed out that artificial intelligence technology has unlimited potential, and the application prospect in the field of linguistics is very broad, and it will become an important driving force for a new round of scientific and technological revolution and industrial transformation. To study the usability of high-tech technologies in language learning, we should sift through the current literature and its comprehensive definitions, working principles, related applications, and concepts. With the time and demands it brings, the education system is constantly in need of renewal. The current generation is a generation bombarded with high-tech information technology. As a result, adapting to new technologies is simpler for this generation as they have focused their lives on digital technology (Lee, 2000). Our students are born into a digital life and are even called "digital natives" (Prensky, 2001). Every digital technology or technological innovation, even if they are not specifically designed for education, has penetrated the classroom, which includes blogging, instant messaging, podcasting and even virtual environments. Among them, the emerging technology of virtual world AR has emerged, which continues to gain momentum in education (Atwood-Blaine &Hoffman, 2017). Some researchers mentioned that a virtual environment is designed to improve the accessibility of language learning content, and it should help and serve students. A virtual environment makes the possibilities extensive and engaging, while realistic situational teaching is relatively more limited. For example, create 3D adventures based on educational learning projects that can be adjusted based on students' demonstrated experience and previous knowledge (Agudo and Rico, 2011; Wortley and Lai, 2017), and can also create a richer learning experience for students.

The opportunity for learners to apply the loop by immersing themselves in the virtual specialized use of English. In the environment, immerse yourself in the foreign language knowledge you have learned, and truly realize the practical use of learning. PlayStation VR, HTC Vive and Oculus Rift. Some of the world's leading 3D interactive software companies have invested heavily in building virtual reality education service platforms to bring 3D virtual reality technology from the field of scientific research into the classroom. Virtual reality platforms demonstrate value beyond other teaching methods in teaching English for specialized purposes. China's special-purpose foreign language teaching has been plagued by the language environment for a long time. The use of a virtual reality platforms allows learners to experience entertainment and education at the same time, which is conducive to getting rid of the boring and tediousness of traditional special-purpose English learning and realizing the concept of edutainment. At present, the application research of virtual reality technology in the teaching of English for special purposes is very limited, and it is of great significance for virtual reality technology to better serve the teaching of English for special purposes at home and abroad, compare to the advantages and disadvantages of relevant research methods and technical routes, and analyze the application research trends of virtual reality technology in English for special purposes in the future.

CLASSROOM INTERACTION AND LEARNERS' RETICENCE

Many of the journals about classroom interaction mentioned reticence, which is a common problem faced by ESL/EFL teachers in classrooms, especially in those with mainly Asian students. The act of being silent, and reluctant to participate or speak using the target language has always been considered as the main source of frustration, and failure for both instructors and students (Zhang & Head, 2009). It is a major obstacle for students to develop oral proficiency in the English language as compared to the development of reading and listening skills (Jenkins, 2008). One reason for being reticent in class may be because of the student's personality, which is understandable. Respecting students' personalities is an important way to promote their learning enthusiasm. Teachers should not force students to talk and simply judge a student based on their performance or whether they are active or not in class.

Since the teacher is always the one who determines what and when students are going to speak in classrooms (Garton 2002; Walsh 2002), researchers believe that the teacher is one of these situational variables, and using teacher-student individual interaction appropriately may help learners reduce reticence in the classroom. In most countries, ESL/EFL classes are in a large and formal setting, where students get together for pedagogical reasons. As Gil (2002) said, in such an institutional setting, a teacher is a person institutionally invested with not only the most talking rights but also the power to control both the content and procedure, discussion topic, and who may participate. As we can see, the main form of the ESL class is teacher fronted, which indicates that it is important for teachers to think about how to conduct their lessons and how their interactions with students can influence learners' communicative behavior in the classroom. Within in the learning process, many students, especially younger ones, experience fear. Some students are fearful of being asked questions by the teachers in the classroom. This fear can stem from many points. For example, the desire to be correct, not wanting to provide the wrong answer, not wanting to appear as less intelligent than peers or simply being afraid of backlash or punishment if an incorrect response is provided to a posed question. Many students fear the anger of teachers or professors and even their parents. An example of why a student might be afraid to participate in the learning process can be seen in the following scenario: Student A is unable to correctly answer a question posed by his or her teacher, the teacher is then upset with student A. Student A is then made to stand for the duration of the class. Student A also receives subsequent punishment at home for failing to perform as deemed adequate by his or her parents. Although never personally witnessed before in

America, in many other countries, such a scenario is common and happens regularly. Fear is a negative deterrent and blockade to the learning process and can often hamper a student's enthusiasm or interest if the fear of failure and the potential punishments are deemed too great by the individual. Many students are able to overcome this or adapt to fear-based learning, but many others are demoralized by it, discouraged, and simply give up. Overall, the learning process is drastically impacted in a negative way. Given the aforementioned statements, it is clear to understand why many students are afraid of being asked questions by teachers or are fearful of speaking in class. Many teachers don't realize that they are interacting with students when they ask students questions, instead, they consider this to be knowledge check, simply wanting to check the students learning progress and validate whether or not the student can provide the correct answer or response. Teachers need to treat this "question and answer forum" as a teacher-student individual interaction, with the purpose of helping students understand what teachers have taught. It is vital that teachers and educators use every opportunity and interaction as a tool and moment to increase the overall understanding and acquisition of the target language. A student's ability to retort the correct response does not mean that they have necessarily grasped true understanding of a given concept or idea. It is only through true interactions between students and teachers that a student's ability can be accurately measured.

PEDAGOGICAL IMPLICATION

The type of classroom interaction that different instructors employ will largely depend on their own teaching philosophy and training. For example, when teaching grammar, some teachers prefer to use the grammar-translation method and teach English through the students' native language by teacher-student individual interaction. Other teachers may use a more communicative method in which grammar constructions are not overtly explained or drilled. There are some articles mentioned about Community Language Learning (CLL), which is a strategy to create classroom interaction. A CLL teacher avoids lecturing and allows students to correct and learn from each other. Some teachers advocate "the Silent Way," a strategy where the teacher says as little as possible, and the students are encouraged to "discover" the language on their own. However, there are many teachers who think "the Silent Way" is not a very effective teaching method because students cannot produce a language without enough input and exposure, especially for beginners. The second opinion holds greater value because for the learners who are at the beginning level, they cannot make any improvement just by talking among themselves and without learning from the teachers. Another reason is about getting feedback, which is a key part of classroom interaction. In order to improve, students should get feedback and correction. During accuracy exercises, teachers can correct students right away, while during fluency exercises teachers may want to simply listen and jot down any glaring mistakes. Instructors can give feedback orally or in writing. Sometimes teachers may care about students' self-esteem and do not want to correct an individual student in front of other students, while at other times it is better to offer general suggestions and corrections for the entire class without pointing out which students have made such mistakes. When giving feedback, teachers have to keep in mind the students' personality and cultural context, as some students may not be comfortable receiving individual correction in front of their peers. ESL instructors have to keep in mind that language is changing all the time and you should not simply use "right" or "wrong" to judge students' answer. As William (2001) concluded in his case study, teachers need to point out the times that they, as learners, understand something new from some of the students. Teachers could say something like, "I never thought about it that way" or "you provide us a different perspective to think about it" to show respect and create a classroom environment where teachers and students learn together.

Dr. James J. Asher states that "good language learners achieve fluency faster when they are immersed in activities that involve them in situational language use", "Good language learners often start their language learning with a period of silence as they watch the effect of language on others" and "Good language learners show comprehension by successfully accomplishing language-generated tasks" (Margaret, Barbara, and Elisabeth, 2003). It is very important that a language teacher demonstrates or models first and gives students some time to understand before having them practice. Through this process of modeling and presentation, students have the opportunity to build a foundational understanding of how the presented language is used in a given scenario. Placing students in situational learning environments, where they can use the target language, causes them to connect language with real events and circumstances, leading to a deeper understanding of the language and its appropriate usage. Carreiro and Media mentioned that instructors need to think about if the goal of the activity is fluency or accuracy before deciding on what type of classroom interaction you want to use for a particular lesson activity. Thinking about the goal of the activity provides a good starting point for teachers to design the interaction activities before the class. With a clear focus on desired outcomes, teachers can better set up lessons and interactions that will effectively align with and match the desired outcomes and goals. In fluency-oriented activities, teachers will want the students to be able to speak without much interruption. The point of fluency activities is to encourage the students to use as much language as they know in order to communicate fluidly without halting. In this case, teachers can use pair-work or group work to provide opportunities for students to practice their listening and speaking. The main concern of accuracy-oriented activities is the opposite. Instructors want students to focus on a particular point, such as grammar or vocabulary, and practice to get it correct. In accuracy exercises, fluency is not as important as pronouncing or using the target vocabulary or grammar correctly, which indicates that teachers can use teacher-student individual interaction to figure out the errors that students would make in a specific grammar or vocabulary point and help students to correct the errors as well as providing helpful feedback.

Talking about active students and quiet students in the classroom interaction, First, teachers should not have a bias toward the students who seems quieter than other students. Some researchers pointed out that if students do not participate in the discussion or verbally share ideas, instructors will usually think that they do not have the desire to learn. This would, consequently, influence the instructors' judgment when assessing the students' language performance since many classroom evaluations of learning relies on student actions and behaviors which are easily observed and measured (Lai and Lim, 2021). From my experience, many instructors care more about the active students who always have a hand-up to answer questions in the class, resulting in that there are some parents who request teachers to ask their kids more questions even if their kids do not want to talk in class because parents think that only in this way teachers can pay more attention on their children. Pointing out students randomly to answer questions is a way to check whether the students have focused on the class and can recite the knowledge, and it is a process of interaction between teachers and students. Instructors should realize that they are actually communicating with students, it should not be a reason to punish the students, or it most likely will make students stressed and even produce resistance to learning. Allowing students to make motions can strengthen students' memorization and also will strengthen students' listening. Using motions is just like using visual/audio tools (picture and sound) and can lead to students acquiring the language quicker. For example, waving when saying hello or goodbye is a common practice in western cultures. This simple gesture, when used in a learning setting, quickly allows students to understand the idea of greeting someone or leaving them. The motion of waving combined with the words "hello" or "bye" become universally understood and synonymous with this event and students quickly grasp and understand the concepts and ideas behind it.

Technology-enhanced education has played a pivotal role in student learning and development (Weisberg, 2011; Worthley and Lai, 2017; Lai, 2016: 2018; Anthony, Rosliza and Lai, 2019. Lai and Liew, 2021; Lai and Tong, 2022). Molka-Danielsen and Deutschmann (2009) as well as Worthley and Lai (2017) propose several ways of teaching in Metaverse world based on social constructivism, active learning and action learning. Teaching in virtual environment, teachers need to face great challenges when preparing and implementing teaching activities. Deutschmann et al. (2009) analyse teacher practices in this virtual environment by considering three main concerns: preparatory issues, task design and the teacher's role in fostering learner autonomy [Deutschmann et al., (2009), p.27]. Role-play-based hands-on activities are one of the better ways to teach online, where students can apply and consolidate knowledge previously acquired in the course. In language courses, each course level is divided into units, and each unit covers a lot of content. Teachers aim to have role-playing activities in each unit as a practical assessment of what has been learned. For students who are quieter in real-world classrooms, this style of teaching may also inspire them to express themselves.

CONCLUSION

Through the review of literature and studies focused on classroom interaction as well as personal experience and professional observation, one can easily arrive at the realization that classroom interaction is more than just an occasion for letting students speak. It is far more important than the simple repetition of words and requires a multi-pronged approach in order to truly be effective in an ESL learning environment. It needs to be conceptualized and structured to support learning opportunities for developing the various components of English learning. When using teacher-student interaction, it is important that teachers guide learners systematically and patiently, introducing activities that are integrated and sequenced and allow students to raise their awareness of the knowledge, skills and strategies needed for various types of interaction and discourse. There are also many factors that instructors need to be concerned with when using different types of classroom interaction, such as the students' personality and identity that were discussed and highlighted in this paper. In the metaverse world, from a technical standpoint, we looked at creating immersive and interactive role-playing scenarios that language learners can engage in, implementing automated avatars capable of taking virtual tours, leading students who are quiet in real-world classrooms to dare to participate, which is also the way to present classroom interaction. Virtual world learning is a very active research direction today. Research into the application of virtual reality technology in English for Specialized Purposes is still in its infancy, with a lot of potential development space. The extension of virtual reality technology in the field of special-purpose foreign language education will greatly compensate for the lack of contextual limitations in foreign language teaching. Through the analysis and collation of related research work, it can be seen that at this stage, the integration of international mainstream virtual reality technology and foreign language teaching is still based on the design and development of three-dimensional learning software, and the development of learning tools for limited and qualitative research-based evaluation and testing, although the results of evaluation and testing have verified the positive role of virtual reality technology in foreign language teaching to a certain extent, but there is still a lack of verification work based on big data support and interactive design improvement of virtual reality foreign language learning platform.

REFERENCES

Agudo, J. E. & Rico, M. (2011). Language learning resources and developments in the Second Life metaverse. *International Journal of Technology Enhanced Learning 3*(5):496-509

Allwright, R. (1984). The importance of interaction in classroom language learning. *Applied Linguistics*, *5*(2), 156–171. doi:10.1093/applin/5.2.156

Anthony, N. T. R., Rosliza, A. M., & Lai, P. C. (2019). The Literature Review of the Governance Frameworks in Health System. *Journal of Public Administration and Governance*, *9*(3), 252–260. doi:10.5296/jpag.v9i3.15535

Atwood-Blaine, D., & Huffman, D. (2017). *Mobile Gaming and Student Interactions in a Science Center: The Future of Gaming in Science Education,* (Vol. 15). Science and Mathematics Education.

Deutschmann, M., Panichi, L., & Molka-Danielsen, J. (2009). Designing oral participation in second life – a comparative study of two language proficiency courses. *ReCALL*, *21*(2), 206–226. doi:10.1017/S0958344009000196

Ellis, R. (2009). *The Study of Second Language Acquisition,* (2nd ed.). Oxford University Press.

Enright, D. S., & McCloskey, M. L. (1985). Yes, Talking!: Organizing the Classroom to Promote Second Language Acquisition. *TESOL Quarterly, 19*(3), 431–453. http://doi.org.proxy.seattleu.edu/10.2307/3586272Gass.

Garton, S. (2002). Learner initiative in the language classroom. *ELT Journal*, *56*(1), 47–55. doi:10.1093/elt/56.1.47

Gil, G. (2002). Two complementary modes of foreign language classroom interaction. *ELT Journal*, *56*(3), 273–279. doi:10.1093/elt/56.3.273

Jenkins, J. R. (2008). Taiwanese Private University EFL Students' Reticence in Speaking Language. *Taiwan Journal of TESOL*, *5*(1), 61–93.

Lai, P. C. (2016). *SMART Healthcare @ the palm of your hand.* ResearchAsia.

Lai, P. C. (2018). Research, Innovation and Development Strategic Planning for Intellectual Property Management. *Economic Alternatives.*, *12*(3), 303–310.

Lai, P. C., & Liew, E. J. (2021). Towards a Cashless Society: The Effects of Perceived Convenience and Security on Gamified Mobile Payment Platform Adoption. *AJIS. Australasian Journal of Information Systems*, *25*. doi:10.3127/ajis.v25i0.2809

Lai, P. C., & Lim, C. S. (2019).. . *The Effects of Efficiency, Design and Enjoyment on Single Platform E-payment Research in Business and Management*, *6*(2), 19–34.

Lai, P. C., & Tong, D. L. (2022). *An Artificial Intelligence-Based Approach to Model User Behavior on the Adoption of E-Payment. Handbook of Research on Social Impacts of E-Payment and Blockchain.* IGI Global.

Lee, K. W. (2000). English Teachers' Barriers to the Use of Computer-Assisted Language Learning. *The Internet TESL Journal, 6*(12).

Margaret, S., Barbara, A., & Elisabeth, P. (2003). *Total Physical Response*. TPR.

Naheed F. (2005) Interactive Approach in English Language Learning and Its Impact on Developing Language Skills, 94-101.

Prensky, M. (2001). Digital Natives, Digital Immigrants. *On the Horizon*, *9*(5), 1–6. doi:10.1108/10748120110424816

Walsh, S. (2002). 'Construction or obstruction: Teacher talk and learner involvement in the EFL classroom'. *Language Teaching Research*, *6*(1), 3–23. doi:10.1191/1362168802lr095oa

Wang, Q. Y., & Castro, C. D. (2010, June). Class Interaction and Language Output. *English Language Teaching*, *3*(2). doi:10.5539/elt.v3n2p175

Weisberg, M. (2011). Student attitudes and behaviors towards digital textbooks. *Publishing Research Quarterly*, *27*(2), 188–196. doi:10.100712109-011-9217-4

Williams, J. A. (2001). Classroom conversations: Opportunities to learn for ESL students in mainstream classrooms. *The Reading Teacher*, *54*(8), 750–757. http://login.proxy.seattleu.edu/login?url=http://search.proquest.com/docview/203275419?accountid=28598

Wortley, D. J., & Lai, P. C. (2017) The Impact of Disruptive Enabling Technologies on Creative Education. *3rd International Conference on Creative Education*, Mar 3-4

Zhang X. Q., Head K. (2009) Dealing with learner reticence in the speaking class. *ELT Journal*, *64*(1). doi:10.1093/elt/ccp0181

Zhao, C. (2012, November 17). Classroom Interaction and Second Language Acquisition: *The more interactions the better?* doi:10.3968/j.sll.1923156320130701.3085

Chapter 12
Adoption of Metaverse in South East Asia:
Vietnam, Indonesia, Malaysia

Hock Leong Chan Julian
HELP University, Malaysia

Tinfah Chung
https://orcid.org/0000-0002-3993-6637
HELP University, Malaysia

Yuyang Wang
Chinese Academy of Sciences, Beijing, China

ABSTRACT

Despite massive support for blockchain startups, countries in Southeast Asia have yet to fully launch themselves into metaverse. The appeal of playing lucrative crypto games such as Axie Infinity, where active players can earn up to US$15 per day, may provide these countries the incentive to kickstart many of the business opportunities in metaverse gaming. Metaverse is a potential hype. The international audience has started riding on the back of metaverse's success. Countries in Southeast Asia (SEA) are at with their respective development within the metaverse space. It is important to recognise the level of development of metaverse in these countries. Among those who felt negative about the Metaverse, the most common reason was their concerns about data security and privacy. This research intends to understand the development of metaverse in countries in SEA – Vietnam, Indonesia and Malaysia. Through understanding the development of metaverse in these countries, these countries can properly position themselves to reap the most benefit out of this hype.

DOI: 10.4018/978-1-6684-5732-0.ch012

BACKGROUND

The word "Metaverse" is a combination of two words – meta and universe. Metaverse refers to both current and future integration of digital platforms focused on virtual and augmented reality. It is widely hyped as the next frontier and seen as a potential business and financial opportunity for the tech industry and other sectors. It was first used in literature by Neal Stephenson in his 1992 dystopian novel *Snow Crash*. In the book, the Metaverse is presented as the ultimate evolution of the internet — a kind of virtual reality where any virtual interaction can have a direct impact on the real world too.

The book pretty much sums up what the Metaverse is. It is a physically persistent virtual space where there are virtual avatars, digital social interactions, and gaming among many unique things that we associate with Metaverse today. *Snow Crash* also underlines how the Metaverse in the story affects developments in the real world of the protagonist, including a conspiracy that turns people, whose brains are connected to the virtual world, insane. Since the release of the book, several other books, films, and television shows have dabbled with the concept to varying degrees, including Steven Spielberg's well-appreciated movie *Ready Player One* (2018), which was adapted from Ernest Cline's 2011 novel of the same name. The common theme in all, is that the Metaverse is a virtual reality wherein, depending on the advancement of the era, people will be able to do everything they do in real life.

Innovative technologies and services either succeed or fail depending on the user's hype flow. The attention of users, media, and researchers are important as it can act as a benchmark on amount of attention being poured towards a certain topic. In the past year, people are warming up to virtual worlds like Decentraland and Sandbox (www.sandbox.game) (Sandbox Team). Household names like Samsung and Nike have set up shop in such spaces. The Ultimate Fighting Championship (UFC) is also planning virtual fights, and K-pop labels are promoting virtual stars. As with any innovation, skepticism quickly emerges. When it comes to the Metaverse, although the idea is fledging accepted, there is still a long way to go. Therefore, user-centric factors and acceptance of the technology also plays an important role in whether the Metaverse hype can be sustain or otherwise. The six user-centric factors are Avatar, Content Creation, Virtual Economy, Social Acceptability, Security and Privacy, and Trust and Accountability.

As with any new technology, the enabler is also a factor of whether a technology can be sustained or fall. The eight-enablers of Metaverse are extended reality, user interactivity, (human-computer interaction), artificial intelligence, blockchain, computer vision, iot and robotics, edge and cloud computing, and future mobile networks.

Despite massive support for blockchain start-ups through initiatives such as the Blockchain Innovation Initiative, developing Southeast Asia has yet to fully launch themselves into Metaverse. Central Banks in the region have issued guidelines to prevent cryptocurrency exchanges from advertising to the public. Although investors are still free to buy cryptocurrencies, however, central banks in these countries discourage their use for speculative trading. The appeal of playing lucrative crypto games such as Axie Infinity, where active players can earn up to US$15 per day, may provide these countries the incentive to kickstart many of the business opportunity in Metaverse gaming (Kan, https://www.bnbchain.org/en/blog/sustainable-gamefi-to-play-or-to-earn/).

Statement of the Problem

Metaverse is a potential hype where users or creators alike are able to reap benefits from it, both socially and economically. The international audience has started riding on the back of Metaverse's success.

But it also begs to see where developing countries in Southeast Asia (SEA) are at with their respective development within the Metaverse space.

It is important to recognise the level of development of Metaverse in these countries as it may use this information to properly position themselves to reap the most benefit out of the hype. Among those who felt negative about the Metaverse, the most common reason was their concerns about data security and privacy. This response may reveal a lack of understanding rather than genuine concern (Seo, Kim, Park, Park, & Lee, 2018). In fact, data security and privacy are two key arguments in favour of blockchain technology (Dinh & Thai, 2018). Decentralization ensures that transactions can take place without users revealing personal information or storing it on company servers.

Research Aims and Objective

This research intends to understand the development of Metaverse in Southeast Asian (SEA) countries – Vietnam, Indonesia and Malaysia. Through understanding the development of Metaverse in these countries, these countries can properly position themselves to reap the most benefit out of this hype. It also has the potential for these countries to become a key player globally if it is found that the country does actually have competitive advantage in this space and necessary development incentives should be rolled out aptly.

Significance of the Study

This research will contribute towards a new segment of Metaverse research where little to no literatures has been found. Most literatures on Metaverse are on defining what is Metaverse and identifying the prospect of Metaverse. A survey of content analysis in articles from newsclip, periodicals and magazines will be undertaken here.

In 2026, it is estimated that 25% of people worldwide will spend at least one hour a day in the metaverse for digital activities including work, shopping, education, social interaction or entertainment. Additionally, almost a third of global businesses are projected to have products and services ready of the metaverse by then.

This research is important because it aims to identify the current state of Metaverse in the Southeast Asian countries where development in this arena is still lagging compared to the United States of America (USA) or China. By properly identifying the current status, necessary policies can be introduced to properly catapult this industry into a lucrative input for the country.

In presenting this status report and roadmap for advancement, attention will be specifically directed to the following four features that are considered central components of a viable Metaverse.

(1) Realism. Is the virtual space sufficiently realistic to enable users to feel psychologically and emotionally immersed in the alternative realm?
(2) Ubiquity. Are the virtual spaces that comprise the Metaverse accessible through all existing digital devices (from desktops to tablets to mobile devices), and do the user's virtual identities or collective personal remain intact throughout transitions within the Metaverse?
(3) Interoperability. Do the virtual spaces employ standards such that (a) digital assets used in the reconstruction or rending of virtual environments remain interchangeable across specific imple-

mentations and (b) users can move seamlessly between locations without interruption in their immersive experience?

(4) Scalability. Does the server architecture deliver sufficient power to enable massive numbers of users to occupy the Metaverse without compromising the efficiency of the system and the experience of the users?

In order to provide context for considering the present state and potential future of 3D virtual spaces, the article begins by presenting the historical development of virtual worlds and conceptions of the Metaverse. This history incorporates literary and gaming precursors to virtual world development as well as direct advances in virtual world technology, because these literary and gaming developments often preceded and significantly influenced later achievements in virtual world technology. Thus, they are most accurately treated as important elements in the technical development of the 3D spaces rather than as unrelated cultural events.

Figure 1. Metaverse potential market opportunity worldwide 2021, by scenario (in trillion U.S. dollars)
Source(s): Goldman Sachs; World Bank; United Nations (https://www.statista.com/statistics/1286718/metaverse-market-opportunity-by-scenario/)

Figure 2. Estimated metaverse use case among consumers and businesses worldwide in 2026
Source(s): Gartner (https://www.statista.com/statistics/1290160/projected-metaverse-use-reach-global-consumers-businesses/)

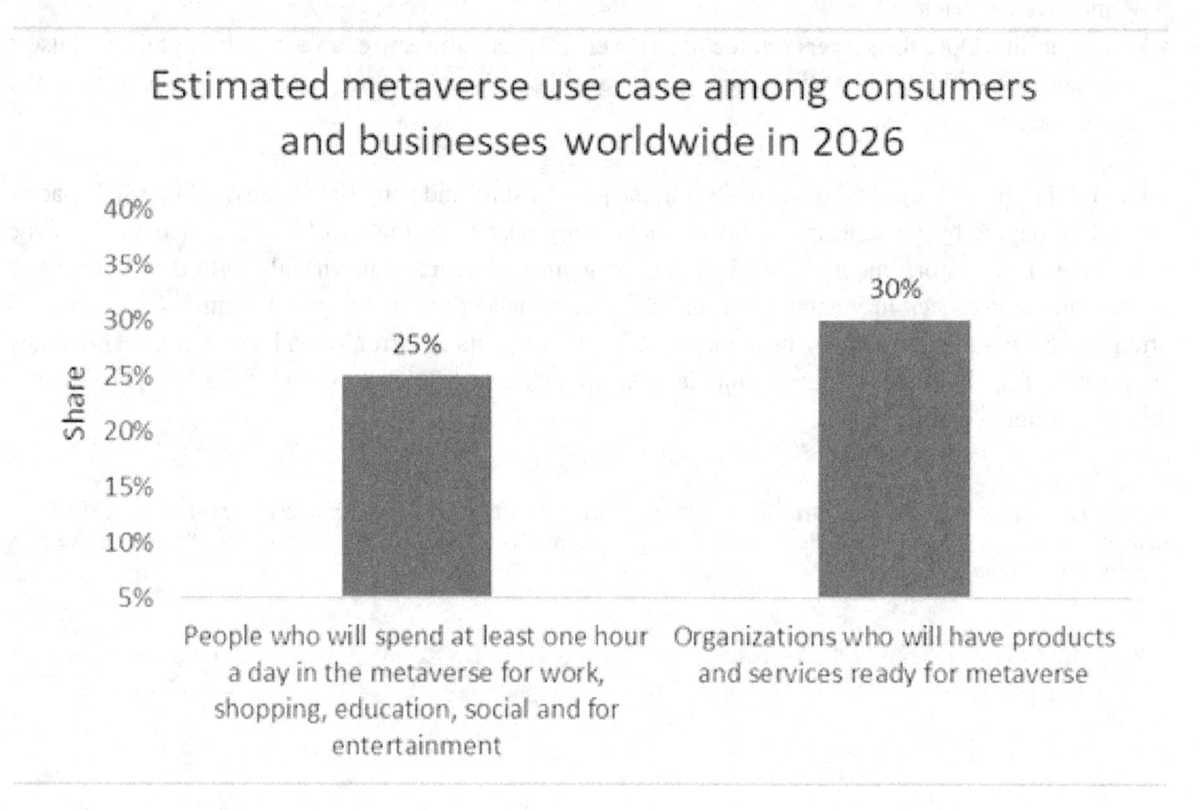

LITERATURE REVIEW

9 out of 10 articles simply define Metaverse and a combination of "Meta" and "Verse" (universe). But what does it really mean? The taxonomy of this term was thoroughly investigated by Park & Kim, A Metaverse: Taxonomy, Components, Applications, and Open Challenges, (2021). The research has elaborated the term Metaverse extensively whereby it is a transcendence of meta and universe. It encapsulates the idea of a three-dimensional virtual world where interactions are free to occur and the avatars and characters in the world are engaged in social, political, economic and cultural activities.

Being an interconnected web of social where it supports multiusers, this technology provides a foundation for seamless way to communicate in real-time (Mystakidis, 2022). Metaverse is also includes multiple somatosensory enabling digital and electronic devices. Among them, the concepts such as Augmented Reality (AR), Virtual Reality (VR) and Cross Reality (XR) are often discussed alongside Metaverse (Lee, et al., 2021). These technology that enables Metaverse have also developed extensively following the hype around Metaverse. For example, XR system enabled by stereoscopic displays are able to convey the perception of depth.

Despite the recent hype on Metaverse, Metaverse was not a new concept. It dates back as early as 1920s, there the first analogue precursor to VR, the Link Trainer, was used to train large cohorts of military plane pilots. In 1968, a mechanical AR heads-up display developed by Ivan Sutherland was introduced, though the term "Augmented Reality" was only coined later by Myron Krueger in 1980s.

Fast forward to the 20th century, electronic devices such as headsets or head-mounted displays starts to garner attention. Consumer-grade wireless, stand-alone VR headsets such as Microsoft HoloLens, Magic Leap and Smart Glasses have also emerged and available for consumers.

Virtual Reality is a key component under the umbrella of Metaverse. Schuemie, Van Der Straaten, Krijn, & Van Der Mast, (2001) have conducted empirical research on the causes of **presence**, although sometimes not consistent, has provided insights on factors that are important for creating presence. To create the feeling of presence while overcoming geospatial barriers, VR technology have come to fill this gap. VR have the ability to create vividness, interactivity and user characteristics, which are supported by empirical evidence as presented in Schuemie et al (2001). Another enabling technology is AR. In 1985, Myron Krueger studies Human-computer Interaction (HCI), which laid the foundations of Augmented Reality (AR) (Krueger, Gionfriddo, & Katrin, 1985). Multiple experiments in alternate modes of HCI was done and found that the premise is that interaction is a central, not peripheral issue in computer science. Among the applications of this technology found was interface described is a deliberately informal one. AR can become a medium of interaction for users. However, it was described as "unnecessary" due to the practicality. It soon proved wrong where the idea of holographic images has been used in movies such as Star Trek, Star Wars and in the Marvel Cinematic Universe (MCU). Holograms acts to serve as a visual simulant.

To measure the adoption of Metaverse, Lee J., A Study on Metaverse Hype for Sustainable Growth, (2021) has used hype flow to identify the probability of the success or failure of innovative technologies. Hype flow analyses attention of users, media and researchers' involvement in a technology. The result yields that there is indeed significant hype about Metaverse amongst Korean. While there are many reasons explaining the adoption (or the lack of) Metaverse, application of Metaverse is also identified to study the feasibility and "usefulness" of Metaverse. Research has forecasted that Metaverse will be applied to Smart City, Entertainment and Game Industry, Remote Office, virtual meeting, digital sightseeing, digital tourism, digital exhibition, psychotherapy, education, economy, culture and social (Ning, et al.).

METAVERSE

Regarded as the next iteration of the internet, the metaverse is where the physical and digital worlds come together. As an evolution of social technologies, the metaverse allows digital representations of people, avatars to interact with each other in a variety of settings. Whether it be at work, in an office, going to concerts or sports events, or even trying on clothes, the metaverse provides a space for endless, interconnected virtual communities using virtual reality (VR) headsets, AR glasses, smartphone apps, or other devices.

Figure 3. Extended reality (XR) headset shipments worldwide from 2016 to 2025 (in millions)
Source(s): Gartner (https://www.statista.com/statistics/1290160/projected-metaverse-use-reach-global-consumers-businesses/)

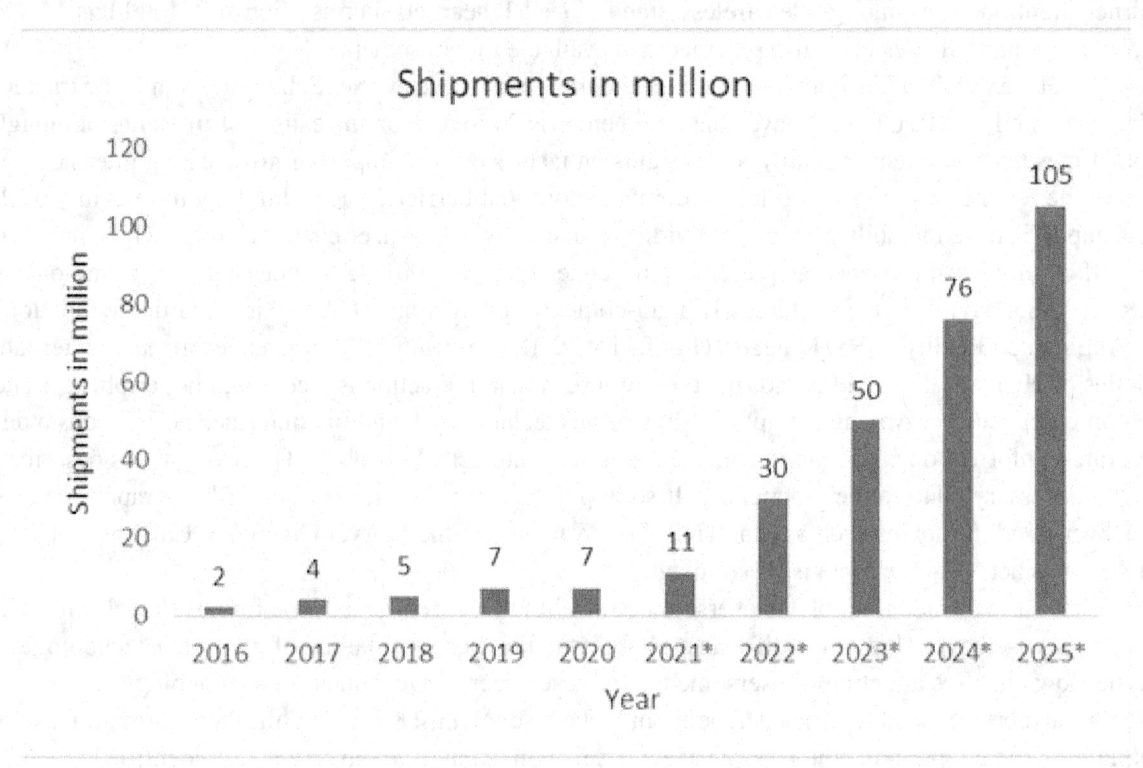

Figure 4. Investment in augmented and virtual reality (AR/VR) technology worldwide in 2024, by use case (in billion U.S. dollars)
Source(s): IDC (https://www.statista.com/statistics/1098345/worldwide-ar-vr-investment-use-case/)

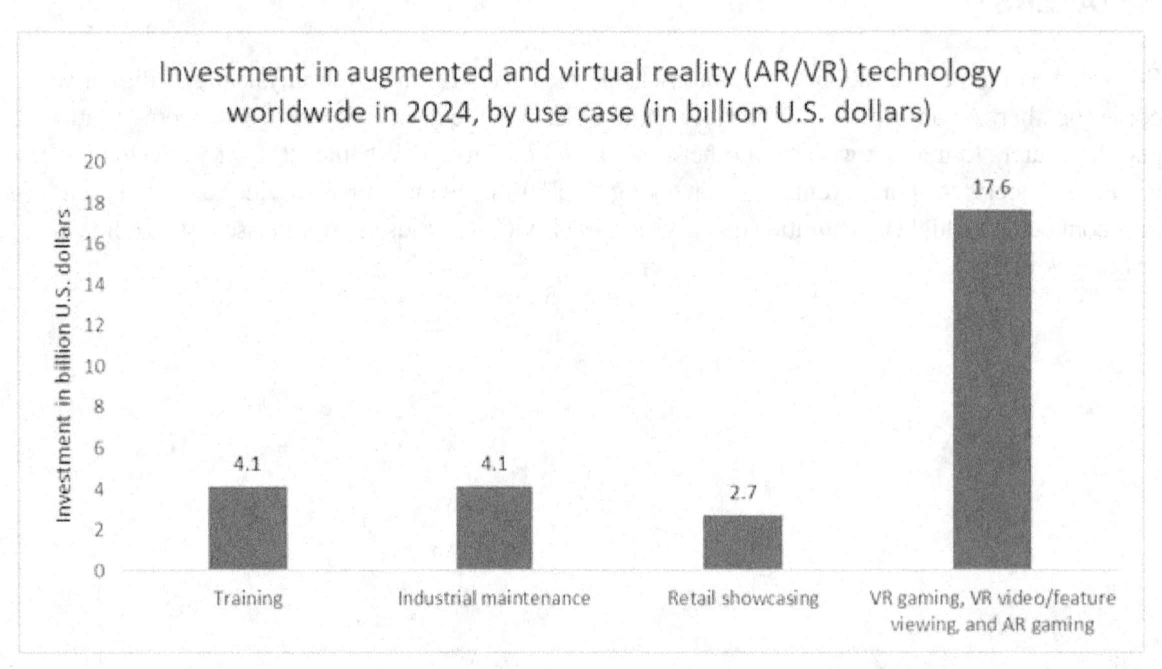

Online Game Makers, Social Networks vie for Metaverse Leadership

Online game makers and existing social networks may vie for leadership of the burgeoning US$800 billion Metaverse economy, on the convergence of megatrends of games, social and user-generated content. Facebook's user scale and VR investments could give it an edge as the market develops, while game engine vendors Unity and Epic may see heightened software demand.

Roblox, Epic Leading but Virtual Worlds Aplenty

Roblox (www.roblox.com), Microsoft's Minecraft and Epic Games' Fortnite (epicgames.com/fortnite) appear to be early leaders in the race for Metaverse leadership but there's ample time for other game makers and social networking companies to tweak existing services or launch new ones to capitalize on the market's growth (Ball, 2021). Other game makers have been able to attract large, active user bases in online titles such as Activision's Call of Duty Warzone and World of Warcraft, EA's the Sims, Take-Two's GTA Online and Nexon's MapleStory and Dungeon&Fighter Online. These companies could seek to add additional social features and make user-generated content to become a larger part of their experiences to capture Metaverse demand.

Figure 5.

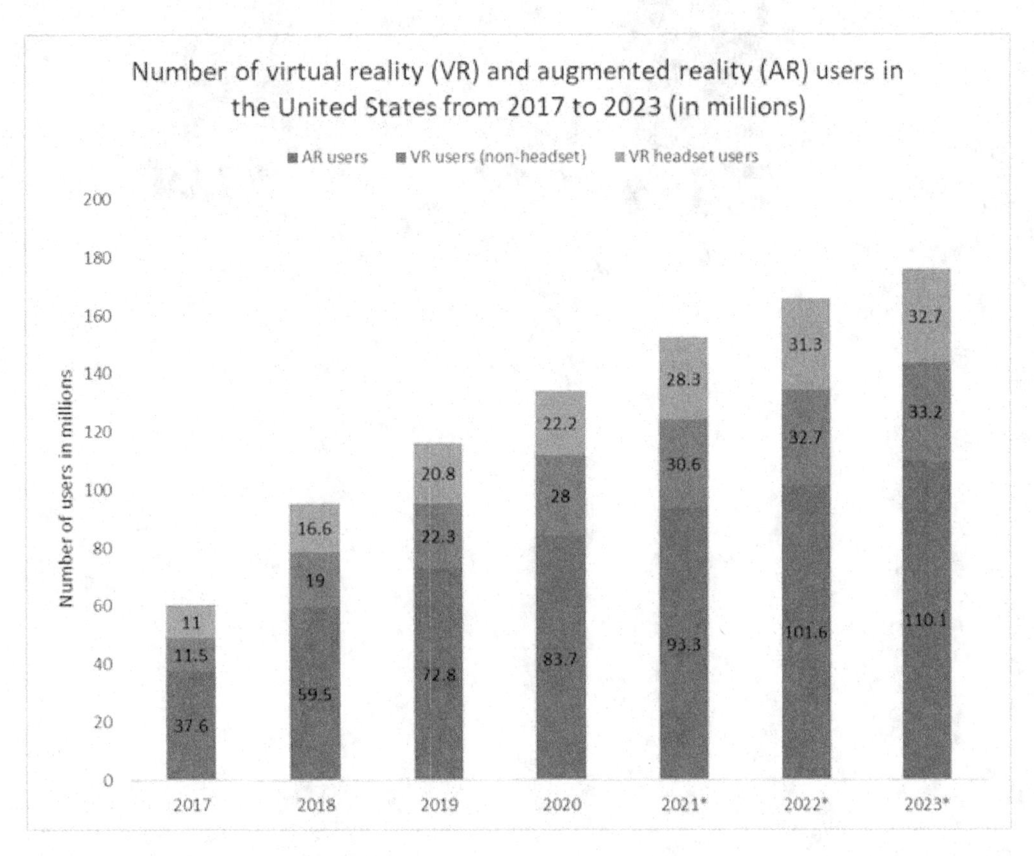

Games that successfully pivot towards virtual 3D worlds can capture a greater share of engagement and user growth, accelerating sales growth.

Figure 6. Share of gamers in the United States who have participated in non-gaming activities or events within video games in the last 12 months in 2021
Source(s): Activate (https://www.statista.com/statistics/1275473/us-gamers-participating-non-gaming-activities

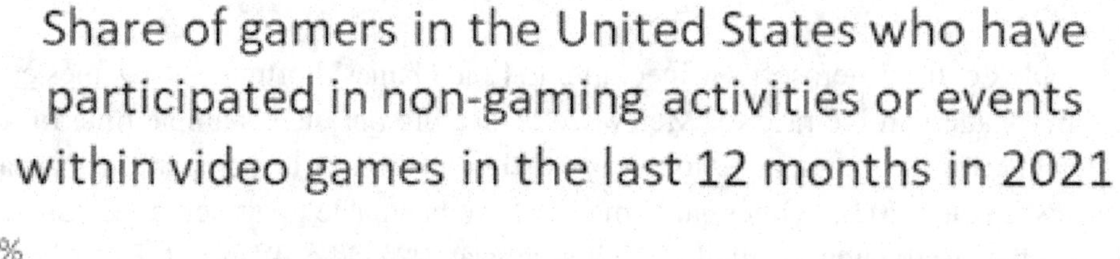

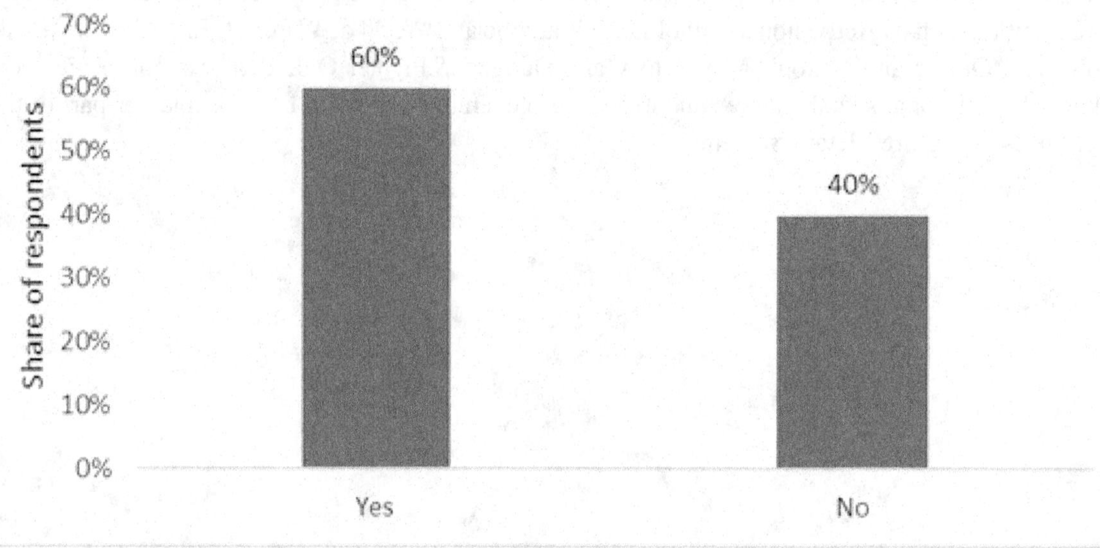

Figure 7. Share of gamers in the United States who have participated in select non-gaming activities or events within video games in the last 12 months in 2021
Source(s): Activate (https://www.statista.com/statistics/1275467/us-gamers-participating-select-non-gaming-activities-within-video-games/)

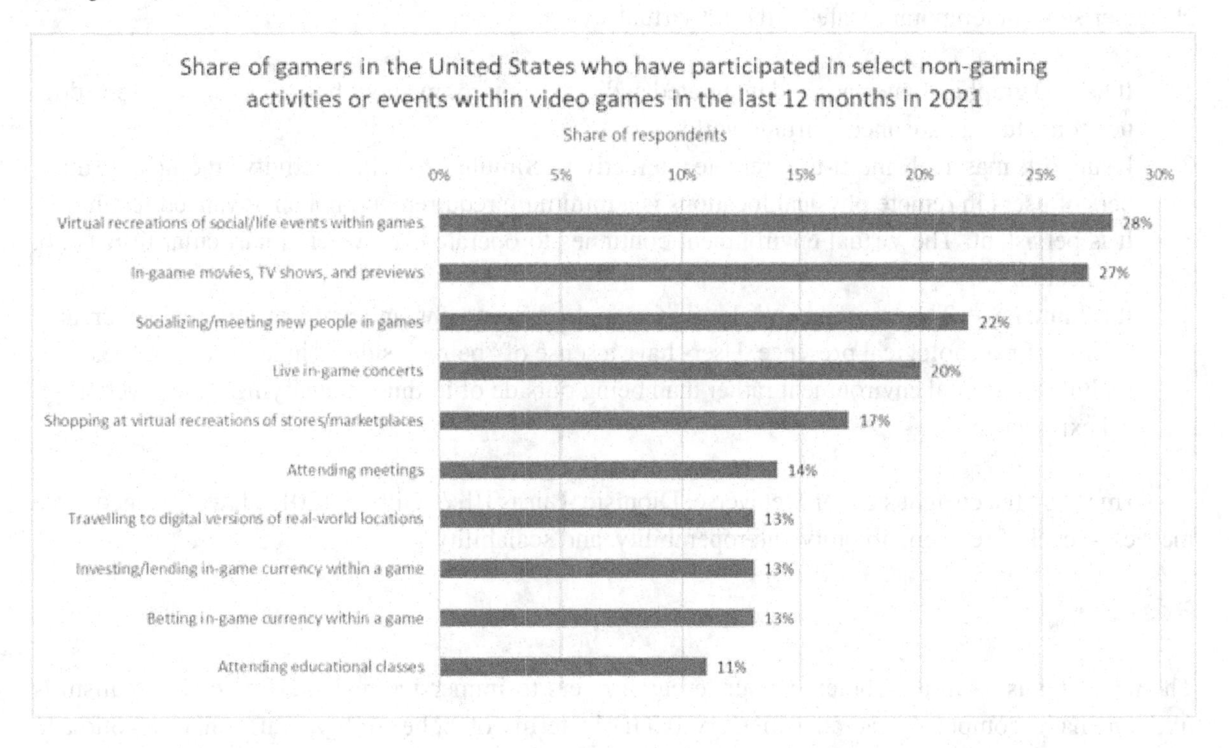

Today, fostering meaningful job opportunities for the masses remains as popular as ever with policymakers and government policies. However, different from the traditional job market where there is a top-down hierarchy, Metaverse is not. The Metaverse comprises an interconnected series of virtual worlds in which humankind can recreate, interact and transact.

Given the amorphous nature of the Metaverse, it can be hard to envisage what a virtual world in which millions clock in and out to earn their crust might resemble. As it happens, though, there is already work being performed in fledging Metaverses. In the play-to-earn (P2E), also known as GameFi sector, virtual pet roams freely (Austria, De Jesus, Marcelo, Ocampo, & Tibudan, 2022). With their respective Metaverses, players can collect tokens and other in-game assets that spawn and trade them for fiat money. This is also unique where the NFTs of these items are inherently valuable due to its immutability. Workers from developing countries such as the Philippines can earn up to $30 per day for performing tasks on behalf of their owners such as walking their pet. It is a simple economy where all the participants benefit commensurate with their interests and financial expectations.

For users who have a large following, or real-life celebrities venturing into this space, much of it will involve marketing and event planning in real-life too. Coders, designers, testers will also be employed to for frontend and backend support.

CURRENT STATE OF METAVERSE

Defining "Metaverse", Dionisio, Burns III, & Gilbert, (2013) has identified five essential features that characterise a contemporary state-of-the-art virtual world.

1. It has 3D graphical interface and integrated audio. An environment with a text interface alone does not constitute an advanced virtual world.
2. It supports massively multi-user remote interactivity. Simultaneous interactivity among large numbers of users in remote physical locations is a minimum requirement, not an advanced feature.
3. It is persistent. The virtual environment continues to operate even when a particular user is not connected.
4. It is immersive. The environment's level of spatial, environment, and multisensory realism created a sense of psychological presence. Users have a sense of "being inside", "inhabiting", or "residing within" the digital environment rather than being outside of it, thus intensifying their psychological experience.

To measure the current state of Metaverse, Dionisio, Burns III, & Gilbert, (2013) have engaged three metrics – level of realism, ubiquity, interoperability, and scalability.

Realism

The term "realism" in the context of our research refers to immersive realism. Immersive realism is like cinematic computer-generated imagery (CGI) in terms of believability rather than devotion to detail (though a sufficient degree of detail is expected). Realism in the Metaverse in sought to produce psychological and emotional user's engagement in the virtual environment. A virtual environment is perceived to be more realistic based on the level it transports a user into an environment and on the transparency of the boundary between the user's physical actions and those of his or her avatar. The virtual world realism is neither purely additive nor, visually speaking, strictly photographic. In many cases, strategic rendering choices can yield better returns than merely adding polygons, pixels, objects, or bits generally across the board.

Across all perspectives on realism, instrumentation through which human beings interact with the environment, through their senses and their bodies remains constant, particularly through their face and hand. This realism is approach through perception and expression.

In this research, we treat the subject at a historical and survey level, examining current and future trends. Glencross, Chalmers, Lin, Otaduy, & Gutierrez have written extensively on the (perceived) realism of virtual worlds.

Ubiquity

The idea of ubiquity in the virtual worlds are derived from the prime criterion that a fully realized Metaverse must provide a milieu for human culture and interaction seamlessly as with the physical world. This notion propels a psychologically compelling prospect for a user.

The default setting, the real world, is the benchmark of ubiquity. We unavoidably dwell in, move around, and interact in the space we call "real world" all the time. Next our presence in the real world, under normal circumstances, is universally recognisable. The features we individually hold, is unique (face, body, voice, fingerprints, retina) but augmented by a small universally applicable set of artifacts, such as our signature, key documents (birth certificates, passports, licenses, etc). Our identity is further extended by what we produce and consume: books, music, or movies that we like; the food that we cook or eat; and memorabilia that we strongly associate with ourselves or our lives (Turkle, 2007).

Despite occasional fluctuation in perception of the real world, such as when we are dreaming – the real world's literal ubiquity persists regardless of what our senses say (and whether or not are we even alive). This is true also when our identities are doubted, inconclusive, forged, impersonated, or stolen, in the end, these situations are borne out of errors, deceptions, and falsehoods: there always remains a real "me" with a core package of artifacts (physical, documentary, etc) that are authoritatively represent "me", which can be proven by identifying the unique feature we individually hold.

To become an alternate medium for human interaction and activity, virtual worlds is essential to support a certain function of aspects of real-world ubiquity – presence and availability. Without one of these two functions, the virtual world will feel "unreal". If it is not seamless, such as artificial impedances, or undue inconveniences involved with identifying our individual identity and the information that we create or use within or across virtual worlds, these creates a lesser believable environment and the virtual world losses a certain degree of investment or immersion within them. We thus divide the remainder of this sections between these two aspects of ubiquity: ubiquitous availability and access, and ubiquitous persona and presence.

Interoperability

Functionality-wise, the context in the virtual world is slightly different from the mainstream idea of interoperability. Interoperability is the ability of distinct systems or platforms the exchange information or interact with each other seamlessly and, when possible, transparently. Interoperability also implies some type of consensus or convention which subsequently become standards or benchmark when formalised.

Generally, interoperability refers to the capability to operate across platforms and make use of information. In the context of virtual worlds, interoperability is viewed merely to enable the technology required to function ubiquitously. However, it is not simply limited to serve this function. Interoperability remains as a key feature of virtual worlds in its own right, because it drives investment into the Metaverse, similar to the way singular capitalized Internet is borne from a layer of standards which allows disparate heterogeneous networks and subnetworks to communicate transparently with each other.

The desired behaviour is analogous to travel and transport in the real world. As our bodies move between physical locations, our identity seamlessly transfers from point to point with no interruption of experience. Our possessions can be sent from place to place, and under normal circumstances, they do not substantially change in doing so. Thus, real-world travel has a continuity in which we and our things remain largely intact in transit. We take this continuity for granted in the real world where it is indeed a matter of course. With virtual worlds, however, this effect ranges from highly disruptive to completely non-existent.

The importance of having a single Metaverse is connected directly to the long-term endpoint of having virtual worlds offer a milieu for human sociocultural interaction that, like the physical world, is psychologically rich and compelling. Such integration immediately makes all compatible virtual world implementations, regardless of lineage, parts of a greater whole. With interoperability, especially in terms of a transferable avatar, users can finally have full access to any environment without the disruption of changing login credentials or losing one's chain of cross-cutting digital assets.

Scalability

As with the features discussed in the previous sections, virtual worlds have scalability concerns that are similar to those that exist with other systems and technologies, while also having some distinct and unique issues from the virtual world perspective. Based on this article's prime criterion that a fully realized Metaverse must provide a milieu for human culture and interaction, scalability thus be the most challenging virtual world feature of all, as the physical world is of enormous and potentially infinite scale on many levels and dimensions (Bainbridge, The Scientific Research Potential of Virtual Worlds, 2007). Three dimensions of virtual world scalability have been identified in the literature (Liu, Bowman, Adams, Hurliman, & Lake, 2010).

(a) Concurrent Users/Authors: The number of users interacting with each other at a given moment.
(b) Scene complexity: The number of objects in a particular locality and their level of detail or complexity in terms of both behaviour and appearance.
(c) User/Avatar Interaction: The type, scope and range of interactions that are possible among concurrent users (e.g., intimate conversations within a small space vs large-area crowd-scale activities such as "the wave").

In many respects, the virtual world scalability problem parallels the computer graphics rendering problem: what we see in the real world is the constantly updating result of multitudes of interactions among photons and materials governed by the laws of physics – something that computers can only approximate and never absolutely replicate. Virtual worlds add further dimensions to these interactions, with the human social factor playing a key role (as previously expressed in the first and third dimensions). Thus, it is no surprise that most of the scientific problems listed by Zhao (2011) focus on theoretical limits to virtual world modeling and computation; because in the end, what else is a virtual world but an attempt to simulate the real world in its entirety, form its physical makeup all the way to the activities of its inhabitants?

THE USE OF METAVERSE ACROSS DIFFERENT INDUSTRIES

Metaverse integrates the most advanced technologies such as 5G, cloud computing, computer vision, blockchain, artificial intelligence, etc., and has applications in numerous fields such as video games, art, and business. In the next five years, Metaverse will have an impact in the real world for the following sectors: financial services, automotive and manufacturing, real estate, education, and retail. The metaverse offers many different attractions for different types of users. A late 2021survey by Tidio of covering 1,050 internet users indicates that over half of the respondents would join the metaverse for work possibilities such as virtual workspaces and networking. 48% stated art and live entertainment as the main reason to join the metaverse, and 44% stated investment into cryptocurrency and non-fungible tokens.

Figure 8. Metaverse potential consumer expenditure total addressable market in the United States as of 2022, by segment (in billion U.S. dollars) Source(s): Morgan Stanley; PwC US; BEA; US Census Bureau; NCES
Source(s): https://www.statista.com/statistics/1288655/metaverse-consumer-expenditure-tam-united-states/)

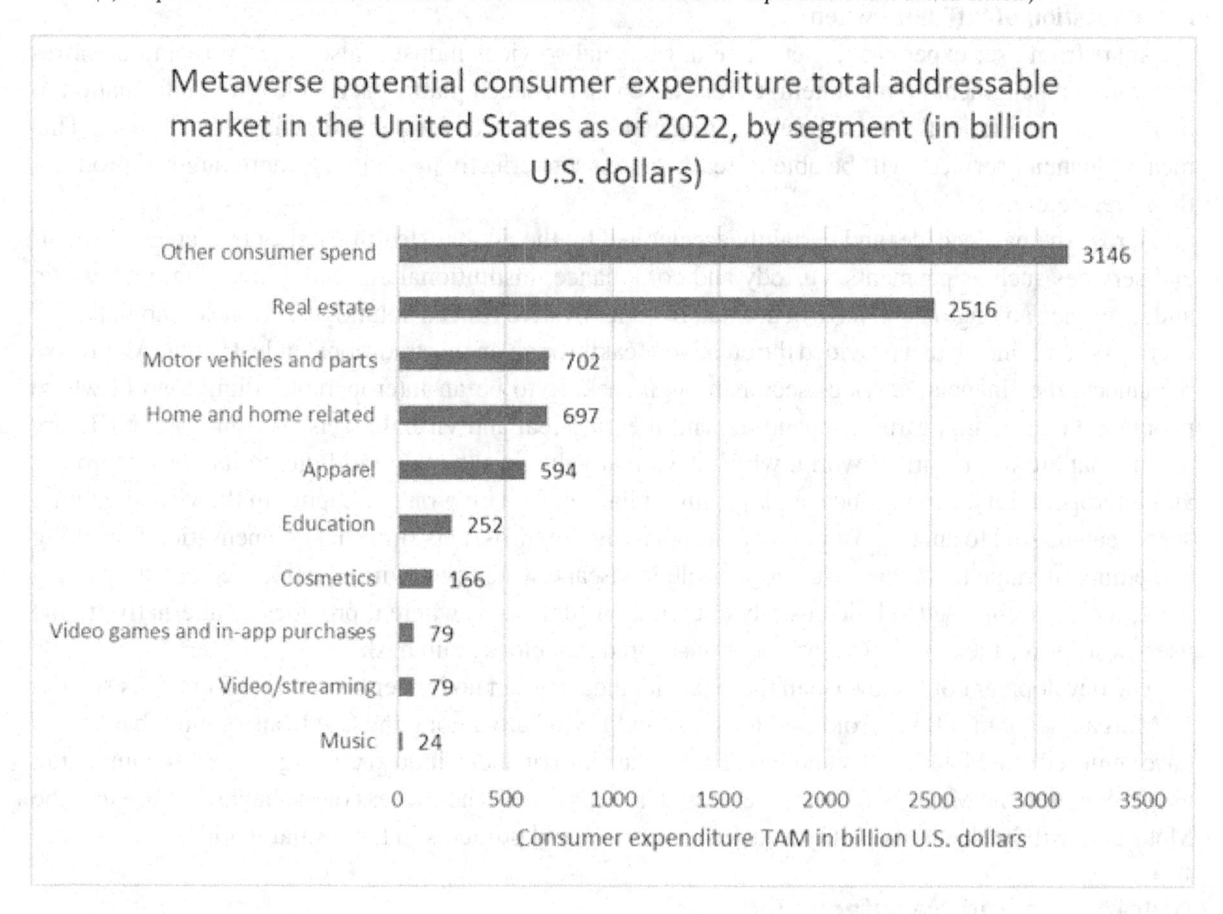

Financial Services

If the Metaverse develops like how nonfungible tokens (NFTs) have, it could be a massive growth opportunity for financial service players. NFT sales alone hit $2 billion in the first quarter of 2021, more than a 20-time increase from the previous quarter (Versprille, 2022). Looking out five years from now, we imagine a place where one has entire user-generated ecosystems, with the Metaverse giving rise to virtual societies that transact and engage in a decentralized manner. One can expect to see:

The development of **virtual-to-physical redemptions** and **financial systems that underpin payments and financing in the ecosystem NFT**s providing users digital ownership and driving the emergence of new asset classes for trading.

Augmented reality/virtual reality (AR/VR) sophistication narrowing the differences between online and offline, driving greater participation and engagement in financial services, and ultimately fueling the convergence of traditional financial services and the new era of innovation.

Traditional financial services players are already leaning into the potential of the Metaverse. **In Asia, banks are creating virtual spaces for branding, education and product development**. Universal banks are using the Metaverse for training and education for both internal stakeholders and external clients. Payment **players are aggressively developing crypto propositions and partnerships to maximize their positioning in the Metaverse**. The **integration of the Metaverse and financial services is therefore not a question of "if" but "when**."

Apart from user experience, Metaverse in financial services industry also able to streamline infrastructure, and augment human intelligence. These smart fintech platforms enable client information to be much more efficiently and robustly aggregated and analysed, further leveraging connectivity. This means financial services will be able to reach more users effectively and with more targeted products than ever before.

Across the past decade, and certainly accelerated by the Global Health Crisis, other areas of financial services such as payments, custody and compliance, institutional and retail investing, real estate, and insurance among others, have benefitted from innovative fintech solutions. It is the combination of Metaverse and fintech that provided the creative breakthrough on this traditional industry. The Metaverse reconnects the financial services sector through seeking to be an interoperable, digital world where people are conversing, earning, spending, and merging real and virtual assets. To illustrate, NFTs are tokens that are in the virtual world, which have real value in the real world due to its non-fungibility and interoperability. For instance, a player might have picked up a rare 'weapon' in the virtual gaming scene, can be sold to another user in real life. The seller in turns, gets financial compensation for selling something of value to another user. A possible landscape whereby a 'virtual bank', regulated by a real bank, will be recognised to hold custody of certain virtual items, where it provides an alternative to the users who stores their NFTs in a wallet or unsecured in a blockchain hash.

The development of Metaverse in the financial industry can no longer be ignored. Big banks such as JP Morgan, Kookmin Bank from South Korea, and HSBC are among the few front-running banks that have ventured into Metaverse technology. The venture into financial industry by big banks is monumental, as it proves that the Metaverse has its merits and is recognised. The success of the banks venture into the Metaverse will be due to its ability to offer its services and products in the virtual world.

Automotive and Manufacturing

A shared online space that spans across dimensions, powered by a combination of VR, AR and MR (mixed reality), will enable the automotive and manufacturing industry to be deeply entrenched in the Metaverse. Creating an industrial Metaverse is possible if it organically integrates cyber-physical systems; digital twins; 5G-powered AR, VR, and AI computer vision; low latency remote control and other applications. With an industrial Metaverse, the future of factories will not only use AR/VR for on-site auxiliary installations or skills training, but also create an immersive and virtual experience of people working together in the virtual world, guided by AI to verify results and correct errors in real-time without needing to be on-site.

In addition, the company's product design, development process, trial production testing, operation management, marketing, and other operations can be simulated and verified within the virtual community before transferring it to the physical world for actual production.

Another application would be for decision-making and results to be recorded through the blockchain, as the basis for assessment and auditing in both the virtual and physical world.

In the automotive industry, cars have evolved from a simple getting from point A to B to a mobile space that integrates work, entertainment and other functions. Autonomous driving releases attention and hands for the driver and passengers to do work or enjoy an entertainment experience that is traditionally constrained to the office or home environment.

In summary, the Metaverse will leverage on VR and AR in automobile, layered onto existing technologies such as smart cockpits, voice recognition and AI.

In the next five years, we expect the Metaverse and automobiles to develop in tandem. In addition to the virtual drive engine, more perceptive technologies such as touch and taste could be integrated, and possibly even developing a unique transaction method via blockchain in the Metaverse for the automotive industry. For example, a virtual car is built in the Metaverse and mapped to the real car and vice versa. Whether the driver is using the car in the physical or virtual world, driving behaviours, technology upgrades, modification preferences, and corresponding data generated can be shared between the two dimensions for a seamless experience.

In the near future, we anticipate **more brands stepping in to break the boundaries between virtual and reality to create more innovative, meaningful interactions with consumers**.

Real Estate

Land and real estate are important assets in the real world, which brings to question its value in the Metaverse. We expect the Metaverse to mirror and adopt characteristics of the real world, and the concept of supply and demand will affect the value and price of virtual land and real estate.

Like in the real world, location is a key factor affecting the value of land in the Metaverse (Maurer, 2022). The two factors that determine the value of the location are the distance from the center of the Metaverse and the quality of neighbours. The buying behaviour of many international celebrities has sparked heated discussions on the topic of virtual real estate investment.

However, in China, the trend may not pick up to the level of interest as in the West due to differences in understanding and perception around the Metaverse. The Metaverse attempts to mirror the rationally quantifiable physical world that is rooted in Western ideologies and culture, which does not align with the East. It would still take some time for the East to accept the concept of a rationally quantifiable world. Therefore, **investments into virtual land and real estate in China** are still speculative and will be accompanied by risks. As the concept gains popularity, it remains to be seen if price fluctuations will stabilize or continue to skyrocket and crash unpredictably.

Education

While the remote education experience has already improved greatly with campus digital twins, we expect the Metaverse to naturally drive spatial changes throughout the education industry in the next five years. Pursuing knowledge will no longer be confined to words, images and lectures available on demand, but reimagined to include an immersive experience accompanied by digital records in the Metaverse.

Education in the Metaverse era is by no means confined to a realistic learning experience. We see the Metaverse as the natural experimental playground for the education sector. Through collaboration and the establishment of a standard framework to create a shared digital ecosystem, the Metaverse will realize the sharing of high-quality educational resources on a global scale, benefiting a wide range of audiences of all ages and social classes, and making education truly a lifelong endeavour.

Companies that have begun their foray into the Metaverse help us imagine unlimited possibilities in the future of learning, illustrating an immersive experience of studying space and history in the Metaverse.

With many "top players" already declaring their focus on the education sector, unlocking the infinite possibilities of learning in the Metaverse is only a matter of time.

The construction of Metaverse can help promote children's education, serious games, and preschool education. Metaverse can contribute to education through immersive, simulating of realistic scenes to promote the understanding of learning content and avoid the harm of reality experiment.

Retail and Consumer Brands

In the era of **digital consumption, brands continue to evolve their approach to establishing direct communication channels with consumers**. From the early days of building a website, to embracing e-commerce, opening social media accounts, and even live broadcasting, **the Metaverse now offers a new concept for brands to experiment with.**

The prevalence and growing importance of virtual characters or avatars have in recent times presents various business opportunities in the retail sector. Many consumers now expect a combination of in-store and digital experiences, and the Metaverse offers the ability to engage with brands and products using a personalized avatar (Burden, Deploying Embodied AI Into Virtual Worlds, 2009). Whether it's trying out clothes, daily necessities, test-driving vehicles, or just elevating the browsing experience in a virtual store, the possibilities are endless.

The **well-known simulation game The Sims 4** already gives us an insight into the Metaverse. **Consumer brands are actively collaborating with game developers** to embed their products in all aspects of the game. Since its launch in 2014, consumer brands have launched item expansion packs for gamers to interact with the brands in the virtual world to raise brand awareness and cultivate brand love.

Beyond marketing products, **brands have also organized or participated in interactive activities with consumers**. The Metaverse has become the new playground, especially for **luxury fashion brands**, with some l**aunching their new collection in the virtual world and others partnering up with developers to create their own bespoke game**.

In the near future, we anticipate **more brands stepping in to break the boundaries between virtual and reality to create more innovative, meaningful interactions with consumers.**

Economy

Blockchain technology, decentralisation and the development and the rise of new industries within the Metaverse can effectively drive economic development.

Method

Business in general is describing an activity and institution that produces products and services in daily life (Amirullah, 2005). Moreover, according to Bukhori Alma, business is a total number of businesses covering Agriculture, Manufacturing, Building, Distribution, Transportation, Communication, Service and Goods business, and Government, whereby they are engaged in making and marketing goods and services to consumers (Bukhori, Pengantar Bisnis, 1993). According to Louis E. Boone, business consists of all activities of profit seeking by providing goods and services needed for the economic system. Some business produces tangible goods, while others provide services (Boone, Kurtz, & Berston, 2007).

According to (Sudarmo, 1996), business can be grouped as follows:

a) Extractive, namely businesses that carries out activities in the mining sector or excavates mining materials contained in the bowels of the Earth
b) Agrarian, namely businesses that involves in Agriculture
c) Industry, namely businesses that are engaged in industries
d) Services, namely businesses that are engaged in producing intangible products.

While business operations are the sector that is responsible for the quality of products and costs of a business or company, business operations make effort to regulate and optimize costs and materials in order to produce a unit of product or service.

USE OF METAVERSE ACROSS DIFFERENT REGIONS

Metaverse in Asia

The Metaverse is a technology that can support a country's development, such as the creative economy and digital economy, as well as drive targeted industries. The government offer opportunities for businesses to have comprehensive Metaverse access because of the high cost of access equipment. Most business sectors are not yet very technologically competent. The digital skills of Southeast Asian countries are not comparable to the industrial developed countries.

Figure 9. Value of fintech investments in the Asia-Pacific region from 2017 to 2021 (in billion U.S. dollars)
Source(s): KPMG; Pitchbook (https://www.statista.com/statistics/1010275/value-of-fintech-investments-asia-pacific/)

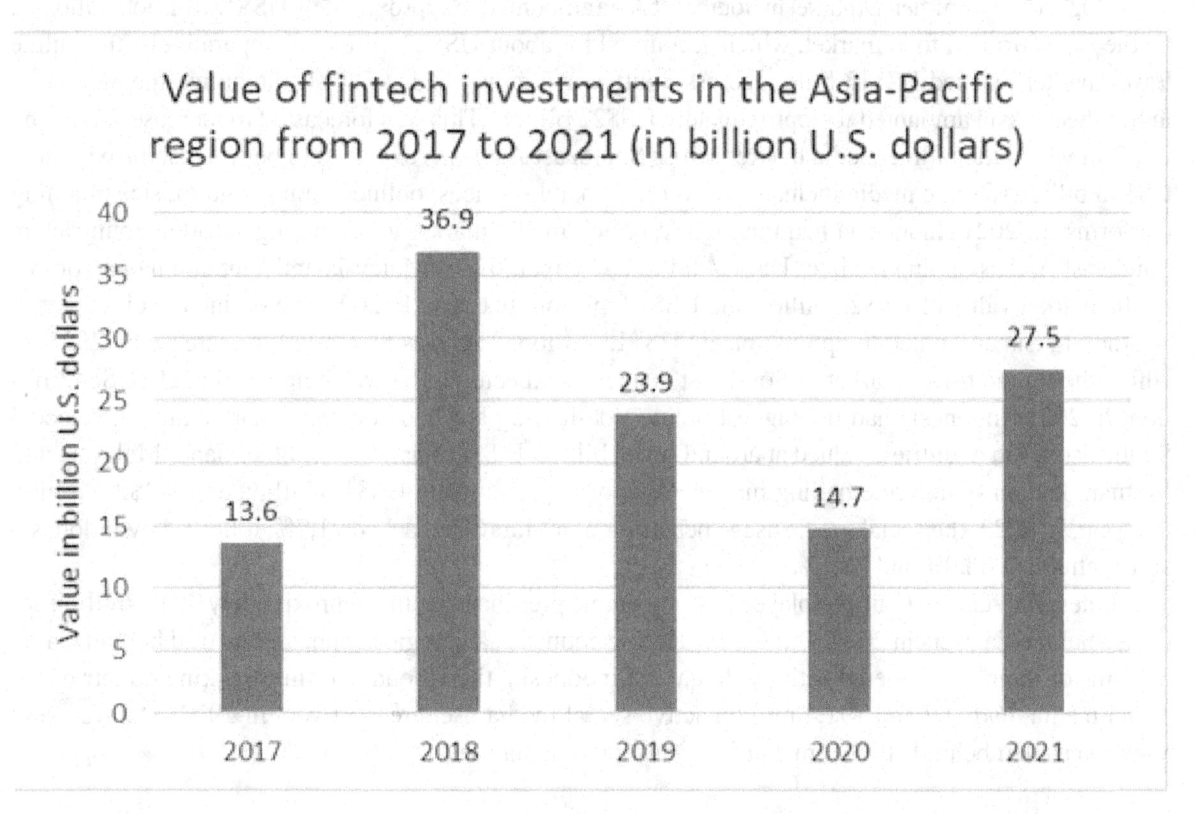

The number of information technology professionals in Southeast Asian countries is still limited, while demand in the IT field over the next 5-10 years is projected to continue growing.

In addition, the Metaverse requires high-speed internet for an immediate and continued response, but in Southeast Asian countries' 5G service areas do not yet span nationwide. A clear law that governs the system is needed because the Metaverse is an open system that allows users around the world to engage in collaborative and simultaneous activities. Oversight and legal controls should be clear, to apply to Southeast Asian countries.

Transactions on the Metaverse require digital assets, but Southeast Asian countries have not yet determined which digital assets can be used for payment of goods and services. Users also require data security to be protected, and they must be careful to avoid data loss.

As of June 2021, Malaysia, Indonesia and Vietnam had internet penetration within the Southeast Asian region of 89%, 76.8% and 77.4%. In 2021, the number of internet users in Southeast Asia amount to an estimated 495.95 million. The online audience is projected to reach 558.48 million users by 2025. In 2021, the number of mobile internet users in Southeast Asia amounts to an estimated 479.04 million. They are projected to reach 551.00 million users by 2025. In 2021, the internet economy size across Southeast Asia was valued at approximately US$174 billion. This was forecasted to increase dramatically by 2025, in which the internet economy in Southeast Asia was expected to reach US$363 billion. In 2021, Indonesia had the largest internet economy size in selected countries across Southeast Asia, of which the internet economy value reached US$70 billion. Comparatively, Malaysia and Vietnam had an internet economy value of US$21 billion for that same year.

In 2021, the e-commerce market in Southeast Asia amounted to approximately US$120 billion. This was forecasted to increase significantly by 2025, in which the e-commerce market in Southeast Asia was expected to be worth US$234 billion.

In 2021, the e-commerce market in Southeast Asia amounted to approximately US$120 billion, followed by the transport and food market, which accounted for about US$22 billion. Comparatively, the online travel market reached US$13 billion across Southeast Asia in 2021. In 2021, the online media market in Southeast Asia amounted to approximately US$22 billion. This was forecasted to increase further by 2025, in which the online media market across Southeast Asia was expected to be worth approximately US$43 billion. Online media includes video-on-demand services, online gaming, and music streaming platforms. In 2021, Indonesia had the highest online media market value among selected countries in Southeast Asia, standing at about US$6.4 billion. Comparatively, Malaysia and Vietnam had an online media market value of US$2.4billion and US$3.9 billion that year. In 2021, the online travel market in Southeast Asia amounted to approximately US$13 billion. This was forecasted to increase by 2025, in which the online travel market in Southeast Asia was expected to be worth approximately US$43 billion. In 2021, Indonesia had the biggest online ride-hailing and food delivery market among selected Southeast Asian countries, valued at around seven billion U.S. dollars. On the other hand, Malaysia and Vietnam had an online ride hailing market value of approximately US$1.8 billion and US$2.4 billion that year. In 2022, the social media usage penetration in Malaysia reached 91.7%, followed by Indonesia and Vietnam at 68.9% and 78.1%.

Chinese citizens certainly displayed a strong online presence, having approximately 983.3 million active social media users in 2022. Other Asia-Pacific countries and regions appeared to trail behind China in terms of their social media activity. India and Indonesia, the second and third ranking countries for social media, had 467 and 191.4 million active social media users respectively in 2022. Malaysia and Vietnam trailed behind at 30.25m and 76.93m social media users.

Leading Platforms

Facebook reigns worldwide and in the Asia Pacific region as the most-used social media site, having displayed an increased number of daily active users. However, Asian social networks, such as WeChat, have also gained immense popularity throughout the Asia Pacific region. Acting not only as a messaging and social media service, but WeChat is also a platform for online shopping and payment services. The social networking site, owned by Tencent, was the fifth most used social media platform globally. Interestingly, almost half of the most popular social media platforms globally were Asian social networks.

A Connected Region

The Asia Pacific region has gone through a technological evolution and alongside the region, its citizens have willingly adopted this change. This is conveyed through the internet penetration rate, which has more than doubled since 2010 in Asia. A better internet infrastructure has catalyzed the emergence of social media throughout the Asia Pacific region. As such, surveyed citizens across the region firmly stated the likelihood of them lessening their social media use was slim.

In 2022, Facebook is the most popular social media platform in all Southeast Asian countries, with the highest share in Timor-Leste at 99.56 percent. YouTube, Twitter, Instagram, and Pinterest were other platforms that had significant shares in Southeast Asian markets in 2022.

In Asia, countries with technological prowess have the first-mover advantage such as South Korea and China, are moving ahead of the other Asian countries. South Korea's capital Seoul has taken the lead and announced a five-year plan to transform itself into "Metaverse Seoul." Through the Metaverse platform, users will be able to virtually file administrative complaints, attend cultural events, and visit tourist sites including remodelling of destroyed landmarks. Seoul's success would boost South Korea's already impressive reserves of soft power.

Meanwhile, in China, tech giants like Tencent and Alibaba are advancing their "digital collectibles" feature – an equivalent to non-fungible tokens (NFTs) – to pave the way for a fully-fledged Metaverse. The Chinese government has maintained a cautious stance on the Metaverse. Nevertheless, as a telecommunications service provider in many countries, China has the potential to command global Metaverse applications and alter the geopolitical landscape.

Metaverse in Indonesia

In Indonesia, the concept of big data and Metaverse technology is still at its adolescence. It has attracted numerous researchers, academics and business practitioners into this arena from the economic incentives it provides. These promised opportunities, however, faces obstacles in adoption such as talent labour and imperfect network infrastructure. Regardless, the Indonesian government have showed serious effort into pushing for development in this sector through providing supports and training (Big data and metaverse toward business operation in Indonesia).

Internet users in Indonesia experienced an exponential growth (Include results from financial inclusion paper, graph). This increase was due to increase in Internet accessibility through mobile devices including mobile phones, laptops and tablets. The metaverse development in Indonesia was strongly support by its government, through its Ministry of Communication and Informatics (KOMINFO), where the government supports local companies to develop the Indonesia Metaverse ecosystem. This regional cooperation promises technological and economic development. Metaverse Indonesia has begun to build from sectors where the user of the ecosystem is most adaptable in adopting digital innovation.

One of the national subsidiaries developing Metaverse intensively in the country is the WIR Group, which presented Indonesia's prototype to the global community at the G20 President's culmination event in Bali in November 2022. To develop Metaverse in Indonesia, the company encourages participants to collaborate with global and domestic companies, including partnerships with Meta and Microsoft to develop hardware for augmented reality and virtual reality eyeglasses.

Metaverse in Malaysia

Preliminary actions have been executed in Malaysia to applied learning based Metaverse applications. In Malaysia, the Arabic Learning Principles (ALP) and the new culture of reading Quran using 3D Metaverse was used to teach Arabic language learning to Muslim students (Basha, Khaleel, Mnaathr, & Rozinah, 2013). The platform comprised of traditional Arabic architecture for Mosques, such as a congregational prayer, Quran recitation and supplication and the use of a compass to determine the direction of Kaaba, etc.

Metaverse in Vietnam

Vietnam is seen as a hub for the latest innovations, with the increasing popularity of the metaverse, non-fungible tokens (NFTs), and Web3.

Vietnamese Metaverse pioneer project The Parallel made headlines towards the end of 2021, for announcing that it will expand the metaverse beyond just gaming and entertainment by building communities that "enjoy to earn."

According to Chainalysis Global Cryptocurrency Adoption Index, this goal appears to be extremely feasible, as Vietnam have the highest cryptocurrency adoption in the world, with huge transaction volumes on peer-to-peer platforms. Over the last two years, the country also saw the rapid rise of blockchain-based startups and platforms being birthed locally.

As a prime example, Vietnam is home to Sky Mavis, the developer behind the popular mobile game Axie Infinity. Within just three years of starting up, Sky Mavis has garnered unicorn status and has also accumulated a total of 2 million Axie players as of October 2021.

Another company poised for growth in Vietnam is Web3 gaming studio Sipher, which secured $6.8 million in an October 2021 seed round to help further develop the World of Sipheria game. The financing round, co-led by Arrington Capital, Hashed and Konvoy Ventures, is expected to help Sipher develop tools for fascinating, fun, and intriguing blockchain-based gaming experiences in the metaverse.

TechWire Asia reports that the country's investments in technology firms have increased by a factor of eight between 2016 and 2019, reaching a peak of $861 million in 2019.

Both Hanoi and Ho Chi Minh City have established a strong ecosystem for technology startups focused on e-commerce, fintech, artificial intelligence (AI), food-tech, enterprise solutions, and information technology services. Blockchain-based startups are also growing in size and number.

Acknowledging the country's potential as a regional tech hub, the government launched the $32 million National Innovation Centre (NIC) in January 2021. NIC's mandate is to construct a start-up and innovation ecosystem that contributes to the country's economic development based on scientific and technological progress.

Within the next few years, the government also expects an influx of foreign investments, as a result of the country's generous tax incentives and the opportunity of 100% foreign ownership of their companies, as opposed to just 50% or less in other Asian countries. Vietnam is also seeking to expand its infrastructure to prepare for the vast flow of information in the Metaverse, as evidenced by the effort to engage domestic and foreign telecom enterprises as well as the expansion of Cloud computing infrastructure and Data centers by cloud service providers. Additional worthwhile news is the Ministry of Information and Telecommunications ("MIC") setting up a steering committee to promote the research and development of 6G mobile technology while the country is still in the phase of commercial testing for the 5G network. These technologies are expected to provide highly reliable, high speed, and low latency networks needed for the Metaverse.

With such incentives from the Government, local tech giants and start-ups are investing heavily into the virtual world. Recently, Viettel Group – the largest telecommunications group in Vietnam – has taken its initial steps, through its affiliate, towards developing a Metaverse platform by analyzing the available 5G platform, business models, and technology trends. VinFast, the global electric vehicle brand owned by Vingroup and Vietnam's largest private conglomerate, launched a collection of NFTs (VinFirst NFTs) as part of the EV reservation process to attract consumers such as Vietnamese customers.

Especially in May 2022, Vietnam Blockchain Association, the country's first entity in the crypto space, launched in Hanoi to allow blockchain experts to collaborate in promoting the development of Vietnam's digital economy. Not long after its inception, the Association announced cooperation with Binance – the world's leading corporation in blockchain development technology – on the blockchain research/application in Vietnam. Such cooperation aims to promote the development of blockchain technology in Vietnam and connect it with other major technology corporations worldwide.

Most importantly, a comprehensive legislative reform by amending and promulgating a large number of legal instruments is also a prominent undertaking in virtual technology. In particular:

First, on May 4th, 2022, the MIC released the Draft Law on E-Transaction ("Draft Law") for public comment, which caught a lot of attention from tech businesses. The Draft Law proposes new regulations and requirements for digital signatures, digital identities, and other topics. As transactions in the Metaverse, like online purchases, are made by cryptocurrency or by connecting a digital wallet to a bank account, it is necessary to have advanced digital payment confirmation such as digital signatures, digital identities, e-contracts, etc.

Second, several legislations regarding various aspects of the Metaverse's digital and virtual worlds, such as data privacy, cyber security, consumer protection, and intellectual property, are also on the way. These upcoming regulations include the Draft Decree detailing the Law on Cybersecurity ("CSD"); the Draft Decree on Personal Data Protection ("PDPD"); the Draft Decree amending Decree No. 72/2013/ND-CP on the Management, Provision, and Use of Internet Services and Online Information ("Draft Decree 72"); the Draft Amendment of Law on Consumers' Rights Protection ("LCRP"); and the Amended Intellectual Property Law, among others.

Third, it is also worth noting that Vietnamese lawmakers are creating even more draft regulations in technology, such as the Draft Law amending Law on Telecommunication; the Draft Law on Digital Technology Industry; and the Draft Revised Law on Radio Frequency. All of which are expected to be considered and approved in 2023.

Such prompt and comprehensive developments would bring a promising future for the Metaverse in Vietnam. With multiple legislations underway, lawmakers, businesses, and lawyers will be very busy catching up with such a tight legislative agenda.

Leading Benefits of the Metaverse Worldwide 2021

Figure 10. Leading benefits of the metaverse worldwide in 2021
Source(s): Tidio (https://www.statista.com/statistics/1285117/metaverse-benefits/)

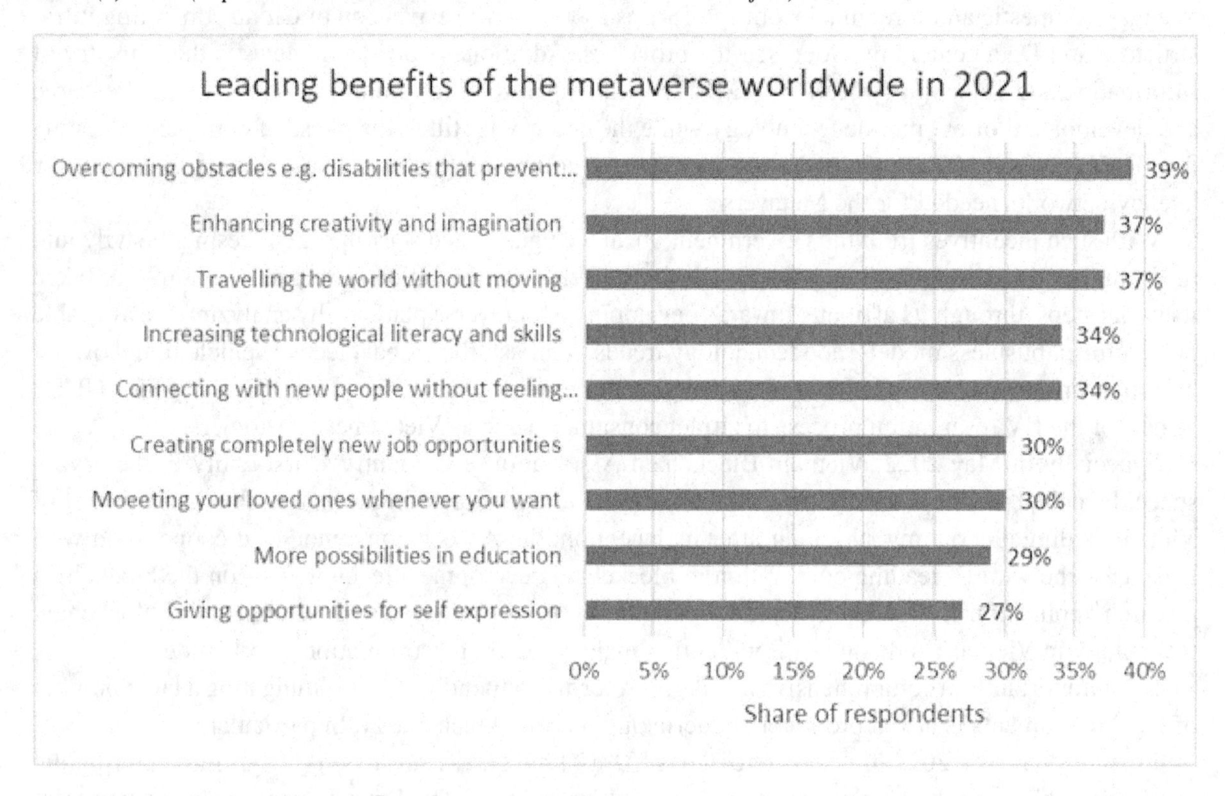

Removing Physical and Geospatial Barriers

Live Events, Social Ads can Double Market

The ability to bring live events such as concerts, film showings and sports into 3D virtual worlds represent additional opportunities for game makers as they elevate online experiences into 3D social worlds to capitalize on the Metaverse opportunity. Game makers including Epic Games and Roblox have hosted concerts inside of their games already, while Unity is investing in opportunities to bring live sports content and tools into its 3D development kit (Han, Heo, & You, 2021).

Figure 11. Gaming Growth Aided by 3D Virtual Worlds
Source(s): Bloomberg Intelligence; Newzoo; IDC

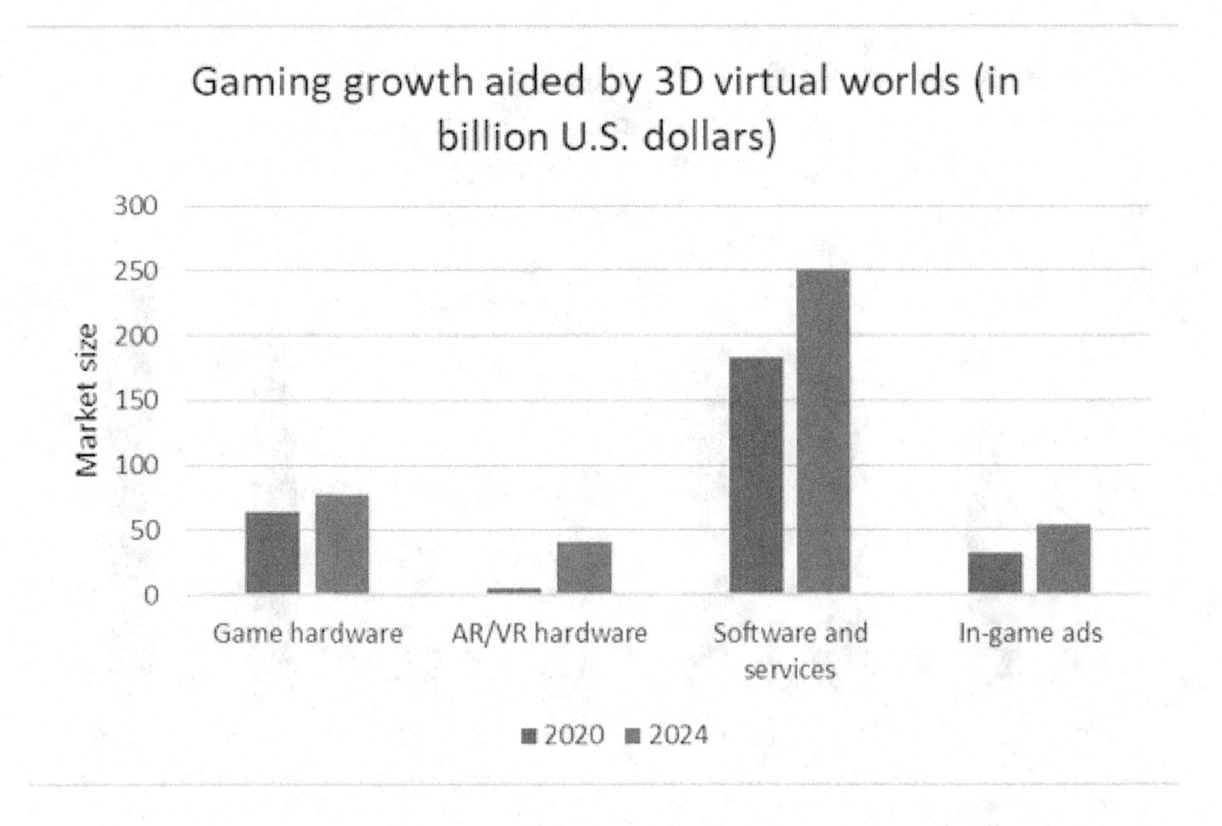

Revenue from live entertainment businesses that can become part of the Metaverse concept – films, live music and sports – may exceed US$200 billion in 2024, roughly flat vs. 2019, as these businesses slowly recover from the Covid-19 pandemic, based on data from PWC, Statista and Two Circles and our projections.

Figure 12. Live Nation Concerts Revenue (in million U.S. dollars) on Recovery Path
Source(s): Company Filings; Consensus Estimates

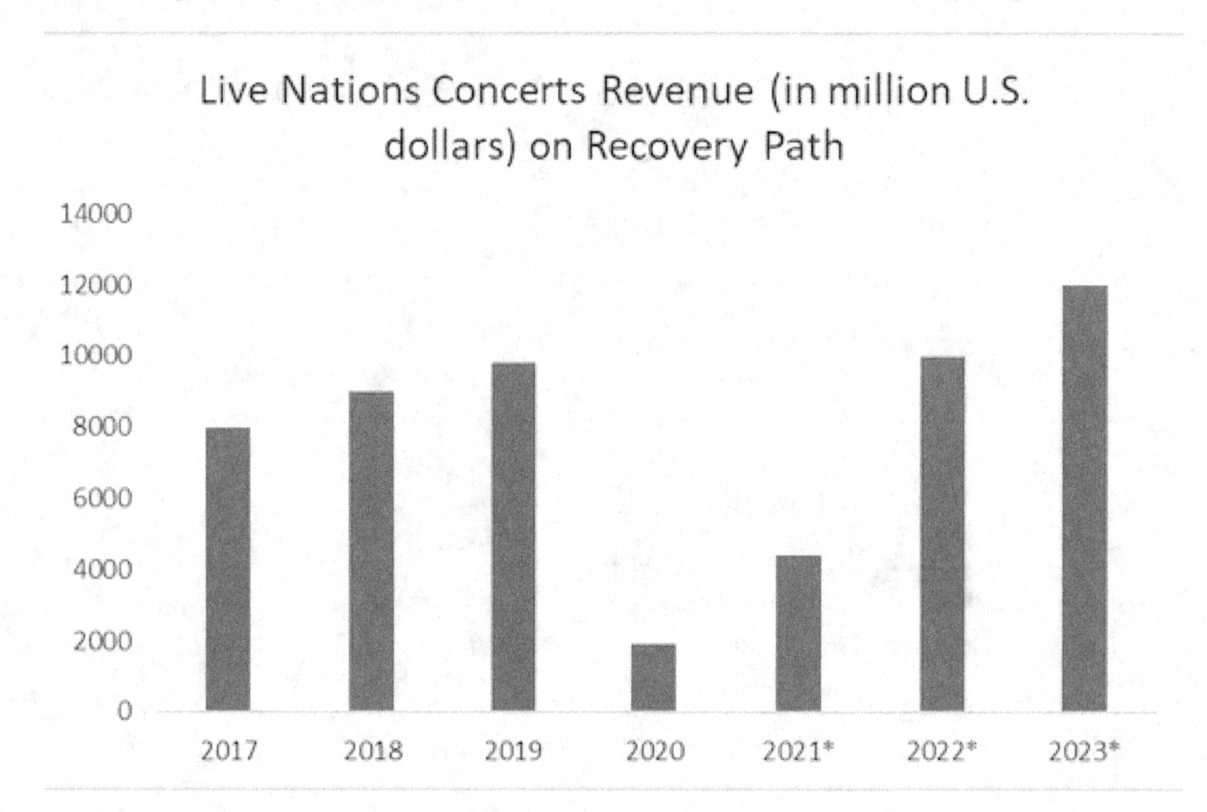

Sustainability

Metaverse overcoming geospatial barriers, allows users to attend workplaces, schools and social gatherings virtually rather than physically. Dropping out daily commute time will mean less time stuck in traffic and lesser carbon footprints. Technology has already substituted less efficient resources use for more efficient resource use in countless ways. Metaverse may provide humanity with more technological options than ever before.

Furthermore, sustaining biodiversity can be further achieved through the use of Metaverse technology. Metaverse technology provides more effective communicating method about sharing the wonder of biodiversity. Immersion in virtual worlds can provide unprecedented ways to interact with aspects of biodiversity that would otherwise be difficult to reach, because they are either geographically remote to large population centres, microscopic and thus inherently difficult to experience, or because they cover time scales that are difficult to capture, such as evolutionary trajectories.

Immersive experiences can replace some visits to sensitive sites, which might help safeguard certain biodiversity hotspots. Metaverse can be harnessed to facilitate the process of environmental science, for example by permitting seamless integration of researchers from various continents and by thus ensuring a greater diversity of voices inn the scientific discourse.

The process of scientific inquiry itself might be improved by building complex simulations and scenarios, for example, of global environmental change or by including human interactions and behaviours.

Such immersive experiences can also assist researchers to better understand the consequences of climate change or of other facets of global change, leading to adjustments in behaviour.

Creation of New Possibilities

Gaming, AR, VR create US$413 Billion Primary Market

The primary Metaverse revenue opportunity for video-game makers consists largely of existing gaming software and services market as well as rising sales of gaming hardware, based on our analysis. Within this primary market opportunity that may reach US$412.9 billion in 2024 from US$274.9 billion in 2020, software and services revenue as well as in-game advertising revenue accounts for about 70% of the total market size. Although this is the existing market for online game makers, those that are successful in capturing a higher share of users and engagement through the elevation of existing games into virtual worlds can garner a higher share of sector sales.

Gaming hardware, including gaming PCs and peripherals and AR/VR hardware such as Facebook's Oculus, accounts for the remainder of the primary market opportunity.

Augmented Reality/Virtual Reality

Figure 13. Augmented Reality (AR), Virtual Reality (VR) and Mixed Reality (MR) market size worldwide from 2021 to 2024 (in billion U.S. dollars)
Source(s): BCG; Mordor Intelligence (https://www.statista.com/statistics/591181/global-augmented-virtual-reality-market-size/)

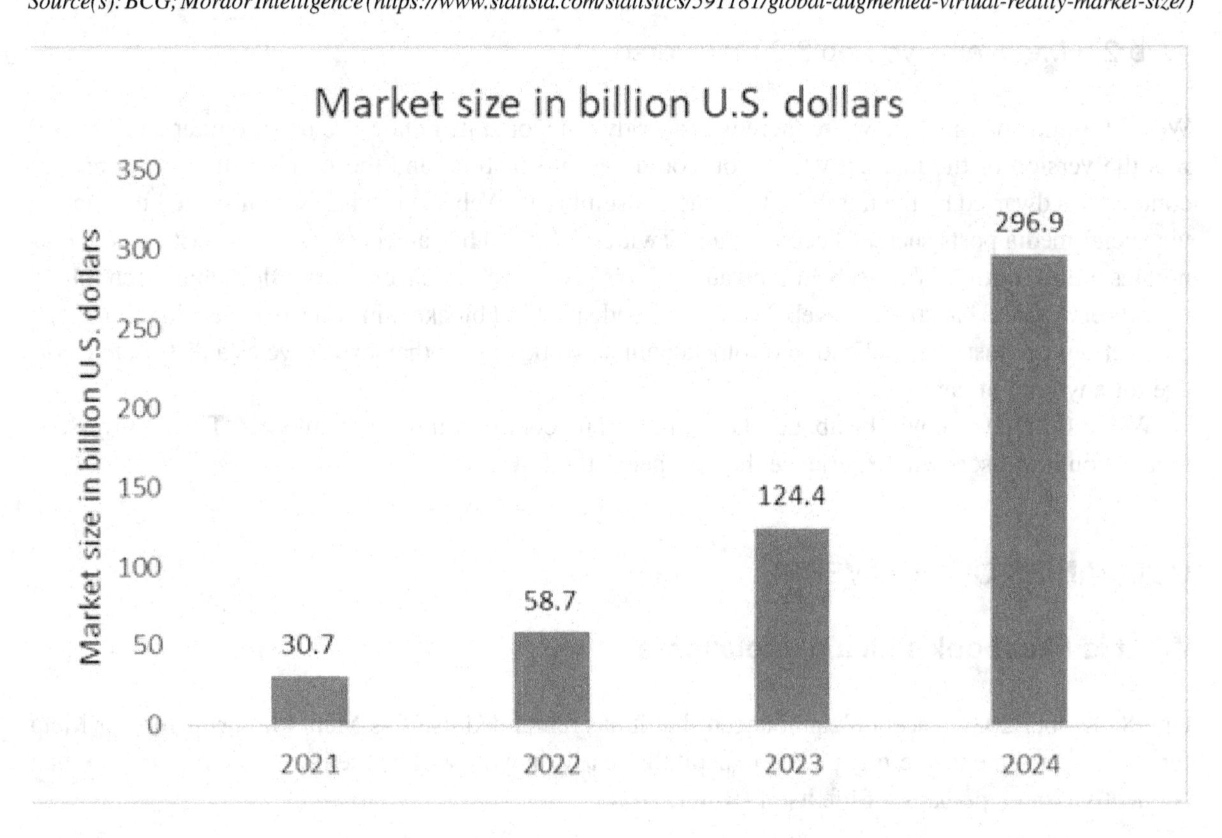

The global augmented reality (AR), virtual reality (VR) and mixed reality (MR) market is forecast to reach US$30.7m in 2021, rising to close to US$300m by 2024

AR technology integrates digital information with the physical environment, live and in real time. Through the addition of graphics, sounds, haptic feedback, or even smell to the natural world as it exits, AR can combine real life with a super-imposed image or animation using the camera on a mobile device or AR headset. As part of the wider extended reality (XR), the global AR market size is expected to grow considerably in the coming years.

Path Analysis of a US$800 Billion Market in 2024 Propelled by Double-Digit Growth According to Bloomberg Intelligence

The Metaverse market may reach $783.3 billion in 2024 from an estimated US$478.7 billion in 2020 propelled by a compound annual growth rate of 13.1%, based on data from Newzoo, IDC, PWC, Statista, Two Circles, and our projections. As video game makers continue to elevate existing titles into 3D online worlds that better resemble social networks, their market opportunity can expand to encapsulate live entertainment such as concerts and sports events as well as fighting for a share of social-media advertising revenue. The total Metaverse market size may reach 2.7x that of just gaming software, services and advertising revenue. According to Goldman Sachs, the Metaverse is an $8 trillion global market opportunity.

Online game makers including Roblox (www.roblox.com), Microsoft, Activision Blizzard, Electronic Arts, Take-Two, Tencent, NetEase and Nexon may boost engagement and sales by capitalizing on the growth of 3D virtual worlds.

Web 2.0 Metaverse vs web 3.0 Metaverse

Web 3.0 builds on the idea where there was already a major shift in how we use the internet. Web 1.0 was the version of the internet where you could get information, and the number of people creating content was dwarfed by the number of people consuming it. Web 2.0 is when we all started interacting via social media posts such as Facebook and Twitter. Web 2.0 has an centralised authority where corporates are in control. With web 3.0, the authority is given back to the creators. Blockchain technology would serve as the backbone of web 3.0. The core idea behind blockchain is decentralisation. Financial transactions are just one application of blockchain networks where they can serve as a distributed storage for any kind of data.

Web 3.0 Metaverse will be about who will own and control tomorrow's internet. The Metaverse is focused on how users will experience the internet of the future.

COMPANIES OF METAVERSE

What is Facebook's idea of Metaverse?

On 28 October 2021, Facebook announced that it has rebranded itself as Meta Platforms Inc., or Meta for short — a name it carefully picked to capitalise early on what will be the inevitable future of human connectivity and, perhaps, of life itself (Ritter, 2021).

Metaverse Competitive Landscape

Figure 14. Metaverse Competitive Landscape
Source(s): Bloomberg Intelligence; Newzoo; IDC; PwC; Two Circles; Statista

Metaverse Competitive Landscape

Online Game Makers	Design Software Vendors	Social Networking	Gaming, AR & VR Hardware	Live Entertainment
Roblox	Unity	Facebook	Facebook	Live Nation
Epic Games	Epic Games	Tencent	Lenovo	Theme Parks
Microsoft	Adobe		HP	Sports Theme
Activision Blizzard	Autodesk		Logitech	
Electronic Arts	Ansys		Acer	
Take-Two			Valve	
Tencent			Razer	
NetEase				
Nexon				
Valve				

"The defining quality of the Metaverse will be a feeling of presence — like you are right there with another person or in another place. Feeling truly present with another person is the ultimate dream of social technology. That is why we are focused on building this," said CEO Mark Zuckerberg in his Founder's Letter following the announcement. Zuckerberg showed a world that can be defined as a much higher level of virtual reality (VR) and augmented reality (AR). In the virtual space of the Metaverse, everything people do in the real world is replicated.

A VR headset, or any other wearable gadget specifically designed for the purpose, will function as a gateway into this world. Facebook's idea of Metaverse includes virtual avatars of people playing games, holding meetings, attending workshops, exercising, studying and socialising besides all kinds of activities that can be done in reality

A larger and much grander Metaverse can be extrapolated from this idea. It is, however, noteworthy that Zuckerberg believes it could take around 10 years for the Metaverse to become mainstream. To think about it, that's a short time.

Figure 15. Meta (formerly Facebook Inc.) revenue and net income from 2007 to 2021 (in million U.S. dollars)
Source(s): Facebook; Meta Platforms (https://www.statista.com/statistics/277229/facebooks-annual-revenue-and-net-income/)

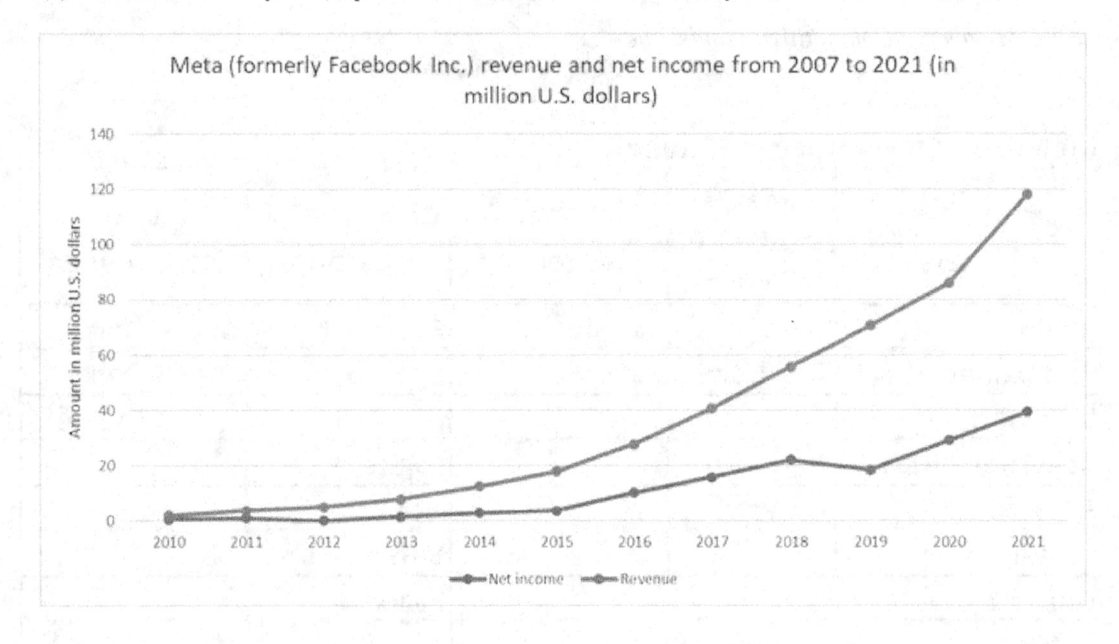

Figure 16. Annual Meta (formerly Facebook Inc.) net income from 2008 to 2021 (in million U.S. dollars)
Source(s): Facebook; Meta Platforms (https://www.statista.com/statistics/1289490/annual-facebook-net-income/)

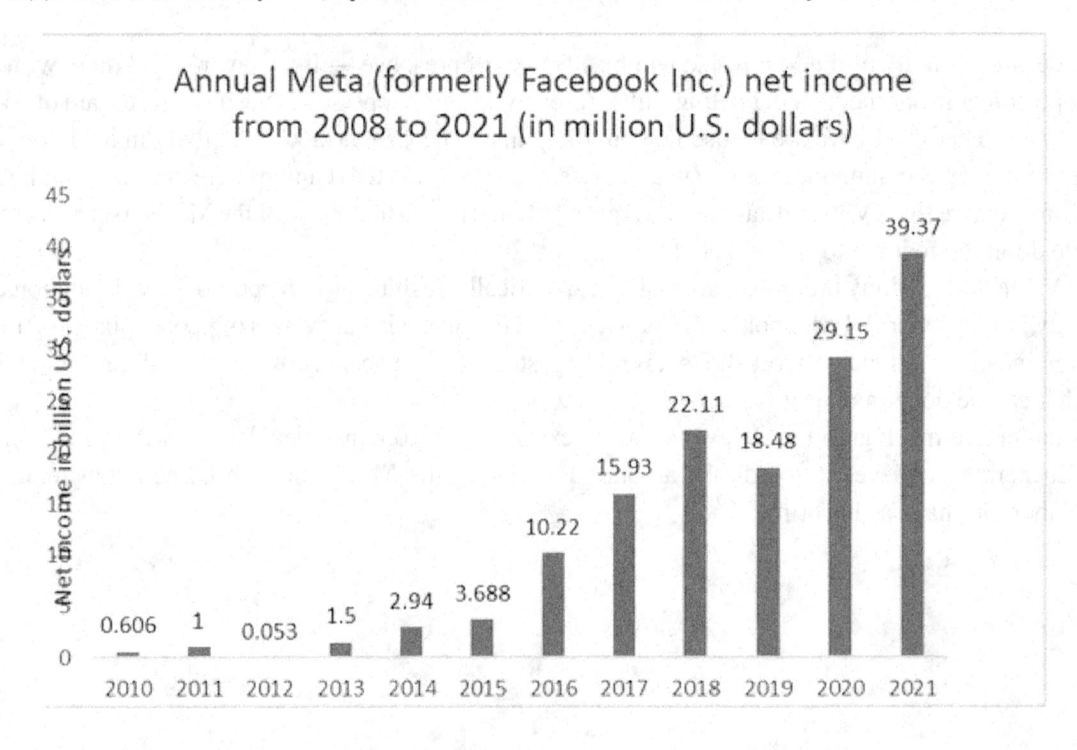

Towards the end of 2021, Facebook Inc, which owns substantial social media platforms Facebook and WhatsApp rebranded and changed its name to Meta. This strategic rebranding came as part of a step towards the metaverse

In 2021, Meta's revenue was over US$117 billion, up by over US$31 billion on the previous year. Within the last decade, the company has increased its overall revenue by over US$114 billion. The majority of Meta's profit come from its advertising revenue, which generated US$114.93 billion in 2021.

Meta's total Family of Apps revenue for 2021 amounted to US$115.66 billion. Additionally, Meta's Reality Labs, the company's VR division, generated US$2.27 billion. Meta's marketing expenditure for 2021 amounted to just over US$14 billion, up from US$11.6 billion over the previous year.

Increasing audience base despite privacy misgivings

Meta's user numbers have continued to grow steadily throughout past years, in the 4th quarter of 2021, there was a total of 3.59 billion worldwide users across all of Meta's platforms. For the same time frame, the company recorded 427 m monthly active users across Europe.

Download of Meta's app Oculus, for which virtual reality headsets are required, increased greatly from 2020 to 2021, reaching a total of 10.62m downloads by the end of 2021. Up until 2021, downloads had grown in a steady manner but from 2020 to 2021, they more than doubled.

Figure 17. Global revenue generated by Meta (formerly Facebook Inc.) as of 4th quarter 2021, by segment (in million U.S. dollars)
Source(s): Facebook; Meta Platforms (https://www.statista.com/statistics/1288912/meta-quarterly-global-revenue-by-segment/)

Global revenue generated by Meta (formerly Facebook Inc.) as of 4th quarter 2021, by segment (in million U.S. dollars)

	Advertising	Other	Family of Apps revenue	Reality labs
2020 Q4	27187	168	27355	717
2021 Q1	25439	198	25637	534
2021 Q2	28580	192	28772	305
2021 Q3	28276	176	28452	558
2021 Q4	32639	155	32794	887

User numbers have increased despite data security issues and past controversy such as the Cambridge Analytica scandal in 2018. There remains scepticism surrounding the idea of metaverse in which Meta aims to immerse itself. Of surveyed audits in the United States, the majority said they were concerned about their privacy if Meta were to succeed in creating the metaverse.

Microsoft – Anyone Else?

Microsoft is another major Big Tech player trying to build a Metaverse. Its idea is called Mesh. One of its most interesting features is what the company calls "Holoportation". Simply put, its users will be able to project their holographic selves to other users.

Its virtual and augmented realities can be best experienced with HoloLens devices — a unique gadget which can make the experience of 'Holoportation' altogether different. However, Mesh can also be accessed via VR headsets, mobile phones, tablets or PCs.

Initially, the projections will be in the form of animated avatars. Eventually, it will become a photorealistic lifelike projection. Something like how characters in Star Wars interact with each other via hologram projection. Mesh is expected to roll out for Teams in 2022. Clearly, Microsoft has its focus on how routine work can be made more immersive in a Metaverse.

While the two front runners have been rapidly developing and investing heavily into Metaverse, several other publicly listed companies such as Amazon, Disney, NVIDIA, and Snapchat have also ventured and catching up to the front runners. Since 2018, Amazon have been developing a "new VR shopping experience" and trying to use Metaverse to create a virtual shopping space where shoppers can interact with digital products by building a kind of virtual "Amazon shopping mall" in Metaverse to asset its dominant position in the market. Disney and Snapchat have focused on another dimension of the Metaverse, where they aim to provide quality experience on entertainment. Tilak Mandadi, Disney's Chief Technology Officer (CTO), said building a "theme park Metaverse" will be the next step in the evolution of Disney's theme parks. Snapchat have introduced custom avatars and filters to fill the world with digital content. Currently, Snapchat has launched the Bitmoji service, which allows user to pose in physical snapshots and create their own 3D Bitmoji avatars.

Other Metaverse platforms – Roblox, Fortnite, Augmented Reality

More advanced Metaverse platforms include Roblox and Fortnite. The former is particularly interesting.

Launched in 2006, Roblox has become so famous that over 50 percent of American children under the age of 16 played it in 2020, but it is more than just a game. For comparison's sake, imagine a fair where stalls allow visitors to play games. Roblox is like a huge fair in the virtual world. Its users can create their own games on the platform and monetise them to make real money by exchanging the virtual currency known as Robux that they earn on the platform.

Recently, Venture Beat reported that at the GamesBeat Summit Next event, the company's chief of technology, Dan Sturman, said that Roblox is creating a Metaverse around its players. That's apparent because companies, such as Nike and the National Football League (NFL) entered the Metaverse with NIKELAND — a virtual playspace — and Roblox store, respectively. While Nike announced its partnership on 18 November, the NFL came out with its news ten days later.

Fortnite, on the other hand, is one of the biggest 'competitors' in the Metaverse space. Launched by Epic Games in 2017, it began as an online multiplayer game and has now become a larger social media space with acclaimed musicians, such as Ariana Grande, performing concerts on its platform.

What about the East?

Figure 18. Leading fintech startups in the Asia-Pacific region in 2021, by total funding (in million U.S. dollars)
Source(s): Crunchbase (https://www.statista.com/statistics/1258820/apac-top-fintech-startups-by-total-funding/)

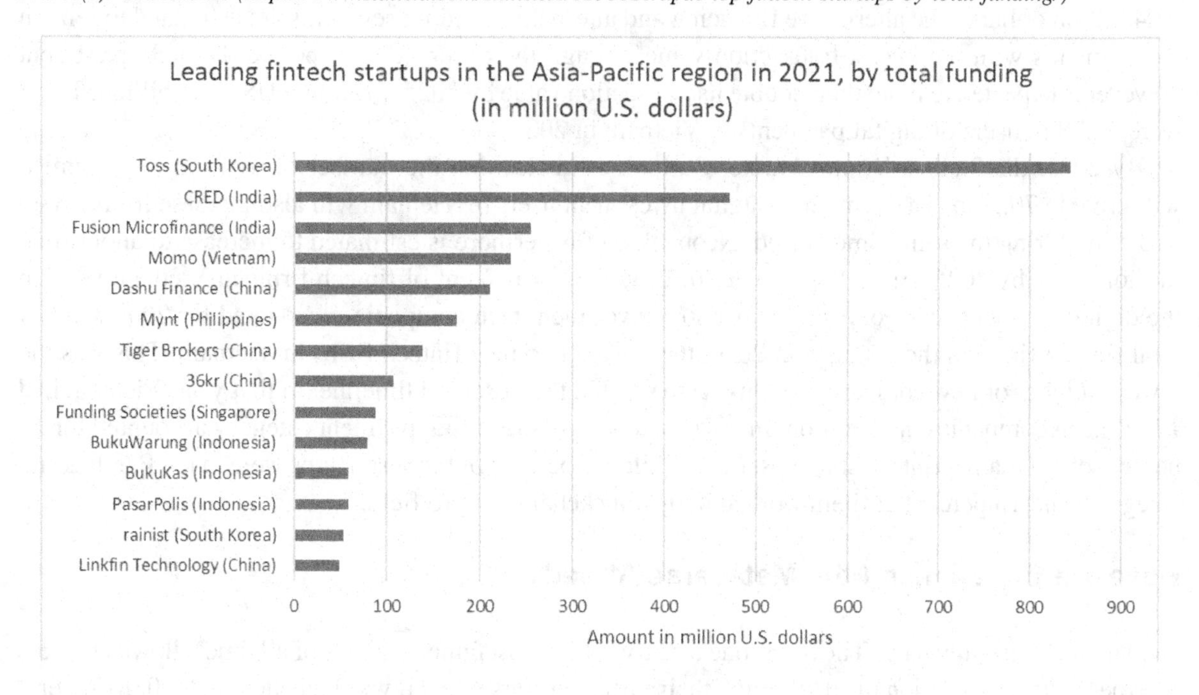

In 2021, the total investments into fintech across the Asia-Pacific region amounted to US$27.5 billion. This was a massive increase from 2020, in which US$14.7 billion was invested into the fintech sector in the Asia-Pacific region. Further to the East, among the develop countries: China, Japan and South Korea, have also invested heavily into this industry. In China, the two biggest companies – Tencent, have introduced a whole series of investments in the Metaverse ecosystem, including the AR development platform, while Alibaba, created a lab for Metaverse, categorised in four divisions:

L1: Holographic construction,
L2: Holographic simulation,
L3: virtual and real fusion, and
L4: Virtual and real linkage.

This shows serious commitment have been into development of the Metaverse space.

South Korean mobile finance platform Toss' total funding amounted to US$844 m as of August 2021. Indian members-only credit card management and bill payments platform CRED followed in second with total funding of US$471 m. In South Korea, under Metaverse Alliance, the Korean Information and Communications Industry Promotion Agency has formed 25 organisations and companies into the "Metaverse Alliance" to build the Metaverse ecosystem under the leadership of the private sector through government and business collaboration, realizing an open Metaverse platform in various fields of reality and virtuality. Another notable company that has ventured into the Metaverse is Samsung. They have launched the "Samsung Global Metaverse fund" with the sole aim of building and development of the Metaverse.

Transaction value in the digital payments segment of Vietnam reached US$14.37 billion in 2021. By 2025, the Statista Digital Market Outlook estimates transaction value in this segment to reach around 23.4 billion dollars. The alternative financing and alternative lending segments are estimated to remain the segments with the lowest transaction value through the observed time period. Digital investment however is expected to more than double its transaction value by 2025, to around US$5.44 billion. There were 51.78 m users of digital payments in Vietnam in 2021.

By 2025, the Statista Digital Market Outlook estimates that the number of users in this segment will grow to 70.91m. The number of digital investment users in Vietnam will also increase from 1.68m to 3.54m during the same time period. Neobanking furthermore is estimated to increase to almost one million users by 2025. As of September 2021, around 21 percent of fintech firms in Vietnam were in the digital payments category. In comparison, investment tech companies accounted for 20 percent of total fintech firms in the country. In 2020, there were two new fintech firms in Vietnam. This was the lowest number of new companies in this sector within the observed timeline. In that year, Vietnam had 141 fintech companies in operation. In 2020, start-ups in the digital payment category accounted for 31 percent of Vietnam's fintech start-ups. Meanwhile, 17 percent of these start-ups were in the P2P lending category, and 13 percent of them worked in the blockchain/ crypto field.

Have We Experienced the Metaverse Already?

Avid gamers certainly have. The Sims, one of the world's most famous games of all time, allowed players to experience a simulation of life through their virtual avatars when it was launched in 2000. Its several sequels and spin-offs have only made the whole "living in virtual reality" more immersive.

The Sims gave a glimpse of what a future Metaverse experience could be like. In 2003, Linden Lab, an American tech company, launched an application known as Second Life. Both fans of virtual reality and experts' credit Second Life for technically planting the seeds of the Metaverse.

Second Life allows its users to experience exactly what its name suggests — a second life. Real people can create their 3D virtual avatars on the platform and do almost everything they can in the real world.

A world in itself, users of Second Life can interact with other users and form relationships through their virtual identities. They can organise, attend and express themselves through music and art, as well as use virtual objects in the environment and even contribute to building things within Second Life such as virtual buildings. They can attend religious seminars, concerts and business meetings.

Moreover, there are educational institutions and even embassies. In 2007, Maldives became the first country to open a virtual embassy in Second Life.

Users can buy and sell virtual items in Second Life using the platform's own closed-loop digital currency — Linden Dollar, which should not be confused with a cryptocurrency.

Several games, including the multiplayer online role-playing World of Warcraft released in 2004, allow players to interact with each other. This is a form of social network where communities are developed around games.

AR and VR Shaping the Metaverse

The arrival of VR headsets, such as Oculus, has given a major boost to the idea of the Metaverse. Games tailored for VR headsets are already making lots of money. According to Fortune Business Insights, the global virtual reality gaming market is projected to grow to USD 53.44 billion in 2028 from USD 7.92 billion in 2021 (Wood, 2021).

A step further is AR, which is where the user's perception of reality is enhanced with the use of technology. It is something like getting more data about an object in the real world while viewing it from an AR device or headset.

A classic way to understand this is to revisit the Marvel Cinematic Universe (MCU) films, featuring Iron Man. Robert Downey Jr.'s Tony Stark (aka our favourite Iron Man) repeatedly displayed tech that is beyond imagination even now. There are several scenes of him wearing the Iron Man suit and getting information and multiple visualisations on his heads-up display (HUD) of things that he is seeing in real time. To a great degree, that is AR. However, these are films after all; Iron Man goes into realms of insanely advanced tech, which will take some more decades to become reality.

However, the speed at which the real world is trying to make VR/AR and Metaverse a reality is indeed something. Apple Inc. is already into AR and claims to have "the world's largest AR platform." Multiple AR apps on the App Store can be used to do anything — from making Snapchat videos more fun to exploring 3D projects in AR.

Nreal, a Chinese AR company, created Nreal Light glasses in 2020 and will start shipping the cheaper Nreal Air glasses, starting December 2021 across China, Japan and South Korea. Moreover, consumer augmented reality glasses look like regular luxury fashion shades and make the experience of enhanced reality more exciting.

FUTURE OF THE METAVERSE

The fact is that the Metaverse will become as real and common as the internet. As we can see, it is but a matter of time.

When Epic Games founder Tim Sweeney was asked by CNN what he thought about the future of the Metaverse, he said, "I think it will take a decade or more to really get to the end point, but I think that is happening."

At the same time, Sweeney said, "The Metaverse isn't going to be created by one company. It will be created by millions of developers each building out their part of it."

So, in other words, the Metaverse is still being constructed brick by brick and everyone will have a hand in its creation.

Negatives

The metaverse could lead to some unintended consequences for users and society at large. Nearly half of 1,050 survey respondents by Tidio perceived addiction to a simulated reality or virtual world as the biggest threat from the metaverse, followed by problems related to privacy and mental health.

The development of Metaverse is still in its infancy, and its business model is not mature. Due to the open issues such as intractability, computing power pressures, ethical constraints, privacy risks, and addiction risks in the different worlds, and the fact that Metaverse development stood still by current technology, research interest in the Metaverse is at an Ebb Stage after 2013.

Pokémon Go creator John Hanke called Metaverse a "dystopian nightmare" in the title of his blog published on 10 August 2021 on the website of his company, Niantic.

While drawing attention to the core plot of the famous novels by Gibson and Cline, Hanke wrote, "A lot of people these days seem very interested in bringing this near-future vision of a virtual world to life, including some of the biggest names in technology and gaming. But in fact, these novels served as warnings about a dystopian future of technology gone wrong."

One of the biggest red flags comes from the inventor of AR — Louis Rosenberg. Writing for Big Think, Rosenberg says that it could be a "dystopian walk in the neighbourhood."

"After all, the shared experience we call 'civilized society' is quickly eroding, largely because we each live in our own data bubble, everyone being fed custom news and information (and even lies) tailored to their own personal beliefs. This reinforces our biases and entrenches our opinions. But today, we can at least enter a public space and have some level of shared experience in a common reality. With AR, that too will be lost," writes Rosenberg.

By the time Metaverse becomes mainstream, maybe we will have systems like Starlink widely available to deliver high-speed data to the remotest corners of the planet. But even though the reach increases, Metaverse might still struggle to entice people.

As Second Life creator Philip Rosedale told Time, "If you live a comfortable life in New York City and you're young and healthy, you probably are going to choose to live there. If I offer you the life of an avatar, you're just not going to use it very much. On the other hand, if you live in a rural location with very little social contact, are disabled or live in an authoritarian environment where you don't feel free to speak, then your avatar can become your primary identity."

Interaction Problem

As a medium between the virtual world and the real world, the interaction technology of the Metaverse needs to meet the following conditions: (1) The interactive device is lightweight, convenient to use, wearable and portable and (2) The transparency of the interactive medium enables the users to ignore the traces of technology and better immerse themselves in the virtual world.

Among the technologies include somatosensory technology, XR (VR, AR, MR) technology, and brain-computer interface. Interaction problem has been greatly overcome in the recent years due to high penetration of mobile devices. Smart devices include smart phones, PDA, handheld consumer devices with Internet access, and the accompanying suites of accessible services. These devices act as a multipurpose information appliance that have a one-to-one binding with the user, offering ubiquitous services and access.

The massive growth in the consumer devices market was mainly drive by smart phones, where the global smart phones market grew 90% in the third quarter of 2010 alone, with vendors shipping more than 81 million smart phones.

Social Privacy

Digital footprints in the social Metaverse can be tracked to reveal the user's real-world identity, and other sensitive information, such as location, shopping preferences and even financial details. The importance of privacy plays a crucial role in shaping the social Metaverse. Applying privacy-preserving schemes is much easier in traditional social networks, as users can decide with whom to share their social media content. On the other hand, privacy control is not possible in the social Metaverse, as users cannot change the virtual properties if the constructed virtual world, which makes user privacy preservation challenging.

Take the following example: If you are navigating in a mall in the social Metaverse, and an avatar follows your avatar and records all the things you buy and your travel history, this information can be used to perform social engineering attack that can violate your privacy in the real world. On the other hand, we cannot just turn off who can follow our avatars in the Metaverse as we can do in the traditional social media. Another example of privacy in the Metaverse: You want to have as much privacy in your home that you have created as you do in the real world. However, the current virtual social networks allow other avatars to navigate freely around the map, including your house, and you cannot call the police, as you would do if someone invaded your house.

REFERENCES

Alsop, T. (2021). Metaverse Potential Market Opportunity Worldwide 2021, by scenario. (https://www.statista.com/statistics/1286718/metaverse-market-opportunity-by-scenario/)

Alsop, T. (2022). Investment in AR/VR Technology Worldwide in 2024, by use case. *Statista*. (https://www.statista.com/statistics/1098345/worldwide-ar-vr-investment-use-case/)

Alsop, T. (2022). U.S. Metaverse Potential Consumer Expenditure TAM 2022, by segment. *Statista*. (https://www.statista.com/statistics/1288655/metaverse-consumer-expenditure-tam-united-states/)

Alsop, T. (2022). Virtual Reality and Augment Reality Users U.S. 2017-2023. *Statista*. (https://www.statista.com/statistics/1017008/united-states-vr-ar-users/)

Alsop, T. (2022). XR/AR/VR/MR Market Size 2021-2028. *Statista*. (https://www.statista.com/statistics/591181/global-augmented-virtual-reality-market-size/)

Alsop, T. (2022). XR Headset Shipments 2016-2025. *Statista*. (https://www.statista.com/statistics/1290160/projected-metaverse-use-reach-global-consumers-businesses/)

Alvin, K. (2021). Sustainable GameFi: To Play or To Earn? *Statista*. (https://www.binance.org/en/blog/sustainable-gamefi-to-play-or-to-earn/)

Amirullah. (2005). *Pengantar Bisnis*. Graha Ilmu.

Austria, D., De Jesus, S. B., Marcelo, D. R., Ocampo, C., & Tibudan, A. J. (2022). Play-to-Earn: A Qualitative Analysis of the Experiences and Challenges Faced by Axie Infinity Online Gamers Amidst the COVID-19 Pandemic. *International Journal of Psychology and Counselling*, 391-424.

Bainbridge, W. S. (2007). The Scientific Research Potential of Virtual Worlds. American Association for the Advancement of Science, 471-476. doi:10.1126cience.1146930

Ball, M. (2021). Payments, Payment Rails, and Blockchains, and the Metaverse. https://www.matthew-ball.vc/all/Metaversepayments.

Basha, A. D., Khaleel, A. I., Mnaathr, S. H., & Rozinah, J. (2013). Applying Second Life in New 3D Metaverse Culture to Build Effective Positioning Learning Platform for Arabic Language Principles. *International Journal of Modern Engineering Research*, 1602-1605.

BlaiseS.De JesusD.MarceloD. R.OcampoC.IsJ.Tibudan, A. (2022). Play-to-Earn : A Qualitative Analysis of the Experiences and Challenges Faced By Axie Infinity Online Gamers Amidst the COVID-19 Pandemic. doi:10.6084/m9.figshare.18856454.v1

Bloomberg. (2021). NFT Market Surpassed $40 Billion in 2021, New Estimate Shows. *Bloomberg*. https://www.bloomberg.com/news/articles/2022-01-06/nft-market-surpassed-40-billion-in-2021-new-estimate-shows.

Boone, L., Kurtz, D., & Berston, S. (2007). *Contemporary Business*. Wiley.

Bukhori, A. (1993). *Pengantar Bisnis*. Alfabeta CV.

Burden, D. (2009). Deploying Embodied AI into Virtual Worlds. *Knowledge-Based Systems*, *22*(7), 540–544. doi:10.1016/j.knosys.2008.10.001

Decentraland, M. (2021). Property. https:// Metaverse.properties/buy-indecentraland

Dinh, T. N., & Thai, M. T. (2018). AI and Blockchain: A disruptive integration. *Computer*, *51*(9), 48–53. doi:10.1109/MC.2018.3620971

Dionisio, J. D., Burns, W. G. III, & Gilbert, R. (2013, June). 3D Virtual Worlds and the Metaverse: Current Status and Future Possibilities. *ACM Computing Surveys*, *45*(3), 38. doi:10.1145/2480741.2480751

Glencross, M., Chalmers, A., Lin, M., Otaduy, M., & Gutierrez, D. (n.d.). *Exploiting Perception in High-Fidelity Virtual Environments.*

Han, J., Heo, J., & You, E. (2021). Analysis of Metaverse Platform as a New Play Culture: Focusing on Roblox and ZEPETO. *2nd International Conference on Human-Centered Artificial Intelligence*.

Krueger, M. W., Gionfriddo, T., & Katrin, H. (1985). Videoplace - An Artifical Reality. *Chi '85 Proceedings*, (pp. 35-40).

Lee, J. (2021). *Blockchain 3.0 Era and Future of Cryptocurrency; Future Info Graphics*. Research Center for New Industry Strategy.

Lee, J. Y. (2021). A Study on Metaverse Hype for Sustainable Growth. *International Journal of Advanced Smart Convergence*, 72-80.

Lee, L.-H., Braud, T., Wang, L., Xu, D., Kumar, A., Bermejo, C., & Hui, P. (2021). All One Needs to Know About MetaverseL A Complete Survey on Technological Singularity, Virtual Ecosystem, and Research Agenda. *Journal of Latex Class Files*, 1-66.

Lik-Hang, L., Braud, T., Zhou, P., Wang, L., Xu, D., Lin, Z., & Hui, P. (2021). All One Needs to Know about Metaverse: A Complete Survey on Technological Singularity, VIrtual Ecosystem, and Research Agenda. *Journal of LATEX Class*, 1-66.

Liu, H., Bowman, M., Adams, R., Hurliman, J., & Lake, D. (2010). Scaling Virtual Worlds: Simulation Requirements and Challenges. *Winter Simulation Conference*, (pp. 778-790). 10.1109/WSC.2010.5679112

Maurer, M. (2022, January 7). Accounting Firms Scoop Up Virtual Land in the Metaverse. *The Wall Street Journal*. https://www.wsj.com/articles/accounting-firms-scoop-up-virtual-land-in-the-metaverse-11641599590

Mystakidis, S. (2022). Metaverse. *Encyclopedia*, 486-497. doi:10.3390/

Ning, H., Wang, H., Lin, Y., Wang, W., Dhelim, S., Farha, F., & Daneshmand, M. (n.d.). A Survey on Metaverse: The State-of-the-art, Technologies, Applications, and Challenges. 1-34.

Ritter, M. (2021). *Cryptocurrencies Rally on Facebook's 'Meta' Rebrand - CNN*. CNN Business.

Robertson, H. (2021). *Jefferies Says the Metaverse Is as Big an Investment Opportunity as the Early Internet*. Insider.

Sandbox. (2020). The Sandbox Whitepaper. *The Sandbox*. https://installers.sandbox. game/The_Sandbox_Whitepaper_2020.pdf [Accessed:2021-01-30]

Sang-Min, P., & Young-Gab, K. (2021). A Metaverse: Taxonomy, Components, Applications, and Open Challenges. IEEE, 4209-4251.

Schuemie, M., Van Der Straaten, P., Krijn, M., & Van Der Mast, C. (2001). Research on Presence in Virtual Reality: A Survey. *Cyberpsychology & Behavior*, *4*(2), 183–201. doi:10.1089/109493101300117884 PMID:11710246

Seo, J., Kim, K., Park, M., & Park, M. (2018). *An Analysis of Economic Impact on IOT Industry under GDPR. Mobile Information Systems. 1-6.*

Smart, J. M., Cascio, J., & Paffendorf, J. (2007). *Metaverse Roadmap Overview*. Acceleration Studies Foundation.

Studio, R. (2021). Anything you can imagine, make it now! *Roblox*. https://www.roblox. com/create

Sudarmo. (1996). *Prinsip Dasar Manajemen*. BPFE-Yogyakarta.

Swan, M. (2015). *Blockchain: Blueprint for a New Economy*. O'Reilly Media.

Turkle, S. (2007). *Evocative Objects, Things we Think With*. MIT Press.

Versprille, A. (2022, January 6). NFT Market Surpassed $40 Billion in 2021, New Estimate Shows. *Bloomberg*. https://www.bloomberg.com/news/articles/2022-01-06/nft-market-surpassed-40-billion-in-2021-new-estimate-shows?leadSource=uverify%20wall

Wall Street Journal. (2022). Accounting Firms Scoop Up Virtual Land in the Metaverse. *The Wall Street Journal*. https://www.wsj.com/articles/accounting-firms-scoop-up-virtual-land-in-the-Metaverse-11641599590/.

Wood, L. (2021). Global Gaming Industry to Cross $314 Billion by 2026 - Microsoft, Nintendo, Twitch, and Activision Have All Reached New Heights in Player Investment, Helped by COVID-19. *Research and Markets*. (https://finance.yahoo.com/news/global-gaming-industry-cross-314-183000310.html)

Zhao, Q. (2011). 10 Scientific Problems in Virtual Reality. *Communications of the ACM*, *54*(2), 116–119. doi:10.1145/1897816.1897847

Chapter 13
Metaverse or Not Metaverse:
A Content Analysis of Turkish Scholars' Approaches to Edufication in the Metaverse

Hatice Büber Kaya
Kırklareli University, Turkey

Tuğba Mutlu
Yozgat Bozok University, Turkey

ABSTRACT

This chapter examines how Turkish educational researchers used the Metaverse concept during the Covid-19 pandemic. A thorough examination of the term metaverse is conducted from its emergence, evolution, past and current usage within the academic circles, along with the terms edufication and gamification. Document analysis method is chosen to do a systematic literature review on academic journals. Search was conducted with added keywords to the Metaverse within the literature in Turkish and English. Although the term is a relatively new entrant to the academic research, it is continuously evolving along with the technological advancements its applications to the industry and social life; hence, the concept and its usage will be a continuous subject of further research. The result of the analysis showed that Turkish researchers mainly prefer 'three-dimensional virtual world' instead of the term Metaverse, and the ones found to use Metaverse mainly misuse the term out of its internationally accepted definition.

INTRODUCTION

The term metaverse was first mentioned and described in the novel published in 1992 "Snow Crash" by Neal Stephonson as a three-dimensional (3D) virtual world (VW) where people existed and interacted with each other as avatars (Joshua, 2017). Since then, many virtual worlds like CitySpace, Active Worlds, SecondLife and Roblox were created (Dionisio et al., 2013). Before October 2021 announcement of the founder of Facebook Mark Zuckerberg, the novel and the movie based on the novel "Ready Player One"

DOI: 10.4018/978-1-6684-5732-0.ch013

made great contribution to make the term concrete in people's brains (Sparkes, 2021). In fall of 2020, 'Omniverse' was introduced by Jensen Huang, CEO of NVIDIA as future of the internet (NVIDIA, 2021)and the Roblox company went public with the first metaverse Initial Public Opening (IPO) (Roblox Corporation, 2020). Finally, but not last, in October 2021 Facebook attracted the crowds' attention on the term by renaming itself Meta, while emphasizing that the new era will be under the umbrella term metaverse, and Meta is intended to be the Player One among the other giant technology companies that already informed to have been investing in the Metaverse. Before long, South Korea announced that the City Seoul joined metaverse and some universities mentioned that they are working on their Metaverse campus (Albawaba, 2021). Thenceforward, the studies dated before October 2021 on metaverse, which were only a few to be named, became popular like never before and new studies have been produced in an enormous speed all over the world (Damar, 2021; Narin, 2021).

Covid-19 pandemic has been (and looks like it will continue to be) a universal problem and with regards to education in the pandemic environment, and many countries addressed this problem almost in the same way. They all have switched to distance education rapidly, regardless of to what extent and in what sense their communities were prepared for such a drastic change. While distance education inevitably became the new normal globally, issues and advancements about metaverse and the possibilities it may bring to the education became very popular not only in the society but also in in the academic community globally. Turkish researchers have a considerable amount of research on the education in the Metaverse to deliver a gameful experience to users. Therefore, from what perspective scholars who grew up within Turkish culture have taken this topic worth addressing academically. In this chapter, we will cover how the term metaverse is used, by investigating content of the articles published by Turkish scholars on Metaverse with regards to edufication during the Covid-19 pandemic.

BACKGROUND

Metaverse

The Metaverse is a hard to define and a complex concept. The term has extended way beyond Stephenson's 1992 3D virtual world, by encapsulating physical word 'things' that interact with virtual environments. In the 2007 report of the Acceleration Studies Foundation's (ASF) project named MetaVerse Roadmap (MVR) (Smart et al., 2007), the term's definition was given a good starting point: *"The Metaverse is the convergence of 1) virtually-enhanced physical reality and 2) physically persistent virtual space. It is a fusion of both, while allowing users to experience it as either."*(p.4).

Gamification and Edufication

Nick Pelling was the first one that mentioned gamification in 2002, however the term was brought to the literature by Marczewski in 2010 as *"The application of gaming metaphors to real life tasks to influence behavior, improve motivation and enhance engagement."* (Marczewski, 2013, p. 4). Gamification is simply applying fun factor of games to any tasks including education while edufication is built up via analogy with education, defined as enriching a non-educational game with educational experiences (Becker & Parker, 2014; Hans Hwang, 2017; Oranje et al., 2019).

Regarding the Metaverse, it is a space that is not specific to education just like the real world, there are educational aspects in it with numerous other aspects. The education experts want to take place in this era as soon as possible while taking advantage of game-factor not by adding games on top of their educational experience but adding educational experiences into the Metaverse. Therefore 'edufication' seems to meet the intended meaning more than 'gamification'.

Problems of the Meaning of Metaverse

While the term metaverse has already been misused as a synonym of VWs; which are including but not limited to the above-mentioned VWs (Park & Kim, 2022), it is not only more than one VW but an embracing term that includes all those VWs, VR and AR technologies and more as a container. Hence, the key feature of the Metaverse is building a web of interconnected multiuser platforms that allow users to interact both in real-life and immersive virtual environments at the same time with real-time and dynamic communication (Mystakidis, 2022). Inevitably, each application plays an important role on both building the Metaverse and embodying its possibilities (Suzuki et al., 2020).

Converting independent VWs into an integrated verse has its own challenges for each research area. Complex issues like ethical problems, fintech infrastructure of it, legal regulations, and more are being and to be discussed in every aspect. Educational researchers have their own discussions as well. According to Dionisio (2013) it has four aspects to be considered: realism, ubiquity, interoperability, and scalability. These multiple-user platforms need to allow learners to use their own avatar through multiple technological devices to make their interaction more sensible and realistic while the objects are transitive between real-world and multiple VWs. Therefore, the learners will feel almost in an additional dimension of real-life, not in a virtual one.

Regarding the education technology, the Metaverse opens up an almost unlimited number of possibilities adding edutaintment's advantages on top of it. Similarly, to the term itself, using the term metaverse in edufication was stuck around SecondLife in earlier studies (Kemp & Livingstone, 2006; Park & Kim, 2022). On the contrary, it is now used as a substitution of almost every technology (AR, VR, VW etc.) used in education. Due to wide scope of the Metaverse, all those technologies take part in it however none of them is enough on its own, moreover it is not essential to use AR or VR to make a metaverse application (Kemp & Livingstone, 2006).

RESEARCH METHODOLOGY FOR THIS STUDY

Research Questions

In conducting this research, the following questions were used to guide the systematic analysis of the literature.

1. How Turkish educational researchers used metaverse concept during the Covid-19 pandemic?
2. Is the term Metaverse being misused by Turkish educational researchers as specified within the international literature?

Literature Search and Evaluation

Materials and Procedure

In order to obtain an insight about the Turkish scholar's point of view about the term 'Metaverse', it was decided to make a *systematic literature review* by using *document analysis* method which is a procedure for both data collection and analysis. Document analysis method involves collecting the data from the documents and evaluating them systematically. The method uses interrelated stages to be followed for raising the data to a conceptual level in order to find an answer in scientific research (Corbin & Strauss, 2008; Lune & Berg, 2017). In this case the data is not a raw data but a list of documents that meet the criteria of this research, documents to be appraised from a specific point of view within a specific period of time.

Inclusion Criterion

To be able to obtain more concise view of the Turkish educational researchers' perception of the term, it was decided to base this research on academic journals. It may seem as excluding those other valuable researches; like theses, conference proceedings, books and book chapters, but keeping an intensive focus on the most current research as they often considered to be.

Time period

The COVID-19 pandemic, also known as the coronavirus pandemic, is an ongoing global pandemic of coronavirus disease 2019 (COVID-19) caused by severe acute respiratory syndrome coronavirus 2 (SARS-CoV-2). The novel virus was first identified from an outbreak in Wuhan, China, in December 2019. (Wikipedia, n.d.)

On January 30, 2020; The World Health Organization (WHO) declared Covid-19 as a Public Health Emergency of International Concern and pandemic on March 11, 2020 (World Health Organization, n.d.). As of April 2022, the world is in the third year of pandemic and according to the WHO's statement dated on March 11, 2022; the pandemic is not over yet (Rocel Ann Junio, 2022).

The Covid-19 pandemic radically changed our lives in all manners: socially, psychologically, and economically. While the world has been expecting educational technologies and gamification to be the remedy for the limitations of both traditional education methods and distance learning within two-dimensional web; the pandemic threw all word in distance learning environments without examining who is ready for it. After the first surprise, scholars had turned their eyes to metaverse and its possibilities to see how it may be used to make distance learning more realistic, to be able to increase its effectiveness and accessibility. Moreover, Lee (2021), expressed that the increasing popularity of the Metaverse is also caused by the lack of socialization of people due to Covid-19 pandemic.

Therefore, the main goal of this study was to analyze the research articles published by Turkish researchers in the field of edufication to be able to explore how they use the term Metaverse in their research during the Covid-19 pandemic period. Although the pandemic is not over yet, this study was made in April 2022 and therefore, the research articles that were published between March 2020 and April 2022 was included in this research.

The Point of View

Throughout the history, adapting to a new situation was a challenge for people and finding solutions to the problems faced on this way differs by culture. Matsumoto (2006) on the other hand, explains how cultures were shaped at first, examining how the universal problems are being solved differently by different groups of people, and then these solutions evolve to become the cultures of these groups themselves. Although many different definitions might be found, a shared system of how people behave when encountered with a universal problem, shapes the culture. Moreover, behavior of the individual members of the group are also shaped by people's ways of life and living conditions, which may not always be their personal choice but depending on the group they were born in.

As it was mentioned earlier; the emergence of distance education and the Metaverse coming into the spotlight occurred close to each other. Exactly similar to the case in the other parts of the world, Turkish scholars got excited by the possibilities of the Metaverse's in the field of education. Consequently, in this research, content of the articles published by Turkish scholars on Metaverse with regards to edufication was investigated and analyzed.

Literature Identification

To identify the literature, a literature search was made by using the following keywords to be added to "metaverse" in both Turkish and English: "education", "game", "gamification", "edutaintment", "edufication", "game-based learning", "educational games" in April 2022. TR Dizin, DergiPark and Tubitak Harman databases were searched, using both English and Turkish keywords because these databases only include the researches published in Turkey. Google Scholar web page was searched, using only Turkish keywords and only in Turkish pages. The searches yielded a total of 531 results which were given in *Table 1. Search results.*

Table 1. Search results

Keyword	TR Dizin	DergiPark	Tubitak Harman	Google Scholar
Metaverse	1	18	21	110
Metaverse + Education (Turkish Eğitim)	0	5	0	88
Metaverse + Education (English)	0	7	4	40
Metaverse + Learning (Turkish Öğrenme)	0	0	0	39
Metaverse + Learning (English)	0	9	0	35
Metaverse + Gamification (Turkish Oyunlaştırma)	0	0	0	5
Metaverse + Gamification (English)	0	1	0	2
Metaverse + Edutaintment (English)	0	0	0	0
Metaverse + Edufication (English)	0	0	0	0
Metaverse + Game+ Education (Turkish oyun + eğitim)	0	1	0	65
Metaverse + Game + Learning (Turkish Oyun + öğrenme)	0	0	0	32
Metaverse + Game+ Education (English)	0	2	0	18
Metaverse + Game + Learning (English)	0	1	0	17
Metaverse + Game-based Learning (English)	0	2	0	2
Metaverse + Educational Games (Turkish Eğitsel oyunlar)	0	0	0	3
Metaverse + Educational Games (English)	0	3	0	0
TOTAL	1	49	25	456

After the duplicates and the non-journal articles that could not be filtered on Google Scholar search were eliminated, 59 articles were identified.

The 59 articles' titles, abstracts and keywords were read to identify which may be classified as educational research. After carefully resolving of the manuscripts, computer science (3), sociology and psychology (8), law (1), business (23), philosophy (3), e-sports (1), arts (5), and architecture (1) articles were excluded; 15 articles were deemed relevant to educational research and their full-text were obtained for quality assessment.

The full-text articles were skimmed for further assessment, four of the articles were excluded because they were reviews of international research, included the education field however, they were not showing the Turkish scholar's point of view. One of the articles was excluded because, it was a social media posts' classification which included educational posts. Although the authors put 'education' in their keywords list, it is only one of their classifications, it could not be classified as educational research as a whole. Two of the articles were excluded because they were published in magazines instead of a peer reviewed journal. The reason that these articles got to this stage is because the magazines were published by universities in Turkey, and they were classified as a journal in google scholar database. After the quality assessment stage had ended, 7 articles were deemed relevant to this research, and they were examined deeply for coding. It is important to note that there was one article that the main discipline of the research seems different than educational research, however it was classified as educational research by both authors of this research only because it is about the perceptions of the students of their discipline about the Metaverse era.

DATA EXTRACTION AND ANALYSIS

The following analysis consists mainly of coding, held in its most common description as analyzing raw data and putting into a conceptual understanding. The act associated with the study is coding and concepts created through this act are called codes. It should be emphasized clearly that coding is a thorough and in-depth study that includes analysis of the data utilizing techniques i.e. questioning the data, comparing and contrasting the data in order to be able to develop the concepts from the data (Corbin & Strauss, 2008). Coding can be described as mining of the data similar to a mineral mine where one needs to dig deep into the ground to get the real value of the resources. The act of digging in the metaphor of data versus mine can consist of reviewing transcribed fieldnotes and dividing and disassembling them while taking their connections into account. The next stage of the analysis is set upon how the collected data is differentiated and then combined based on its properties and dimensions (Miles & Huberman, 1994).

How the Metaverse was Defined in the Chosen Articles

The articles were examined in terms of the definitions they provided for the term Metaverse within their concept and following codes were driven from the articles (7) to provide a projection of the term: imitation of real world (2), synonym of Virtual World (3), 3D- AR and VR services (2), digital era (2), and universe of many VWs (1). This section briefly introduces the definitions of the term Metaverse in the full text of selected articles. The articles were presented chronologically.

The first article included in this research, which was published by Demirbağ (2020), used the term Metaverse as an inspirational concept of 3D virtual worlds and defined as: 'Interactive and multiuser environments, where the real world is imitated with 3D VR technology'. This definition is consistent with their inference from review of literature, that is the advancement of VWs was ruled by the developments in both sci-fi literature and game industry, and these advancements led the researchers to define Metaverse – 3D VW as above mentioned since it is very similar to Stephenson's definition.

Damar (2021) conducted a bibliometric study on Metaverse, and defined Metaverse as a 3D virtual shared world. His definition depends on the AR and VR services and equipment, which are introduced as essential peripherals of this technology. Although the definition seems to be quite narrow at first glance, the article was concluded with the acceptance that the term is complex and will take many years to develop and understand thoroughly.

Sağlık and Yıldız (2021), conducted a study on Web 2.0 tools in education. Although they use Metaverse and its components (VR, AR, VWs and so.) often in the study, they did not define Metaverse as a part of their research. Therefore, this study was not included in the coding process.

Dursun and Yetimova (2022), stated that the Metaverse era and obligatory distance education period during the pandemic, caused Turkish students' perceptions about their future work to shift towards digital business areas from the traditional ones. They used the Metaverse as a synonym to 'internet of the World' in the Industry 4.0 era. This definition seems vague since it might stand either for 'a net of Virtual Worlds' or 'a World similar to internet'. Apart from this vagueness, other meanings might have been meant. The reason could be the translation problems, since Dursun and Yetimova's article is written in Turkish. It might have been translated not accurately while the international literature was synthesized. The same situation appears also for Altan and Özmusul's article (2022), in which the Metaverse term was used as a synonym to Digital era, telecommunications era and artificial intelligence. In this article, accompanying the previously mentioned situation, there might be a desire to use a new and popular world in the article for SEO (search engine optimization) purposes.

Çelik (2022), stated in his article that the new fictional universes had drawn the digital era population in with new communication technologies. The metaverse was defined as a VW in which all worlds are gathered where all digital elements are combined, and hence a virtual living space was created. Çelik also suggested that Web 3.0 technologies including AR and VR will be existed without intermediaries on the metaverse-based digital environments. This definition seems to be the closest to the definition that was given on MVR (Smart et al., 2007).

While Türk and Darı's article (2022) is about sociology of communication, it was included into the coding process of this research because Türk and Darı stated that socialization is shaped by educational institutions along with other institutions. Educational institutions, unlike many other units of society, subject individuals to the modern formation process by using the innovations in communication tools. In this way, the experiences of the individual outside the educational institutions (i.e. to have fun, work, relax) also pass through this formation filter. Moreover, they defined the Metaverse as a special 3D universe where all the actions of socialization including but not limited to shopping, education, entertainment and working, are carried out through special clothes and apparatus. Hence it is a new and developing concept, as they suggested, its definition is being built day by day.

LIMITATIONS AND FUTURE RESEARCH DIRECTIONS

This research is limited with:

- The date of literature search 26.04.2022,
- The TR Dizin, DergiPark and Tubitak Harman ana Google Scholar web databases of search,
- The keywords that authors chose,
- The search result of 59 research in which 10 of them included in the study.

Possible future studies may include the theses, proceedings, and book chapters in the field for deeper understanding.

CONCLUSION

The metaverse is considered as a new and complex concept containing all aspects of real life since it is a hybrid reality in which people are expected to be confused about the boundaries between real life and Metaverse. As a natural consequence of this situation, it offers almost endless possibilities of research in all fields. This review article is conducted to throw a light to the misuse of Metaverse within the Turkish educational academic literature, which seems to be affected from colloquial misuse of the term. With the Covid-19 pandemic, people had to do practice many activities in the virtual environment i.e. socialization with friends, taking courses, attending concerts, working from home, having distant education, most of which had to happen in two dimensions online. This obligatory situation made the Metaverse concept in people's heads more concrete than ever before and a spark went off for the new dimension of the internet. The giant technology companies invested in this idea to support people to be a part of this hybrid reality that allows them to switch between VWs easily. When this idea is brought to life, most of the daily activities can be performed within itself.

To return to the research questions of this study, Turkish educational researchers mostly used metaverse concept during the Covid-19 pandemic as suggested in international research, synonymously with virtual worlds and/or imitation of real world. Also, some of the researchers defined the components of Metaverse (AR, VR etc.) instead of the term itself.

Although the journal articles were chosen because it was suggested as the most current research indicator in international research (Alasuutari, 2010; Hargens, 2000), it was concluded that this is not the case for edufication researchers in Turkey. While screening the first 531 research for inclusion, both authors noted that edufication researchers in Turkey are more likely to share their research as books and book chapters. This may be because edufication researchers produce publications on the interaction of different disciplines, as Clemens et al. (1995) suggested *"...Books generate conversations across subfields and disciplines..."*(Clemens et al., 1995, p. 433).

Regarding the edufication researchers, it is important to note that, Turkish edufication researchers were observed to use the term "Three-dimensional virtual world" in Turkish instead of the term Metaverse, which explains the narrow search results.

REFERENCES

Alasuutari, P. (2010). The rise and relevance of qualitative research. *International Journal of Social Research Methodology*, *13*(2), 139–155. doi:10.1080/13645570902966056

Albawaba. (2021). Seoul Becomes First City to Join the Metaverse. *Albawaba*. https://www.albawaba.com/business/seoul-becomes-first-city- join-metaverse-1454641

Altan, M. Z., & Özmusul, M. (2022). Geleceğin Türkiye'sinde Öğretmen Refahı: Öğretmenlik Meslek Kanununun Kayıp Parçası [Teacher Welfare in Future Turkey: The Missing Part of the Teaching Profession Law]. *Ahmet Keleşoğlu Eğitim Fakültesi Dergisi*, *4*(1), 24–42.

Becker, K., & Parker, J. (2014). An overview of game design techniques. In *Learning, education and games,* (pp. 179–198). ETC Press.

Çelik, R. (2022). Metaverse Nedir? Kavramsal Değerlendirme ve Genel Bakış [What Is The Metaverse? Conceptual Evaluation And Overview]. *Balkan and Near Eastern Journal of Social Sciences*, *8*(1), 67–74.

Clemens, E. S., Powell, W. W., McIlwaine, K., & Okamoto, D. (1995). Careers in print: Books, journals, and scholarly reputations. *American Journal of Sociology*, *101*(2), 433–494. doi:10.1086/230730

Corbin, J., & Strauss, A. (2008). *Basics of qualitative research: Techniques and procedures for developing grounded theory,* (3rd ed.). SAGE. doi:10.4135/9781452230153

Damar, M. (2021). Metaverse Shape of Your Life for Future: A bibliometric snapshot. *Journal of Metaverse, 1*(1), 1–8.

Demirbağ, İ. (2020). Üç boyutlu sanal dünyalar [Three-dimensional virtual worlds]. *Açıköğretim Uygulamaları ve Araştırmaları Dergisi, 6*(4), 97–112.

Dionisio, J. D. N., Burns, W. G. III, & Gilbert, R. (2013). 3D virtual worlds and the metaverse: Current status and future possibilities. *ACM Computing Surveys, 45*(3), 1–38. doi:10.1145/2480741.2480751

Dursun, T., & Yetimova, S. (2022). Güncel İletişim Teknolojilerinin Sunduğu Dijital İş İmkânlarının (Digital Business) Öğrenciler Tarafından Nasıl Algılandığı Üzerine Bir Saha Araştırması [Field Research On How Students Perceive The Digital Business Offered By Current Communication Technologies]. *Third Sector Social Economic Review, 57*(1), 366–391.

Hargens, L. L. (2000). Using the literature: Reference networks, reference contexts, and the social structure of scholarship. *American Sociological Review, 65*(6), 846–865. doi:10.2307/2657516

Hwang, H. (2017). The "Edufication" of Gaming. *LinkedIn*. https://www.linkedin.com/pulse/edufication-gaming-hans-hwang

Joshua, J. (2017). Information Bodies: Computational Anxiety in Neal Stephenson's Snow Crash. *Interdisciplinary Literary Studies, 19*(1), 17–47. doi:10.5325/intelitestud.19.1.0017

Kemp, J., & Livingstone, D. (2006). Putting a Second Life "metaverse" skin on learning management systems. *Proceedings of the First Second Life Education Workshop, Part of the 2006 Second Life Community Convention, 20*.

Lee, B.-K. (2021). The Metaverse World and Our Future. *Review of Korea Contents Association, 19*(1), 13–17. https://www.koreascience.or.kr/article/JAKO202119759273785.pdf

Lune, H., & Berg, B. L. (2017). *Qualitative research methods for the social sciences*. Pearson.

Marczewski, A. (2013). *Gamification: a simple introduction*. Andrzej Marczewski.

Matsumoto, D. (2006). Culture and nonverbal behavior. In V. Manusov & M. L. Patterson (Eds.), *The SAGE handbook of nonverbal communication,* (pp. 219–235). Sage Publications, Inc., doi:10.4135/9781412976152.n12

Miles, M. B., & Huberman, A. M. (1994). *Qualitative data analysis: An expanded sourcebook,* (2nd ed.). Sage Publications, Inc.

Mystakidis, S. (2022). Metaverse. *Encyclopedia, 2*(1), 486–497. doi:10.3390/encyclopedia2010031

Narin, N. G. (2021). A Content Analysis of the Metaverse Articles. *Journal of Metaverse, 1*(1), 17–24. https://dergipark.org.tr/en/download/article-file/2167699

NVIDIA. (2021). *GTC Spring 2021 Keynote with NVIDIA CEO Jensen Huan*. YouTube. https://www.youtube.com/watch?v=eAn_oiZwUXA

Oranje, A., Mislevy, B., Bauer, M. I., & Jackson, G. T. (2019). Summative game-based assessment. In D. Ifenthaler & Y. J. Kim (Eds.), *Game-based assessment revisited,* (pp. 37–65). Springer International Publishing., doi:10.1007/978-3-030-15569-8_3

Park, S.-M., & Kim, Y.-G. (2022). A Metaverse: Taxonomy, components, applications, and open challenges. *IEEE Access : Practical Innovations, Open Solutions*, *10*, 4209–4251. doi:10.1109/ACCESS.2021.3140175

Roblox Corporation. (2020, November 19). Registration statement on form S-1. *SEC.gov*. https://www.sec.gov/Archives/edgar/data/0001315098/000119312520298230/d87104ds1.htm

Sağlık, Z. Y., & Yıldız, M. (2021). Türkiye'de Dil Öğretiminde Web 2.0 Araçlarının Kullanımına Yönelik Yapılan Çalışmaların Sistematik İncelemesi [A Systematic Review of Studies on the Use of Web 2.0 Tools in Language Teaching in Turkey] [JRES]. *Eğitim ve Toplum Araştırmaları Dergisi*, *8*(2), 418–442. doi:10.51725/etad.1011687

Smart, J. M., Cascio, J., & Paffendorf, J. (2007). *Metaverse Roadmap Overview*. https://www.metaverseroadmap.org/overview/

Sparkes, M. (2021). What is a metaverse. *New Scientist*, *251*(3348), 1–18. doi:10.1016/S0262-4079(21)01450-0

Suzuki, S. N., Kanematsu, H., Barry, D. M., Ogawa, N., Yajima, K., Nakahira, K. T., Shirai, T., Kawaguchi, M., Kobayashi, T., & Yoshitake, M. (2020). Virtual experiments in metaverse and their applications to collaborative projects: The framework and its significance. *Procedia Computer Science*, *176*, 2125–2132. doi:10.1016/j.procs.2020.09.249

Türk, G. D., & Darı, A. B. (2022). Metaverse'de bireyin toplumsallaşma süreci [The Socialization Process Of The Individual In The Metaverse]. *Stratejik ve Sosyal Araştırmalar Dergisi*, *6*(1), 277–297.

Wikipedia. (n.d.). COVID-19 pandemic. *Wikipedia*. https://en.wikipedia.org/wiki/COVID-19_pandemic

World Health Organization. (2020). Listings of WHO's response to COVID-19. *WHO*. https://www.who.int/news/item/29-06-2020-covidtimeline

Yadav, R. & Moon, S. (2022, March 11). Opinion: Is the pandemic ending soon? *WHO*. https://www.who.int/philippines/news/detail/11-03-2022-opinion-is-the-pandemic-ending-soon

ADDITIONAL READING

Çelik, R. (2022). Metaverse Nedir? Kavramsal Değerlendirme ve Genel Bakış [What Is The Metaverse? Conceptual Evaluation And Overview]. *Balkan and Near Eastern Journal of Social Sciences*, *8*(1), 67–74.

Damar, M. (2021). Metaverse Shape of Your Life for Future: A bibliometric snapshot. *Journal of Metaverse*, *1*(1), 1–8.

Demirbağ, İ. (2020). Üç boyutlu sanal dünyalar [Three-dimensional virtual worlds]. *Açıköğretim Uygulamaları ve Araştırmaları Dergisi*, *6*(4), 97–112.

Gökkaya, Z. (2014). Yetişkin Eğitiminde Yeni Bir Yaklaşım: Oyunlaştırma [A New Approach Of Adult Education: Dramatization]. *Hayef Journal of Education*, *11*(1), 71–84.

Hursen, C., & Bas, C. (2019). Use of Gamification Applications in Science Education. *International Journal of emerging technologies in Learning, 14*(1).

Mystakidis, S. (2022). Metaverse. *Metaverse Encyclopedia*, *2*(1), 486–497. doi:10.3390/encyclopedia2010031

Sparkes, M. (2021). What is a metaverse. *New Scientist*, *251*(3348), 1–18. doi:10.1016/S0262-4079(21)01450-0

KEY TERMS AND DEFINITIONS

Augmented Reality (AR): A computer-generated synchronous or asynchronous enhanced appearance of real-life spaces with sounds, graphics.

Avatar: An image that was created by user electronically, it may be realistic or manipulated, depending on the user's wishes.

Edufication: Enriching a non-educational game with educational experiences.

Gamification: Applying fun factor of the games into any real-life tasks to transfer advantage of gameful experiences ie. motivation to engage.

Metaverse: A web of interconnected multiuser platforms that allow users to interact both in real-life and immersive virtual environments at the same time with real-time and dynamic communication.

Three-dimensional Virtual World (3D VW): A three-dimensional digital space that represents a world with some realistic or figmentive characteristics.

Virtual Reality (VR): An immersive hybrid reality that contains fictions and imagination which was created using technology.

Chapter 14
Effects of Digital Technologies on Academic Performance of Nigerian Adolescents

Desmond Onyemechi Okocha
https://orcid.org/0000-0001-5070-280X
Bingham University, Nigeria

Julson Nimat James
Bingham University, Nigeria

Terhile Agaku
Bingham University, Nigeria

ABSTRACT

The rapid development of technology has penetrated almost all sectors of the society and makes any form of resistance almost impossible. The incorporation of digital technology into adolescent's daily life, as well as its impact on their cognitive, emotional, and social development, is growing by the day. They can use technology to play, explore, and learn in a variety of ways. This is because their brains are so adaptable, these learning opportunities represent a vital growth stage throughout this time period, which helps and encourages them to improve their communication skills and knowledge. No one will deny the numerous benefits that accrue from digital technology usage such as the internet and social network, which is an instant hit site that launches individuals into the world beyond imagination. A growing number of adolescents believe that happiness is linked to direct and indirect interactions with digital environments and technology that facilitate and mediate communication. This study investigated the effects of digital technology on adolescents in Nigeria.

DOI: 10.4018/978-1-6684-5732-0.ch014

INTRODUCTION

Technology has become part of man as it has evolved from the mastery of fire by man to super computers and electronics of this generation. People, regardless of their region, have found technology to be useful. The rapid development of technology has penetrated almost all sectors of the society and makes any resistance almost impossible. No one can deny the various advantages that a social network like Facebook provides, which is an immediate hit site that immerses users in a world beyond their wildest dreams. It assists in resolving lost contacts as long as the name survives in the internet world, telling people about others, and providing updates on significant life events. Technology is also used for the purpose of educational interactions, educating and enlightening one another on academics social and moral issues (Eberendu, 2015).

According to Queen Rania Foundation, studies have consistently revealed that digital technology is associated with moderate learning gains: on average, an additional four months' progress. However, there is considerable variation in impact. Technology approaches are used to supplement other teaching, rather than replace more traditional approaches. It is unlikely that particular technologies bring about changes in learning directly, but some have the potential to enable changes in teaching and learning interactions. For example, they can support teachers to provide more effective feedback or use more helpful representations, or they can motivate students to practise more.

Adolescents today are growing up in an increasingly digital world, with technology and computing becoming an inextricable aspect of their personal life and the society at large. Adolescents with digital access, unlike earlier generations, live both online and offline lives, and utilize digital technology not only to explore the world around them but also to discover their positions in it. Adolescents' key gateways to digital worlds are devices that allows them access to platforms and services powered by data analytics, machine learning, and other forms of artificial intelligence (AI). A growing number of adolescents believe that happiness is linked to direct and indirect interactions with digital environments and technology that facilitate and mediate communication. It is worth noting that digital experiences and habits develop during childhood and adolescence can have a beneficial or bad impact on the trail to maturity (Holly et al., 2022).

Digital technology is widely accepted in Nigeria. Uzuegbunam (2019) asserts that Nigeria is one of the African countries with the highest and consistently growing internet penetration. He further noted that a growing number of Nigerian adolescents have access to digital technologies, particularly mobile phones and the internet. Many of them have access to either a shared smartphone, personal smartphones, or just feature phones bought by their parents, guardians, or older relatives.

It is on record that the electronic digital computer made its first appearance in Nigeria in 1963, The Information and Communication Technology (ICT) in Schools was launched in December, 2004 and revised in 2010 to provide opportunities for secondary school students to mainly build their capacity on ICT skills and make them learn through computer aided learning process in connection with the analysis of the 1962/1963 national census data. Smartphone adoption in Nigeria is predicated upon the advent of mobile telecommunication in Nigeria in 2011 as this development allowed for the exposure to mobile phones and internet penetration in Nigeria.

Digital technology has evoked a seismic shift in the ways children learn, play and communicate in Nigeria. New technologies have permeated and transformed life in the 21st century. Children of this generation have been exposed with digital technologies and are the most regular users of new online and digital services (OECD, 2016). In our society today, almost nine out of ten adolescents use diverse digital devices. They utilize it for a variety of things such as business, social networking, research, and assignments. But the question still remains, how many of these adolescents make effective and efficient use of these digital devices? Against this backdrop, the study evaluates the effects of digital technologies on the academic performance of adolescents in Jos South Local Government Area of Plateau State.

OBJECTIVES OF THE STUDY

The broad objective of the study is to assess the effects of digital technologies on the academic performance of adolescents in Jos South Local Government Area of Plateau State. Specifically, however, the study aims to:

1. Evaluate the effects of digital technologies on the academic performance of adolescents
2. Determine the level of adolescent's usage of digital technologies

CONCEPTUAL CLARIFICATIONS

Digital Technology

The concept of digital technology was invented to help man in achieving certain goals and objectives. In line with the Nigerian Communication Commission (NCC), digital technology refers to electronic tools, systems, processes, equipment, and resources that generate, store, or analyse data so as to fulfil the user-defined goals. Computers, television, radio, social media, online gaming, multimedia, mobile phones, are all well-known examples of digital technology.

Olajide, Afolabi and Ajayi (2019) argued that digital technology is a branch of scientific or engineering knowledge that deals with the practical use of digital or computerized devices, methods, systems. Digital technology also refers to electronic tools, systems, devices and resources which generate, store or process data. Well known examples include social media, online games, multimedia and mobile phones (https://www.education.vic.gov.au/school/teachers/teachingresources/digital/Pages). Digital technology is also concerned with the use of computer and technology assisted strategies to support learning within schools. Approaches in this area vary widely, but generally involve: technology for students, where learners use programmes or applications designed for problem solving or open-ended learning; or technology for teachers, such as interactive whiteboards or learning platforms.

Adolescents

Adolescent is often connected to puberty and the cycle of physiological changes that advances to reproductive maturity in young adults. Adolescence is a transitional phase of growth and development between childhood and adulthood. It is during this time that knowledge and skills are developed, emotions are learned or controlled, and qualities and abilities are acquired in order to enjoy the adolescent years and adult roles. Quoting Britannica (2019), Olajide, Afolabi and Ajayi (2019) argued that adolescence is a transitional phase of growth and development between childhood and adulthood. The implication is that children between this phase of growth and development are referred to as adolescents. The term "adolescence" is a sociological construct, like other developmental phases in human growth and development, but unlike others, it breeds a lot of ambiguity.

Adolescence can also be described as the period of life between childhood and adulthood which corresponds roughly to the teenage years (thirteen to eighteen years).

The World Health Organization (WHO) describes adolescence as the stage between childhood and adulthood, from ages 10 to 19. It is said to be a unique stage of human development and an important time for laying the foundations of good health. Adolescents usually experience rapid physical, cognitive and psychosocial growth. This affects how they feel, think, make decisions, and interact with the world around them. During this phase, adolescents establish patterns of behavior – for instance, related to diet, physical activity, substance use, and sexual activity – that can protect their health and the health of others around them, or put their health at risk now and in the future. Adolescents are characterized by rebellion, commencement of search for identity, concern about 'looking good, increased peer group alliance, increased sex drive, increased aggressive drive and unpredictable attitudes and opinions (Olajide, Afolabi and Ajayi, 2019).

THEORETICAL FRAMEWORK

The study is anchored on the Uses and Gratification theory. The theory was propounded by Katz and Blumer (1974) to explain why people use certain types of media, what needs do they have to use them, and what gratifications do they get from using them. The Uses and Gratifications Theory (UGT) proposes that people choose to consume certain kinds of media because they expect to obtain specific gratifications as a result of those selections. Beneficiaries make planned, purposeful choices about the media messages they open themselves which has a positive or negative impact in their lives.

Shraddha (2018) argues that human needs and gratification can be divided into five broad categories such as:

1. Affective needs: Affective needs refer to the emotional satisfaction and pleasure people derive from viewing soap operas, television series, and movies. People identify with the characters and empathize with the emotions they display. When they cry, the audience cries with them, and when they laugh, the audience laughs with them.
2. Cognitive needs: People turn to the media for knowledge and to meet their mental and intellectual demands. People mostly watch news to satisfy this urge. Quiz shows, educational shows, children's arts and crafts shows, documentaries, how-to films (DIYs), and so on are all examples. The Internet is also being utilized to obtain information in order to meet this demand.
3. Needs for social integration: Each person's need to socialize with others, such as family and friends, is a social integrative need. People utilize social networking sites such as Facebook, Myspace, and Twitter to mingle and interact. People also use media to improve their social relationships by providing them with subjects to discuss with their friends and family. People can also benefit from the media since it provides them with subjects and ideas to discuss with their friends and family, thereby improving their social interaction skills.
4. Integrative requirements of a person: The demands for self-esteem and respect are known as personal integrative needs. People require reassurance in order to build their position, trustworthiness, strength, authority, and other attributes, which is accomplished through the use of media. They use

the media to watch advertisements and learn about current trends in order to adapt their lifestyle and blend in with others.

5. Tension free needs: When people are stressed or bored, they listen to music or watch television to reduce their stress. People may have numerous tensions in their lives that they do not want to face, so they turn to the media for relief.

The uses and gratification theory is relevant to this study because it explains the academic benefits adolescents seek to gain from the use of digital technology and its impact to their academic and social life. It offers explanations to suggest that students use digital technologies for varying reasons. In other words, users select media based on how well each one helps them meet specific needs or goals.

REVIEW OF RELATED LITERATURE

Impact of Digital Technologies on Education

One of the undeniable advantages of the advent and use of digital devices is that it has tremendously aided learning and made education more interesting. This is obvious in the use of digital learning platform which is defined as a new kind of classroom learning infrastructure enabled by advances in theory, research, and one-to-one computing initiatives. This system is designed to operate in a teacher-led classroom as the major carrier of the curriculum content and to function as the primary instructional environment (Dede and Richards 2012).

Olajide, Afolabi and Ajayi (2019) opined that learning and knowledge acquisition has been enhanced through the societal embracement of digital device. This is more so that self-learning via the use of digital devices has become the order of the day for quick, easy and vast knowledge acquisition. Sofela (2012), in his work titled 'The Effect of social media on Students,' highlighted the impact of digital technology on students, stressing on the positive impact in the student's academic performance, although it comes with some negative effects. As such, there is need for the students to create a balance between social media and their academics in order to prevent setbacks.

Higgins et al. (2012) provide a summary of research findings from studies with experimental and quasi-experimental designs, which have been combined in meta-analyses to assess the impact of digital learning in schools. Their search identified 48 studies which synthesised empirical research of the impact of digital tools and resources on the attainment of school age learners (5–18-year-olds). They found a consistent but small positive association between digital learning and educational outcomes. Harris (2009) identified statistically significant findings, positively associating higher levels of ICT use with school achievement at each Key Stage in England, and in English, maths, science, foreign languages and design technology. Meyer (2005) identified a link between high levels of ICT use and improved school performance. They found that the rate of improvement in tests in English at the end of primary education was faster in ICT Test Bed education authorities in England than in equivalent comparator areas.

Olajide, Afolabi and Ajayi (2020) conducted a survey of opinions of a wide spectrum of the society in what digital technology has brought to the nation on the effects of technology on the nation, especially the attendant effects on the adolescents. The research design adopted for the study was a descriptive survey. The population of this study comprised five (5) secondary schools from Ondo West Local Government. The study targeted adolescents from these schools. These five secondary schools were selected to represent the secondary schools in Ondo State. The findings of the study showed that, adolescents' addiction and exposure to digital technology can have a negative effect on their academic performances.

Digital Platforms Used Among Adolescents

The following are common digital platforms used among adolescents:

- **Smart phones and mobile phones**: Smart phones have proven to be extremely valuable to people in recent years; having one smart phone device is seen as a major requirement for adolescents. The first and most vital function of those devices is to enable individuals speak with others. Whether they are close or far away, adolescents can connect with one another through verbal conversations as well as written texts and messages. They use these devices to search for information as well as for entertainment and amusement and they use the phones as cameras to take pictures of individuals. These devices are also used for paying of bills, purchasing groceries and other items. The phones also used to take pictures which individuals are able to transfer easily from one place to another by following the maps on their phones (Capaldo, Flanagan, and Littrell 2008).
- **Digital Camera**: The primary purpose of a camera is for taking pictures of locations, objects, articles, things, and individuals. In educational institutions the main purpose of digital camera is to take pictures of individuals as well as things. It is additionally considered as efficient variety of digital technology. Individuals take pictures of field trips, places, activities, experiments, meetings, presentations, seminars, conferences etc. Pictures enable individuals to get meaningful and important strategies for learning and communicating. (Ten Technologies which would change Our Lives, 2015)
- **Social Media:** In recent years, social networking platforms such as Facebook, WhatsApp Twitter, YouTube and Instagram have gained popularity. They use a wide range of digital technologies to allow users to communicate via text, photographs, and video, as well as form social groups. Facebook, as one of the virtual social networking sites created by Mark Zuckerberg and friends in 2004, has allowed people to stay in touch with one another, share information more easily and increase business marketing opportunities. (Mohd Azul et al. 2019). Facebook is the most popular social media among adolescents and society (Boyd and Ellison, 2008). A study has proven that adolescents nowadays are more comfortable to interact through the Facebook than face to face (Robin, B., 2008). This is because the use of Facebook allows a person to interact with anyone around the world without being limited by time and geographical distance. Through the Facebook social site, all information related to current issues, education and technology can be accessed quickly and easily. Facebook has become a well-known communication channel for sharing information and communicating fast. It is not surprising, then, that the number of Facebook users has risen, particularly among youngsters.
- **WhatsApp** is another popular social media used among adolescent nowadays. It is derived from the English word "what's up "When compared to other social media platforms for connecting with family and friends, WhatsApp offers simplicity and privacy (Jamiah et al. 2016). This program allows a user to communicate or tell a narrative as if there is a genuine presence among the individuals that interact with one another.
- **Instagram** was established as a social networking program in October 2010. Adolescents love to upload and share personal photographs and videos; Instagram is a picture and video sharing app

that allows users to save and share their photographs and videos digitally. Instagram is a photo-sharing program that allows users to share photographs, alter them with the 20 filters available, and then publish them to the Instagram website to be shared with other Instagram users.

- **YouTube** is the next most popular social media application among teenagers. YouTube is a video-sharing website where users can publish, download, watch, comment on, and share video clips, including video blogs. YouTube is a video-based social media platform that disseminates information, entertainment, and news.

Students' Exposure/ Addictiveness to Digital Technologies

Monaha and Schumacher (2000) argued that social media addiction is the excessive use of the internet and the failure to control this usage which seriously harms a person's life. In recent years, it has been observed that students have unrestricted access to digital tools. (Peter & Valkenburg, 2009). Students use computers to send and receive information from all over the world. Even when they do not have personal phones, they still use that of their friends, older siblings at home or even that of their parents. Thanks to the development and distribution of similarly advanced cellular phones. Some schools are so well-equipped that internet connections are provided both in the classroom and in the library. Teenagers have become much more accustomed to this lifestyle in recent years than previous generations, as it is all they know (Lewis, 2008). Teenagers increasingly rely on digital technologies for the majority of their daily activities, as compared to previous generations who relied on television and newspapers.

Peer Pressure

Peer influence can be positive or negative. Coping well with peer influence is about getting the right balance between being yourself and fitting in with your group. Hartney (2022) argued that peer pressure is the process by which members of the same social group influence other members to do things that they may be resistant to, or might not otherwise choose to do. Usually, the term 'peer pressure' is used when people are talking about behaviors that are not considered socially acceptable or desirable, such as experimentation with alcohol or drugs.

For example, they may pressure you into doing something you are uncomfortable with, such as shoplifting, doing drugs or drinking, taking dangerous risks when driving a car, or having sex before you feel ready. Peer Pressure Statistics indicates that 75% of adolescents have tried alcohol due to peer pressure. 28% of those who gave in to peer pressure improved their social status. 70% of teen smokers began as a result of peer pressure. 33% of teen boys feel pressured to have sex.

A recent study on digital technology usage among adolescents revealed that students get to use smartphones or create a social media account even if they do not want to as a way of

"Fitting in" with peers. In fact, it promotes a feeling of self-esteem and wellbeing in students, that if you are not part of it, you become dissatisfied. At such they tend to go any length to get involved in what is in vogue that it has become a fundamental role in students. A student stated research study by Dr. Danah Boyd at Berkeley "If you're not on Myspace, you don't exist" (Boyd, 2007)

Furthermore, because of the disconnect between real life and what people post on social media, they are only exposed to a highly edited 'highlight reel' of other people's lives. This effectively creates the impression that others' lives are more intriguing, perfect, or thrilling than our own. Peer pressure is often seen during the adolescence stage of a teenagers because they often seek comfort among their peers and intend to do what their peers does without knowing if it is good or bad for them. Adolescence is a period in the life of an individual that is transitory, when a child reaches the point in changing its childhood to adulthood (Adeniyi and Kolawole, 2015). Adolescence social environment could affect teenagers in their adolescence, because mostly in this period teenagers tend to communicate more by their peers. As children grow and reach adolescence, teenagers become more dependent with their peers than their family especially in making choices and enhancing their moral values in life (Uslu, 2013).

Peer pressure could easily affect the self-esteem of students that an important factor adolescence. Individuals adapt attitudes towards a certain aspect that they encountered or they are aware of (Uslu, 2013). In many events students fantasize and visualize what they dream to become through their colleagues' atmosphere. Eventually, they pursue their choices through the influence of peer pressure (Owoyele and Toyobo, 2008).

Influence of Digital Technologies on Adolescents Academic Performance

Digital technologies have a huge influence on the academic performance of students particularly adolescents. Odofin (2014) argued that teachers use technology to improve the effectiveness of instructions in the class, motivates students, school attendance, students' participation, and is used as both a learning and communication tool. Technology forms the bases for life learning and problem solving. Students can use technologies in their academics, works, and other areas of interests.

Adomi, Okiy and Roteyan (2004) averred that digital technology when used for the purpose of learning enhances students' engagement in learning by searching the web for information to complete individual and group assignments, communicating via the e-mail, allowing them to connect with their peers, and for entertainment. The emergence of digital technology has been a thing of joy, interest and entertaining to members of the communities. These technologies such as the internet and the assessment of database would help secondary school students to communicate with one another, to be able to search for educational information that will help them in their school tasks and be able to know what is happening around the world (Idowu, Idowu and Adawgouno, 2004).

Most teenagers in the secondary schools today, are fully aware of the benefits that can be derived from the use of technology, e.g. the internet. Studies have shown that students constitute the highest users of the internet in African countries, and most of them use it for E-mail communication. These teenagers and young adults usually send e-mail to communicate with their friends or partners. The effect of these affects all aspects of their lives. In

Nigeria, digital technology may be an invaluable instrument for learning and teaching (Ojedokon and Owolabi, 2005). The developed countries of Europe and America have less problems meeting the educational resources needed by their students. Educational sources such as books, CD-ROM, projector, and the internet are in abundance. The emergence of digital technology in developing countries has bridged the digital gap. This enables students and adults to access and participate globally in sharing of information resources. Hence, secondary school students are now enjoying the existence of digital technology in enabling them to use and tap the available educational resources open online in making new friends internationally, distance learning also known as the E-Learning, sport and entertainment. Secondary school students are in the age of digital technology of acquitting themselves with information that could either affect their academic engagement or disengagement behavior. (Adeogun, 2005).

Adolescents in Nigeria today are occupied with social media networks and technological social lives. How will this affect their studies? It is estimated that even those students who do graduate high school, one out of three does not possess the knowledge and skills that would lead him or her to the next level, such as college or an advanced trade school (Bowen, 2008). The top academic areas that many school professionals are concerned about are mostly lacking in them. This is because instead of studying they do the contrary; they live in a fast-paced technological world with many different types of communication happening all at the same time. For example, he or she may be on the computer playing games, while also talking on the phone, sending instant messages to a friend, and emailing someone else all at the same time (Williams, 2008). Teens and teenagers have taken cognizance of such websites because they allow them to communicate with their classmates, share information, reinvent their personas, and display their social lives. Adolescents' use of English and grammar is influenced by their use of digital technological tools. Students are accustomed to using abbreviated forms of writing words in chat rooms, they forget and use the same in class. They use words like '4' instead of for, 'U' instead of you, 'D' instead of the, and so on, which could affect their grades.

Many arguments can be made about the possible risks of adolescent usage of digital technology, it is important to point out the benefits of these tools as well to the adolescents. The introduction and use of digital gadgets in the twenty-first century has greatly helped learning and made education more entertaining for teens. High school students use these sites as tools to obtain information.

Many schools have started to use these sites to promote education, keep students up to date with assignments, and offer help to those in need (Boyd, 2007). In general, the Internet and social networking sites can be a positive influence on adolescents. It provides an outlet for teens to express themselves in their own unique ways. In addition, they serve both as a meeting place for teens to interact with other like-minded people and as showplaces applying for college visit profiles of that college's students to view pictures and read blogs of past students to determine whether the college would be a good fit (Boyd & Ellison, 2007). It helps them to develop and sustain helpful relationships. They create their own identities through self-expression, learning and talking. The internet helps adolescents in learning how to type faster and the ability to carry out multiple tasks at a time.

Negative Effects of Digital Technologies on Adolescents

Digital technologies, or social media platforms such as Facebook, Twitter, Snapchat, and Instagram are designed to bring people together; yet they may have the opposite effect in some cases because most of the time they are active on social media platforms for posting and sharing messages, photos, videos and gaming. This behaviour may isolate them from the social life. It also does not allow them to spend enough time with their family members, and because of its nature, social media as a digital technology actively encourages social comparisons, as it is rife with records that may easily be exploited as indicators of obvious social achievement (e.g., friends, likes, shares, fans and so forth).

Odofin (2014) rightly observed that digital technology has disconnected children from their parents, peers and academic activities, making it difficult to acquire social skills, communication skills, emotional skills, and basic problem solving skills which may have adverse effect on the child's moral, attitude, psychological well-being, habits and learning behavior, since so much time and attention could be wasted by the child in engaging in the internet. This may include:

1. Digital technology can be a source of distraction to the child: the child may disengage from learning activities or learn less when he or she uses technologies such as tablets, IPhone, IPad, Androids, Phones, Laptops, watching television screen or using other devices during lectures. The uses of

these devices distract the child from learning thereby reducing his attention span and academic engagement.

2. Digital technology can disengage the child from social interactions: spending too much time and using modern electronics devices may cause social withdrawal or hinder social skill development through decrease in face-to-face interaction with people at home and in school.
3. Technology can increase examination malpractice in the classroom: technology now makes it easier for students to engage in examination malpractices, coping and pasting other peoples work. It can really create an avenue for students to cheat.
4. Sleep deprivation: Excessive use of electronic devices relates to sleep deprivation, which affects growth and development in children and adolescents. Risky sexual behavior: sexual content and increased availability of the internet and electronic devices may increase risky sexual behavior.
5. Risky sexual behavior: Sexual content and increased availability of the internet and electronic devices may increase risky sexual behavior.
6. Aggressive behavior: Violet content of videos and online games may have adverse effects on the behavior of children, adolescents and adults.
7. Other social and psychological problems: Excessive use of the internet and electronic devices can be associated with a range of social and psychological problems such as psychological well-being, poor self-confidence, and reduced academic performance (Nwosu, 2019).
8. All students may not have access to technological tools: Since some of these devices are expensive, some students may not have the means to afford them. For such students, the teachers can recommend the use of library or work in groups with their peers and share resources.
9. The quality of information they get online may not be reliable: students may not be able to identify reliable educational sources from unreliable sources. Hence, they should be properly guided on how to differentiate between reliable and unreliable source of information.

These measurements are troublesome in and of themselves, since if teenagers do not receive enough 'likes' on a comment or a photo they have shared, or if someone has more likes or friends than them, they may feel inferior.

RESEARCH METHOD

The research method adopted for this study is the survey research design. Survey research design involves the use of questionnaire or interview to generate data for a particular study particularly in social sciences (Okoro, 2003). The survey research design was employed with questionnaire as instrument for data collection on a sample size of 120 respondents randomly selected. This was to ensure that every member of the population had an equal chance of being selected for the study.

Data Presentation, Interpretation and Analysis

This section deals with the analysis of data. The data collection tool administered by the researcher is a set of questionnaires. The data was collected from some communities and a total of 120 students were included in the overall data collected.

Table 1. Demographic Characteristics of Respondents

Sex	Frequency	Percentage (%)
Male	65	54.2
Female	55	45.8
Total	120	100
Age	**Frequency**	**Percentage (%)**
10-12	5	4.2
12-14	50	41.7
14-16	65	54.2
	120	**100**

Source: Field Survey, 2022

Data in table 1 shows the demographic data of the respondents: 54.2% of the respondents were males and 45.8% were females. 4.2% of the respondents were between 10-11 years, 41.7% of the respondents were between 12-14 years, 54.2% of the respondents were between the ages of 15-17. This implies that most of the respondents aged 14-16 were male.

Table 2. Effects of digital technology on academic performance of adolescents

S/N Statement	SA	A	SD	D
1. Addiction to digital technologies is a problematic issue that affects my academic life	12 (10%)	30 (25%)	40 (33.3%)	38 (31.7%)
2. Digital technologies distract me from my studies	12 (10%)	30 (25%)	40 (33.3%)	38 (31.7%)
3. Hours spent on Digital Technologies can never be compared to the number of hours I spend reading	12 (10%)	30 (25%)	40 (33.3%)	38 (31.7%)
4. There is no improvement in my grades since I engaged in the use of Digital Technologies	60 (50%)	40 (33.3%)	10 (8.3%)	10 (8.3%)
5. I usually have unlimited access to Digital Technologies and this has not affected my academic performance	60 (50%)	40 (33.3%)	10 (8.3%)	10 (8.3%)

Source: Field Survey, 2022

Data in table 2: shows the effects of digital technology on academic performance of adolescents, 10%strongly agreed that addiction to digital technology such as Laptops and television is a problematic issue that affects their academic life; 25% agreed while 33.3% strongly disagreed and 31.7% disagreed. 10% strongly agreed digital technology distracts them from their studies; 25% agreed while 33.3% strongly disagreed and 31.7% disagreed. 10% strongly agreed that hours spent on digital technology can never be compared to the number of hours they spent reading; 25% agreed while 33.3% strongly disagreed and 31.7% disagreed. The implication is that digital technologies have no effect on academic performance of adolescents in Jos South Local Government Area.

Table 3. Adolescents usage of digital technologies

S/N Statement	SA	A	SD	D
1. I use digital technologies for my academic activities and these have improved my academic performance	45 (37.7%)	38 (31.7%)	7 (5.8%)	30 (25%)
2. We make use of digital technologies to disseminate knowledge in with my class mates	45 (37.7%)	38 (31.7%)	7 (5.8%)	30 (25%)
3. I rely on information gotten from the internet to do my assignment	50 (41.7%)	28 (23.3%)	15 (12.5%)	27 (22.5%)
4. The usage of digital technologies for my assignment has helped to improve my grades	37 (30.8%)	23 (19.2%)	25 (20.8%)	35 (29.2%)
5. Engaging in academic forum through digital platforms reduces my rate of understanding	23 (19.2%)	37 (30.8%)	25 (20.8%)	35 (29.2%)

Source: Field Survey, 2022

Data in Table 3 shows adolescents 'usage of digital technology, 37.5% strongly agreed that they use digital technology for their academic activities and this has improved their academic performance; 31.7% make use of digital technology to disseminate knowledge with their classmates while 5.8% strongly disagreed and 25% disagreed. This shows that the majority of the respondents' performance use digital technology to disseminate knowledge. 41.7% strongly agreed that they rely on information gotten from digital technology to do their assignments; 23.3% agreed while 5.8% strongly disagreed and 22.5% disagreed. This shows that the majority of the respondents rely on digital technology for their assignment. 19.2% strongly agreed that engaging in academic forum through digital technology reduces their rate of understanding; 30.8% agreed while 20.8% strongly disagreed and 29.2% disagreed. This shows that the majority of the respondents' who engage in academic forum through digital technology reduces their rate of understanding. The implication is that adolescents use digital technologies to disseminate knowledge to their classmates as affirmed by 45 respondents of the entire population.

DISCUSSION OF FINDINGS

The study was conducted to evaluate the effects of digital technologies on the academic performance of adolescents. In response to this objective, data evidence revealed that adolescents' use of digital technology has no negative effects on their academic performance of students in Jos South Local government Area. This is in line with the Higgins et al. (2012) submission that a positive relationship exists between digital learning and educational outcomes. The findings also agree with Kelvin (2014) submission that the impact of digital technology on students' performances in schools by transforming the classroom into an interactive learning environment, providing a sense of competence and confidence in students' computer skills and helping them to get information in multiple ways.

However, the findings also contradict the findings of Olajide, Afolabi and Ajayi (2020) who conducted a survey of opinions of a wide spectrum of the society in what digital technology has brought to the nation brought to the nation, particularly the consequences for youth. The findings of the study showed that, adolescents' addiction and exposure to digital technology can have a negative effect on them such as distracting them from their academic work, taking most of their productive time, loss of concentration in school activities which sometimes causes them to exhibit strange and bad behaviours within the society.

The study was also conducted to determine the level of adolescent's usage of digital technologies. In response to this objective, data evidence revealed that adolescents use digital technologies especially the use of digital technologies to disseminate knowledge to class mates. This was affirmed by 45 respondents of the entire population. This finding tallies with Katz and Blumer (1974) uses and gratification theory (UGT) which postulates that people choose to consume certain kinds of media because they expect to obtain specific gratifications as a result of those selections.

RECOMMENDATIONS

Technology can create great benefits for adolescents in numerous areas of their needs ranging from education to entertainment, shopping, health and many more. To unlock the best benefits from the digital world adolescents should be

1. Equipped with a comprehensive set of digital skills.
2. Using digital technology for educational purposes rather than for fun and entertainment alone to prevent setback
3. Properly monitored by teachers, parents and guardian on the use of digital technology

CONCLUSION

Digital technology has become an important part of everyday life for many adolescents. It brings both the good and bad experiences to the adolescents. Compared to adults, adolescents use it for several purposes such as education, entertainment, social interaction and it is a place where they turn to for almost everything. Most of these tools are easy to use and adolescent can use them on their own to get the best experiences. So, it is no surprise that majority of adolescents are heavy users of digital technology especially social media sites.

The result from the findings of this study showed that, adolescents' addiction and exposure to digital technology does have negative effects on them.

REFERENCES

Adeniyi, M. A., & Kolawole, V. A. (2015). The influence of peer pressure adolescents'social behaviour. *University of Mauritus Research Journal, 21*. https://www.ajol.info/index.php/umrj/article/view/1220

Adeogen, M. (2005). The digital Divide and University Education Systems in Sub-Saharan *African. Afr. J. Lib. Arch & Information Science, 13*(1), 11-20.

Adomi, T. T., Okiy, R. B., & Roteyan, J. O. (2004). A Survey of Cyber Cafes Delta State Nigeria. Electronic Library 21(5), 487-495.

American Psychological Association. (2020). Developing adolescents: a reference for professionals. First Street publishers.

Archer, K. & Savage, R. (2014). Examining the effectiveness of technology use in classrooms: A tertiary meta-analysis. *Computers & Education: an international journal. 7*(8), 140-149

Boyd, D., & Ellison, N. (2007). Social network sites: Definition, history, and scholarship. *Journal of Computer-Mediated Communication, 13*(1), 1–11. doi:10.1111/j.1083-6101.2007.00393.x

Capaldo, C., Flanagan, K., & Littrell, D. (2008). Teacher interview 7377 – introduction to technology in schools. *Philaliteracy.* https://philaliteracy.org/wp-content/uploads/2014/05/2013-What-are-the-different-types-of-technology-you-use-in-your-cla
ssroom.pdf

Dede, C., & Richards, J. (2012). Digital teaching platforms: customizing classroom learning for each student. Teachers college press.

Eberendu, A. C. (2015). Negative impacts of technology in Nigerian society. *International journal of business and management review 3*(2), 23 – 29.

Goleman, D. (1994). *Emotional intelligence.* Bantam.

Harris, S. (2009). Lessons From e-learning: transforming teaching. *Proceedings of the International Conference on e-Learning.*

Hartney, E. (2022). What Is Peer Pressure? Types, Examples, and How to Deal With Peer Pressure. *Very Well Mind.* https://www.verywellmind.com/what-is-peer-pressure-22246

Higgins, S., Xiao, Z., Katsipataki, M., (2012). The impact of digital technology on learning: a summary for the education endowment foundation.

Holly, L., Wong, B., Agrawal, A., Awah, I., Azelmat, M., Kickbusch, L., & Ndili, N. (2022). Opportunities and threats for adolescent well- being provided by digital transformations. *NCBI.* www.ncbi.nlm.nih.gov › pmc › articles › PMC7366938

Idow, P.A, Idowu, A.O. & Adawgouno. (2004). A comparative survey of Information and communication technologies in High Educational Institutions in Africa: a case study from Nigeria and Mozambi. *Journal of Information Technology impact, 4*(6), 67-74.

Katz, E., Blumer, J. G., & Gurevitch, M. (1974). Uses and gratifications research. *Public Opinion Quarterly, 37*(4), 509–524. doi:10.1086/268109

Lewis, S. (2008). Where young adults intend to get news in five years. *Newspaper Research Journal, 29*(4), 36–5. http://findarticles.com/p/ articles/mi_qa3677/ is_200810/ ai_n39229321/. doi:10.1177/073953290802900404

Meyer, K. A. (2005). Planning for cost-efficiencies in online learning. *Planning for Higher Education, 33*, 19–30.

Mihalyi, C. (nd). Adolescence. *Britannica.* https://www.britannica.com/science/adolescence

Museum Of Applied Arts and Sciences. *(2017).* Museum of applied arts and sciences. Retrieved 20th *NSW Government.* https://maas.museum/about/

Ndukwe, E. (2011). The telecommunication revolution in Nigeria. *A Convocation Lecture.*

Nigerian Communication Commission. (2019). Study on young children and digital technology: a survey across Nigeria. *NCC.* www.ncc.gov.ng › Technical Regulation › Research & Development

Nwosu, N. (2019). *Developmental Psychology and Education.* Published by Agatha Series Ltd.

Odofin, T. (2014). the influence of digital technology on secondary school students academic engagementbehaviour in nigeria. *The Educational Psychologist, 14*(1). https://www.journals.ezenwaohaetorc.org/index.php/ncep/article/viewfile/1847/1879

OECD (2016). Trends Shaping Education 2016. *OECD Publishing Paris.* doi:. doi:10.1787/trends_edu-2016-en

Ojedokon, A. A & Owolabi, E.O (2005).Internet access competence and the use of internet for teaching and research activities by university of Botswana Academic staff. *Africa Journal of Library, archives and information science, 13*(1), 43-53

Olajide M. S., Afolabi, O.P. & Ajayi, A.A. (2020). *Impact of digital technology on adolescents in Nigeria.* https://www.researchgate.net/publication/343796926

Olweus, D. (1996). Bullying at school: Knowledge base and an effective intervention program. In C. Ferris & T. Grisso (Eds.), *Understanding aggressive behaviour in children.* New York Academy of Sciences. doi:10.1111/j.1749-6632.1996.tb32527.x

Osazee-Odia, O. U. (2016). Reflections on smartphones: A new paradigm of modernity in Nigeria. *IISTE New Media and Communication Journal, 56*(1), 39–52.

Owoyele, J. W., & Toyobo, O. M. (2008). Parental will, peer pressure, academic ability and school subjects selection by students in senior secondary schools. *Social Sciences, 3*(8), 583–586. http://docsdrive.com/pdfs/medwelljournals/sscience/2008/583-586.pdf

Reed, P., Hughes, A., & Phillips, G. (2013). Rapid recovery in sub-optimal readers in Wales through a self- placed computer-based reading programme. *British Journal of Special Education, 40*(4), 162–166. doi:10.1111/1467-8578.12040

Shannon, C. E., & Weaver, W. (2003). *The mathematical theory of communication.* University of Illinois Press.

Shraddha, C. (2018) Uses and Gratifications Theory. *Businesstopia*, https://www.businesstopia.net/mass-communication/uses-gratifications-theory

Taye, C. & Samuel, O. (2019). From news breakers to news followers: the influence of Facebook on the coverage of the January 2010 crisis in Jos. *Bingham journal of humanities, social and management sciences, 1,* 7-9.

Uslu, M. (2013). Relationship between degrees of self-esteem and peer pressure in high school adolescents. *International Journal of Academic Research Part B,* 117122.https://www.researchgate.net/publication/314524207

Uzuegbunam, C. (2019). A child-cantered study of teens' digital lifeworld's from a Nigerian perspective. https://blogs.lse.ac.uk/parenting4digitalfuture/2019/10/16/a-child-centred-study-of-teens-digital-lifeworlds-from-a-nigerian-perspective/

World Health Organization. (2020). Orientation program me on adolescent health for health-care providers. *WHO.* https://www.who.int/health-topics/adolescent-health#tab=tab_1

ADDITIONAL READING

Liao Y. C., Chang, H. W., & Chen, Y. W. (2008). Effects of computer applications on elementary school students' achievement: a meta-analysis of students in Taiwan. *Computers in the schools [sic], Jan.*

Martin, B. (2011). 20 years ago today, the World Wide Web was born. *TNW Insider.* https://thenextwep.com/insider/2011/08/06/20-years-ago-today-the-world-wide-web-opened-to-the-public/

Rubin, A. (2002). The uses-and-gratifications perspective of media effects. In J. Bryant & D. Zillmann (Eds.), *Media effects: Advances in theory and research,* (2nd ed., pp. 525–548). Lawrence Erlbaum Associates, Inc.

Strasburger, V. C., Jordan, A. B., & Donnerstein, E. (2010). American Academy of Paediatrics [Edition Pearson Education Limited: England]. *Health Effects of Media on Children and Adolescents, 1*(4), 756–767.

Chapter 15
A Hybrid SEM-ANN Approach for Intention to Adopt Metaverse Using C-TAM-TPB and IDT in China

Yuyang Wang
Chinese Academy of Sciences, Beijing, China

Tinfah Chung
https://orcid.org/0000-0002-3993-6637
HELP University, Malaysia

Eik Den Yeoh
https://orcid.org/0000-0001-6232-2345
HELP University, Malaysia

ABSTRACT

The study aims to identify factors affecting university students' intention to adopt the metaverse due to the 4th industrial revolution and the COVID-19 pandemic in China, based on the integration of C-TAM-TPB model and IDT theory. A questionnaire survey was conducted on university students for data collection, as the metaverse is expected to be actively used or developed by them in future. A sample of 441 valid data was analysed by T-test and SEM-ANN analysis. Results show that subjective norm, attitude, compatibility, perceived usefulness, and relative advantages significantly affect Chinese university students' metaverse adoption intention, except for perceived behavioural control. Subjective norm holds the highest influence, while compatibility ranks the lowest. Perceived risk negatively moderates the relationship between relative advantage and adoption intention. There is no significant difference exists in different gender groups and experience groups for Chinese university students' metaverse adoption intention.

DOI: 10.4018/978-1-6684-5732-0.ch015

INTRODUCTION

Background

As a new favourite buzzword attracted the worldwide attention, "Metaverse" is not entirely new. All the ingredients for it existed, even before official metaverse applications appeared. Virtual reality, gaming and social media could be some of the early applications of the metaverse (Kharpal, 2022). To turn the idea of the metaverse into reality and offer an immersive metaverse virtual experience, cutting-edge but not new technologies are incorporated, including blockchain, augmented reality (AR) and virtual reality (VR), 3D reconstruction, artificial intelligence (AI), and the Internet of things (IoT) (Howell, 2022). The idea of "metaverse" was first used by Neal Stephenson, an American science fiction novelist in his novel "Snow Crash" describing an immersive 3D virtual environment in 1992. A character named "Avatar" is featured (Sourin, 2017). The metaverse has become a very grand scene, as everyone is "Avatar" in the most fascinating future virtual world constructed by their imagination and self-construction, via sci-fi works resonates, capital speculates.

The prolonged isolation due to the COVID-19 pandemic has also accelerated the interest in developing metaverses (Caminiti, 2022). There is an increased demand for more interactive ways to connect with others as people were tired of the web-based lifestyle (Thomala, 2022a). Since 2020, internet leading tech-companies already have their 3D digital world up and running, such as Roblox, Facebook (parent Meta), Microsoft, and Nvidia (Kayyali, 2022). It is expected it to grow more popular in 2022 (Mileva, 2022). The boundary between the virtual world and the real world is broken down. 3D Virtual spaces that let users have remote meetings, catch up, mingle, and collaborate more engaging and fun using their own "Avatars".

"Metaverse" is a term with no concrete definition, but broadly defined as an 3D virtual space, where digital representations of people live in, interact, and socialize with an immersive virtual experience (Dionisio et al., 2013; Kharpal, 2022). It promises to provide a space for persistent and boundless virtual communities with the help of virtual reality (VR) headsets, augmented reality (AR) glasses, smartphone apps, or other devices (Johnson, 2022a). The AR and VR market will accelerate at a compound annual growth rate of over 35% through 2020-2024 considering the impact of COVID-19, although high development costs associated with AR and VR applications might hamper the market growth to some extent (Technavio, 2020).

Metaverse is the future of social network (Ma, 2022). It has very broad space, and is an emerging thing that has not yet taken shape. Technology is nothing without practical application scenarios to create any value in capital market. It is worth nothing that China's official medias such as "People's Daily", "Xin Hua Net", "People's Posts and Telecommunications News", and "Economic Daily" have also focused on the market frenzy of metaverse and warned people to be rational. No matter how exciting the concept of metaverse is, it needs to be implemented and practiced step by step, otherwise it is just a bubble.

Motivation and Problem Statement

The metaverse has been described as many things: the future of the internet, the next evolution of social connections, and an unprecedented business opportunity (Fly and Grünberg, 2022). However, it is important to recall that the definition of metaverse will undoubtedly continue to shift as most of its functionalities are still under development in the long run. The use cases of metaverse would expand beyond gaming and social media platforms into various applications, such as digital identity management, remote workspaces, and decentralized governance (Howell, 2022). Mass adoption of the virtual world will take a long time, given major technological and regulatory hurdles (Chittum, 2022).

But whether future or fad, one thing is certain: China will not miss out on shaping this new ecosystem as an area in which it can beat the West (The Economist, 2022). These will obviously be the areas that China big technology giants will follow first, such as 5G mobile networks, payment, VR, social media, gaming, and WeChat like online service, which has laid good foundations and can be extended to build into the metaverse (Mok, 2022).

Chinese users are rather open-minded about new technologies. With most daily activities being supported by technology, the concepts of digital worlds are widely accepted in China. According to a survey conducted in 2021, over 70 percent of Chinese netizens stated that they had a sufficient understanding about metaverse (Thomala, 2022a). Morgan Stanley estimated $8 trillion (about ¥52 trillion) metaverse market in China alone (Chittum, 2022). Chinese technology giants from Tencent to Alibaba, Byte Dance/TikTok, NetEase and Baidu have also started creating their versions of fully-fledged metaverse as front-runners in metaverse, but taken a more cautious approach amid tighter regulation in China compared to U.S. companies (Kharpal, 2022).

The vision is that the metaverse will be a virtual body with Chinese characteristics. Globally, business mode on mobile internet and smartphones are maturing. Things are moving so fast; metaverse has struck a chord with young Chinese users. It is critical for China's technology companies to embrace the metaverse to find new ways to engage the youngest generation (Gen-Z who born from 1995 to 2009) of internet users, especially for the on-campus university students who are main active end-users and developers of metaverse technology in future. Market research and cultivation should be carried out simultaneously with the construction of metaverse related infrastructure and the development of software and hardware.

In academic, a few studies have also reviewed the acceptance in metaverse, such as intention to use metaverse system in medical education (Almarzouqi et al., 2011), utilizing the metaverse for learner-centered constructivist education (Suh and Ahn, 2022), using intentions toward early users of metaverse platforms (Park and Kang,2021), intention to use the virtual world metaverse (Oh et al., 2021), metaverse adoption in higher institutions of gulf area (Akour et al., 2022), literature reviews on consumer behaviour research and virtual commerce application design for promoting user purchase in metaverse (Shen et al., 2021). Practical metaverse research in China are rare. Therefore, this study focuses on factors affecting university students' intention to adopt metaverse due to the 4th industrial revolution and the COVID-19 pandemic in China, using a conceptual framework that integrates the models of C-TAM-TPB and IDT.

Aims and Objectives

China can properly position itself to become a key player globally and reap the most benefit out of the metaverse hype by well recognising and understanding its users of metaverse, as China indeed have competitive advantage in this space and necessary development incentives by the government should be rolled out aptly. It necessitates a deeper understanding of factors may impact Chinese university students' intention/behaviour towards metaverse adoption for successful market cultivation and development strategies making. The objectives of this study are as followed:

RO1: To establish a tailored Chinese characteristic model illustrating the relation between key factors and adoption intention of metaverse based on the integrated C-TAM-TPB and IDT approach.

RO2: To examine key factors influencing on-campus university students' intention of metaverse adoption in China.

Research Questions

The following research questions are formulated to address the research aim:

RQ1: What is the model for the adoption of metaverse using the integrated C-TAM-TPB and IDT approach in China?

RQ2: Which is the most significant factor influencing the adoption intention of metaverse by Chinese on-campus university students?

The study includes seven sections. Following the introduction section, a literature review on "post-pandemic internet and metaverse", "Theoretical foundations – Combined-Technology acceptance and planed behaviour (C-TAM-TPB) model, and Diffusion of Innovation Theory (IDT)" followed. Conceptual framework was established and hypotheses were developed based on previous studies. A questionnaire survey was conducted on on-campus university students for data collection as the metaverse is expected to be actively used or developed by them in future. In data analysis part, SEM-ANN two-step hybrid approach was used to enhance the model's predictive performance. Results and discussion presented further. Implications, recommendation, limitation and future study trends were discussed at the end.

LITERATURE REVIEW

"Post-pandemic" Internet and Metaverse

By 2021, there were an estimated 4.9 billion internet users around the world, while Asia accounts for almost 60% global internet users and Eastern Asia is home to 1.16 billion internet users. As of February 2022, China was ranked first with 1.02 billion internet users of the estimated 1.4 billion of population, more than triple the amount of the U.S. (Johnson, 2022b). Digital content such as apps, social media, and online platforms have long become part of everyday life. This scenario became reality with the outbreak of the global COVID-19 pandemic: nationwide lockdowns, social distancing measures, stay-at-home mandates forced people around the world in many cases having to rely on the internet as their lifeline for work, social life, and entertainment. What if daily life changed abruptly and digital needs had to adapt in time accordingly?

According to google trend reports (trends.google.com/trends), there has been a severe increase of searches on "Metaverse" in the google search engine since the time Mark Zuckerberg announced that Facebook will be renamed "Meta" on 28th, October 2021. Overcoming obstacles that prevented them from doing something in real life was perceived as the biggest benefit of the metaverse, while enhancing creativity and imagination was ranked second. The metaverse also holds promise for upskilling, connecting with new people without feeling awkward, and creating new job opportunities (Johnson, 2022c).

The metaverse represents a significant economic opportunity as the global market size of the metaverse in 2020 was $4.69 billion and is estimated to grow by 43.3% annually between 2020–2027 (Sudan et al., 2022). Despite its current hype and attendant risks, the metaverse could potentially offer benefits in multiple sectors, including financial services, automotive and manufacturing, real estate, retail, healthcare, education and training, and urban development (iiMedia Research, 2022). Main reasons for global internet users joining the metaverse were gaming or creative activities in 2021 (Clement, 2022b). While the Metaverse is still in its early stages of development, many businesses are already looking into its possibilities. Globally, 17% companies in the computer and IT sector have already invested in the metaverse as a business opportunity as of March 2022 (Clement, 2022a).

The existing trends in the metaverse clearly show that it is a thriving ecosystem with many technological foundations powering the metaverse. Looking at the history of the metaverse shown in Table 1, as Blockchain (enable digital currencies and NFTs), 3D reconstruction, AR, VR, AI and IoT technologies advance and mature, it would likely see exciting new features in 3D virtual world from the era of Web 2.0 to Web 3.0 (J.P.Morgan, 2022). It is those technologies, together with social and economic drivers flourishing and collaborating leading to the explosive development of the metaverse or "Metanomics" called by J.P. Morgan.

In China, as mentioned, despite social and economic drivers, there are more political drivers weaving deeply into the long-term survival of the metaverse relevant companies and investors, and users. China is running its own metaverse way with multiple regulatory and functional departments who put strict regulations in internet platform economy for impelling and prompting them to develop healthily and sustainably. National Press and Publication Administration (NPPA) issued the new 2021 Anti-Addiction Notice, which further restrict the playing time of online games for minors. New anti-monopoly laws for internet platforms were proposed by Anti-monopoly Commission, while a landmark personal information protection law has been passed in 2021.

Table 1. Web 2.0 versus Web 3.0 approach to the metaverse

		Web 2.0	Web 3.0
Example virtual worlds		Second Life; Roblox; Fortnite; World of Warcraft	Decentraland; The Sandbox; Somnium Space; Cryptovoxels
Platform characteristics	Organizational structure	Centrally owned; Decisions are based on adding shareholder value	Community governed, generally through a foundation decentralized autonomous organization (DAO); Native tokens are issued and enable participation in governance; Decisions are based on user consensus
	Data storage	Centralized	Decentralized (game assets)
	Platform format	PC/Console; VR/ AR hardware; Mobile/app	PC; VR/ AR hardware; Mobile/app coming soon
	Payments infrastructure	Traditional payments (e.g., credit/debit card)	Crypto wallets
User interaction	Digital assets ownership	Leased within platform where purchased	Owned through non-fungible tokens (NFT)
	Digital assets portability	Locked within platform	Transferable
	Content creators	Game studios and/or developers	Community; Game studios and/or developers
	Activities	Socialization; Multi-player games; Game streaming; Competitive games (e.g., esports)	Play-to-earn games; Experiences (Same activities as Web 2.0, see box on left)
	Identity	In-platform avatar	Self-sovereign and interoperable identity; Anonymous private-key-based identities
Commercials	Payments	In-platform virtual currency	Cryptocurrencies and tokens
	Content revenues	Platform; app store; developer	Peer-to-peer; developers (content creators) directly earn revenue from sales; Users can earn through play or participation in platform governance; Royalties on secondary trades of NFTs to creators

Source: J.P.Morgan, 2022

Theoretical Foundations: Combined TAM-TPB and IDT

Technology Acceptance Model (TAM)

The Technology Acceptance Model (TAM) was proposed by Davis (1989) and Davis et al. (1989) based on Theory of Reasoned Action (TRA) (Fishbein and Ajzen,1975). TAM has been widely used as the theoretical basis for many previous studies of user technology acceptance, as it works by using two main variables, namely perceived usefulness, and perceived ease of use (Venkatesh and Davis, 2000). As one of the most influential research models determining information system adoption, TAM only provides general and wide scope of predicting technology acceptance. Further information is needed regarding its use in specific fields of human–computer interactions (Davis et al.,1989; Lee et al., 2011). In recent years, TAM has been used to examine users' adoption of World-Wide-Web context (Moon and Kim, 2001; Page-Thomas, 2006), educational context (e.g., online courses, e-learning systems, mobile learning) (Baby and Kannammal, 2020; Chaveesuk and Chaiyasoonthorn, 2022; Al-Emran et al., 2018), health informatics (Ammenwerth, 2019; Kamal et al.,2020), online finance (e.g., tax filing systems, banking

(Ahmad, 2018; Gor, 2015), and autonomous vehicle (Koul and Eydgahi, 2018; Dirsehan and Can, 2020). It shows that TAM are good at assessing the use of the compulsory system which users must use.

Combined TAM-TPB Model

Taylor and Todd (1995) integrated subjective norm and perceived behavioral control derived from TPB (Ajzen, 1991) to TAM, and proposed a new model named C-TAM-TPB has high fitness in explaining using behaviors toward new technology (here is computing resources center) for both experienced and inexperienced users. TPB extends TRA (Fishbein and Ajzen, 1975) by adding perceived behavioral control. It assumes that consumers make rational decisions by evaluating the costs and benefits of different actions by examining three variables (attitude, subjective norm, and perceived behavioral control) (Ajzen, 1991). Obviously, C-TAM-TPB theory is more suitable for optional technology adoption behavior (e.g., metaverse adoption in the study) by covering users' personal characteristics and technology properties to explain user behaviors using new technologies (Chen, 2013).

Diffusion of Innovation Theory (IDT)

As proposed by Rogers (2003), Diffusion of Innovation Theory (IDT) signifies a group communication phenomenon by how an innovation (e.g., new idea, practice, or object) diffuses/spreads over time, and finally is adopted by members of a social system. IDT provides five factors (relative advantage, compatibility, complexity, trialability, and observability) influencing innovation adoption. It is the most appropriate theory for investigating technology adoption in higher education (Al-Mamary et al., 2016).

Previous studies have revealed that relative advantage and compatibility may provide more explanation for consumer intention to new technology adoption intention (Agag and El-Masry, 2016; Lu et al., 2011). Accordingly, we also include relative advantage and compatibility representing IDT divers in this study. Some researchers propose that relative advantages is similar to TAM's perceived usefulness (Lee and Kozar, 2008; Wu and Wang, 2005). However, it needs to be emphasized that relative advantages explicitly contain a comparison between the innovation and its precursor, while perceived usefulness does not (Amaro and Duarte, 2015; Shin, 2010).

The study proposed a conceptual framework that measures users' metaverse adoption intention as depicted in Figure 1. The integration of C-TAM-TPB and IDT could provide an even stronger explanation to information technologies than either standing alone.

Figure 1. Conceptual framework of the study

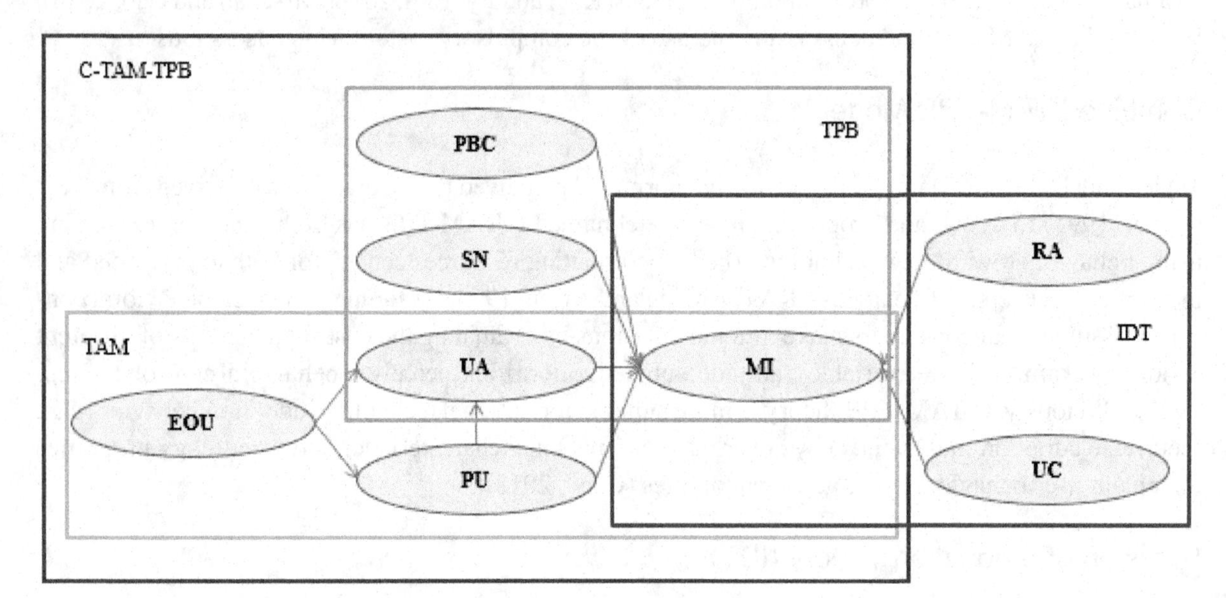

Research Model and Hypotheses

Six constructs, namely perceived usefulness (PU), perceived ease of use (EOU), attitude (UA), perceived behavioural control (PBC), subjective norm (SN) and metaverse adoption intention (MI) were taken from C-TAM-TPB model, relative advantage (RA) and compatibility (UC) were derived from IDT theory, with perceived risk (PR) as a moderator. It thus tested the validity and applicability of the proposed model based on the following hypotheses as shown in Figure 2.

Figure 2. Research model of the study

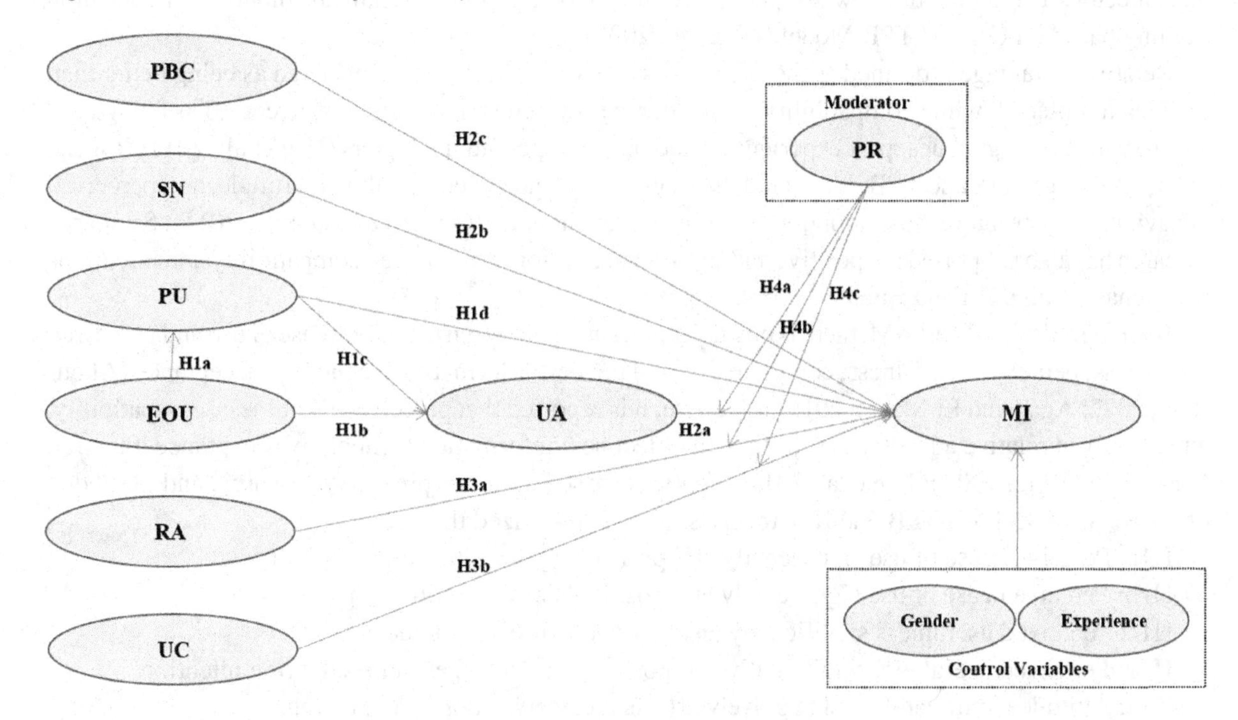

Constructs of C-TAM-TPB and IDT

TAM shapes technological innovation, which is usually affected by perceived ease of use and perceived usefulness of technology as mentioned (Davis, 1989). The former is defined as the degree to which a person believes that using a particular system would be free of effort, while the latter is the degree to which an individual believes that a particular system would enhance his or her job performance within an organizational context (Davis, 1989). Perceived ease of use has a positive effect on perceived usefulness to information systems using, and perceived usefulness positively influence behavioral intention based on TAM related models (Dirsehan and Can, 2020; Manis and Choi, 2019; Yang and Su, 2017). Positive relationships exist between perceived ease of use and attitude towards using technology innovations (Venkatesh and Davis, 2000). Perceived usefulness has positive relationship with attitude and behavioral intention to use blended learning (Nadlifatin et al., 2020), mobile learning (Alhumaid et al., 2021), user acceptance of internet (Jiang et al., 2000), intention to use internet system (Gor, 2015), big data adoption (Soon et al.,2016), and metaverse system adoption (Akour et al.,2022).

Both perceived usefulness and perceived behavioral control are significant determinants of innovation adoption (Awa et al., 2015). Attitude, subjective norm, and perceived behavioral control affect behavioral intention in Taiwan based on C-TAM-TPB model (Safeena et al., 2013). Similarly, it is also supported by (Nadlifatin et al. (2020) with respect for measuring Taiwanese group but Indonesian group. Yang and Su (2017) and Alhumaid et al. (2021) also propose that users' perceived behavior control, subjective norm, and attitude (exerting the greatest influence on behavioral intention), which directly and positively influence their behavioral intention based on the C-TAM-TPB model. However, it is also found that perceived usefulness, subjective norm, and perceived behavioral control do not have significant relationship with fee collectors' using intention towards personal digital assistant (PDA) within one month after education training based on C-TAM-TPB Model (Jen et al., 2009).

Relative advantage is defined as the degree to which an innovation is considered as being better than the idea it replaced, while compatibility is the degree to which an innovation is perceived as being consistent with existing values, past experiences, and needs of potential adopters (Hsu et al., 2011; Rogers, 2003). Based on IDT and TPB, it is found that relative advantage, compatibility, attitude, and perceived behavioural control affect the adoption of online travel shopping (Amaro and Duarte, 2015). Similarly, research has also supported the positive and significant relationship between compatibility and intentions to purchase online (Li and Buhalis, 2006).

By integrating IDT to TAM, there is positive relationship between perceived ease of use and perceived usefulness; perceived usefulness, compatibility, and the new information technology acceptance (Akour et al., 2022; Agag and El-Masry, 2016). Likewise, it is reported that perceived usefulness, compatibility, and perceived relative advantages significantly affect new information technology acceptance intention (Chang and Tung, 2008a; Lee et al.,2011). To recap, based upon the preceding research, and grounded on TAM, TPB, C-TAM-TPB and IDT theories, it is hypothesized that:

H1a: Perceived ease of use significantly and positively affects perceived usefulness.

H1b: Perceived ease of use significantly and positively affects attitude.

H1c: Perceived usefulness significantly and positively affects attitude.

H1d: Perceived usefulness significantly and positively affects metaverse adoption intention.

H2a: Attitude significantly and positively affects metaverse adoption intention.

H2b: Subjective norm significantly and positively affects metaverse adoption intention.

H2c: Perceived behavioral control significantly and positively affects metaverse adoption intention.

H3a: Relative advantage significantly and positively affects metaverse adoption intention.

H3b: Compatibility significantly and positively affects metaverse adoption intention.

Moderation Effect of Perceived Risk

Perceived risk should be considered as an important factor primarily due to the uncertainty and potential negative outcomes of adopting a new/emerging technology innovations (Esteves and Curto, 2013; Heart, 2010; Luo, et al., 2010) or service which will influence users' decision-making (Featherman and Pavlou, 2003). It has been introduced into models integrating between TAM, TPB and IDT constructs, that perceived risk negatively influences intention and attitude (Amaro and Duarte, 2015; Rogers, 2003; Takele and Sira, 2013). Perceived risk is negatively related to purchasing online (Nittala, 2015). Similarly, perceived risk negatively affects big data adoption (Soon et al., 2016). Research looking at perceived risk associated with metaverse system adoption are rare. In view of these previous findings, the study is concerned in examining the moderation effect of perceived risk. Therefore, it is hypothesized that:

H4a: Perceived risk moderates the relationship between attitude and metaverse adoption intention.

H4b: Perceived risk moderates the relationship between relative advantage and metaverse adoption intention.

H4c: Perceived risk moderates the relationship between compatibility and metaverse adoption intention.

RESEARCH METHODOLOGY

Data Collection

Data for this study is collected from on-campus university students in Beijing Municipality, Shandong province, and Guangxi province using a questionnaire survey. Primary reasons for on-campus university students as target audience are this group is main active end-users and developers of metaverse technology as mentioned. To ensure that the study does not pose any ethical concerns, the respondents are informed that the participation is voluntary and they are free to withdraw at any point. Respondents are also informed that the data collected will not be used for any purpose other than for the research study. Further, all survey data is kept in a secure place and can be made available where needed.

Measurement scales are developed for the questionnaire survey data collection based on those previous studies depict in Table 2. All 27 items of 9 variables are measured on a 7-point Likert scale with a "1" representing "Strongly disagree" to a "7" representing "Strongly agree". A group of 45 from the target audience participate the pilot test for ensuring content validity. Accordingly, factor analysis is run to ensure construct validity and get the final measure items by using pilot data.

Table 2. Measurement scales

Variables	Items	Sources
EOU	1. I think it is easy to learn to use metaverse system. 2. I think I can easily use metaverse system for different purposes. 3. I think I can interact with metaverse system with no effort.	Davis & Wiedenbeck, 2001
PU	1. I think metaverse system is useful in my daily life. 2. I think metaverse system could improve my social performance. 3. I think adopting metaverse system help me live my life.	Davis & Wiedenbeck, 2001
UC	1. I think adopting metaverse system is compatible with my standards. 2. I think adopting metaverse system is compatible with the way I like to work. 3. I think adopting metaverse system is compatible with my lifestyle.	Agag & El-Masry, 2016
RA	1. I think metaverse system provides me better experience than the old online world. 2. I think metaverse system provides me more opportunities. 3. I think metaverse system enjoys me more.	Amaro & Duarte, 2015
UA	1. I like to use metaverse system. 2. I am interested in using metaverse system. 3. I feel comfortable with metaverse system.	Singh, 2015
SN	1.People who influence my behaviour using metaverse system. 2. Whether to using metaverse system will be influenced by my friends. 3. Whether to using metaverse system will be influenced by my family.	Singh, 2015
PBC	1. I can't master everything that appears during metaverse system adoption. 2. I can't arrange time on my own to using metaverse system. 3. I think using metaverse system causes disturbance to my life/work/studies.	Yang & Su, 2017
PR	1. There is too much uncertainty over the metaverse system. 2. Using metaverse system is risky. 3. I feel apprehensive about using metaverse system.	Soon et al., 2016 Amaro & Duarte, 2015
MI	1. I intend to use metaverse system. 2. I expect that I will use metaverse system. 3. I plan to use metaverse system.	Dirsehan & Can, 2020

Demographic Description

Target sample size is 500, 441 valid samples are collected, suitable for both factor analysis and further SEM analysis in this model (27 items from 9 constructs) despite being a little higher than required sample size, after excluding outliers and incomplete or invalid responses (Siddiqui, 2013). Regarding the demographic characteristics, 250 are males (56.7%) and 191 are females (43.3%). 359 respondents (81.4%) have experienced in virtual world before, while 82 respondents have not (18.6%). According to the results of T-test, there is no significant difference exists in different gender (Sig.=0.540>0.05), experience (Sig.=0.449>0.05) for metaverse adoption intention.

Due to China's ongoing and fast-paced economic development, and a cultural inclination towards technology, Chinese people also had shown general rise in digital literacy levels and interest in adopting metaverse. 230 respondents (52.2%) present that they only heard of the term metaverse, while 128 respondents (29.0%) present that they have basic understanding, 65 respondents (14.7%) indicate that they have good understanding, and 18 respondents (4.1%) show that they have strong understanding. 47 respondents (10.7%) present that they don't have willingness to use metaverse-empowered applications to socialize in metaverse, while 284 respondents (64.4%) show that they have willingness, and 110 respondents (24.9%) indicate that it is depends on the situations. To some extent, the results of the study keep consistency with previous research which show over 70% of Chinese internet users had a sufficient understanding about metaverse, and over 60% of respondents clearly intended to socialize in the metaverse for more personal online interactions (Thomala, 2022a).

EFA and CFA Test

Discriminant and convergent validity during an EFA by SPSS 24.0 are fully determined and satisfied in this study as shown in Table 3-4. Correlation values between constructs mostly around 0.45 (<0.7) are perfectly significant and fully satisfied (Achchuthan and Velnampy, 2016). The KMO value is .908

Table 3. Inter-construct correlation matrix

Coefficient	UA	RA	UC	MI	PU	EOU	SN	PBC	PR
UA	1.000								
RA	.544***	1.000							
UC	.505***	.441***	1.000						
MI	.545***	.471***	.477***	1.000					
PU	.449***	.425***	.426***	.466***	1.000				
EOU	.506***	.454***	.515***	.552***	.486***	1.000			
SN	.400***	.345***	.345***	.531***	.394***	.462***	1.000		
PBC	.232***	.218***	.309***	.230***	.360***	.369***	.214***	1.000	
PR	.298***	.263***	.397***	.264***	.340***	.365***	.210***	.447***	1.000
Mean	3.9788	4.1950	4.6901	4.3507	4.5586	4.4626	4.8080	4.7785	4.5820
SD	1.24359	1.18141	1.15479	1.19867	1.05345	1.14423	1.39658	1.19493	1.04779

Note: n=441; *** Sig. Level P<0.001

Table 4. KMO and Bartlett's Test

Kaiser-Meyer-Olkin Measure of Sampling Adequacy		.908
Bartlett's Test of Sphericity	Approx. Chi-Square	9143.552
	df	351
	sig	.000

Table 5. Rotated Factor Matrix

	Factor								
	UA	RA	UC	MI	PU	EOU	SN	PBC	PR
UA1	.793								
UA2	.804								
UA3	.806								
RA1		.787							
RA2		.813							
RA3		.851							
UC1			.799						
UC2			.842						
UC3			.812						
MI1				.807					
MI2				.801					
MI3				.758					
PU1					.781				
PU2					.816				
PU3					.765				
EOU1						.793			
EOU2						.810			
EOU3						.773			
SN1							.889		
SN2							.900		
SN3							.899		
PBC1								.834	
PBC2								.863	
PBC3								.867	
PR1									.834
PR2									.863
PR3									.867
Cronbach's a	.901	.877	.898	.896	.845	.907	.959	.881	.873
Eigenvalue	10.501	2.716	1.932	1.473	1.333	1.253	1.142	1.110	1.027
Variance (%)	38.894	10.061	7.157	5.455	4.937	4.642	4.231	4.112	3.803

Extraction Method: Principal Component Analysis.
Rotation Method: Varimax with Kaiser Normalization.
a. Rotation converged in 10 iterations.

greater than 0.6, and the Bartlett's test reached statistical significance (p=.000<0.05), well support the factorability of the correlation matrix. Further, factor rotation is conducted for better interpretation of factor analysis as shown in Table 5. Factor loadings are ranging from 0.758 to 0.900 (>0.6) (Hair et al., 1998) indicate that 9 constructs are well differentiated from each other, while items within a single construct are highly correlated. Moreover, Cronbach's α values range from 0.845 to 0.959 meeting a 0.7 level, which indicates that the scale has a high internal consistency and is reliable (Pallant, 2005).

Moreover, all results display evidence at satisfactory levels of the reliability and validity by further conducting a CFA using SEM-AMOS 24.0 as shown in Table 6. The t-values for all the standardized factor loadings of items are significant at a 0.001 level. Composite reliability values (CR) range from 0.8468 to 0.917 (>0.7) (Fornell and Lacker, 1981), while the average variance extracted (AVE) values range from 0.6489 to 0.8897 meeting a 0.5 level.

Table 6. CFA results

Variables	Item	Factor loading	SE	t-value	Cronbach's α	AVE	CR
PBC	PBC1	1			0.881	0.7176	0.8838
	PBC2	1.015	0.049	20.714			
	PBC3	0.926	0.049	18.898			
SN	SN1	1			0.959	0.8897	0.917
	SN2	1.136	0.026	43.692			
	SN3	1.125	0.033	34.091			
PU	PU1	1			0.845	0.6489	0.8468
	PU2	1.094	0.066	16.576			
	PU3	1.071	0.066	16.227			
EOU	EOU1	1			0.907	0.7558	0.9028
	EOU2	1.016	0.043	23.628			
	EOU3	1.008	0.043	23.442			
RA	RA1	1			0.877	0.7083	0.8788
	RA2	1.285	0.066	19.469			
	RA3	1.188	0.065	18.277			
UC	UC1	1			0.898	0.7462	0.8981
	UC2	0.915	0.036	26.083			
	UC3	0.948	0.042	22.571			
UA	UA1	1			0.901	0.7125	0.8814
	UA2	1.089	0.045	24.200			
	UA3	0.995	0.047	21.170			
MI	MI1	1			0.896	0.7333	0.8918
	MI2	0.988	0.043	22.976			
	MI3	0.945	0.045	21.000			

Structural Model

There is n = 24 observed variables; 66 distinct parameters to be estimated in total. As λ =66<24*25/2 = 300, the model can be considered as an overidentified one used for structural equation modeling (SEM) (Byrne, 2001). The SEM results show that GFI =.920, NFI = .945, CFI = .972, RMSEA = .047, RMR=0.120. The $\chi 2$ of 461.340 with 233 degrees of freedom showed a ratio of 1.980:1 (<3:1). All fitting indicators reach the ideal indice as shown in Table 7. Therefore, the model as shown in Figure 3 is statistically assessed to be valid and finally can be applied.

Further, as shown in Figure 3 below, the variance of constructs explained by antecedents can be used to evaluate the research model. In this study, EOU explained 36.0% of variance in PU; PU and EOU explained 40.0% of variance in UA; PBC, SN, PU, UA, RA and UC explained 51.0% of variance in MI.

Table 7. Model fitness statistics

Index	X2/df	GFI	NFI	CFI	RMSEA	RMR
	1.980	0.920	0.945	0.972	0.047	0.120

Figure 3. Model Path Diagram

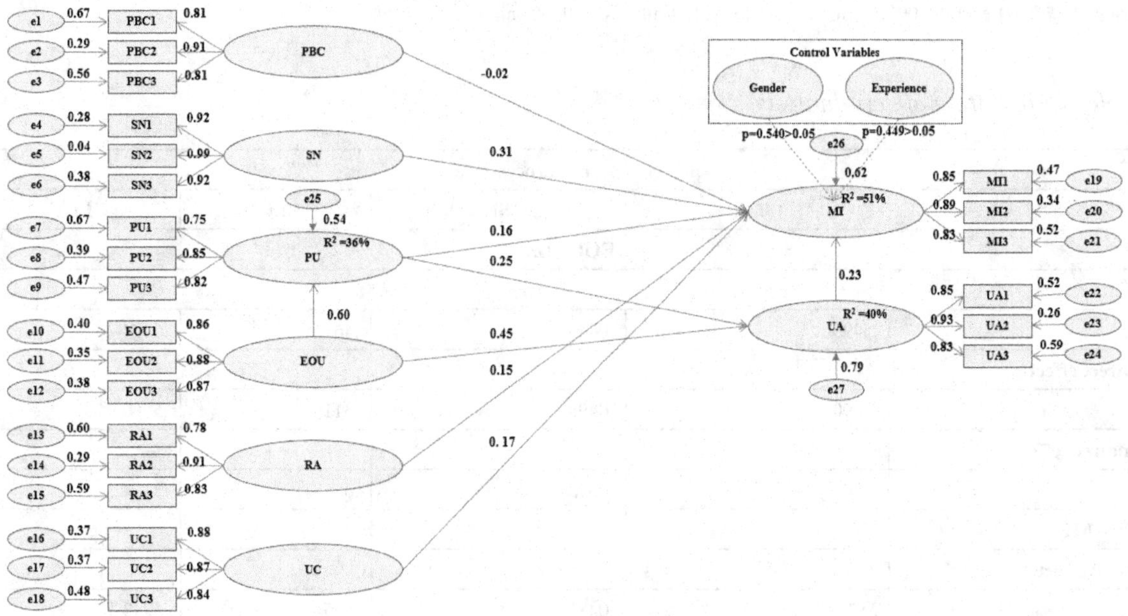

Hypothesis Testing and Results

The results of SEM depict path estimate (β) and statistical significance (p-value) of β weights for 9 path relationships as shown in Table 8. H1, H2a, H2b, and H3 are all supported with a significant predicting level (p<0.01), while H2c is unsupported (p=0. 0.608 >0.05) statistically. Thus, EOU significantly influence PU (β = .512). As antecedents of UA, PU and EOU both significantly affect UA (β = .311, β = .484, respectively); PU, UA, SN, RA, UC have significant relationship with MI (β =. 0.199, β =0.224, β = 0.276, β =. 0.174, β = 0.167 respectively), but PBC does not have as mentioned above. Further, at 95% level, PU show partial mediation effects on the relationship from EOU to UA, while UA also partially mediates the relationship between PU and MI using Process v5.0 as shown in Table 9.

Table 8. Coefficient estimate results

				Path Estimate	Standardized RegressionWeights	S.E.	C.R.	P	Hypothesis
H1a	PU	←	EOU	0.512	0.599	0.047	10.894	***	Support
H1b	UA	←	EOU	0.484	0.453	0.063	7.630	***	Support
H1c	UA	←	PU	0.311	0.249	0.075	4.172	***	Support
H1d	MI	←	PU	0.199	0.162	0.064	3.089	0.002**	Support
H2a	MI	←	UA	0.224	0.228	0.050	4.500	***	Support
H2b	MI	←	SN	0.276	0.308	0.040	6.898	***	Support
H2c	MI	←	PBC	-0.022	-0.022	0.043	-0.513	0.608	Not Support
H3a	MI	←	RA	0.174	0.148	0.059	2.965	0.003**	Support
H3b	MI	←	UC	0.167	0.168	0.051	3.265	0.001**	Support

Note: *** Sig. Level P‹0.001. ** Sig. Level P‹0.01. Repetitions: 5000; Confidence interval: 95%

Table 9. Mediating effects analysis

Bootstrapping (Process v5.0)				
	Effect	**SE**	**LLCI**	**ULCI**
EOU - UA				
Total effect				
	.5495	.0448	.4615	.6374
Direct effect				
	.4090	.0494	.3119	.5060
Indirect effect				
	.1405	.0375	.0667	.2164
PU - MI				
Total effect				
	. 5298	.0481	.4354	.6243
Direct effect				
	.3148	.0487	.2190	.4106
Indirect effect				
	.2151	.0468	.1324	.3142

*** Sig. Level P‹0.001. Repetitions: 5000; Confidence interval: 95%

Process v5.0 by Andrew Hayes is also used to examine the moderating effect of PR on the relationship between MI and UA (H4a), MI and RA (H4b), and MI and UC (H4c) as shown in Table 10. Both H4a and H4c are unsupported statistically. For H4b, Sig RA*PR = 0.0000(<0.05), Coefficients RA*PR = -0.1292 (<0) which show that PR negatively moderates the relationship between RA and MI. Further, Figure 4 depicts the result of simple slopes analysis.

Table 10. Coefficient estimate results (RA with MI by PR)

	Coefficients	Std.Error	t	Sig
Constant	4.3927	.0506	86.9944	.0000
RA	.4506	.0454	8.2597	.0000
PR	.1591	.0507	6.6546	.0011
RA*PR	-.1292	.0322	-2.9144	.0000

Sig. Level P‹0.05. Repetitions: 5000; Confidence interval: 95%

Figure 4. Simple Slopes Analysis

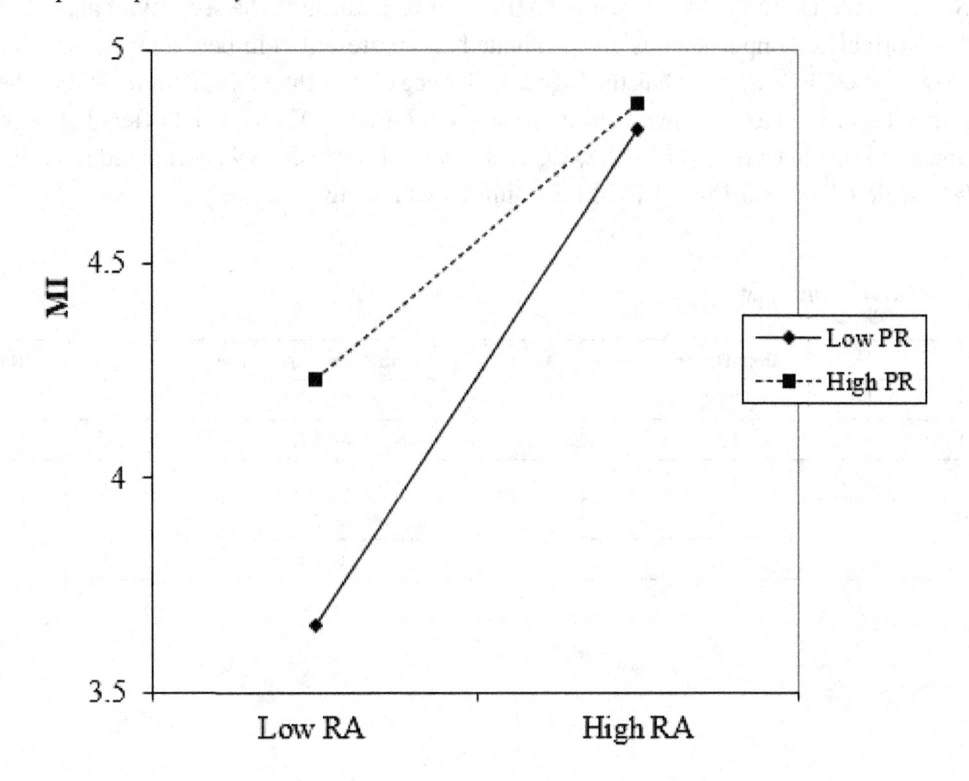

ANN Results

ANN modeling is more advanced than SEM, due to its non-linear mapping and self-learning ability (Raut et al., 2019). The study conducts ANN analysis as a compensatory balance of SEM. As a structured neural network model, SEM-ANN model can not only explain the causal relationship and influence degree between each network node, but also improve the goodness of fit of the model by using the non-linear fitting ability of neural network. Thus, it can provide further understanding of metaverse adoption intention in the study as ANN models are highly robust and adaptable. The structure of the ANN model is given in Figures 5-7, using only the significant predictors obtained from Amos-SEM, thus the topology of ANN model becomes more reliable with improved interpretability. Multi-layered model with an input layer, output layer, and hidden layers mapping relations between the aforementioned two layers.

Data is parted as 90% for training and 10% for testing, while10-fold cross validation is applied to prevent ANN model from overfitting (Raut et al., 2019). MLP (Multilayer Perceptron) with a BP (Back Propagation) algorithm which is used to train the connection weights among the network nodes in the model. The Sigmoid function is accurate for two-hidden layers, while Softmax function is used for output layer activation (Akour et al., 2022; Lee et al., 2020). Root mean square error (RMSE) and decision coefficient (R^2) are used for evaluating the performance of the ANN model, while the degree of significance of SN, PU, RA, UC, and UA that predict MI is evaluated through the sensitivity analysis as stated in Table 10. Normalized importance is also evaluated and represented in percentage form (Raut et al., 2019). The outcomes show that PU has the higher influence on UA than EOU, while SN has the highest influence on MI and UC has the lowest influence on MI. Further, ANN model offers better predictive power of the relationship from SN, PU, RA, UC and UA to MI (R^2 =53.3%) compared to AMOS -SEM (R^2 = 51%), while RMSE= 0.1468/0.1556 for training and testing separately.

Table 11. Sensitivity analysis

ANN	Importance	Normalized Importance	Ranking
UA	.250	82.7%	2
RA	.128	42.2%	4
UC	.117	38.7%	5
PU	.202	66.9%	3
SN	.302	100.0%	1

Figure 5. ANN analysis

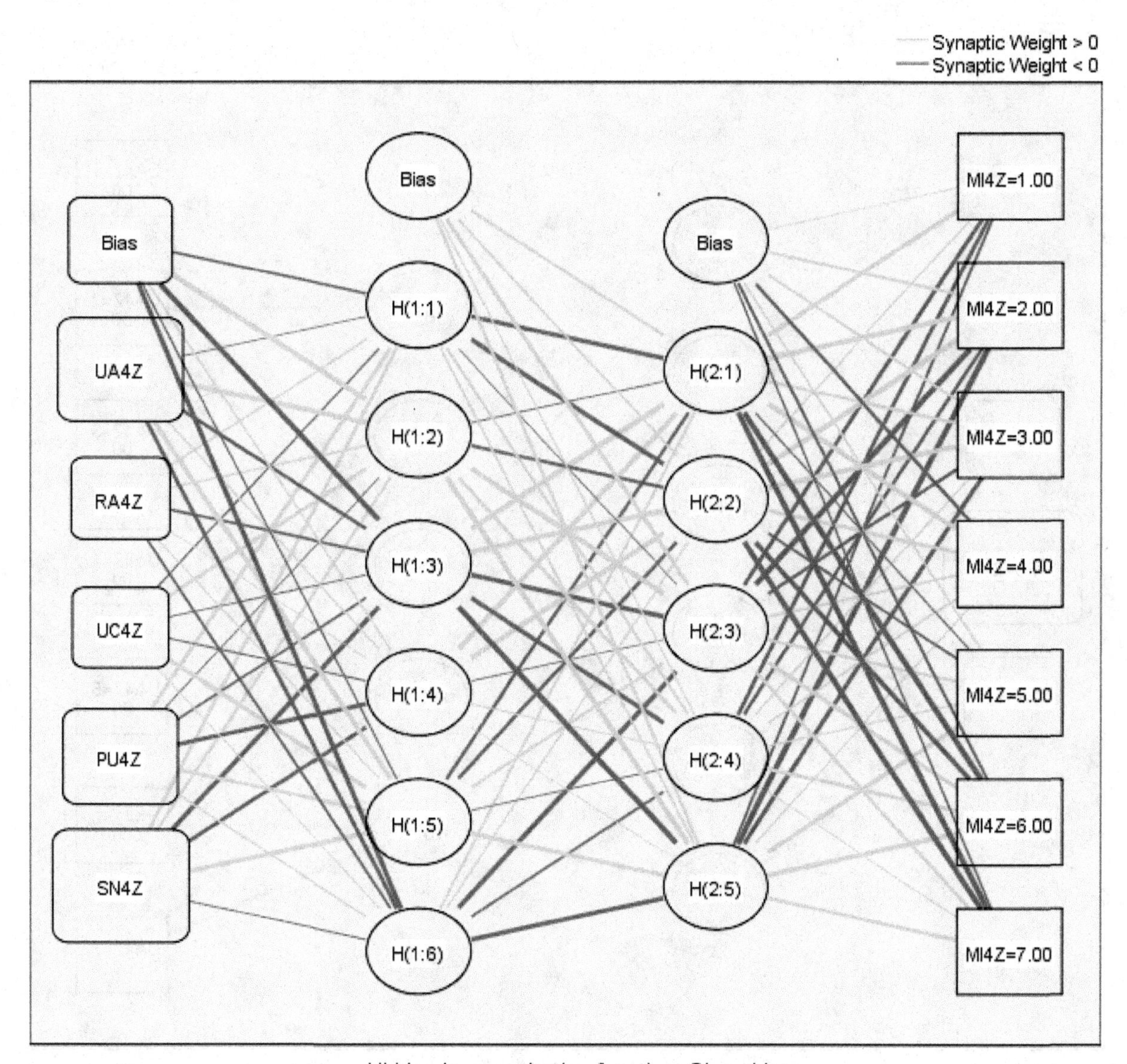

Figure 6. ANN analysis

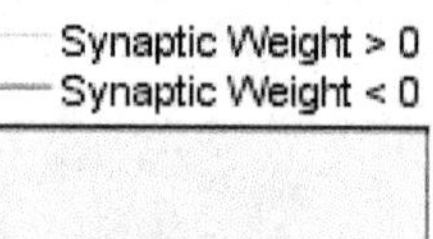

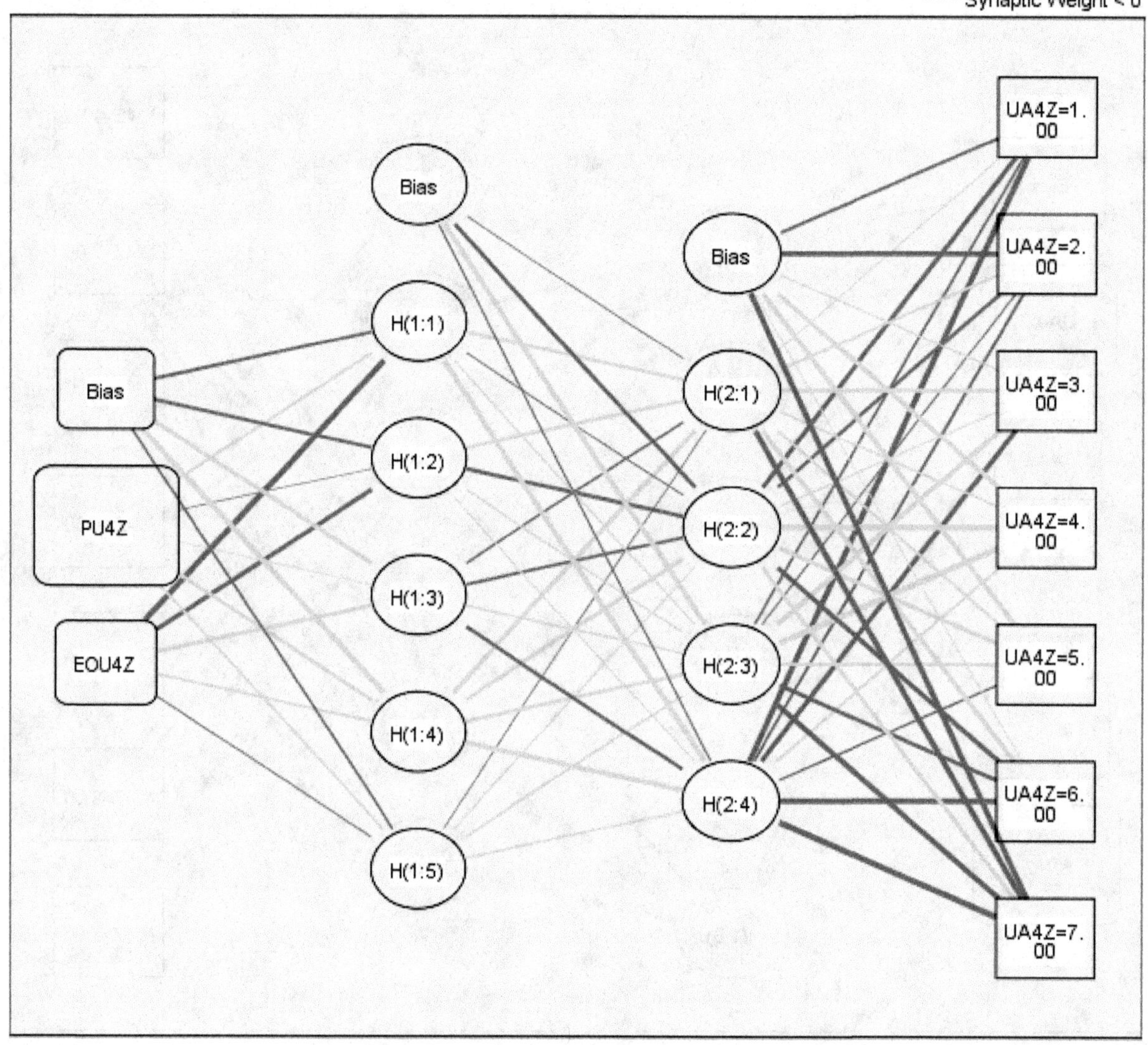

Hidden layer activation function: Sigmoid

Output layer activation function: Softmax

Figure 7. ANN analysis

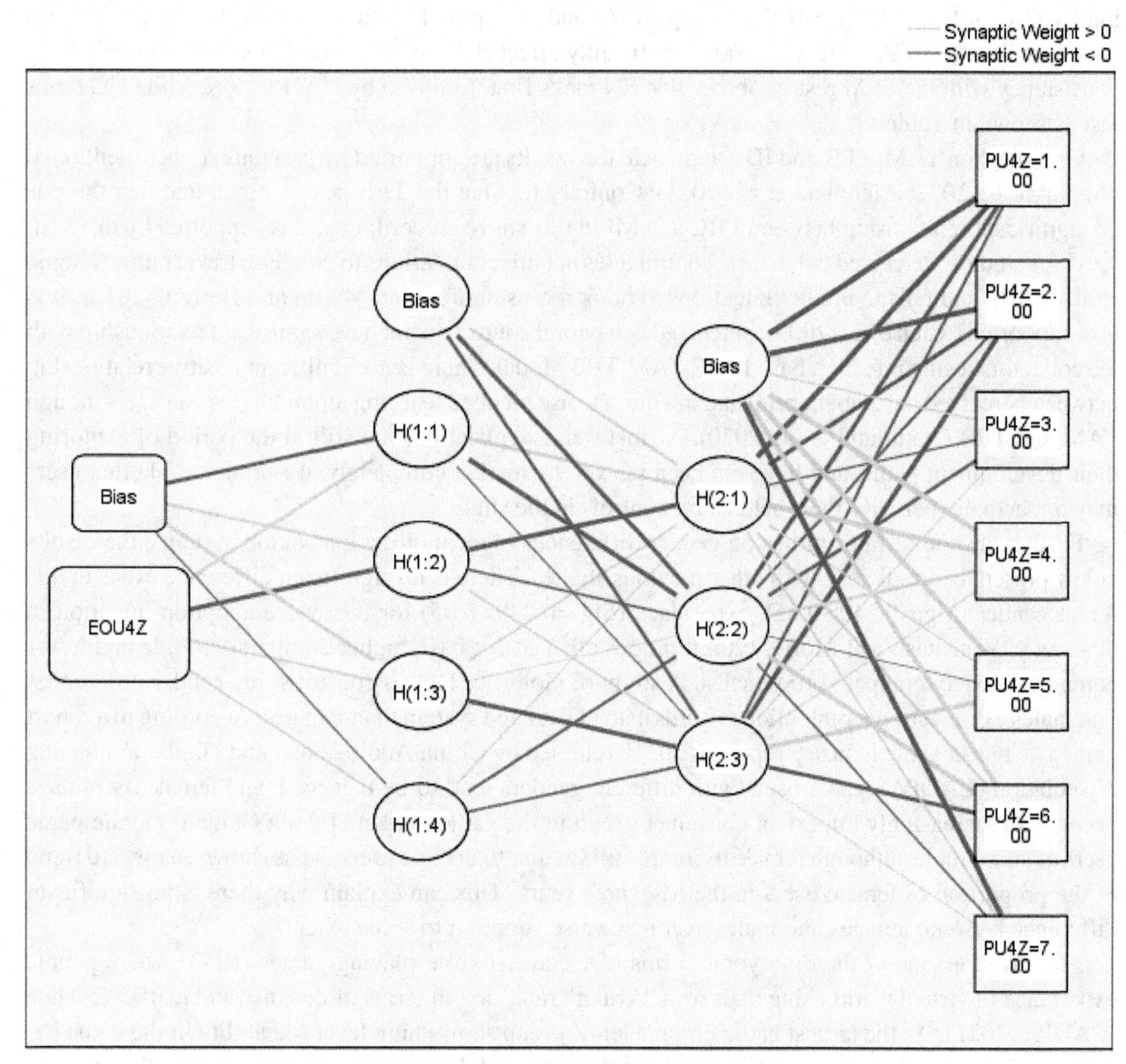

DISCUSSION AND CONCLUSION

This study uses two-step hybrid SEM-ANN analysis (Akour et al., 2022; Raut et al., 2019) to identify factors influencing Chinese on-campus students' metaverse adoption intention. As shown from the results, the hypothesis (H1-H4) and the proposed expanded C-TAM-TPB-IDT framework are verified and discussed.

SEM results of this study show that SN (p=0.31) has the biggest influencing effect on MI, following by UA (p=0.23), UC (p=0.17), PU (p=0.16) and RA (p=0.15). EOU (p=0.45) and PU (p=0.25) significantly affect UA. EOU (p=0.60) significantly affect PU. Accordingly, ANN results largely keep consistency with the SEM results above, that SN ranks first, followed by UA, PU, RA, while UC ranks last as shown in Table 11.

Grounded on TAM, TPB and IDT approach, the results are supported by previous studies mentioned (Akour et al., 2022; Alhumaid et al., 2021). Contrary to what the TPB posits, it is found that there is no significant relationship between PBC and MI in this study. Accordingly, it is supported by different previous studies. Perceived behavioral control does not affect intentions to purchase travel online (Bigné et al., 2010), and fails to predict intentions to book rooms online (San Martín and Herrero, 2012). It is also supported by Jen et al. (2009), perceived behavioral control do not have significant relationship with fee collectors' using intention based on C-TAM-TPB Model. There is no significant positive relationship between perceived behavior control and intention to use blended learning among Indonesia users though TAM and TPB (Nadlifatin et al., 2020). As metaverse applications are still in the period of exploring their development paths, and have not been put on the market completely, it is conceivable that users may not actually perceive their behavioral controls in the study.

Further, as one of rare studies on gender differences in technology innovation systems, the results below present contradictory with other previous studies. There is no significant difference exists in different gender (Sig.=0.540>0.05), experience (Sig.=0.449>0.05) for metaverse adoption intention in this study. Venkatesh and Morris (2000) propose that males have higher computer attitude and lower computer anxiety compared to females. A study of Goh (2011) also confirms this gender differences that males exhibit stronger intention towards using IT related system than females. According to a report named "China's game industry report in 2021" released by China Audio-Video and Digital Publishing Association (CADPA), game users with different genders tend to be balanced, and female users have become an increasingly important consumer group in the game market. Taking Chinese mobile game users as an example, although female users are still smaller than male users, it has shown an upward trend of the proportion of female users in the past three years. This can explain why there is no significant difference between females and males over metaverse adoption to some extent.

Moreover, as one of the embryonic forms of metaverse, role-playing games (RPG) where people experience in virtual world using their own "Avatar" rank first in terms of quantity and market revenue (CADPA, 2021). As the largest game group, Gen-Z group shows high level sociability in the game experience. They conduct voice chat while playing games, and don't even playing without online friends. This largely explains that more than 80% respondents have experienced in virtual world, and why no significant difference exists in respondents with experience and without experience in this study, as experiencing in virtual world that is already widely adopted among Gen-Z group.

Additionally, research looking at perceived risk associated with metaverse system adoption are rare. In the study, it is explainable that PR moderates the relationship between RA and MI, but UA, UC and MI by considering the concept of RA, as those perceived uncertainty and negative outcomes of new technology could lower users' perception and evaluation of new technology compared to the old technology it replaced.

IMPLICATIONS

Theoretical Implication

The study will contribute towards a new segment of metaverse research where rare literatures has been reviewed. Most literatures on metaverse are on how to define metaverse and identify its prospect. The integration of TAM, TPB, and IDT in the study not only provide worthy insights into IT acceptance models to deal with the technology adoption and infusion issues, but also how future users and developers adopt metaverse system. The study also increases the current literature of gender differences in technology innovation systems adoption.

Practical Implication

Firstly, trial and error costs a lot for Chinese technology companies as the government implement strict rules, and close check on the practices of them. They must maintain great sensitivity and vigilance, so understanding and recognizing metaverse users becomes particularly important though marketing survey in the early stage.

Secondly, as an out and out "Topic king" in the A-share market, it takes time to promote the gradual maturity to be down-to-earth of metaverse from a new concept to mature industry. "Gold-rush" or "Shilly-shally"? To take a rational view of metaverse industry and provide capital support to relevant technology companies and markets, it is more important to well know users. The study can also provide reference for investors.

Thirdly, a deeper understanding of the current situation of its residents/users will help Chinese government to make more appropriate long-term policy planning, regulation and formulation of laws and regulations. Timely and appropriately launching long-term development strategy and enterprise subsidies and preferential policies to maintain the enthusiasm and confidence of capital investors, companies, and users in the metaverse industry, while reminding them to keep rationality under strict supervision.

RECOMMENDATIONS

According to a survey conducted in 2021 among Chinese netizens as mentioned, over 60% of respondents expressed their willingness to use metaverse-empowered social networking services (Thomala, 2022a). The prevailing of the metaverse-sociality is not a difficult as "herd behavior" appears when the user group reaches a certain scale. The metaverse era ushers in new cast of both business advantages and ordeals. It is believed that metaverse will have bright win-win future through some corporation and sacrifice among users, developers, governments, and businesses as followed:

Firstly, Game IP avatars gained the highest level of interest as Chinese users were most enthusiastic about characterizing by their own avatars and socialize in metaverse (Thomala, 2022b). However, it is frightened that and the number of game fraud cases has increased. The metaverse giants should also put anti-risk mechanism plan and design as specific strategic objectives for achieving long term sustainable development of metaverse related applications.

Secondly, 70% of respondents worried that metaverse might lead to an addiction to the virtual world (Thomala, 2022c). The game publishing and working committee of the China audio and digital association jointly issued the "Convention on the Prevention of Addiction and Self-Discipline in the Online Game Industry" with 213 member units, formulated several targeted provisions, and strengthened the compliance awareness and self-discipline in the game industry in 2021. Metaverse companies have social responsibility to comprehensively consider users' concerns and their pain points, and take the construction of anti- addiction mechanism as the top priority.

Thirdly, although China has a good development momentum in VR/AR terminal hardware, virtual idols, Sandbox games and other prototype industries of the metaverse with expanding market scale, these products all precede the concept of metaverse. We should guard against the risk of speculation, capital kidnapping and financial fraud in business, as economic risks may spread from the virtual world to the real world.

LIMITATIONS AND FUTURE STUDIES TRENDS

Continued studies are required to provide further understanding about other potential factors that may have impacts on the sustainable development of the metaverse system, users' acceptance behaviour and to provide useful guidelines for developers, governments, and businesses.

Firstly, when we experience in metaverse, we need both supported software and hardware components (physical devices and sensors). Hardware components are not considered in the study. Future research may focus on the comfort of wearing those devices, and related interactions between those devices and users, as those factors could directly influence users' metaverse adoption behaviour, even their intention.

Secondly, consumers evaluate the value of an acquired product or service while simultaneously considering price. These conditions exist in the VR hardware market as the cost of VR hardware ranges from $15 to over $1000 where multiple factors must be considered at different price levels (Manis and Choi, 2019). Future study should consider how to include the price of those hardware components in the emerging VR hardware components market.

Thirdly, in the era of Web 3.0, Non-Homogenous Token (NFT), Decentralized Organization (DAO), Decentralized Finance (DFI) are developing rapidly based on blockchain technology. In the world of metaverse, NFT are used to trade and build users' own ownership world. Future study should consider how metaverse will ushering the international monetary and financial system, especially the Central Bank Digital Currency (CBDC).

REFERENCES

Achchuthan, S., & Velnampy, T. (2016). A quest for green consumerism in Sri Lankan context: an application of comprehensive model. *Asian economic and social society*, *6*(3), 59-76.

Agag, G., & El-Masry, A. A. (2016). Understanding consumer intention to participate in online travel community and effects on consumer intention to purchase travel online and WOM: An integration of innovation diffusion theory and TAM with trust. *Computers in Human Behavior*, *60*, 97–111. doi:10.1016/j.chb.2016.02.038

Ahmad, M. (2018). Review of the technology acceptance model (TAM) in internet banking and mobile banking. *International Journal of Information Communication Technology and Digital Convergence*, *3*(1), 23–41.

Ajzen, I. (1991). The Theory of Planned Behavior. *Organizational Behavior and Human Decision Processes*, *50*(2), 179–211. doi:10.1016/0749-5978(91)90020-T

Akour, I. A., Al-Maroof, R. S., Alfaisal, R., & Salloum, S. A. (2022). A conceptual framework for determining metaverse adoption in higher institutions of gulf area: An empirical study using hybrid SEM-ANN approach. *Computers and Education: Artificial Intelligence*, *3*, 100052. doi:10.1016/j.caeai.2022.100052

Al-Emran, M., Mezhuyev, V., & Kamaludin, A. (2018). Technology Acceptance Model in M-learning context: A systematic review. *Computers & Education*, *125*, 389–412. doi:10.1016/j.compedu.2018.06.008

Al-Mamary, Y. H., Al-nashmi, M., Hassan, Y. A. G., & Shamsuddin, A. (2016). A critical review of models and theories in field of individual acceptance of technology. *International Journal of Hybrid Information Technology*, *9*(6), 143–158. doi:10.14257/ijhit.2016.9.6.13

Alhumaid, K., Habes, M., & Salloum, S. A. (2021). Examining the factors influencing the mobile learning usage during COVID-19 Pandemic: An Integrated SEM-ANN Method. *IEEE Access : Practical Innovations, Open Solutions*, *9*, 102567–102578. doi:10.1109/ACCESS.2021.3097753

Almarzouqi, A., Aburayya, A., & Salloum, S. A. (2022). Prediction of User's Intention to Use Metaverse System in Medical Education: A Hybrid SEM-ML Learning Approach. *IEEE Access : Practical Innovations, Open Solutions*, *10*, 43421–43434. doi:10.1109/ACCESS.2022.3169285

Amaro, S., & Duarte, P. (2015). An integrative model of consumers' intentions to purchase travel online. *Tourism Management*, *46*, 64–79. doi:10.1016/j.tourman.2014.06.006

Ammenwerth, E. (2019). Technology acceptance models in health informatics: TAM and UTAUT. *Studies in Health Technology and Informatics*, *263*, 64–71. PMID:31411153

Awa, H. O., Ojiabo, O. U., & Emecheta, B. C. (2015). Integrating TAM, TPB and TOE frameworks and expanding their characteristic constructs for e-commerce adoption by SMEs. *Journal of Science and Technology Policy Management*, *6*(1), 76–94. doi:10.1108/JSTPM-04-2014-0012

Baby, A., & Kannammal, A. (2020). Network Path Analysis for developing an enhanced TAM model: A user-centric e-learning perspective. *Computers in Human Behavior*, *107*, 106081. doi:10.1016/j.chb.2019.07.024

Bigné, E., Sanz, S., Ruiz, C., & Aldás, J. (2010). Why some internet users don't buy air tickets online. In *Information and communication technologies in tourism 2010,* (pp. 209–221). Springer. doi:10.1007/978-3-211-99407-8_18

Byrne, B. M. (2001). Structural Equation Modeling with AMOS, EQS, and LISREL: Comparative approaches to testing for the factorial validity of a measuring instrument. *International Journal of Testing*, *1*(1), 55–86. doi:10.1207/S15327574IJT0101_4

CADPA. (2021). Report of China's game industry in 2021. *CADPA*. https://www.sohu.com/a/509131223_120840?scm=&spm=smpc.channel_164.tpl-author-box-pc.14.1655048642096nc9B6Jk_324

Caminiti, S. (2022). No one knows what the metaverse is and that's what's driving all the hype. CNBC. https://www.cnbc.com/2022/01/25/no-one-knows-what-the-metaverse-is-and-thats-driving-all-the-hype.html

Chang, S., & Tung, F. (2008). An empirical investigation of students' behavioral intentions to use the online learning course websites. *British Journal of Educational Technology*, *39*(1), 71–83.

Chaveesuk, S., & Chaiyasoonthorn, W. (2022). COVID-19 in Emerging Countries and Students' Intention to Use Cloud Classroom: Evidence from Thailand. *Education Research International*. doi:10.1155/2022/6909120

Chen, C. C. (2013). The exploration on network behaviors by using the models of Theory of planned behaviors (TPB), Technology acceptance model (TAM) and C-TAM-TPB. *African Journal of Business Management*, *7*(30), 2976–2984. doi:10.5897/AJBM11.1966

Chittum, M. (2022). Morgan Stanley Sees $8 Trillion Metaverse Market — In China Alone. Blockworks. https://blockworks.co/morgan-stanley-sees-8-trillion-metaverse-market-eventually/

Clement, J. (2022a). Leading business sectors worldwide that have already invested in the metaverse as of March 2022. *Statista*. https://www.statista.com/statistics/1302091/global-business-sectors-investing-in-the-metaverse/

Clement J. (2022b). Main reasons for global internet user interest in the metaverse 2021. *Statista*. https://www.statista.com/statistics/1296982/metaverse-interest-reasons-worldwide/

Davis, F. D. (1989). Perceived usefulness, perceived ease of use, and user acceptance of information technology. *Management Information Systems Quarterly*, *13*(3), 319–340. doi:10.2307/249008

Davis, F. D., Bagozzi, R. P., & Warshaw, P. R. (1989). User acceptance of computer technology: A comparison of two theoretical models. *Management Science*, *35*(8), 982–1003. doi:10.1287/mnsc.35.8.982

Davis, S., & Wiedenbeck, S. (2001). The mediating effects of intrinsic motivation, ease of use and usefulness perceptions on performance in first-time and subsequent computer users. *Interacting with Computers*, *13*(5), 549–580. doi:10.1016/S0953-5438(01)00034-0

Dionisio, J. D. N., III, W. G. B., & Gilbert, R. (2013). 3D virtual worlds and the metaverse: Current status and future possibilities. *ACM Computing Surveys*, *45*(3), 1-38.

Dirsehan, T., & Can, C. (2020). Examination of trust and sustainability concerns in autonomous vehicle adoption. *Technology in Society*, *63*, 101361. doi:10.1016/j.techsoc.2020.101361

Dirsehan, T., & Can, C. (2020). Examination of trust and sustainability concerns in autonomous vehicle adoption. *Technology in Society*, *63*, 101361. doi:10.1016/j.techsoc.2020.101361

Dirsehan, T., & Can, C. (2020). Examination of trust and sustainability concerns in autonomous vehicle adoption. *Technology in Society*, *63*, 101361. doi:10.1016/j.techsoc.2020.101361

Esteves, J., & Curto, J. (2013). A Risk and Benefits Behavioral Model to Assess Intentions to Adopt Big Data. *Journal of Intelligence Studies in Business*, *3*(3), 37–46. doi:10.37380/jisib.v3i3.74

Featherman, M. S., & Pavlou, P. A. (2003). Predicting e-services adoption: A perceived risk facets perspective. *Human-Computer Studies*, *59*(4), 451–474. doi:10.1016/S1071-5819(03)00111-3

Fishbein, M., & Ajzen, I. (1975). *Belief, attitude, intention and behavior: an introduction to theory and research*. Addison-Wesley.

Fly, Y. J., & Grünberg, L. (2022). What Will China's Metaverse Look Like? *Thediplomat*. https://thediplomat.com/2022/03/what-will-chinas-metaverse-look-like/

Fornell, C., & Lacker, D. F. (1981). Evaluating structural equation models with unobservable variables and measurement error. *JMR, Journal of Marketing Research*, *18*(1), 39–50. doi:10.1177/002224378101800104

Goh, T. T. (2011). Exploring Gender Differences in SMS-Based Mobile Library Search System Adoption. *Journal of Educational Technology & Society*, *14*(4), 192–206.

Gor, K. (2015). Factors influencing the adoption of online tax filing systems in Nairobi, Kenya. *The Strategic Journal of Business and Change Management*, *2*(77), 906–920.

Hair, J. F., Anderson, R. E., Tatham, R. L., & Black, W. C. (1998). *Multivariate data analysis,* (5th ed.). Prentice-Hall.

Heart, T. (2010). Who is Out There? Exploring the Effects of Trust and Perceived Risk on SaaS Adoption Intentions. *The Data Base for Advances in Information Systems*, *41*(3), 49–67. doi:10.1145/1851175.1851179

Howell, J. (2022). Key Technologies That Power The Metaverse. *Blockchains*. https://101blockchains.com/technologies-powering-metaverse/

Hsu, J. S. C., Huang, H. H., & Linden, L. P. (2011). Computer-mediated Counter-Arguments and Individual Learning. *Journal of Educational Technology & Society*, *14*(4), 111–123.

iiMedia Research (2022). Report on user behavior analysis of China' s meta universe industry in 2021. iiMedia Research. Retrieved from https://report.iimedia.cn/repo3-0/43053.html?acPlatCode=IIMReport&acFrom=recomBar_1020&iimediaId=85146

Jen, W., Lu, T., & Liu, P. T. (2009). An integrated analysis of technology acceptance behaviour models: Comparison of three major models. *MIS REVIEW: An International Journal*, *15*(1), 89–121.

Jiang, J., Hsu, M., Klein, G., & Lin, B. (2000). E-commerce user behaviour model: An empirical study. *Human Systems Management*, *19*(4), 265–276. doi:10.3233/HSM-2000-19406

Johnson, J. (2022a). Metaverse - statistics & facts. *Statista*. https://www.statista.com/topics/8652/metaverse/#dossierContents__outerWrapper

Johnson, J. (2022b). Countries with the highest number of internet users 2022. *Statista*. https://www.statista.com/statistics/262966/number-of-internet-users-in-selected-countries/

Johnson, J. (2022c). Benefits of the metaverse worldwide 2021. Statista. Retrieved from https://www.statista.com/statistics/1285117/metaverse-benefits/

Kamal, S. A., Shafiq, M., & Kakria, P. (2020). Investigating acceptance of telemedicine services through an extended technology acceptance model (TAM). *Technology in Society*, *60*, 101212. doi:10.1016/j.techsoc.2019.101212

Kayyali, A. (Feb 6, 2022). How the Metaverse will Alter the World's Direction. *Insidetelecom*. https://www.insidetelecom.com/how-the-metaverse-will-alter-the-worlds-direction/

Kharpal, A. (Feb 14, 2022). China's tech giants push toward an $8 trillion metaverse opportunity. Tech drivers. *CNBC*. https://www.cnbc.com/2022/02/14/china-metaverse-tech-giants-latest-moves-regulatory-action.html

Koul, S., & Eydgahi, A. (2018). Utilizing technology acceptance model (TAM) for driverless car technology adoption. *Journal of Technology Management & Innovation*, *13*(4), 37–46. doi:10.4067/S0718-27242018000400037

Lee, Y., & Kozar, K. A. (2008). An empirical investigation of anti-spyware software adoption: A multitheoretical perspective. *Information & Management*, *45*(2), 109–119. doi:10.1016/j.im.2008.01.002

Lee, Y. H., Hsieh, Y. C., & Hsu, C. N. (2011). Adding innovation diffusion theory to the technology acceptance model: Supporting employees' intentions to use e-learning systems. *Journal of Educational Technology & Society*, *14*(4), 124–137.

Li, L., & Buhalis, D. (2006). E-commerce in China: The case of travel. *International Journal of Information Management*, *26*(2), 153–166. doi:10.1016/j.ijinfomgt.2005.11.007

Lu, Y., Yang, S., Chau, P. Y., & Cao, Y. (2011). Dynamics between the trust transfer process and intention to use mobile payment services: A cross-environment perspective. *Information & Management*, *48*(8), 393–403. doi:10.1016/j.im.2011.09.006

Luo, X., Li, H., Zhang, J., & Shim, J. P. (2010). Examining multi-dimensional trust and multifaceted risk in initial acceptance of emerging technologies: An empirical study of mobile banking services. *Decision Support Systems*, *49*(2), 222–234. doi:10.1016/j.dss.2010.02.008

Ma, W. (2022). China's tech giants in the metaverse. *CNBC*. https://www.cnbc.com/2022/02/14/china-metaverse-tech-giants-latest-moves-regulatory-action.html

Manis, K. T., & Choi, D. (2019). The virtual reality hardware acceptance model (VR-HAM): Extending and individuating the technology acceptance model (TAM) for virtual reality hardware. *Journal of Business Research*, *100*, 503–513. doi:10.1016/j.jbusres.2018.10.021

Mileva, G. (2022). 50+ Metaverse Statistics-- Market Size & Growth. *Influencer Marketing*. https://influencermarketinghub.com/metaverse-stats/

Mok, C. (2022). Metaverse is the future of social network;. *CNBC*. https://www.cnbc.com/2022/02/14/china-metaverse-tech-giants-latest-moves-regulatory-action.html

Moon, J. W., & Kim, Y. G. (2001). Extending the TAM for a World-Wide-Web context. *Information & Management*, *38*(4), 217–230. doi:10.1016/S0378-7206(00)00061-6

Morgan, J. P. (2022). Opportunities in the metaverse. *JP Morgan*. https://www.jpmorgan.com/content/dam/jpm/treasury-services/documents/opportunities-in-the-metaverse.pdf

Nadlifatin, R., Miraja, B. A., Persada, S. F., Belgiawan, P. F., Redi, A. P., & Lin, S.-C. (2020). The Measurement of University Students' Intention to Use Blended Learning System through Technology Acceptance Model (TAM) and Theory of Planned Behavior (TPB) at Developed and Developing Regions: Lessons Learned from Taiwan and Indonesia. *International Journal of Emerging Technologies in Learning*, *15*(09), 219–230. doi:10.3991/ijet.v15i09.11517

Oh, J. H. (2021). A Study on Factors Affecting the Intention to Use the Metaverse by Applying the Extended Technology Acceptance Model (ETAM): Focused on the Virtual World Metaverse. *The Journal of the Korea Contents Association*, *21*(10), 204–216.

Page-Thomas, K. (2006). Measuring task-specific perceptions of the world wide web. *Behaviour & Information Technology*, *25*(6), 469–477. doi:10.1080/01449290500347962

Pallant, J. (2005). SPSS survival guide. Allen & Unwin.

Park, S., & Kang, Y. J. (2021). A Study on the intentions of early users of metaverse platforms using the Technology Acceptance Model. *Journal of Digital Convergence*, *19*(10), 275–285.

Park, S. M., & Kim, Y. G. (2022). A Metaverse: Taxonomy, components, applications, and open challenges. *IEEE Access : Practical Innovations, Open Solutions*, *10*, 4209–4251. doi:10.1109/ACCESS.2021.3140175

Raut, R. D., Mangla, S. K., Narwane, V. S., Gardas, B. B., Priyadarshinee, P., & Narkhede, B. E. (2019). Linking big data analytics and operational sustainability practices for sustainable business management. *Journal of Cleaner Production*, *224*, 10–24. doi:10.1016/j.jclepro.2019.03.181

Rogers, E. M. (2003). Diffusion of Innovations, 5th Edition. Simon and Schuster.

Safeena, R., Date, H., Hundewale, N., & Kammani, A. (2013). Combination of TAM and TPB in internet banking adoption. *International Journal of Computer Theory and Engineering*, *5*(1), 146–150. doi:10.7763/IJCTE.2013.V5.665

San Martín, H., & Herrero, Á. (2012). Influence of the user's psychological factors on the online purchase intention in rural tourism: Integrating innovativeness to the UTAUT framework. *Tourism Management*, *33*(2), 341–350. doi:10.1016/j.tourman.2011.04.003

Shen, B. Q., Tan, W.M., Guo, J.Z., Zhao, L.S., & Qin. P. (2021). How to Promote User Purchase in Metaverse? A Systematic Literature Review on Consumer Behavior Research and Virtual Commerce Application Design. *applied sciences*, 11, 11087

Shin, D. H. (2010). MVNO services: Policy implications for promoting MVNO diffusion. *Telecommunications Policy*, *34*(10), 616–632. doi:10.1016/j.telpol.2010.07.001

Siddiqui, K. (2013). Heuristics for sample size determination in multivariate statistical techniques. *World Applied Sciences Journal*, *27*(2), 285–287.

Singh, D. P. (2015). Integration of TAM, TPB, and self-image to study online purchase intentions in an emerging economy. *International Journal of Online Marketing*, *5*(1), 20–37. doi:10.4018/IJOM.2015010102

Soon, K. W. K., Lee, C. A., & Boursier, P. (2016). A study of the determinants affecting adoption of big data using integrated Technology Acceptance Model (TAM) and diffusion of innovation (DOI) in Malaysia. *International journal of applied business and economic research,* 14(1), 17-47.

Sourin, A. (2017). Case study: shared virtual and augmented environments for creative applications. In *Research and development in the academy, creative industries and applications,* (pp. 49–64). Springer. doi:10.1007/978-3-319-54081-8_5

Sudan, R., Petrov, O., & Gupta, G. (2022). Can the metaverse offer benefits for developing countries? *Worldbank*. https://blogs.worldbank.org/digital-development/can-metaverse-offer-benefits-developing-countries

Suh, W., & Ahn, S. (2022). Utilizing the Metaverse for Learner-Centered Constructivist Education in the Post-Pandemic Era: An Analysis of Elementary School Students. *Journal of Intelligence*, *10*(1), 17. doi:10.3390/jintelligence10010017 PMID:35324573

Taylor, S., & Todd, P. (1995). Assessing it usage: The role of prior experience. *Management Information Systems Quarterly*, *19*(4), 561–570. doi:10.2307/249633

Technavio. (2020). The increasing demand for AR and VR technology to boost growth. *Technavio*. https://www.businesswire.com/news/home/20200903005356/en

The Economist (2022). Building a metaverse with Chinese characteristics. *The Economist*. https://www.economist.com/china/2022/02/04/building-a-metaverse-with-chinese-characteristics

Thomala, L. L. (2022a). Level of understanding of metaverse in China 2021. *Statista*. https://www.statista.com/statistics/1287390/china-familiarity-with-metaverse/

Thomala, L. L. (2022b). Expectations of metaverse adoptions in China 2021. *Statista*. https://www.statista.com/statistics/1287403/china-most-in-demand-applications-of-metaverse-2021/

Thomala, L. L. (2022c). Major concerns of metaverse in China 2021. *Statista*. https://www.statista.com/statistics/1287437/china-worries-about-the-popularity-of-metaverse/

Venkatesh, V., & Davis, F. D. (2000). A theoretical extension of the technology acceptance model: Four longitudinal field studies. *Management Science*, *46*(2), 186–204. doi:10.1287/mnsc.46.2.186.11926

Venkatesh, V., & Morris, M. (2000). Why don't men ever stop to ask for direction? Gender, social influence and their role in technology acceptance and usage behavior. *Management Information Systems Quarterly*, *24*(1), 115–139. doi:10.2307/3250981

Wu, J. H., & Wang, S. C. (2005). What drives mobile commerce? An empirical evaluation of the revised technology acceptance model. *Information & Management*, *42*(5), 719–729. doi:10.1016/j.im.2004.07.001

Yang, H. H., & Su, C. H. (2017). Learner Behaviour in a MOOC Practice-oriented Course: In Empirical Study Integrating TAM and TPB. *International Review of Research in Open and Distributed Learning*, *18*(5), 35–63. doi:10.19173/irrodl.v18i5.2991

Chapter 16
Customer E-Satisfaction Towards Online Grocery Sites in the Metaverse World

Mohamed Salman Faiz B. Mohamed Jahir Hussain
Sunway University, Malaysia

Evelyn Toh Bee Hwa
https://orcid.org/0000-0002-2953-4909
Sunway University, Malaysia

ABSTRACT

The retail industry has changed drastically whereby in 2022, a further evolution into the metaverse is having a dramatic effect on the future of retail into virtual shopping and digital experiences. This study examines the factors influencing consumers' satisfaction towards online grocery shopping in Malaysia with the evolution of the metaverse. The technology acceptance model (TAM) serves as the underpinning theory for this study to assess the factors influencing people's intention to shop in online grocery websites. The main objective of this study is to establish the dimensions of customer satisfaction which is hypothesized to have a relationship with customer intention to purchase from online grocery sites and testing the robustness of the TAM model. This study was conducted in Klang Valley with 200 responses obtained and tested on SPSS and Smart PLS3.0. The findings of this research papers showing that PEOU, SEC, and PA do not influence customer satisfaction (SAT) and this will provide organizations insights behind customer satisfaction in e-commerce in the metaverse.

INTRODUCTION

The introduction of e-commerce is considered to be the broadest ranging and important area of concurrent expansion in marketing (Barwise et al., 2002). Online retailing is common component of corporations' distribution strategy since the advent of the internet, which has then changed the method consumers buy the products and services (Rose et al., 2012). The advantages of shopping online are that online retailers

DOI: 10.4018/978-1-6684-5732-0.ch016

can provide an array of products with better accessibility and convenience without posing any time or capacity limitations in the purchasing process (Giovanis & Athanasopoulou, 2014). This is supported in the global crisis of the Covid-19 pandemic. During this time, physical stores were obliged to adhere to with the time and capacity limitations as ordered by the government (Achariam, 2021).

The e-commerce market in Malaysia is picking up its pace and set to become one of the greatest in Southeast Asia. Its expansion is outperforming the region's traditional established markets, prior the effects of the pandemic. The rise in e-commerce is put together with ever rising consumer's online shopping (Aprameya, 2020). Despite the drop of 36.6 percent total sales in value of the wholesale and retail trade, Malaysia's online retail sales observed a climb in 28.9 percent back in April 2020. The Department of Statistics Malaysia (DoSM) mentioned that the increase in online retail sales is the result of the development of e-commerce activities in Malaysia. Overall, Malaysia's e-commerce market is expected to record a 24.7 percent of growth in 2020 (TheStar, 2020; Diong & Toh, 2021). The market in Malaysia is estimated to attain MYR 51.6 billion (US$ 12.6bn) by 2024, while growing at a compound annual growth rate (CAGR) of 14.3% between 2020 and 2024 (GlobalData, 2020). In the last two decades, the grocery retail landscape saw a significant transformation with the quick rise in online grocery shopping. Online grocery shopping is inclusive of the buying of food, drinks and ordinary necessities, particularly fast-moving consumer goods (Sonal & Thomas, 2019). Numerous grocery retailers are competing to gain a competitive edge via online grocery retailing by connecting with customers who are keen to purchasing their groceries online (Chintagunta et al., 2012).

The Covid-19 pandemic and Movement Control Order (MCO) was proof that the ideology of online grocery shopping is here to stay with potential expansion, despite it being deemed as a novel idea five years ago (Banoo, 2020). The online grocery market in Malaysia logged the highest progress (2.2x) in online retail penetration in 2020 since 2018 (Afandi, 2020). Due to uncertain times, online grocery platforms have developed as legitimate, gained popularity and serves as a substitute to traditional grocery shopping, during the period where people were advised to remain home to curb the spread of the virus. MyGroser, a local online grocery platform, mentions that they have seen an increase in demand by over 1000% during the past quarter of the year ("*Online grocery platforms*", 2020). During these times, the concept of metaverse further spring boarded into the online grocery platforms. The concept of metaverse is not new. According to Forbes (2022), the metaverse is defined as a fully immersive internet, where the ability to access augmented and virtual reality as well as interact with all sorts of environments using persistent avatars and innovative digital technology. The Metaverse is the future version of the Internet that further blurs the boundaries between reality and virtuality, with the convergence of immersive spaces, social & collaborative experiences, and the creator economy. "Future version of the Internet" emphasizes the fact that this is a new evolution of the Internet and not a paradigm shift or a particular private platform. "That further blurs the boundaries between reality and virtuality" aims to highlight that what we call reality is and increasingly will be, augmented by layers of digital information.

E-commerce in the Metaverse allows shoppers to emerge themselves through an avatar of their choice into an online shop while they might be completely on the opposite side of the world. E-commerce relies heavily on a fundamental grasp on the factors that influence online customer's satisfaction. Customer satisfaction is the result of experiences at different phases of the shopping process; needing something, gathering information about it, evaluating purchasing alternatives, actual purchase decision and post purchase behaviour (Mumtaz et al., 2011; Diong & Toh, 2021; Lai, Toh & Alkhrabsheh, 2020). The e-commerce platform is highly competitive; hence e-retailers have to constantly remain competitive in the market. In order to achieve this, e-satisfaction is the pathway to succeed. As mentioned earlier, the metaverse is the combination of several technological innovations that all operate seamlessly together. From non-fungible tokens (NFTs) to social commerce to augmented and virtual reality, the metaverse brings technology out of the digital arena and into the physical world. As an example, e-commerce giant, Amazon, has incorporated early metaverse technology into its marketplace (Forbes 2022). The use of the newest AR shopping tool, Room Decorator, allows users to use their phones or tablets to see what furniture and other home décor will look in reality. Other advantages are that the users can view multiple products together, and even save the Augmented Reality (AR) snapshots of the chosen room to review later. From the retail front, especially online grocery sites, this is the way forward in the coming years.

Hence, customer experience is increasingly being recognized by researchers and retailers as a key driver of market success and an important factor in customer satisfaction and intent to remain with a business (Sreeram et al., 2017). This is especially true for online retailers because the concept of metaverse poses a few challenges to them when it comes to digitizing the tactile (texture) and aromatic components of products.

Despite having the expansion of new product categories and fresh players in the industry, few studies have been conducted regarding customer satisfaction within the retail food industry. Satisfaction is important to compete in the ever changing and competitive environment of grocery retail (Huddleston, 2009). This is especially so in the context of online grocery retailing. In addition, a consumer's intent to repurchase and spread electronic-word-of-mouth is dependent on the level of satisfaction they acquire (Guzel et al., 2020).

According to Katrina & Benedict (2019), governmental bodies encourage the adoption of e-commerce by allowing more online merchants to display an array of items and categories for customers to pick from. Based on recent studies, it is noted that factors like ease of use and providing information about product assortment can lead to customer satisfaction (Raman, 2019). Despite this, the measures of product assortment have been rarely included in previous studies (Haridasan & Fernando, 2018).

For online platforms in the metaverse, it is critical to configure technology and user interface to meet the demands and desires of the customers. Qualities such as, ease of navigation, accessibility, and content utility has a significant effect on the customer's behaviour. The functionality and capabilities of technology clearly generates differences in purchasing behaviour between traditional and online settings (Bauerová & Klepek, 2018; Lai, Toh & Alkhrabsheh, 2020). For instance, the web design, greatly reduces the time spent shopping for groceries online. With the widespread of technology adaptation, it actually plays a vital role in both physical and online shopping. The acceptance of this adaptation is defined as the behaviour that encourages the use of technology rather than inhibiting it (Huijts et al., 2012).

Customers adaptation on technology is driven by various factors. These, can then be studied through the Technology Acceptance Model (TAM) by Davis (1989). Given the significant investments on e-commerce at this time, a section of fascinating factors has yet emerged to better understand the purchasing interactions via online platforms in metaverse which influences their satisfaction level leading to future purchase intents and customer loyalty.

Obtaining customer e-satisfaction poses a challenge in e-commerce in the metaverse hence the need for this study. This study will provide an extensive view on the variables which customers will respond to, when they perceive towards satisfaction when performing e-commerce transaction, focusing on online grocery shopping. Furthermore, this research will focus on the Malaysian market where there are limited studies conducted with regards to online grocery shopping attitude (Kian et al., 2018). The effects of the pandemic have resorted to people wanting to stay safe and be wary of closed areas for shopping. A previous study indicated that about 48% of participants are shopping online to decrease their chances of being exposed to the coronavirus (Katrina & Benedict, 2019). This fills the gap of past studies, whereby this study, will provide insights understanding customer satisfaction for online grocery shopping experience in the absence of a physical setting. By implementing the notion of experience, the findings of this study will provide managers of grocery stores and future researchers about customers experience, enhancing satisfaction while retaining customers. As such, the objective of this study is as follows:

- To determine whether Perceived Ease of Use (PEOU) contribute to satisfaction towards online grocery shopping site.
- To determine whether Perceived Usefulness (PU) contribute to satisfaction towards online grocery shopping site.
- To determine whether product assortment contribute to satisfaction towards online grocery shopping site.
- To determine whether security contribute to satisfaction towards online grocery shopping site.
- To identify the relationship with satisfaction with website usage and intention to purchase.

The aim of this study is to investigate the effect of PEOU, PU, Security and Product Assortment on customer satisfaction towards online grocery shopping with the intention to purchase. This study is based on a Technology Acceptance Model which is modified to include some variables to explain the current phenomenon.

LITERATURE REVIEW

Technology Acceptance Model (TAM)

The technology acceptance model (TAM) by Davis (1989) is currently the most widely used theory for technology acceptance in the field of information services (IS). TAM has been used in numerous empirical studies and is a simple and reliable model to examine technology adoption behaviours for a wide range of information technology (IT) (Gefen et al., 2003). TAM was chosen for this study in order to explain consumers' adoption of shopping online in the metaverse because of its constant ability to enlighten a significant number of variances between behavioural intention and actual behavioural based in terms of the purchasing of technology linked goods (Grabner-Krauter & Kaluscha, 2003; Lai and Tong, 2022).

Davis (1989), introduced the TAM model back in 1986. In TAM, the users' choices on accepting a new technology is based on two evaluations of projected outcomes; (i) Perceived Ease of Use (PEOU) and (ii) Perceived Usefulness (PU). PEOU is well-defined as to which degree the user trusts the technology system would be effortless, while PU is described as the users' expectation that the usage of new information technology will boost their job performance. Based on previous research, PEOU and PU is vital and in the acceptance of information system (IS) (Singh & Keswani, 2016; Lai and Liew, 2021).

For this study, as it looks at online grocery shopping in the metaverse, some new factors such as product assortment strategies, security and satisfaction have started to emerge because of the pandemic. Both trust and satisfaction are starting points in fostering an effective e-commerce relationship (Kim et al., 2009). Trust and satisfaction influence purchase intention. Lastly, the impact product assortment on customer satisfaction in traditional retailing was the initial component to enter the progressive equation. As a result, shoppers place a high value on product variety.

Perceived Ease of Use (PEOU) and Perceived Usefulness (PU)

PEOU remains the key determinant on technology acceptance. When contemplating the use of technology, PEOU is defined as the attentiveness of mental and physical efforts that a user assumes to get such as the extent to which a specific technological system becomes effortless (Davis, 1989). According to Burton-Jones and Hubona (2005) the simplicity of learning and mastering extensive technologies, such as technology and layouts of online shopping sites, are drivers of what makes it simple to use or move around.

Users are more likely to embrace technology that is believed to be simpler to use (Selamat et al., 2009). According to Teo (2001), easy-to-use system necessitates less effort to learn from the consumers perspective thus increasing the likelihood of acceptance and usage. From the perspective of online shopping customers 'attitude towards utilising the internet were positively influenced by perceived ease of use (Yulihasri & Daud, 2011). This is in line with Childers et al. (2001), whom claimed that online retailers who are able to offer online shopping sites which are transparent and comprehensible, requires less mental effort, allowing consumers to shop their desired way resulting in ease-to-use perceptions in consumers minds leading to a satisfied attitude towards online retailers. Online businesses should focus more on creating a user-friendly website. A study claims that there is a relationship between PEOU and satisfaction in the online shopping setting. Many researchers have looked into PEOU and satisfaction, resulting in a significance between them (Rezaei & Amin, 2013).

PU is defined the extent to which a user feels that technology will improve the performance of an activity (Davis, 1989). The ability to increase shopping performance, productivity and achieve shopping goals are all determinants of making a shopping activity successful (McCloskey, 2003; Barkhi et al. 2008). Customers will acquire positive attitudes towards products or services they perceive will give adequate advantages or qualities toward a solution. Kim et al. (2004) stated that online shopping sites that provide services that assist consumers in making better purchasing selections leads to satisfaction towards online shopping. Consumers with favourable attitudes towards online shopping platforms are usually valuable to online retailers and this can be seen by their overall purchasing efficiency, effectiveness and capability (Childers et al., 2001; Lai, Toh & Alkhrabsheh, 2020).

Customer satisfaction is considered one of the prime components for expansion. Moreover, customer satisfaction in e-commerce context can be clarified by PU, the main component of TAM (Amin et al., 2014). In their research framework, Soud and Fisal (2011) proved that there is a relationship between perceived usefulness and satisfaction. In addition, Akram (2011) also discovered that there is a relationship between perceived usefulness and satisfaction when purchasing online. Hence, the following hypotheses is derived: -

H1: There is a positive relationship with Perceived Ease of Use (PEOU) and Satisfaction (SAT) of Online Grocery Website.

H2: There is a positive relationship with Perceived Usefulness (PU) and Satisfaction (SAT) of Online Grocery Website.

Security (SEC)

Consumers who make purchases online regard security as a critical aspect (Eid, 2011; Lai and Liew, 2021). It is due to the fact that security and privacy concerns are critical in establishing confidence when performing an online transaction. Because online purchasing frequently entails payment by debit or credit cards, customers may shift their focus to the retailer's information as a sort of security. This is especially so for e-commerce in the metaverse (Forbes 2022). Consumers' propensity to visit online retailers and make purchases is closely tied to their trust in delivering personal information and performing credit card transactions Consumers are more likely to purchase a product from a reputable vendor or a well-known brand (Vasic et al., 2019). One of the most important factors impacting the success or failure of e-retailers is their level of assurance (Prasad & Aryasri, 2009). Consumers are afraid of being misled by sellers that would misuse their private information, especially credit card data (Comegys et al., 2009; Lai and Liew, 2021). Security is sectioned into two parts; first is related to concerns with data and transaction security while the second is concerned with consumer legitimacy (Guo et al., 2012). Security in e-commerce is one of the prime elements that consumers evaluate when making purchases. Security in websites will make customers more dependable and satisfied. According to Guo et al. (2012), security was categorized as a technological factor that results in a favorable relationship between customers and online retailers. As such, the following hypothesis is derived:

H3: There is a positive relationship between Security (SEC) and Satisfaction of Online Grocery Websites (SAT).

Product Assortment (PA)

A customer's view of the store is influenced by the range of available products. Fashion, computers and automotive, alongside fast-moving consumer goods all have the propensity to seek out for more product variation (Sahai et al., 2020). Hence, satisfaction and shop preferences are influenced by perceptions of product diversity. When it comes to grocery shopping, the width and depth of the assortment affects store patronage (Putit, Muhammad & Aziz, 2017). Ordinary grocery stores are normally larger than specialty grocery stores, and are more likely to shelf various product range compared to specialty store (Huddleston et al., 2009). According to Paulins & Geistfeld (2003), a store's appealing goods assortments is crucial in achieving its perceived desired status. Retailers should stress on characteristics that can support satisfaction, such as product selection choice, to enhance consumers' lasting trustworthiness to the store. A wide range of products may draw clients, and the customer satisfaction will be higher. The product assortment in e-commerce represents the product mix a business presents before the customer to maximize sales. As there are many choices provided, the company is able to address the need and satisfy the demand of a diverse client via online shopping (Guo et al., 2012). Hence, the following hypothesis is derived:-

H4: There is a positive relationship between Product Assortment (PA) and Satisfaction of Online Grocery Websites (SAT).

Satisfaction (SAT)

Customer satisfaction remains the key indicator of a company's performance, as it affects attitude, product repurchasing and word-of-mouth marketing (Sivadas & Baker-Prewitt, 2000; Diong & Toh, 2021). As such, customer satisfaction influences profitability and long term to customer loyalty. In a study by Giese and Cote (2000), a response, either cognitive or affective to a specific emphasis such as purchasing experience determines customers satisfaction.

Hence, based on this definition, customer's satisfaction, with their shopping experience could be a result of the worth implemented by the shopping adventure. Carpenter and Fairhurst (2005) discovered that both utilitarian and hedonic purchasing advantages have a favourable effect on satisfaction. Customers are more likely to be satisfied with a store, if the experience includes attributes they appreciate. Customer satisfaction is one of the company's most valuable assets. Satisfied customers are more like to use services, have a greater repurchase intention, and are more likely to suggest the product or service to others (Kim, Jin, et al., 2009). Customer satisfaction has been shown to have a beneficial impact on future repurchase intention based on previous research (Revels et al., 2010). A customer's positive online experiences in the metaverse could boost their likelihood of repeating the behaviour. In an online shopping setting, being satisfied with the seller has a significant impact on the customer's purchasing intention (Chu & Zhang, 2016).

H5: There is a positive relationship between Satisfaction of Online Grocery Websites (SAT) and Intention to Purchase (INT) from these sites.

Purchase Intention (INT)

Purchase intention (INT) happens when consumers willingly use internet technology to purchase goods (Ha & Janda, 2014). The behavioural aspect of intention is defined as people who will have a positive or negative attitude towards a product, which will then influence the intention to purchase (Chin & Goh, 2017). Previous research stated that the trust of consumers towards e-commerce is a vital factor towards intentions to purchase (Bianchi & Andrews, 2012). Consumers are more likely to purchase from online stores if they feel satisfied with the services offered by the online shop (Lee et al., 2011). This makes consumer purchase intention a vital factor in determining the online shopping behaviour of consumers active in online grocery shopping. Hence, purchase intention is treated as dependent variable in this research. The diagram below illustrates the relationship between variables.

Figure 1. Proposed Framework

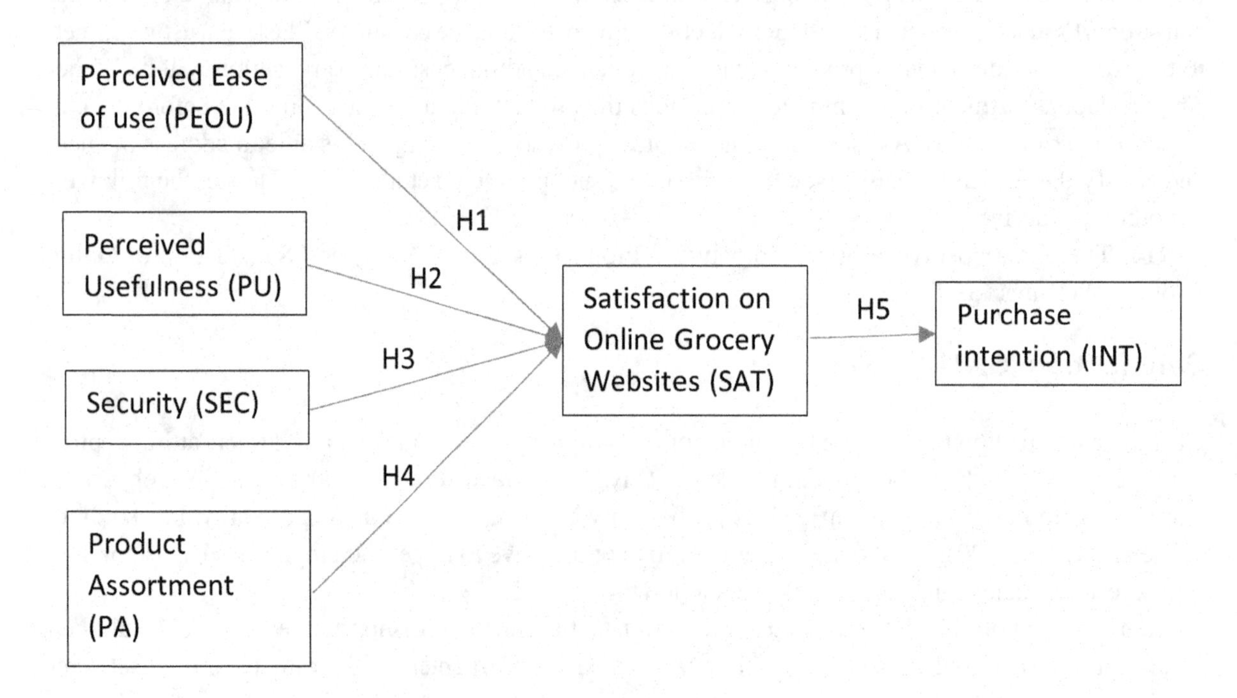

METHODOLGY

For this study, the hypothetic-deductive prism is used to determine the framework and foundations of positivism. This method is a circular process that starts with literature theory, moving to (i) building testable hypotheses, (ii) designing an experiment conceptualising variable such as; recognising factors to manipulate and assess through group assignments, and (iii) perform an empirical study built on the outcomes of the experiment. The results of this empirical investigation can be used to reinforce or enhance a theory (Park et al., 2019).

Research Design

There are various types of quantitative research; causal-comparative research, correlational research, experimental research and survey research. This study implements the correlational research design whereby it determines, to what extent a relation exist between two or more variables within a population. As such this is the case of how PEOU, PU, Product Assortment and Security, leads to satisfaction with website usage leading to the purchase intention.

The data collected for this study was done via an online. As this study was conducted during the Movement Control Order (MCO), with various restrictions implemented, a link was generated allowing it to be shared to individuals via platforms such as WhatsApp and Facebook. This questionnaire was based on a voluntary basis, and the respondents had the right to be a part of this study. To provide a sense of validity of this research, the institution name and email address of both the researcher and the supervisor were included. To maintain anonymity and confidentiality, any contactable information, such as mobile number and/or email address was required.

As the study required respondents who shop for groceries online where the population is nearly finite, hence the rationale behind using convenience sampling. Convenience sampling is a non-probability sampling which targets a particular population; ease of access, geographical closeness, convenience at a specific time or desire to participate (Etikan et al., 2016; Lai, 2020). The convenience sampling method's main goal is to collect data from individual who are easily accessible to the researchers (Etikan et al., 2016; Lai and Lim, 2019).

Respondents of working age (15-64 years old) account for the vast majority of e-commerce consumers in Malaysia (Malaysian Communications and Multimedia Commission, 2018). E-commerce in Malaysia developed its popularity due to the demographic shift and the emergence of millennials and Gen Z (Rafee, 2019). Millennials are born between the years of 1981 to 1996 meanwhile Gen Z, are individuals born from 1997 onwards (Dimock, 2019). Taking into consideration of the legal age majority being 18 years old, this study will be targeted to individuals aged from 18-60 years old (Dorall, 2020). In addition, the population will be targeted at respondents from the Klang Valley area, as this area highly dense has good internet adoption rate (Poh, 2019). A suggested sample size of 100 to 400 is adequate when using the structural equation model (Hair et al., 2010)

Measurement

The survey adopted a 5-point Likert scale, ranging from the extent of "agree" to "disagree", which will allow the participants to pick their choice based on the questions asked. The questions asked will is retrieved from past studies as seen in the table below. Furthermore, the questions are adopted from previous studies with a Cronbach α ranging from 0.7 till 0.9. The Cronbach alpha with a value of 0.7 or greater, indicates suitable consistency (Olaniyi, 2019). Table 1 indicates the items used as well as the Cronbach α for each variable is above the threshold of 0.7.

Table 1. Items measurement

Dimensions		Items	Cronbach α	Source
Perceived Ease of use (PEOU)	PEOU1	I find most online shopping sites easy to use.	0.96	Lim & Ting (2012)
	PEOU2	I find it easy learning to use most online shopping sites.		
	PEOU3	I find it easy to use most online shopping sites to find what I want.		
	PEOU4	I find it easy to become skillful at using most online shopping sites.		
	PEOU5	I find it easier to compare products when shopping at online retailers.		
	PEOU6	I find that most online shopping sites are flexible to interact with.		
	PEOU7	I am able to browse online shopping sites with ease.		
Perceived Usefulness (PU)	PU1	I am able to accomplish my shopping goals more quickly when I shop online.	0.959	Lim & Ting (2012)
	PU2	I am able to improve my shopping performance when I shop online.		
	PU3	I am able to increase my shopping productivity when I shop online.		
	PU4	I am able to increase my shopping effectiveness when I shop online.		
	PU5	I find the website of online retailers useful in aiding my purchase decision.		
	PU6	Shopping from online retailers improves my purchse decisions.		
	PU7	Shopping from online retailers makes it easier for me to satisfy my needs.		
Product Assortment (PA)	PA1	The store offers the assortment of products I am looking for.	0.877	Huddleston et. al. (2009)
	PA2	This store has the right merchandise selection.		
	PA3	This store has an extensive assortment of products.		
	PA4	This store is well-stocked across its different departments.		
Security (SEC)	S1	I feel safe in my stransactios with the online shop.	0.837	Rita et.al. (2019)
	S2	The online shop has adequate security features.		
	S3	This site protects information about my credit card.		
Satisfaction (SAT)	SA1	I am satisfied with this online shop.	0.855	Rita et.al. (2019)
	SA2	the online shop is getting close to the ideal online retailers.		
	SA3	The online shop always meets my needs.		
Purchase intention (INT)	INT1	I'm likely to purchase the products on this website.	0.853	Tang & Neuven, (2013)
	INT2	I'm likely to recommend this website to my friends.		
	INT3	I'm likely to make another purchase from this website if I need the products that I will buy.		

RESULTS AND FINDINGS

The demographics analysis of the respondents is shown in Table 2. This study collected 200 valid responses with an equal distribution of females and males at 50% each. In terms of race, the majority of the participants are Chinese (50%), Malay (25.5%), Indian (16.5%) and others (8%). Majority of the respondents are aged 18 to 25 (59.5%), 26 to 35 (20.0%), 36 to 45 (8%) and 55 and above (8%). In terms of education level, 70% of the participants are degree graduates. Respondents with diploma graduates at 13.5%, Masters degree graduates at 10.5%, high school graduates at 4.5% and PhD graduates at 1.5%.

Table 2. Demographic profile

Variable	Details	Frequency	Percentage
Gender	Female	100	50.0
	Male	100	50.0
Race	Malay	51	25.5
	Chinese	100	50.0
	Indian	33	16.5
	Others	16	8.0
Age	18 – 25	119	59.5
	26 – 35	40	20.0
	36 – 45	16	8.0
	46 – 55	9	4.5
	55 and above	16	8.0
Education Level	Bachelor Degree	140	70.0
	Diploma	27	13.5
	High School	9	4.5
	Masters	21	10.5
	PhD	3	1.5

Before the data is analysed the test for central tendency is conducted as prescribed by Kaur et al. (2018).

Table 3. Descriptive statistics

Variables	Mean	Median	Mode	Skewness	Kurtosis
INT	3.9983	4.0000	5.00	-1.027	0.722
PA	4.5563	4.7500	5.00	-1.983	6.970
PEOU	4.6525	4.8333	5.00	-2.786	14.810
PU	4.4133	4.5000	5.00	-1.716	5.417
SAT	4.0817	4.0000	5.00	-0.939	1.094
SEC	4.8167	5.0000	5.00	-3.534	20.588

Based on the results tabulated in Table 3, it can be concluded that the data is not of normal distribution. The variables are negatively skewed, indicating the mode is greater than the median and mean. Apart from that, with the low sample collected, warrants the necessities to conduct the analysis using the SmartPLS. The first step in conducting the confirmatory factor analysis on Smart PLS is to ascertain the composite reliability and AVE levels.

Composite Reliability and Average Variance Extracted (AVE)

The composite reliability (CR) that measures the internal consistency shows that all variables are above the 0.7 threshold as prescribed by Fornell & Larcker (1981; Hair et al., 2011).

Table 4. Reliability and AVE values

	Cronbach's Alpha	rho_A	Composite Reliability	Average Variance Extracted (AVE)
INT	0.914	0.918	0.946	0.853
PA	0.847	0.859	0.897	0.686
PEOU	0.880	0.889	0.909	0.626
PU	0.902	0.907	0.925	0.673
SAT	0.826	0.832	0.897	0.743
SEC	0.808	0.834	0.886	0.722

The variables are then examined for convergent and discriminant validity to determine their validity. Latent variable should obtain at least 50% of the variance manifest variables (Hussain et al., 2018). Hence, the constructs AVE value should be greater than 0.5. Based on the results tabulated in Table 4, the data in this study are reliable and the AVE is in the acceptable range above 0.5 (0.6 to 0.8). As seen in Table 5, all the item loadings are above the 0.7 threshold as the loadings that are below that threshold have been deleted (PEOU4 and PU1).

Table 5. Outer loadings

	INT	PA	PEOU	PU	SAT	SEC
INT1	0.917					
INT2	0.937					
INT3	0.916					
PA1		0.823				
PA2		0.782				
PA3		0.838				
PA4		0.866				
PEOU1			0.742			
PEOU2			0.867			
PEOU3			0.814			
PEOU5			0.779			
PEOU6			0.759			
PEOU7			0.780			
PU2				0.787		
PU3				0.869		
PU4				0.854		
PU5				0.813		
PU6				0.839		
PU7				0.754		
SAT1					0.878	
SAT2					0.905	
SAT3					0.800	
SEC1						0.794
SEC2						0.893
SEC3						0.858

The R^2 value indicates the variation of one variable to the other. An R^2 value of 0.75 is deemed significant, 0.50 is regarded as moderate and 0.26 is considered weak (CFI Education, 2018). Based on the results tabulated in Table 6, the variables are considered to be moderate and weak. However, despite having a low R^2 value, with the independent variable being significant, it can still consider crucial inferences in the relationship between the variables.

Table 6. R square

	R Square	R Square Adjusted
INT	0.438	0.436
SAT	0.249	0.233

Discriminant validity refers to the amount to which the constructs experimentally differ from one another (Rönkkö & Cho, 2020). The HTMT method is the newest addition to the marketing literature's preferred discriminant validity test (Henseler et al., 2015). In the HTMT test, a value near or greater than 1.0 is considered a discriminant validity infringement. Even though the specific HTMT ratio that would result in a discriminant validity infringement is unknown, the recommended starting point is at 0.85 and 0.90 (Henseler et al., 2015). Based on the results tabulated in Table 7, it can be concluded that the discriminant validity has been achieved and is considered valid for all variable with values between 0.1 and 0.8.

Table 7. HTMT criterion

	INT	PA	PEOU	PU	SAT	SEC
INT						
PA	0.341					
PEOU	0.203	0.757				
PU	0.387	0.691	0.626			
SAT	0.758	0.464	0.356	0.556		
SEC	0.167	0.626	0.821	0.523	0.320	

Multicollinearity occurs when two or more independent variables are significantly associated with one another (Bhandari, 2020). Variance inflation factor (VIF) with high values, implies that the linked independent variable has a high degree of collinearity with other variables in the model. High multicollinearity is an issue since it increases the variance of regression coefficient, leaving them volatile (Li, 2019).

Table 8. Inner VIF

	INT	PA	PEOU	PU	SAT
INT					
PA					2.118
PEOU					2.616
PU					1.695
SAT	1.000				
SEC					1.924

One way to estimate multicollinearity is the variance inflation factor (VIF), which assesses how much the variance of an estimated regression coefficient increases when predictors are correlated. If no factors are correlated, the VIFs will all be 1. A VIF between 5 and 10 indicates high correlation that may be problematic. If the VIF is greater than 10, it is likely that the regression coefficients are under-estimated due to multicollinearity, and should be addressed (Akinwande et al., 2015). Based on the results tabulated in Table 8, the VIF value is well below the suggested value (between 1 and 2.616) hence the issue of multicollinearity does not exist.

Path Coefficients for Hypothesis Testing

Based on the results tabulated in Table 9, 3 out of 5 hypotheses is rejected. Amongst the hypotheses that are rejected are PEOU to SAT ($\beta = -0.048$, $p = 0.651$), SEC to SAT ($\beta = 0.034$, $p = 0.736$) and PA to SAT ($\beta=0.163$, $p = 0.096$). The hypotheses that are accepted are PU to SAT ($\beta = 0.395$, $p = 0.000$) and SAT to INT ($\beta = 0.662$, $p = 0.000$). And, lastly, Therefore, it can be concluded, H1, H3, H4 is rejected meanwhile H2 and H5 is accepted.

Table 9. Path coefficient

Variable	Original Sample (β)	Sample Mean (M)	Standard Deviation (STDEV)	T Statistics (|O/STDEV|)	P Values	Decision
H1: PEOU -> SAT	-0.048	-0.043	0.105	0.453	0.651	Reject
H2: PU -> SAT	0.395	0.402	0.086	4.577	0.000	Accept
H3: SEC -> SAT	0.034	0.033	0.101	0.338	0.736	Reject
H4: PA -> SAT	0.163	0.161	0.098	1.668	0.096	Reject
H5: SAT -> INT	0.662	0.660	0.067	9.888	0.000	Accept

The diagram below illustrates the relationship between the variables in this study

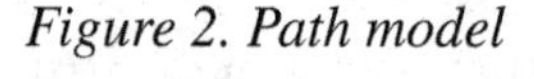

Figure 2. Path model

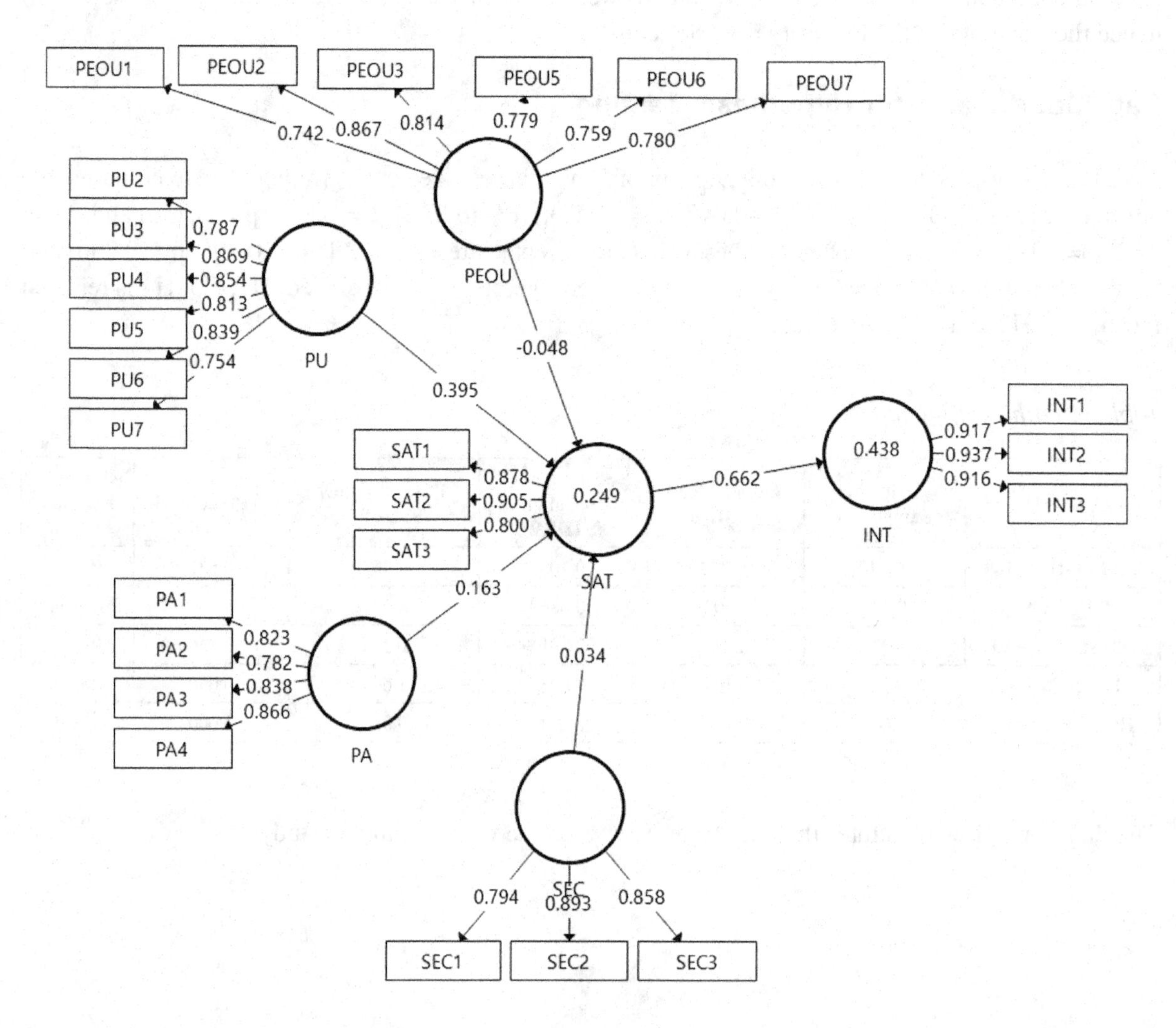

DISCUSSION, LIMITATIONS AND CONCLUSION

The Technology Acceptance Model (TAM), has been widely analyzed in various literature (Davis, 1989). It focuses on the philosophy of information services and describes how people come into acceptance of a technology (Yusuf Dauda & Lee, 2015).

Despite studies determining that PEOU has an impact on customer satisfaction when shopping online, due to easy use of the system and overall user interface, this study has discovered a different finding. In this study, the hypotheses; whether PEOU contributes towards satisfaction towards online grocery shopping sites, has been rejected ($\beta = -0.048$, $p = 0.651$). Previous researchers such as Kim & Chang (2007) and Premkumar & Bhattacherjee (2008) have also reported a nonsignificant relationship between PEOU and SAT. Even though, the notion of PEOU is that a technology should be effortless and simple to understand in order to encourage usage and repeat usage. According to Dalle (2010), the ease of use can be determined by how frequent the system has been utilised and interacted with. Additionally, ease of use also has no substantial effect on skilled online shoppers (Hernandez et al., 2009). In this study, perhaps the respondents have used the online grocery platform before hence have no issues about the interface that they are familiar with. Another argument is that they may already be loyal customers hence hold different expectations in post purchase transactions.

Perceived usefulness (PU) is described as the degree to which customers believe an online site could possibly add value and efficiency to their online shopping experience (Lai & Wang, 2012). PU has been used in various studies for various technological systems such as e-CRM, mobile payments and mobile data services (Zarmpou et al., 2012). PU and SAT were found to have a positive association in this study. Previous studies by Hung et al. (2012), Mohamed et al. (2014), Bölen & Özen (2020) and Diong & Toh (2021) supports the findings of the significant relationship between PU and SAT. The finding indicates that the bigger the PU benefits, the greater the motivation for them to be satisfied, thus leading to a continuous online purchasing. According to Alreck et al. (2009), online vendors who emphasise on factors such as time saving and quicker task completion have a higher chance of satisfying customers. According to Xu et al. (2013), PU is impacted by the information satisfaction and quality of information provided. Thus, PU has the ability to influence the mindset and usage behaviour of new technologies.

Consumers that shop online regard security to be an important element. Security is defined as an online store's ability to manage and maintain the data security related to their transactions (Vasic et al., 2019). In previous studies, security has recorded a good significant effect on customer satisfaction (Harwani & Safitri, 2017, Wang et al., 2019). Unlike previous studies, this study notes that there is no significant relationship between security and satisfaction. A similar finding was recorded in a study conducted by Mustafa (2011), hypothesising there is no security-satisfaction link. In addition, this finding is also aligned with Schaupp and Belanger (2005), who discovered that there is no significance with security and satisfaction. Another reason for the non-significance result is that during the lockdown MCO), people were forced people to shop online for safety reasons and online retailers took extra steps to ensure higher levels of security in websites and apps.

Customer satisfaction has been determined as critical in both online and traditional business (Alam & Yasin, 2010). A high level of satisfaction should be associated with the increase in purchasing intention (INT) or loyalty (Prateek Kalia et al., 2016). In this study, the relationship between customer satisfaction (SAT) and purchase intention (INT) is significant. This goes in line with the findings of past studies by Hsu et al. (2012) and Ali (2016).

A consumer's decision to purchase is definitely influenced by the product assortment or the variety available at the store. For instance, a store's recurring business is based on the product diversity offered (Dunner, Lusch & Carver, 2014; Putit, Muhammad & Aziz, 2017). Customers make purchases online for the perks of having a wide and large range of product selection. An extensive range of selection may be linked to higher satisfaction, since it enhances the chances of a better match between a customer's preferred product and current alternatives (Gao & Simonson, 2015).

Despite studies proving the significance of product assortment and customer satisfaction (Wilson & Christella, 2019, Mofokeng, 2021) this study has proved likewise. In this study, there is no significant relationship between product assortment and customer satisfaction. Thus, this goes in line with another study, that product assortment does not have a link to satisfaction (Sahai et al., 2020). In this study, the respondents did not link the availability of product assortment to satisfaction. The reason for the non-significance result is that perhaps the respondents are not able to differentiate the product offerings from different sites and end up shopping online for better bargains and other conveniences such as payment options, delivery options such as BOPIS (Buy online, pick in store) etc.

This study comes with some limitations. Firstly, the data collection process was done online during the lockdown (MCO) and post MCO period where there was limited conversation between the respondent and researcher. Respondents were probably not able to understand the questionnaire. Also, the sample size of this study is small, hence this could have resulted in the data obtained to differ from previous studies. Data was only collected in the Klang Valley region hence the results for this study cannot be generalised to represent the entire population of Malaysia.

The further directions of this study can be expanded beyond the Klang Valley region (Ariffin, 2021). The recommendations for future studies could also focus on a specific group of individuals, mainly millennials to better understand their purchasing factors on online grocery shopping. Lastly, a good recommendation to better understand the influencing factor is to conduct a comparative study with Singapore. Despite being a small island, Singapore is considered one of the most digitally competitive country (Yip, 2019). Hence with this, a comparative study would give researchers a better understanding on how these demographics from two different nations differ or share similarities amongst one another. Further research can be conducted with the Metaverse application for online shopping.

REFERENCES

Ab Hamid, M. R., Sami, W., & Mohmad Sidek, M. H. (2017). Discriminant Validity Assessment: Use of Fornell & Larcker criterion versus HTMT Criterion. *Journal of Physics: Conference Series*, *890*, 012163. doi:10.1088/1742-6596/890/1/012163

Achariam, T. (2021). Business outlets to operate from 8am to 8pm under new measures to curb Covid-19. *The Edge Markets*. https://www.theedgemarkets.com/article/shops-only-allowed-operate-8am-8pm-under-stricter-mco-measures-announced-today

Afandi, H. (2020). #TECH: Malaysia has most digital consumers in the region. *NST Online*. https://www.nst.com.my/lifestyle/bots/2020/09/627994/tech-malaysia-has-most-digital-consumers-region

Akinwande, M. O., Dikko, H. G., & Samson, A. (2015). Variance Inflation Factor: As a Condition for the Inclusion of Suppressor Variable(s) in Regression Analysis. *Open Journal of Statistics*, *05*(07), 754–767. doi:10.4236/ojs.2015.57075

Alam, S. S., & Yasin, N. M. (2010). An Investigation into the Antecedents of Customer Satisfaction of Online Shopping. *ResearchGate*. https://www.researchgate.net/publication/266495477_An_Investigation_into_the_Antecedents_of_Customer_Satisfaction_of_Online_Shopping

Ali, F. (2016). Hotel website quality, perceived flow, customer satisfaction and purchase intention. *Journal of Hospitality and Tourism Technology*, *7*(2), 213–228. doi:10.1108/JHTT-02-2016-0010

Alreck, P. L., & Dibartolo, G., Diriker, M., & Settle, R. B. (2009). Time Pressure, Time Saving And Online Shopping: Exploring A Contradiction. *ResearchGate*. https://www.researchgate.net/publication/242089354_Time_Pressure_Time_Saving_And_Online_Shopping_Exploring_A_Contradiction

Aprameya, A. (2020). Ecommerce in Malaysia: Growth, Trends & Opportunities. *Capillary Blog*. https://www.capillarytech.com/blog/capillary/ecommerce/ecommerce-in-malaysia-growth/

Ariffin, A. (2021). Top 5 most populous states in Malaysia 2021. *Iproperty.com.my*. https://www.iproperty.com.my/guides/most-populous-states-in-malaysia/

Banoo, S. (2020). E-Commerce - Growing acceptance of online grocery shopping. *The Edge Markets*. https://www.theedgemarkets.com/article/ecommerce-growing-acceptance-online-grocery-shopping

Barkhi, R., Belanger, F., & Hicks, J. (2008). A Model of the Determinants of Purchasing from Virtual Stores. *Journal of Organizational Computing and Electronic Commerce*, *18*(3), 177–196. doi:10.1080/10919390802198840

Barwise, P., Elberse, A., & Hammond, K. (2002). Marketing and the Internet: a research review (pp. 01-801), London Business School.

Bauerová, R., & Klepek, M. (2018). Technology Acceptance as a Determinant of Online Grocery Shopping Adoption. *Acta Universitatis Agriculturae et Silviculturae Mendelianae Brunensis*, *66*(3), 737–746. doi:10.11118/actaun201866030737

Bhandari, A. (2020). What is Multicollinearity? Here's Everything You Need to Know. *Analytics Vidhya*. https://www.analyticsvidhya.com/blog/2020/03/what-is-multicollinearity/

Bianchi, C., & Andrews, L. (2012). Risk, trust, and consumer online purchasing behaviour: A Chilean perspective. *International Marketing Review*, *29*(3), 253–275. doi:10.1108/02651331211229750

Bölen, M. C., & Özen, Ü. (2020). Understanding the factors affecting consumers' continuance intention in mobile shopping: The case of private shopping clubs. *International Journal of Mobile Communications*, *18*(1), 101–129. doi:10.1504/IJMC.2020.104423

Burton-Jones, A., & Hubona, G. S. (2005). Individual differences and usage behavior. *ACM SIGMIS Database*, *36*(2), 58–77. doi:10.1145/1066149.1066155

Carpenter, J. M., Moore, M., & Fairhurst, A. E. (2005). Consumer shopping value for retail brands. *Journal of Fashion Marketing and Management*, *9*(1), 43–53. doi:10.1108/13612020510586398

Education, C. F. I. (2018). R-Squared - Definition, Interpretation, and How to Calculate. *Corporate Finance Institute*. https://corporatefinanceinstitute.com/resources/knowledge/other/r-squared/

Chang, C. (2011). The Effect of the Number of Product Subcategories on Perceived Variety and Shopping Experience in an Online Store. *Journal of Interactive Marketing*, *25*(3), 159–168. doi:10.1016/j.intmar.2011.04.001

Childers, T. L., Carr, C. L., Peck, J., & Carson, S. (2001). Hedonic and utilitarian motivations for online retail shopping behavior. *Journal of Retailing*, *77*(4), 511–535. doi:10.1016/S0022-4359(01)00056-2

Chin, S. L., & Goh, Y. N. (2017). Consumer Purchase Intention Toward Online Grocery Shopping: View from Malaysia. *Global Business and Management Research*, 9.

Chintagunta, P. K., Chu, J., & Cebollada, J. (2012). Quantifying Transaction Costs in Online/Off-line Grocery Channel Choice. *Marketing Science*, *31*(1), 96–114. doi:10.1287/mksc.1110.0678

Comegys, C., Hannula, M., & Váisánen, J. (2009). Effects of Consumer Trust and Risk on Online Purchase Decision-Making: A Comparison of Finnish and United States Students. *International Journal of Management*, *26*(2), 295–308.

Dalle, J. (2010). *The Relationship Between PU and PEOU Towards the Behavior Intention in New Student Placement (NSP) System of Senior High School in Banjarmasin, South Kalimantan, Indonesia*.

Davis, F. D. (1985). *A technology acceptance model for empirically testing new end-user information systems: theory and results*. Massachusetts Institute Of Technology.

Davis, F. D. (1989). Perceived Usefulness, Perceived Ease of Use, and User Acceptance of Information Technology. *Management Information Systems Quarterly*, *13*(3), 319–340. 10.2307/249008 doi:10.2307/249008

Dilmegani, C. (2021). Data Cleaning in 2021: What it is, Steps to Clean Data & Tools. *Clean (Weinheim)*.

Dimock, M. (2019). Defining generations: Where Millennials End and Generation Z Begins. *Pew Research Center.* https://www.pewresearch.org/fact-tank/2019/01/17/where-millennials-end-and-generation-z-begins/

Diong, I. H., & Toh, E. B. H. (2021). The Influences of Reference Groups Towards the Usage of E-Wallet Payment Systems. In *Handbook of Research on Social Impacts of E-Payment and Blockchain Technology* (pp. 428–455). IGI Global.

Dorall, A. (2020). Can An Adult Legally Be In A Relationship With A 17-Year-Old In Malaysia? *TRP*. https://www.therakyatpost.com/2020/09/27/can-an-adult-legally-be-in-a-relationship-with-a-17-year-old-in-malaysia/

Dunner, P. M., Lusch, R. F., & Carver, J. R. (2014). Retailing. (8th Ed). South Western.

Eid, M. I. (2011). Customer satisfaction, brand trust and variety seeking as determinants of brand loyalty. *Journal of Electronic Commerce Research*, *12*(1). doi:10.5897/ajbm11.2380

Eri, Y., Aminul Islam, M., & Ku Daud, K. A. (2011). Factors that Influence Customers' Buying Intention on Shopping Online. *International Journal of Marketing Studies*, *3*(1). doi:10.5539/ijms.v3n1p128

Etikan, I., Musa, S. A., & Alkassim, R. S. (2016). Comparison of Convenience Sampling and Purposive Sampling. *American Journal of Theoretical and Applied Statistics*, *5*(1), 1. C:\Users\salma\Desktop\10.11648\j.ajtas.20160501.11 doi:10.11648/j.ajtas.20160501.11

Fornell, C., & Larcker, D. F. (1981). Evaluating Structural Equation Models with Unobservable Variables and Measurement Error. *JMR, Journal of Marketing Research*, *18*(1), 39–50. doi:10.1177/002224378101800104

Gao, L., & Simonson, I. (2015). The positive effect of assortment size on purchase likelihood: The moderating influence of decision order. *Journal of Consumer Psychology*, *26*(4), 542–549. doi:10.1016/j.jcps.2015.12.002

Gefen, Karahanna, & Straub. (2003). Trust and TAM in Online Shopping: An Integrated Model. *MIS Quarterly*, *27*(1), 51. doi:10.2307/30036519

Giese, J. L., & Cote, J. A. (2000). Defining Consumer Satisfaction. *Academy of Marketing Science Review*, 1–24.

Giovanis, A. N., & Athanasopoulou, P. (2014). Gaining customer loyalty in the e-tailing marketplace: The role of e-service quality, e-satisfaction and e-trust. *International Journal of Technology Marketing*, *9*(3), 288. doi:10.1504/IJTMKT.2014.063857

GlobalData. (2020). COVID-19 Accelerates e-commerce Growth in Malaysia. *GlobalData*. https://www.globaldata.com/covid-19-accelerates-e-commerce-growth-malaysia-says-globaldata/

Grabner-Kräuter, S., & Kaluscha, E. A. (2003). Empirical research in on-line trust: A review and critical assessment. *International Journal of Human-Computer Studies*, *58*(6), 783–812. C:\Users\salma\Desktop\10.1016\s1071-5819(03)00043-0 doi:10.1016/S1071-5819(03)00043-0

Guo, X., Ling, K. C., & Liu, M. (2012). Evaluating Factors Influencing Consumer Satisfaction towards Online Shopping in China. *Asian Social Science*, *8*(13). doi:10.5539/ass.v8n13p40

Gupta, A., Mishra, P., Pandey, C., Singh, U., Sahu, C., & Keshri, A. (2019). Descriptive statistics and normality tests for statistical data. *Annals of Cardiac Anaesthesia*, *22*(1), 67. doi:10.4103/aca.ACA_157_18 PMID:30648682

Guzel, M., Sezen, B., & Alniacik, U. (2020). Drivers and consequences of customer participation into value co-creation: A field experiment. *Journal of Product and Brand Management*, *30*(7), 1047–1061. doi:10.1108/JPBM-04-2020-2847

Ha, H.-Y., & Janda, S. (2014). The effect of customized information on online purchase intentions. *Internet Research*, *24*(4), 496–519. doi:10.1108/IntR-06-2013-0107

Hair, J. F., Black, W. C., Babin, B. J., & Anderson, R. E. (2010). *Multivariate Data Analysis,* (7th ed.). Pearson Prentice Hall.

Hair, J. F., Ringle, C. M., & Sarstedt, M. (2011). PLS-SEM: Indeed a Silver Bullet. *Journal of Marketing Theory and Practice*, *19*(2), 139–152. doi:10.2753/MTP1069-6679190202

Hamid, A., Razak, F. Z. A., Bakar, A. A., & Abdullah, W. S. W. (2016). The Effects of Perceived Usefulness and Perceived Ease of Use on Continuance Intention to Use E-Government. *Procedia Economics and Finance*, *35*, 644–649. doi:10.1016/S2212-5671(16)00079-4

Haridasan, A. C., & Fernando, A. G. (2018). Online or in-store: Unravelling consumer's channel choice motives. *Journal of Research in Interactive Marketing*, *12*(2), 215–230. doi:10.1108/JRIM-07-2017-0060

Haridasan, A. C., & Fernando, A. G. (2018). Online or in-store: Unravelling consumer's channel choice motives. *Journal of Research in Interactive Marketing, 12*(2), 215–230. doi:10.1108/JRIM-07-2017-0060

Harwani, Y., & Safitri. (2017). Security and Ease of Use Effect on Customers' Satisfaction Shopping in Tokopedia. *Journal of Resources Development and Management*. https://www.semanticscholar.org/paper/Security-and-Ease-of-Use-Effect-on-Customers%E2%80%99-in-Harwani-Safitri/c54eb79cd78c575b23640394fec36b96909bfe2d

Henseler, J., Hubona, G., & Ray, P. A. (2016). Using PLS path modeling in new technology research: Updated guidelines. *Industrial Management & Data Systems, 116*(1), 2–20. doi:10.1108/IMDS-09-2015-0382

Henseler, J., Ringle, C. M., & Sarstedt, M. (2015). A new criterion for assessing discriminant validity in variance-based structural equation modeling. *Journal of the Academy of Marketing Science, 43*(1), 115–135. doi:10.100711747-014-0403-8

Henseler, J., Ringle, C. M., & Sinkovics, R. R. (2009). The use of partial least squares path modeling in international marketing. *Advances in International Marketing, 20*, 277–319. doi:10.1108/S1474-7979(2009)0000020014

Hernandez, B., Jimenez, J., & Jose Martin, M. (2009). The impact of self-efficacy, ease of use and usefulness on e-purchasing: An analysis of experienced e-shoppers. *Interacting with Computers, 21*(1-2), 146–156. doi:10.1016/j.intcom.2008.11.001

Chan, H. K., Lettice, F., & Durowoju, O. A. (2012). Decision-making for supply chain integration: supply chain integration. Springer.

Hsu, C. L., Chang, K. C., & Chen, M. C. (2012). The impact of website quality on customer satisfaction and purchase intention: Perceived playfulness and perceived flow as mediators. *Information Systems and e-Business Management, 10*(4), 549–570. doi:10.100710257-011-0181-5

Hu, L., & Bentler, P. M. (1999). Cutoff criteria for fit indexes in covariance structure analysis: Conventional criteria versus new alternatives. *Structural Equation Modeling, 6*(1), 1–55. doi:10.1080/10705519909540118

Huddleston, P., Whipple, J., Nye Mattick, R., & Jung Lee, S. (2009). Customer satisfaction in food retailing: Comparing specialty and conventional grocery stores. *International Journal of Retail & Distribution Management, 37*(1), 63–80. doi:10.1108/09590550910927162

Huijts, N. M. A., Molin, E. J. E., & Steg, L. (2012). Psychological factors influencing sustainable energy technology acceptance: A review-based comprehensive framework. *Renewable & Sustainable Energy Reviews, 16*(1), 525–531. doi:10.1016/j.rser.2011.08.018

Hung, M. C., Yang, S. T., & Hsieh, T. C. (2012). An examination of the determinants of mobile shopping continuance. *International Journal of Electronic Business Management, 10*(1), 29–37. http:// 203.72.2.146/bitstream/987654321/27104/1/An +Examination.pdf

Hussain, S., Fangwei, Z., Siddiqi, A., Ali, Z., & Shabbir, M. (2018). Structural Equation Model for Evaluating Factors Affecting Quality of Social Infrastructure Projects. *Sustainability*, *10*(5), 1415. https://www.forbes.com/sites/bernardmarr/2022/03/21/a-short-history-of-the-metaverse/?sh=503572be5968. doi:10.3390u10051415

Kang, H. (2013). The prevention and handling of the missing data. *Korean Journal of Anesthesiology*, *64*(5), 402. doi:10.4097/kjae.2013.64.5.402 PMID:23741561

Katrina, B., & Benedict, L. (2019). *Why Are Malaysians Shopping Online and What's Their Shopping Behaviour Like?* Janio. https://janio.asia/articles/why-are-malaysians-shopping-online/

Kaur, P., Stoltzfus, J., & Yellapu, V. (2018). Descriptive statistics. *International Journal of Academic Medicine*, *4*(1), 60. doi:10.4103/IJAM.IJAM_7_18

Kian, T. P., Loong, A. C. W., & Fong, S. W. L. (2018). Customer Purchase Intention on Online Grocery Shopping. *International Journal of Academic Research in Business and Social Sciences*, *8*(12). doi:10.6007/IJARBSS/v8-i12/5260

Kim, D. J., Ferrin, D. L., & Rao, H. R. (2009). Trust and Satisfaction, Two Stepping Stones for Successful E-Commerce Relationships: A Longitudinal Exploration. *Information Systems Research*, *20*(2), 237–257. doi:10.1287/isre.1080.0188

Kim, D., & Chang, H. (2007). Key functional characteristics in designing and operating health information websites for user satisfaction: An application of the extended technology acceptance model. *International Journal of Medical Informatics*, *76*(11-12), 790–800. doi:10.1016/j.ijmedinf.2006.09.001 PMID:17049917

Kim, J., Jin, B., & Swinney, J. L. (2009). The role of etail quality, e-satisfaction and e-trust in online loyalty development process. *Journal of Retailing and Consumer Services*, *16*(4), 239–247. doi:10.1016/j.jretconser.2008.11.019

Kim, S., Williams, R., & Lee, Y. (2004). Attitude Toward Online Shopping and Retail Website Quality. *Journal of International Consumer Marketing*, *16*(1), 89–111. doi:10.1300/J046v16n01_06

Lai, P. C. (2020). *Intention to use a drug reminder app: a case study of diabetics and high blood pressure patients SAGE Research Methods Cases*. doi:10.4135/9781529744767

Lai, C. P., & Zainal, A. A. (2015). Consumers' intention to use a single platform e-payment system: A study among Malaysian internet and mobile banking users. *Journal of Internet Banking and Commerce*, *20*(1), 1–13.

Lai, E., & Wang, Z. (2012). An empirical research on factors affecting customer purchasing behavior tendency during online shopping. *IEEE International Conference on Computer Science and Automation Engineering*. 10.1109/ICSESS.2012.6269534

Lai, P. C., & Lim, C. S. (2019). The Effects of Efficiency, Design and Enjoyment on Single Platform E-payment. Research in Business and Management, *6*(2), 19–34.

Lai, P. C., Toh, E. B. H. & Alkhrabsheh, A. A. (2020), Empirical Study of Single Platform E-Payment in South East Asia. *Strategies and Tools for Managing Connected Consumers*, 252-278

Lai, P. C., & Liew, E. J. (2021). Towards a Cashless Society: The Effects of Perceived Convenience and Security on Gamified Mobile Payment Platform Adoption. *AJIS. Australasian Journal of Information Systems*, *25*. doi:10.3127/ajis.v25i0.2809

Lai, P. C., & Tong, D. L. (2022). An Artificial Intelligence-Based Approach to Model User Behavior on the Adoption of E-Payment. Handbook of Research on Social Impacts of E-Payment and Blockchain. IGI Global.

Lee, M. K. O., Shi, N., Cheung, C. M. K., Lim, K. H., & Sia, C. L. (2011). Consumer's decision to shop online: The moderating role of positive informational social influence. *Information & Management*, *48*(6), 185–191. doi:10.1016/j.im.2010.08.005

Lee, T., & Jun, J. (2007). The role of contextual marketing offer in Mobile Commerce acceptance: Comparison between Mobile Commerce users and nonusers. *International Journal of Mobile Communications*, *5*(3), 339. doi:10.1504/IJMC.2007.012398

Leedy, P. D., Ormrod, J. E., & Johnson, L. R. (2010). Practical research: planning and design (9th ed.). New York Pearson Education, Inc.

Li, T. (2019). *Variance Inflation Factor Definition*. Investopedia. https://www.investopedia.com/terms/v/variance-inflation-factor.asp

Li, Y. (2016). Empirical Study of Influential Factors of Online Customers' Repurchase Intention. *IBusiness*, *08*(03), 48–60. doi:10.4236/ib.2016.83006

Lim, W. M., & Ting, D. H. (2012). E-shopping: An Analysis of the Technology Acceptance Model. *Modern Applied Science*, *6*(4). doi:10.5539/mas.v6n4p49

Malaysian Communications and Multimedia Commission. (2018). E-Commerce Consumers Survey 2018. *MCMC*. https://www.mcmc.gov.my/skmmgovmy/media/General/pdf/ECS-2018.pdf

McCloskey, D. (2003). Evaluating electronic commerce acceptance with the technology acceptance model. *Journal of Computer Information Systems*, *44*(2).

Mofokeng, T. E. (2021). The impact of online shopping attributes on customer satisfaction and loyalty: Moderating effects of e-commerce experience. *Cogent Business & Management*, *8*(1), 1968206. doi:10.1080/23311975.2021.1968206

Mohamed, N., Hussein, R., Zamzuri, N. H. A., & Haghshenas, H. (2014). Insights into individual's online shopping continuance intention. *Industrial Management & Data Systems*, *114*(9), 1453–1476. doi:10.1108/IMDS-07-2014-0201

Mumtaz, H., Aminul Islam, M., Ku Ariffin, K. H., & Karim, A. (2011). Customers Satisfaction on Online Shopping in Malaysia. *International Journal of Business and Management*, *6*(10). C:\Users\salma\Desktop\10.5539\ijbm.v6n10p162 doi:10.5539/ijbm.v6n10p162

Mustafa, I.E. (2011). Determinants of e---commerce customer satisfaction, trust, and loyalty in Saudi Arabia. *Journal of Electronic Commerce Research, 12*(1), 78---93.

Olaniyi, A. (2019). Type and Cronbach's Alpha Analysis in an Airport Perception Study. *Scholar Journal of Applied Sciences and Research, 2*(4). https://www.innovationinfo.org/articles/SJASR/SJASR-4-223.pdf

Park, Y. S., Konge, L., & Artino, A. R. Jr. (2019). The Positivism Paradigm of Research. *Academic Medicine, 95*(5), 690–694. doi:10.1097/ACM.0000000000003093 PMID:31789841

Paulins, V. A., & Geistfeld, L. V. (2003). The effect of consumer perceptions of store attributes on apparel store preference. *Journal of Fashion Marketing and Management, 7*(4), 371–385. doi:10.1108/13612020310496967

Prasad, Ch. J. S., & Aryasri, A. R. (2009). Determinants of Shopper Behaviour in E-tailing: An Empirical Analysis. *Paradigm, 13*(1), 73–83.

Kalia, P., Arora, D. R., & Kumalo, S. (2016). E-service quality, consumer satisfaction and future purchase intentions in e-retail. *e-Service Journal, 10*(1), 24. doi:10.2979/eservicej.10.1.02

Premkumar, G., & Bhattacherjee, A. (2008). Explaining information technology usage: A test of competing models. *Omega, 36*(1), 64–75. doi:10.1016/j.omega.2005.12.002

Putit, L., Muhammad, N. S., & Aziz, Z. D. A. (2017). *Retailing*. Oxford Revision Series.

Rafee, H. (2019). Cover Story: Retail market remains steady amid changes in e-commerce and demography. *The Edge Markets.* https://www.theedgemarkets.com/article/cover-story-retail-market-remains-steady-amid-changes-ecommerce-and-demography

Raman, P. (2019). Understanding female consumers' intention to shop online. *Asia Pacific Journal of Marketing and Logistics, 31*(4), 1138–1160. doi:10.1108/APJML-10-2018-0396

Revels, J., Tojib, D., & Tsarenko, Y. (2010). Understanding consumer intention to use mobile services. *Australasian Marketing Journal, 18*(2), 74–80. doi:10.1016/j.ausmj.2010.02.002

Rita, P., Oliveira, T., & Farisa, A. (2019). The impact of e-service quality and customer satisfaction on customer behavior in online shopping. *Heliyon, 5*(10), e02690. doi:10.1016/j.heliyon.2019.e02690 PMID:31720459

Rönkkö, M., & Cho, E. (2020). An Updated Guideline for Assessing Discriminant Validity. *Organizational Research Methods, 109442812096861*. doi:10.1177/1094428120968614

Sahai, P., Sharma, M., & Singh, V. K. (2020). Effect of Perceived Quality, Convenience, and Product Variety on Customer Satisfaction in Teleshopping. *Management and Economics Research Journal, 6*(3), 1–8. doi:10.18639/MERJ.2020.9900021

Sarstedt, M., & Mooi, E. (2014). Regression Analysis. *Springer Texts in Business and Economics*, 193–233.

Schaupp, L.C., and Belanger, F. (2005). A conjoint analysis of online customer satisfaction. *Journal of Electronic Commerce Research, 6*(2), 95---111

Selamat, Z., Jaffar, N., & Ong, H. B. (2009). (PDF) Technology Acceptance in Malaysian Banking Industry. *European Journal of Economics, Finance and Administrative Sciences*.

Singh, P., & Keswani, S. (2016). A Study of Adoption Behavior for Online Shopping: An Extension of Tam Model. *International Journal Advances in Social Science and Humanities, 4*(7).

Sivadas, E., & Baker-Prewitt, J. L. (2000). An examination of the relationship between service quality, customer satisfaction, and store loyalty. *International Journal of Retail & Distribution Management, 28*(2), 73–82. doi:10.1108/09590550010315223

Ringle, C. M., Wende, Sven, & Baker, J. M. (2017). SmartPLS 4. *SmartPLS*. https://www.smartpls.com/documentation/algorithms-and-techniques/bootstrapping

Sonal, K., & Thomas, S. (2019). (PDF) Online grocery retailing – exploring local grocers beliefs. *International Journal of Retail & Distribution Management, 47*(12).

Sreeram, A., Kesharwani, A., & Desai, S. (2017). Factors affecting satisfaction and loyalty in online grocery shopping: An integrated model. *Journal of Indian Business Research, 9*(2), 107–132. doi:10.1108/JIBR-01-2016-0001

Taherdoost, H. (2016). Validity and Reliability of the Research Instrument; How to Test the Validation of a Questionnaire/Survey in a Research. *SSRN Electronic Journal, 5*(3). doi:10.2139/ssrn.3205040

Tang, L.-L., & Nguyen, H. T. H. (2013). Common causes of trust, satisfaction and TAM in online shopping: An integrated model. *Journal of Quality, 20*(5), 483–501.

Teo, T. S. H. (2001). Demographic and motivation variables associated with Internet usage activities. *Internet Research, 11*(2), 125–137. doi:10.1108/10662240110695089

Lumpur, K. (2020). Malaysia's online retail sales up 28.9% in April. *The Star*. Www.thestar.com.my. https://www.thestar.com.my/business/business-news/2020/06/11/malaysia039s-online-retail-sales-up-289--in-april

Vasic, N., Kilibarda, M., & Kaurin, T. (2019). The Influence of Online Shopping Determinants on Customer Satisfaction in the Serbian Market. *Journal of Theoretical and Applied Electronic Commerce Research, 14*(2), 0. doi:10.4067/S0718-18762019000200107

Wang, Y., Seo, J. H., & Song, W.-K. (2019). Factors That Influence Consumer Satisfaction with Mobile Payment: The China Mobile Payment Market. *International Journal of Contents, 15*(4), 82–88.

Williams, C. (2007). Research Methods. *Journal of Business & Economics Research, 5*(3). doi:10.19030/jber.v5i3.2532

Wilson, N., & Christella, R. (2019). An Empirical Research of Factors Affecting Customer Satisfaction: A Case of the Indonesian E-Commerce Industry. *DeReMa (Development Research of Management). Jurnal Manajemen, 14*(1), 21. doi:10.19166/derema.v14i1.1108

Xu, J. D., Benbasat, I., & Cenfetelli, R. T. (2013). Integrating Service Quality with System And information Quality: An Empirical Test in the E-Service Context. *Management Information Systems Quarterly*, *37*(3), 777–794. doi:10.25300/MISQ/2013/37.3.05

Yi, M. Y., Jackson, J. D., Park, J. S., & Probst, J. C. (2006). Understanding information technology acceptance by individual professionals: Toward an integrative view. *Information & Management*, *43*(3), 350–363. doi:10.1016/j.im.2005.08.006

Yip, W. Y. (2019). Singapore is the world's second most digitally competitive country, after the US. *The Straits Times*. https://www.straitstimes.com/tech/singapore-is-the-worlds-second-most-digitally-competitive-country-after-the-us

Yusuf Dauda, S., & Lee, J. (2015). Technology adoption: A conjoint analysis of consumers′ preference on future online banking services. *Information Systems*, *53*, 1–15. doi:10.1016/j.is.2015.04.006

Zarmpou, T., Saprikis, V., Markos, A., & Vlachopoulou, M. (2012). Modeling users' acceptance of mobile services. *Electronic Commerce Research*, *12*(2), 225–248. doi:10.100710660-012-9092-x

Žukauskas, P., Vveinhardt, J., & Andriukaitienė, R. (2018). Research Ethics. *Management Culture and Corporate Social Responsibility*. doi:10.5772/intechopen.70629

ADDITIONAL READING

Abutabenjeh, S., & Jaradat, R. (2018). Clarification of research design, research methods, and research methodology. *Teaching Public Administration, 36*(3), 237–258. doi:10.1177/0144739418775787

Apuke, O. D. (2017). Quantitative Research Methods: a Synopsis Approach. *Kuwait Chapter of Arabian Journal of Business and Management Review, 6*(11), 40–47. doi:10.12816/0040336

Aseri, A. (2021). Security Issues for Online Shoppers. *International Journal of Scientific & Technology*, *10*(3). https://www.ijstr.org/final-print/mar2021/Security-Issues-For-Online-Shoppers.pdf

Byrne, B. M. (2008). *Structural Equation Modeling with EQS: Basic Concepts, Applications, andProgramming*. Psychology Press.

Fraser, J., & Fahlman, D., Arscott, J., & Guillot, I. (2018). Pilot Testing for Feasibility in a Study of Student Retention and Attrition in Online Undergraduate Programs. *The International Review of Research in Open and Distributed Learning, 19*(1). doi:10.19173/irrodl.v19i1.3326

Frey, F. (2017). SPSS (Software). *The International Encyclopedia of Communication Research Methods*, 1–2. doi:10.1002/9781118901731.iecrm0237

Hassan, Z. A., Schattner, P., & Mazza, D. (2006). Doing A Pilot Study: Why Is It Essential? *Malaysian Family Physician : the Official Journal of the Academy of Family Physicians of Malaysia*, *1*(2-3), 70–73. PMID:27570591

Hayes, A. (2021). Descriptive Statistics. *Investopedia*. https://www.investopedia.com/terms/d/descriptive_statistics.asp#:~:text=Descriptive%20statistics%20are%20brief%20descrip
tive

Huck, S. W. (2007). *Reading statistics and research.* Pearson Education Limited.

Jang-Jaccard, J., & Nepal, S. (2014). A survey of emerging threats in cybersecurity. *Journal of Computer and System Sciences, 80*(5), 973–993. doi:10.1016/j.jcss.2014.02.005

McLeod, S. (2019). P-Values and Statistical Significance. *Simply Psychology*. Www.simplypsychology.org. https://www.simplypsychology.org/p-value.html#:~:text=The%20smaller%20the%20p%2Dvalue

Sekaran, U., & Bougie, R. (2010). Research Methods for Business a skill-building Approach, (5th ed.). Haddington: John Wiley & Sons.

Tarka, P. (2017). An overview of structural equation modeling: Its beginnings, historical development, usefulness and controversies in the social sciences. *Quality & Quantity, 52*(1), 313–354. doi:10.100711135-017-0469-8 PMID:29416184

Venkatesh, V., & Davis, F. D. (1996). A Model of the Antecedents of Perceived Ease of Use: Development and Test. *Decision Sciences, 27*(3), 451–481. doi:10.1111/j.1540-5915.1996.tb01822.x

Wong, A. (2020). *4 ways Touch 'n Go eWallet can secure its user accounts better.* SoyaCincau. https://soyacincau.com/2020/03/10/touch-n-go-ewallet-4-security-suggestions-otp/

Chapter 17
The Role of Legal Governance Framework in the Metaverse World

Chin Chin Sia
Taylor's University, Malaysia

ABSTRACT

The book chapter analyses how the current legal framework of self-governance is weak in regulating the metaverse world. The rule of law, as a discourse that emphasizes the legitimacy of governance and appropriate limits on the exercise of power, provides a useful framework as a first step to reconceptualizing and evaluating these tensions in communities at the intersection of the real and the virtual, the social and the economic, and the public and the private. Technologists, practitioners, and regulators must be open to these tensions to appropriately develop the correct legal governance through a mix of user control, industry practice, and regulatory oversight for a just and equitable metaverse world.

INTRODUCTION

One of the principles of governance law requires legal authorization (Risch, 2009) for the exercise of power. Incorporating this insight into metaverse world of self-governance implies that the contracts that underpin participation in metaverse world ought to be enforceable against the providers of those communities as well as the participants (Lastowka, 2010). Nevertheless, there are some shortcomings in the ways that these contracts are drafted. Namely, they are drafted considerably in favor of the providers (Fairfield, 2010), grant wide discretionary powers, and significantly limit any potential liability to the providers. It is argued that while autonomy of contract is critical in metaverse world, there is a severe tendency to reduce state intervention in private governance through self-governance by the major service providers.

Regulation comes in several different forms, each of which affects participants. Regulation also comes from a number of different sources: the moral force of the community, the imposed rule of the provider, and the laws of territorial states. Some conflicts are best illustrated from a position internal to the rules and norms of the metaverse world, while others more visible from an external position. There

DOI: 10.4018/978-1-6684-5732-0.ch017

are overlapping constraints from multiple sources, but what matters is the cumulative effect of the law on its subjects. The interaction between internal and external perspectives and sources of regulation constructs the experience of participating users (Fernandez & Hui, 2022), who are subject to all these forces simultaneously. This book chapter will proceed on the basis that legal framework highlight tensions that can be in different sources of regulation in metaverse world and provide insights that may be relevant to a number of different forms of governance.

The most immediate legal limits on a provider's discretion usually lie in the contractual terms of service that purport to govern most communities within the metaverse world. First, providers are expected to act in accordance with the terms of service since these contractual documents ought to be enforceable against providers and not merely for the benefit of providers. This leads to some serious problems, particularly as most terms of service are drafted in a manner that greatly favors the interests of the provider. Most importantly, terms of service generally include clauses that reserve a wide discretion to the provider. In metaverse world where the value of the rule of law against arbitrary power is significant, clauses that allow absolute discretion should be regarded suspiciously (Suzor, 2010).

The analysis in this chapter also aimed to show that privacy is inadequately handled in the current governance model. It will be interesting to see local governance develop. Different governance models are likely to develop within the various communities in the metaverse world and some will certainly address privacy needs of the individual participants or communities (Leenes, 2007). Whether privacy as a social value will be acknowledged and handled within these diverse forms of governance is a different question.

BACKGROUND

Metaverses are immersive three-dimensional virtual worlds in which people interact as avatars with each other and with software agents, using the metaphor of the real world but without its physical limitations (Park & Kim, 2022). Metaverse is an interconnected web of social, networked immersive environments in persistent multiuser platforms. It enables seamless embodied user communication in real-time and dynamic interactions with digital artifacts (Mystakidis, 2022). Its first iteration was a web of virtual worlds where avatars were able to teleport among them. The contemporary iteration of the Metaverse features social, immersive VR platforms compatible with massive multiplayer online video games, open game worlds and AR collaborative spaces.

The term Metaverse was invented and first appeared in Neal Stevenson's science fiction novel Snow Crash published in 1992. It represented a parallel virtual reality universe created from computer graphics, which users from around the world can access and connect through goggles and earphones (Mayer-Schonberger & Crowley 2006). The backbone of the Metaverse is a protocol called the Street, which links different virtual neighborhoods and locations an analog concept to the information superhighway. Users materialize in the Metaverse in configurable digital bodies called avatars. Although Stevenson's Metaverse is digital and synthetic, experiences in it can have a real impact on the physical self. A literary precursor to the Metaverse is William Gibson's VR cyberspace called Matrix in the 1984 science fiction novel Neuromancer (Fernandez. & Hui 2022).

The novel Metaverse differs from the earlier Metaverse in three ways (Moy & Gadgil 2022). First, the rapid development of deep learning dramatically improves the accuracy of vision and language recognition, and the development of generative models enables a more immersive environment and natural movement. The processing time and complexity were reduced using multimodal models as end-to-end solutions with a multimodal pre-trained model. Second, Metaverse previously served based on PC access and had low consistency due to time and space constraints, but now it is possible to easily access the Metaverse anytime, anywhere due to the mobile devices that could always connect to the Internet. Lastly, the current Metaverse differs from the previous one because the program coding can be done in the Metaverse world, and it is more bonded to real life with virtual currency. Metaverse expands with various social meanings (e.g. conference, fashion, event, game, education, and office) based on immersive interaction (Chin, 2006). Cryptocurrencies serve as an economic bridge between the Metaverse and the real world, giving people deeper social meaning.

In a wide variety of online spaces, there are a wide arrays of potential behaviours and activities which users can engage in (Barker, 2016). This is, in part, dependent upon the particular space – for example, in certain games, it is acceptable to seek to rape and pillage in Sociolotron' as it is an adult sex game (Whitty and Young, 2012).

The modern form of metaverse is far more costly to provide than previous systems. The amount of computer server power, network bandwidth, and programming time required is much greater for graphical interaction as compared to textual worlds. Additionally, there are many more users due to growth in popularity and decreases in the cost of network access. Thus, today's metaverse worlds are run almost exclusively by companies with a profit motive (Risch, 2009).

In order to draw paying users, worlds provide different experiences; some worlds are role-playing games, some are devoted to fantasy combat, some attempt to simulate the real world, and some appeal to children (Cheong, 2022). These worlds have one common denominator: most have some form of virtual money or other trading currency used to acquire virtual "stuff' such as services or information. Many worlds also allow avatars to obtain virtual property and to transfer that property to others. Several also incorporate varying levels of social status, especially worlds that involve combat with computer-controlled enemies; the more enemies killed, the higher the "level" an avatar might achieve (Strikwerda, 2014). Rarer property and higher levels are valuable to those who do not want to spend their own time appropriating them. Those who do not wish to spend the time instead pay to obtain such property, levels, or any other virtual asset they do not have.

WHY METAVERSE REQUIRES THE RULE OF LAW?

Sources of Regulations in the Metaverse

Regulations in the Metaverse are basically formed by a number of different sources: the moral force of the community, the imposed rule of the provider, and the laws of territorial states.

Regulations Through Moral Force of Metaverse Community in Conflict Resolution

Conflicts in a Metaverse world may arise when two participants have agreed on a contract, and one of them fails to perform. For example, in Animal Crossing, participants may buy objects from each other, like the virtual clothes with which they dress their avatars. These merchants are designers of an informational good—a stream of bits that, when rendered through a graphics engine, instantiates a beautiful dress in the virtual world. Buyers contract with these merchants and pay them the virtual currency. If a seller agrees to sell a virtual dress and collects the money, but then fails to deliver the informational good—that is, fails to transfer control over the virtual dress to the buyer—the seller may be in breach of contract and soon embroiled in a conflict with the transaction partner. In virtual worlds that permit their participants to freely trade in virtual objects, commercial transactions between participants are commonplace and, unfortunately, so are conflicts arising from contractual obligations (Garon, 2022).

In addition, conflicts may arise between individuals over aspects of communication, because, as Jack Balkin has explained, most if not all activity in Metaverse worlds begins "as a form of speech." Through their online actions and speech, participants may commit "some form of communications tort," like defamation, fraud or misrepresentation, each of which can lead to serious conflicts with other participants in virtual worlds (Leenes, 2007). Finally, in addition to conflicts between participants—ex contractus or ex delicto—conflicts can arise between the society itself and an individual, when participants in Metaverse worlds violate what we call criminal statutes in the real world—i.e., the societal rules of accepted and prohibited behavior. For example, a participant, by manipulating the code, could hijack another participant's avatar and keep the avatar locked up for a ransom.

Dealing with each of these types of conflict requires a functioning virtual world governance system that addresses these conflicts in two ways.

First, ex post, societal and moral conflict resolution structures, institutions, and processes are necessary to settle conflicts authoritatively. Second, ex ante, by setting general rules, such societal structures, institutions and processes provide guidelines for behavior and create predictable outcomes to conflicts, such that if one adheres to these rules, one will likely have any conflict settled in one's favor. Each conflict resolved as expected further strengthens this predictability

Regulations by the Imposed Rule of the Service Provider

All of these issues arising in online platforms in the Metaverse– are supposedly regulated and dealt with at online platform level under the End User License Agreement (EULA). Such agreements require the consent of user of each online platform – irrespective of gaming or otherwise – before full access to the platform is granted. EULAs are imposed without negotiation or flexibility, and are widely recognised as being contracts of adhesion, compounding the situation in respect of gaps in the regulatory paradigm.

At the level of implementation, service providers possess both building blocks of a functioning governance system—rule setting and rule enforcement. By changing the software code, providers can set rules and thereby constrain behavior in their worlds (Barker, 2016). For example, to keep participants in a Metaverse world from moving instantaneously from one part of the world to another (a feature often called "teleporting"), providers only have to modify the software that implements teleporting. Unlike the real world, Metaverse worlds are creations of the mind that are modeled in software and, as such, are completely changeable. In this respect, code is law in the Metaverse worlds, arguably making them the most "Lessigian" of all spaces of online interaction.

As software code can only limit and not eliminate conflict, service providers must apply more old-fashioned governance mechanisms to deal with conflicts that cannot be constrained through modifications to the underlying code: codifying social norms into written rules and providing effective enforcement of these rules. Enactment of such a means of governance is relatively straightforward. When participants in Metaverse enter into a contract with providers to register for the service—usually through an end user license agreement ("EULA") or the terms of service— they agree to be bound by the rules and regulations in that contract (Suzor, 2010). Providers thus have a mechanism to set the rules they deem necessary

Regulations by Laws of the Territorial States

Territorial states of the real-world regulators may attempt to limit these regulatory dynamics through real-world inter-jurisdictional coordination. Three distinct regulatory options are available to them: generally harmonizing laws regulating Metaverse world across jurisdictions, constraining users' ability to switch providers, or prohibiting users from switching to providers domiciled outside their real-world jurisdiction. By finding common regulatory ground, real-world territorial lawmakers could limit the ability of Metaverse world users to switch to virtual world providers who operate outside the reach of real-world territorial state's regulators. Such coordination does not need to be comprehensive and coverage does not have to be complete in order to reduce competitive dynamics.

Service providers who are already required to follow their jurisdiction's real-world territorial rules, especially those from restrictive real-world territorial jurisdictions, would welcome real-world regulatory coordination, in as much as this coordination is leveling the playing field. However, given wide variances in societal values and differing ideologies, achieving effective coordination may prove difficult.

Real-world territorial regulators may choose a more palatable strategy to constrain regulatory arbitrage: they may restrict users from choosing service providers who operate outside the real-world jurisdictions in which those users live. Users could still choose among providers within their jurisdiction. This tactic could successfully limit regulatory competition around attracting new users and level the playing field for service providers by interdicting users from joining Metaverse worlds outside of their real-world territorial jurisdictions (Gibbons, 2008). Although competition among service providers would continue, direct competitors would all reside in the same jurisdiction and be bound to enforce the same real-world rules in each of their virtual worlds. In this way, the global market of virtual worlds would be broken into national markets along jurisdictional (and thus real-world regulatory) borders.

Dutch case law provides a real case example of a human act made possible by computer simulation that has been brought under the scope of criminal law (Strikwerda 2014). In 2009, Dutch judges have convicted several minors of theft, because they had stolen virtual property in the virtual worlds of online multiplayer computer games. Three minors were convicted of theft for the stealing of virtual furniture in the virtual world of the online multiplayer computer game Habbo. Habbo is a metaverse and consists of a virtual hotel where players have their own room. They can buy virtual furniture for their rooms with 'credits', which have to be obtained with real, non-virtual money. By means of deceit (phishing) the perpetrators obtained the usernames and passwords of other Habbo players, so that they could access the other players' accounts and transfer their virtual furniture to their own Habbo accounts.

Another example of a putative virtual cybercrime is the murder on an avatar. Several media have reported that a 43-yearold Japanese woman hacked into the account of the person behind the avatar her own avatar was married to within the virtual world of the online multiplayer computer game MapleStory and deleted her virtual husband, because it had suddenly divorced her avatar. When the person found out, he called the police. The police investigated the case and even arrested the woman at her home, but she was never formally charged (Gibbons, 2008)

Rationale of the Rule of Law in the Metaverse

The conception of the rule of law requires "that laws be declared publicly in clear terms in advance, be applied equally, and be interpreted and applied with certainty and reliability" in order that the law "be capable of guiding the behavior of its subjects. "It follows that "all laws should be prospective, open, and clear" and that "laws should be relatively stable." (Egliston, & Carter 2021).

These principles, stated in a number of different ways, form the rationale and standard liberal understanding of the rule of law. The emphasis on the law's ability to guide the behavior of its subjects leads to two somewhat separable themes in this conception of the rule of law: an aspiration towards clarity and predictability in legal rules and, to a lesser extent, a set of due process requirements in the application of those rules.

For example, a small, tight-knit community with shared understandings of appropriate behaviour may not need formally articulated rules or restraints on the power of the administrator to eject members deemed to be disruptive or unwanted. In Metaverse communities, however, particularly those that foster a more diverse population and are relatively open-ended, a perceived lack of predictability may be harmful to the interests of participants and imposing limits on private governance may be justified.

Clear Rules

The requirement that rules be clearly expressed and promulgated is familiar in the liberal rule of law discourse, where the emphasis is on the ability of law to guide behaviour and the ability of citizens of Metaverse to plan their lives (Fairfield, 2010). This discourse immediately highlights that the rules in Metaverse communities are often unclear, obscure, and difficult to understand. The contractual terms of service and end user license agreement **("EULA")** documents are usually written in dense legalese and are usually presented in a form that discourages reading.

Changing Rules

Another problematic component of Metaverse community governance is the rate at which legal rules can change and the lack of responsibility that providers have to compensate any participants who may be adversely affected by rule changes (Charamba, 2022). Many providers purport to have the right to modify the terms of service at any time, often without notice to the participants. Changes in these legal rules are rarely highlighted to the participant, who may have substantial difficulty in identifying the changes and their legal effect.

This suggests that the mechanism of changing rules should be investigated, requiring, for example, that providers make clear statements about the effects of any changes and highlight modified sections in the dense legal agreements in order to enable participants to identify and understand impact of rule changes.

Inconsistent Application and Discretionary Enforcement

Perhaps the most troubling aspect of virtual community governance is that the rules on the books, the EULAs and terms of service, sometimes bear almost no resemblance to the rules in force in the community (Barker, 2016). Virtual community contracts are typically drafted in a very risk-averse manner, reserving for the provider almost total power to deal with members of the community. This often includes broad prohibitions on behaviour that is commonplace within the community. In many cases, the provider is not interested in enforcing these contracts as written but will use them as a tool against particular participants as it sees fit, even to the extent of infringement of privacy of users. Essentially, these contracts are designed to reserve a wide range of discretionary powers for the provider, which is a concept that directly contradicts the values of formal legality in therule of law that are generally understood to require that "similar cases be treated similarly."

Resolving the tension between the need for flexibility and the need to avoid the worst effects of inconsistent application of discretionary rules is a difficult task that speaks to the core of the tension between formal and substantive conceptions of justice. In moving away from purely positive accounts of law and responding to the need to allow, but simultaneously constrain the discretionary exercise of governance powers, the next set of values of the rule of law embrace requirements of fairness, equality, and transparency as measures of legitimacy in decision making.

Right to Privacy

The potential threats that the use of networked technologies pose to the privacy of participants has been the subject of much discussion in recent decades (Fernandez & Hui 2022). The growing importance of participation and the increasing computational power and storage capacity of computer networks highlights immediate concerns about the collection, use, and distribution of personal information. Because all actions that occur "within" a Metaverse community are essentially reduced to information flows, they are all easily recorded and stored. Actions that are ephemeral in the corporeal world perversely take on a more tangible form when mediated through virtual information networks.

Information that is not displayed or carried out synchronously must necessarily be processed and stored for later use; personal messages left on bulletin boards and profile pages are kept indefinitely on the provider's server, for example. Even information that is used synchronously, however, is vulnerable to capture, including all actions, searches, information and products browsed, real-time chats, and exchanges between participants are potentially logged and stored (Leenes, 2007).

UN Guiding Principles for Business and Human Rights

UN Guiding Principles for Business and Human Rights contain three chapters, or pillars: protect, respect and remedy. Each defines concrete, actionable steps for governments and companies to meet their respective duties and responsibilities to prevent human rights abuses in company operations and provide remedies if such abuses take place. This denotes that real-world regulators and service providers of the Metaverse world are to meet their duties to protect human rights in the context of business operations in the Metaverse world (Mares, 2011). This includes enacting and enforcing laws that require service providers to respect human rights; creating a regulatory environment that facilitates business respect for human rights in the Metaverse world; and providing guidance to service providers on their responsibilities.

The Guiding Principles also affirms that business enterprises, including service providers, must prevent, mitigate and, where appropriate, remedy human rights abuses that they cause or contribute to.

Service providers must seek to prevent or mitigate any adverse impacts related to their operations, products or services, even if these impacts have been carried out by suppliers or business partners.

RECOMMENDATIONS

Consent may be the single most important aspect of legitimacy in the governance of Metaverse communities. Cyberlaw theory suggests that the main benefit of the autonomy of Metaverse communities is the ability of participants to come together in spaces whose norms differ from those of other communities. At its libertarian extreme, this ideal holds that through consensual participation in a boundless array of potential communities, each community's rules will more closely match the preferences of its participants than any default set of rules could. There must be room for participants who consensually choose to participate in Metaverse communities whose rules may seem strange or arbitrary. Where consensual rules conflict with external values, territorial states continue to have an interest in limiting autonomy. Territorial states will often limit the internal norms that are socially repugnant or that have deleterious effects on people outside of the Metaverse community. Territorial states routinely limit the scope of consent in issues of discrimination, for example, or in content matters such as the sexualized depiction of underage persons. Concerns about sexual play and the exposure and exploitation of children in Metaverse worlds are also increasingly prominent as territorial states begin to consider what type of behaviour is permissible and when regulation is necessary.

Whilst clearly not a perfect model of consensual governance, this approximation at least provides an avenue for territorial courts to examine the internal social norms of the Metaverse community in relation to both external values and contractual terms. Consent provides a useful indication of the internal legitimacy of Metaverse community rules that can then be used as a normative guide as to whether the territorial state ought to support a particular contractual interpretation or not.

FUTURE RESEARCH DIRECTIONS

A future direction of legal governance framework of the Metaverse may be the rather pragmatic realization on the part of real-world regulators that robust governance must derive from those that are governed, not from an outside (real-world) regulatory body.

According to this direction, the best that one could wish for virtual worlds is that they are able to bring about their own governance structures and encourage the development of systems that are participatory and fair. Should real-world regulators follow this path, they may find themselves in the uneasy but promising role of midwives for the birth of self-governance in Metaverse worlds, inculcating the values that they hold dear from real-world governance systems into these nascent attempts at self-governance. This may ensure that democracy's enduring values are encoded in the DNA of each virtual world's governance, thereby facilitating a pragmatic compromise of policy and regulatory challenges faced by Metaverse worlds, including those caused by permeability with the real world.

CONCLUSION

In conclusion, it is important to revisit the adequacy of legal governance frameworks for the protection of human rights in the metaverse world. In so doing, the book chapter questions how we can understand self-governance and corporate responsibility in relation to human rights in metaverse world, it discusses the adequacy of the UN Guiding Principles for Business and Human Rights, and it posits that we should consider a more comprehensive legal framework through corporate responsibility to respect and protect rights of participants in the metaverse world. The scope of the metaverse and its potential social importance will reshape these principles in sometimes unpredictable ways. Technologists, practitioners, and regulators must be open to these shifts to appropriately develop the correct legal governance through a mix of user control, industry practice, and regulatory oversight for a just and equitable metaverse world.

REFERENCES

Barker, K. (2016). Virtual spaces and virtual layers–governing the ungovernable? *Information & Communications Technology Law*, *25*(1), 62–70. doi:10.1080/13600834.2015.1134146

Charamba, K. (2022). *Beyond the Corporate Responsibility to Respect in the Dawn of a Metaverse*. doi:10.2139srn.4043254

Cheong, B. C. (2022). Avatars in the metaverse: potential legal issues and remedies. *International Cybersecurity Law Review*, 1-28.

Chin, B. M. (2006). Regulating Your Second Life-Defamation in Virtual Worlds. *Brook. L. Rev.*, *72*, 1303.

Egliston, B., & Carter, M. (2021). Critical questions for Facebook's virtual reality: data, power and the metaverse. *Internet Policy Review, 10*(4).

Fairfield, J. A. (2010). Castles in the Air: Greg Lastowka's Virtual Justice.

Fernandez, C. B., & Hui, P. (2022). Life, the Metaverse and Everything: An Overview of Privacy, Ethics, and Governance in Metaverse.

Garon, J. (2022). Legal implications of a ubiquitous metaverse and a Web3 future.

Gibbons, L. J. (2008). Law and the emotive avatar. *Vand. J. Ent. & Tech. L.*, *11*, 899.

Lastowka, G. (2010). *Virtual justice*. Yale University Press.

Leenes, R. (2007). Privacy in the Metaverse. In *IFIP International Summer School on the Future of Identity in the Information Society,* (pp. 95–112). Springer.

Mares, R. (Ed.). (2011). *The UN guiding principles on business and human rights: foundations and implementation*. Martinus Nijhoff Publishers.

Mayer-Schonberger, V., & Crowley, J. (2006). Napster's Second Life: The Regulatory Challenges of Virtual Worlds. *Nw. UL Rev.*, *100*, 1775.

Moy, C., & Gadgil, A. (2022). Opportunities in the Metaverse: How Businesses Can Explore the Metaverse and Navigate the Hype vs. Reality. *Erişim Tarihi, 23*(02). https://www. jpmorgan. com/content/dam/jpm/treasury-services /documents/opportunities-in-the-metaverse. pdf .

Mystakidis, S. (2022). Metaverse. *Metaverse. Encyclopedia*, *2*(1), 486–497. doi:10.3390/encyclopedia2010031

Park, S. M., & Kim, Y. G. (2022). A Metaverse: Taxonomy, components, applications, and open challenges. *IEEE Access : Practical Innovations, Open Solutions*, *10*, 4209–4251. doi:10.1109/ACCESS.2021.3140175

Risch, M. (2009). Virtual rule of law. *West Virginia Law Review*, *112*(1), 1–52.

Strikwerda, L. (2014). Should Virtual Cybercrime by Regulated by Means of Criminal Law: Philosophical, Legal-Economic, Pragmatic and Constitutional Dimension. *Information & Communications Technology Law*, *23*(1), 31–60. doi:10.1080/13600834.2014.891870

Suzor, N. (2010). The role of the rule of law in virtual communities. *Berkeley Technology Law Journal*, *25*, 1817.

Whitty & Young. (2012). Rape, pillage, murder, and all manner of ills: Are there some possibilities for action that should not be permissible even within gamespace? In K. Poels & S. Malliet (Eds.), *Moral Issues in Digital Game Play,* (p. 2). ACCO Academic.

Compilation of References

Ab Hamid, M. R., Sami, W., & Mohmad Sidek, M. H. (2017). Discriminant Validity Assessment: Use of Fornell & Larcker criterion versus HTMT Criterion. *Journal of Physics: Conference Series, 890*, 012163. doi:10.1088/1742-6596/890/1/012163

Abraham, J., Higdon, D., & Nelson, J. (2018). Cryptocurrency price prediction using tweet volumes and sentiment analysis. *SMU Data Science Review, 1*(3), 22.

Aburbeian, A. M., Owda, A. Y., & Owda, M. (2022). A Technology Acceptance Model Survey of the Metaverse Prospects. *AI, 3*(2), 285-302.

Achariam, T. (2021). Business outlets to operate from 8am to 8pm under new measures to curb Covid-19. *The Edge Markets*. https://www.theedgemarkets.com/article/shops-only-allowed-operate-8am-8pm-under-stricter-mco-measures-announced-today

Achchuthan, S., & Velnampy, T. (2016). A quest for green consumerism in Sri Lankan context: an application of comprehensive model. *Asian economic and social society, 6*(3), 59-76.

Adeniyi, M. A., & Kolawole, V. A. (2015). The influence of peer pressure adolescents'social behaviour. *University of Mauritus Research Journal, 21*. https://www.ajol.info/index.php/umrj/article/view/1220

Adeogen, M. (2005). The digital Divide and University Education Systems in Sub-Saharan *African. Afr. J. Lib. Arch & Information Science, 13*(1), 11-20.

Adomi, T. T., Okiy, R. B., & Roteyan, J. O. (2004). A Survey of Cyber Cafes Delta State Nigeria. Electronic Library 21(5), 487-495.

Afandi, H. (2020). #TECH: Malaysia has most digital consumers in the region. *NST Online*. https://www.nst.com.my/lifestyle/bots/2020/09/627994/tech-malaysia-has-most-digital-consumers-region

Afzal, S., Arshad, M., Saleem, S., & Farooq, O. (2019). The impact of perceived supervisor support on employees' turnover intention and task performance: Mediation of self-efficacy. *Journal of Management Development, 38*(5), 369–382. doi:10.1108/JMD-03-2019-0076

Agag, G., & El-Masry, A. A. (2016). Understanding consumer intention to participate in online travel community and effects on consumer intention to purchase travel online and WOM: An integration of innovation diffusion theory and TAM with trust. *Computers in Human Behavior, 60*, 97–111. doi:10.1016/j.chb.2016.02.038

Agudo, J. E. & Rico, M. (2011). Language learning resources and developments in the Second Life metaverse. *International Journal of Technology Enhanced Learning 3*(5):496-509

Agur, I. Peria, S. M., & Rochon, C. (2020). Digital Financial Services and the Pandemic: Opportunities and Risks for Emerging and Developing Economies. *International Monetary Fund (MF).*

Ahmad, M. (2018). Review of the technology acceptance model (TAM) in internet banking and mobile banking. *International Journal of Information Communication Technology and Digital Convergence*, *3*(1), 23–41.

Ajzen, I. (1991). The Theory of Planned Behavior. *Organizational Behavior and Human Decision Processes*, *50*(2), 179–211. doi:10.1016/0749-5978(91)90020-T

Akhter, S. H. (2014). Privacy concern and online transactions: The impact of internet self-efficacy and internet involvement. *Journal of Consumer Marketing*, *31*(2), 118–125. doi:10.1108/JCM-06-2013-0606

Akinwande, M. O., Dikko, H. G., & Samson, A. (2015). Variance Inflation Factor: As a Condition for the Inclusion of Suppressor Variable(s) in Regression Analysis. *Open Journal of Statistics*, *05*(07), 754–767. doi:10.4236/ojs.2015.57075

Akour, I. A., Al-Maroof, R. S., Alfaisal, R., & Salloum, S. A. (2022). A conceptual framework for determining metaverse adoption in higher institutions of gulf area: An empirical study using hybrid SEM-ANN approach. *Computers and Education: Artificial Intelligence*, *3*, 100052. doi:10.1016/j.caeai.2022.100052

Alaloul, W. S., Liew, M. S., Zawawi, N. A. W. A., & Kennedy, I. B. (2020). Industrial Revolution 4.0 in the construction industry: Challenges and opportunities for stakeholders. *Ain Shams Engineering Journal*, *11*(1), 225–230. doi:10.1016/j.asej.2019.08.010

Alam, S. S., & Yasin, N. M. (2010). An Investigation into the Antecedents of Customer Satisfaction of Online Shopping. *ResearchGate*. https://www.researchgate.net/publication/266495477_An_Investigation_into_the_Antecedents_of_Customer_Satisfaction_of_Online_Shopping

Alasuutari, P. (2010). The rise and relevance of qualitative research. *International Journal of Social Research Methodology*, *13*(2), 139–155. doi:10.1080/13645570902966056

Albantani, A. M., & Madkur, A. (2019). Teaching Arabic in the era of Industrial Revolution 4.0 in Indonesia: Challenges and opportunities. *ASEAN Journal of Community Engagement*, *3*(2), 12–31. doi:10.7454/ajce.v3i2.1063

Albawaba. (2021). Seoul Becomes First City to Join the Metaverse. *Albawaba*. https://www.albawaba.com/business/seoul-becomes-first-city- join-metaverse-1454641

Alcácer, V., & Cruz-Machado, V. (2019). Scanning the industry 4.0: A literature review on technologies for manufacturing systems. *Engineering science and technology, an international journal, 22*(3), 899-919.

Al-Emran, M., Mezhuyev, V., & Kamaludin, A. (2018). Technology Acceptance Model in M-learning context: A systematic review. *Computers & Education*, *125*, 389–412. doi:10.1016/j.compedu.2018.06.008

Alexander, B. (2020). *Academia next: The futures of higher education.* Johns Hopkins University Press.

Alhumaid, K., Habes, M., & Salloum, S. A. (2021). Examining the factors influencing the mobile learning usage during COVID-19 Pandemic: An Integrated SEM-ANN Method. *IEEE Access : Practical Innovations, Open Solutions*, *9*, 102567–102578. doi:10.1109/ACCESS.2021.3097753

Ali, F. (2016). Hotel website quality, perceived flow, customer satisfaction and purchase intention. *Journal of Hospitality and Tourism Technology*, *7*(2), 213–228. doi:10.1108/JHTT-02-2016-0010

Allam, Z., Sharifi, A., Bibri, S. E., Jones, D. S., & Krogstie, J. (2022). The Metaverse as a Virtual Form of Smart Cities: Opportunities and Challenges for Environmental, Economic, and Social Sustainability in Urban Futures. *Smart Cities*, *5*(3), 771–801. doi:10.3390martcities5030040

Allwright, R. (1984). The importance of interaction in classroom language learning. *Applied Linguistics*, *5*(2), 156–171. doi:10.1093/applin/5.2.156

Al-Mamary, Y. H., Al-nashmi, M., Hassan, Y. A. G., & Shamsuddin, A. (2016). A critical review of models and theories in field of individual acceptance of technology. *International Journal of Hybrid Information Technology*, *9*(6), 143–158. doi:10.14257/ijhit.2016.9.6.13

Almarzouqi, A., Aburayya, A., & Salloum, S. A. (2022). Prediction of User's Intention to Use Metaverse System in Medical Education: A Hybrid SEM-ML Learning Approach. *IEEE Access : Practical Innovations, Open Solutions*, *10*, 43421–43434. doi:10.1109/ACCESS.2022.3169285

Alpha Inc. (2022). Engage. *IPS News.* http://ipsnews.net/business/2021/10/04/in-game-advertising-market-to-reach-usd-18-41-billion-by-2027-is-going-to-boom-wi thrapidfire-inc-playwire-media-llc-atlas-alpha-inc-engage/.

Alreck, P. L., & Dibartolo, G., Diriker, M., & Settle, R. B. (2009). Time Pressure, Time Saving And Online Shopping: Exploring A Contradiction. *ResearchGate.* https://www.researchgate.net/publication/242089354_Time_Pressure_Time_Saving_And_Online_Shopping_Exploring_A_Contradiction

Al-Saedi, K., Al-Emran, M., Ramayah, T., & Abusham, E. (2020). Developing a general extended UTAUT model for M-payment adoption. *Technology in Society*, *62*, 101293. doi:10.1016/j.techsoc.2020.101293

Al-Smadi, A. M., Alsmadi, M. K., Baareh, A., Almarashdeh, I., Abouelmagd, H., & Ahmed, O. S. S. (2019). Emergent situations for smart cities: a survey. *International Journal of Electrical & Computer Engineering (2088-8708), 9*(6).

Alsop, T. (2021). Metaverse Potential Market Opportunity Worldwide 2021, by scenario. (https://www.statista.com/statistics/1286718/metaverse-market-opportunity-by-scenario/)

Alsop, T. (2022). Investment in AR/VR Technology Worldwide in 2024, by use case. *Statista.* (https://www.statista.com/statistics/1098345/worldwide-ar-vr-investment-use-case/)

Alsop, T. (2022). U.S. Metaverse Potential Consumer Expenditure TAM 2022, by segment. *Statista.* (https://www.statista.com/statistics/1288655/metaverse-consumer-expenditure-tam-united-states/)

Alsop, T. (2022). Virtual Reality and Augment Reality Users U.S. 2017-2023. *Statista.* (https://www.statista.com/statistics/1017008/united-states-vr-ar-users/)

Alsop, T. (2022). XR Headset Shipments 2016-2025. *Statista.* (https://www.statista.com/statistics/1290160/projected-metaverse-use-reach-global-consumers-businesses/)

Alsop, T. (2022). XR/AR/VR/MR Market Size 2021-2028. *Statista.* (https://www.statista.com/statistics/591181/global-augmented-virtual-reality-market-size/)

Altan, M. Z., & Özmusul, M. (2022). Geleceğin Türkiye'sinde Öğretmen Refahı: Öğretmenlik Meslek Kanununun Kayıp Parçası [Teacher Welfare in Future Turkey: The Missing Part of the Teaching Profession Law]. *Ahmet Keleşoğlu Eğitim Fakültesi Dergisi*, *4*(1), 24–42.

Altun, D. (2021). *Sanal Ve Artirilmiş Gerçeklikle Dönüşen Yeni Nesil Sosyal Medya Mecrasi: Metaverse* [The Next Generation Social Media Channel Transforming with Virtual and Augmented Reality: Metaverse.]. Uluslararası İşletme ve Pazarlama Kongresi.

Alvin, K. (2021). *Sustainable GameFi: To Play or To Earn?* https://www.binance.org/en/blog/sustainable-gamefi-to-play-or-to-earn/

Alvin, K. (2021). Sustainable GameFi: To Play or To Earn? *Statista*. (https://www.binance.org/en/blog/sustainable-gamefi-to-play-or-to-earn/)

Aly, M., Khomh, F., & Yacout, S. (2021). What do practitioners discuss about iot and industry 4.0 related technologies? characterization and identification of iot and industry 4.0 categories in stack overflow discussions. *Internet of Things*, *14*, 100364. doi:10.1016/j.iot.2021.100364

Amaro, S., & Duarte, P. (2015). An integrative model of consumers' intentions to purchase travel online. *Tourism Management*, *46*, 64–79. doi:10.1016/j.tourman.2014.06.006

American Psychological Association. (2020). Developing adolescents: a reference for professionals. First Street publishers.

Amirullah. (2005). *Pengantar Bisnis.* Graha Ilmu.

Ammenwerth, E. (2019). Technology acceptance models in health informatics: TAM and UTAUT. *Studies in Health Technology and Informatics*, *263*, 64–71. PMID:31411153

Ante, L. (2021). Non-fungible token (NFT) markets on the Ethereum blockchain: Temporal development, cointegration and interrelations.

Anthony, N., Rosliza, A. M., & Lai, P. C. (2019). The Literature Review of the Governance Frameworks in Health System. *Journal of Public Administration and Governance*, *3*(9), 252–260.

Aprameya, A. (2020). Ecommerce in Malaysia: Growth, Trends & Opportunities. *Capillary Blog*. https://www.capillarytech.com/blog/capillary/ecommerce/ecommerce-in-malaysia-growth/

Arcenegui Almenara, J., Arjona, R., Román Hajderek, R., & Baturone Castillo, M. I. (2021). Secure combination of iot and blockchain by physically binding iot devices to smart non-fungible tokens using pufs. *Sensors, 21* (9), 3119. doi:10.3390/s21093119

Archer, K. & Savage, R. (2014). Examining the effectiveness of technology use in classrooms: A tertiary meta-analysis. *Computers & Education: an international journal. 7*(8), 140-149

Ariffin, A. (2021). Top 5 most populous states in Malaysia 2021. *Iproperty.com.my*. https://www.iproperty.com.my/guides/most-populous-states-in-malaysia/

Ariffin, K. A. Z., & Ahmad, F. H. (2021). Indicators for maturity and readiness for digital forensic investigation in era of industrial revolution 4.0. *Computers & Security*, *105*, 102237. doi:10.1016/j.cose.2021.102237

Atwood-Blaine, D., & Huffman, D. (2017). *Mobile Gaming and Student Interactions in a Science Center: The Future of Gaming in Science Education,* (Vol. 15). Science and Mathematics Education.

Austria, D., De Jesus, S. B., Marcelo, D. R., Ocampo, C., & Tibudan, A. J. (2022). Play-to-Earn: A Qualitative Analysis of the Experiences and Challenges Faced by Axie Infinity Online Gamers Amidst the COVID-19 Pandemic. *International Journal of Psychology and Counselling*, 391-424.

Averbek, G. S., & Türkyilmaz, C. A. (2022). Sanal Evrende Markalarin Geleceği: Yeni İnternet Dünyasi Metaverse Ve Marka Uygulamalari. [The Future of Brands in the Virtual Universe: The New Internet World Metaverse and Brand Applications.] Sosyal Bilimlerde Multidisipliner Çalışmalar Teori, Uygulama Ve Analizler, 99.

Awa, H. O., Ojiabo, O. U., & Emecheta, B. C. (2015). Integrating TAM, TPB and TOE frameworks and expanding their characteristic constructs for e-commerce adoption by SMEs. *Journal of Science and Technology Policy Management*, *6*(1), 76–94. doi:10.1108/JSTPM-04-2014-0012

Aziz, M. Y. (2014). Business intelligence trends and challenges. In *The Fourth International Conference on Business Intelligence and Technology (BUSTECH) 2014 Proceedings,* (pp. 1-7).

Baby, A., & Kannammal, A. (2020). Network Path Analysis for developing an enhanced TAM model: A user-centric e-learning perspective. *Computers in Human Behavior*, *107*, 106081. doi:10.1016/j.chb.2019.07.024

Bagozzi, R. P., & Yi, Y. (2012). Specifications, evaluation, and interpretation of structural equation models. *Journal of the Academy of Marketing Science*, *40*(1), 8–34. doi:10.100711747-011-0278-x

Bag, S., Gupta, S., & Kumar, S. (2021). Industry 4.0 adoption and 10R advance manufacturing capabilities for sustainable development. *International Journal of Production Economics*, *231*, 107844. doi:10.1016/j.ijpe.2020.107844

Baihui, Z. (2019). Research on interactive design of social sharing in Pan entertainment mobile live broadcast [Doctorat dissertation]. Jiangnan University.

Bailenson, J. N. (2021). Nonverbal overload: A theoretical argument for the causes of Zoom fatigue.

Bainbridge, W. S. (2007). The Scientific Research Potential of Virtual Worlds. American Association for the Advancement of Science, 471-476. doi:10.1126cience.1146930

Balica, R. Ş., Majerová, J., & Cuțitoi, A. C. (2022). Metaverse Applications, Technologies, and Infrastructure: Predictive Algorithms, Real-Time Customer Data Analytics, and Virtual Navigation Tools. *Linguistic and Philosophical Investigations*, 21.

Ball, M. (2021). Payments, Payment Rails, and Blockchains, and the Metaverse. https://www.matthewball.vc/all/metaversepayments.

Ball, M. (2021). Payments, Payment Rails, and Blockchains, and the Metaverse. https://www.matthewball.vc/all/Metaversepayments.

Banoo, S. (2020). E-Commerce - Growing acceptance of online grocery shopping. *The Edge Markets.* https://www.theedgemarkets.com/article/ecommerce-growing-acceptance-online-grocery-shopping

Bao, H., & Roubaud, D. (2022). Non-Fungible Token: A Systematic Review and Research Agenda. *Journal of Risk and Financial Management, 1*(1), 44–46. doi:10.3390/jrfm15050215

Barker, K. (2016). Virtual spaces and virtual layers–governing the ungovernable? *Information & Communications Technology Law, 25*(1), 62–70. doi:10.1080/13600834.2015.1134146

Barkhi, R., Belanger, F., & Hicks, J. (2008). A Model of the Determinants of Purchasing from Virtual Stores. *Journal of Organizational Computing and Electronic Commerce*, *18*(3), 177–196. doi:10.1080/10919390802198840

Barteit, S., Lanfermann, L., Bärnighausen, T., Neuhann, F., & Beiersmann, C. (2021). Augmented, mixed, and virtual reality-based head-mounted devices for medical education: Systematic review. *JMIR Serious Games*, *9*(3), e29080. doi:10.2196/29080 PMID:34255668

Barwise, P., Elberse, A., & Hammond, K. (2002). Marketing and the Internet: a research review (pp. 01-801), London Business School.

Basha, A. D., Khaleel, A. I., Mnaathr, S. H., & Rozinah, J. (2013). Applying Second Life in New 3D Metaverse Culture to Build Effective Positioning Learning Platform for Arabic Language Principles. *International Journal of Modern Engineering Research*, 1602-1605.

Basil, W. (2022). *The" Metaverse." The arrival of the future of 3D research, learning, life & commerce.*

Bauer, T., Antonino, P. O., & Kuhn, T. (2019, May). Towards architecting digital twin-pervaded systems. In *IEEE/ACM 7th International Workshop on Software Engineering for Systems-of-Systems (SESoS) and 13th Workshop on Distributed Software Development, Software Ecosystems and Systems-of-Systems (WDES),* (pp. 66-69). IEEE. 10.1109/SESoS/WDES.2019.00018

Bauerová, R., & Klepek, M. (2018). Technology Acceptance as a Determinant of Online Grocery Shopping Adoption. *Acta Universitatis Agriculturae et Silviculturae Mendelianae Brunensis*, *66*(3), 737–746. doi:10.11118/actaun201866030737

Baumeister, R. F., & Leary, M. R. (1995). The need to belong: Desire for interpersonal attachments as a fundamental human motivation. *Psychological Bulletin*, *117*(3), 497–529. doi:10.1037/0033-2909.117.3.497 PMID:7777651

Becker, K., & Parker, J. (2014). An overview of game design techniques. In *Learning, education and games,* (pp. 179–198). ETC Press.

Bécue, A., Praça, I., & Gama, J. (2021). Artificial intelligence, cyber-threats and Industry 4.0: Challenges and opportunities. *Artificial Intelligence Review*, *54*(5), 3849–3886. doi:10.100710462-020-09942-2

Beibei, D. (2021). Research on the influence and Countermeasures of webcast on College Students' Socialist Core Values Education. *Science and technology communication,13*(16), 142-145. doi:.1674-6708.2021.16.050. doi:10.16607/j.cnki

Belfiore, M. P., Urraro, F., Grassi, R., Giacobbe, G., Patelli, G., Cappabianca, S., & Reginelli, A. (2020). Artificial intelligence to codify lung CT in Covid-19 patients. *La Radiologia Medica*, *125*(5), 500–504. doi:10.100711547-020-01195-x PMID:32367319

Belk, R. W. (1992). Attachment to Possessions. In I. Altman & S. M. Low (Eds.), *Place Attachment,* (pp. 37–62)., doi:10.1007/978-1-4684-8753-4_3

Benford, S., & Giannachi, G. (2011). *Performing mixed reality*. MIT Press.

Bergström, I., Azevedo, S., Papiotis, P., Saldanha, N., & Slater, M. (2017). The Plausibility of a String Quartet Performance in Virtual Reality. *IEEE Transactions on Visualization and Computer Graphics*, *23*(4), 1352–1359. doi:10.1109/TVCG.2017.2657138 PMID:28141523

Bervell, B. B., Kumar, J. A., Arkorful, V., Agyapong, E. M., & Osman, S. (2022). Remodelling the role of facilitating conditions for Google Classroom acceptance: A revision of UTAUT2. *Australasian Journal of Educational Technology*, *38*(1), 115–135.

Bhandari, A. (2020). What is Multicollinearity? Here's Everything You Need to Know. *Analytics Vidhya*. https://www.analyticsvidhya.com/blog/2020/03/what-is-multicollinearity/

Bhattarai, M., Jin, Y., Smedema, S. M., Cadel, K. R., & Baniya, M. (2021). The relationships among self-efficacy, social support, resilience, and subjective well-being in persons with spinal cord injuries. *Journal of Advanced Nursing*, *77*(1), 221–230. doi:10.1111/jan.14573 PMID:33009842

Bianchi, C., & Andrews, L. (2012). Risk, trust, and consumer online purchasing behaviour: A Chilean perspective. *International Marketing Review*, *29*(3), 253–275. doi:10.1108/02651331211229750

Bianchi, D., Dickerson, A., & Babiak, M. (2019). *Trading volume and liquidity provision in cryptocurrency markets. SSRN*. Electronic Journal., doi:10.2139/SSRN.3239670

Bibby, L., & Dehe, B. (2018). Defining and assessing industry 4.0 maturity levels–case of the defence sector. *Production Planning and Control*, *29*(12), 1030–1043. doi:10.1080/09537287.2018.1503355

Bibri, S. E. (2022). The Social Shaping of the Metaverse as an Alternative to the Imaginaries of Data-Driven Smart Cities: A Study in Science, Technology, and Society. *Smart Cities*, *5*(3), 832–874. doi:10.3390martcities5030043

Bigné, E., Sanz, S., Ruiz, C., & Aldás, J. (2010). Why some internet users don't buy air tickets online. In *Information and communication technologies in tourism 2010,* (pp. 209–221). Springer. doi:10.1007/978-3-211-99407-8_18

Bingjie, Z. (2021). Webcast: a relational media. *Southeast propagation*, (09), 52-55. doi:.dncb. cn35-1274/j.2021.09.016. doi:10.13556/j.cnki

Blaise, S., de Jesus, D., Marcelo, D. R., Ocampo, C., Is, J., Colleges, L., & Tibudan, A. J. (2022). *Play-to-Earn: A Qualitative Analysis of the Experiences and Challenges Faced By Axie Infinity Online Gamers Amidst the COVID-19 Pandemic.* doi:10.6084/m9.figshare.18856454.v1

Bloomberg. (2021). NFT Market Surpassed $40 Billion in 2021, New Estimate Shows. *Bloomberg*. https://www.bloomberg.com/news/articles/2022-01-06/nft-market-surpassed-40-billion-in-2021-new-estimate-shows.

Bölen, M. C., & Özen, Ü. (2020). Understanding the factors affecting consumers' continuance intention in mobile shopping: The case of private shopping clubs. *International Journal of Mobile Communications*, *18*(1), 101–129. doi:10.1504/IJMC.2020.104423

Bolton, S. J., & Cora, J. R. (2021). Virtual Equivalents of Real Objects (VEROs): A type of non-fungible token (NFT) that can help fund the 3D digitization of natural history collections. *Megataxa*, *6*(2), 93–95. doi:10.11646/megataxa.6.2.2

Boone, L., Kurtz, D., & Berston, S. (2007). *Contemporary Business*. Wiley.

Bostrom, N. (2003). Are You Living in a Computer Simulation? *The Philosophical Quarterly*, *53*(211), 243–255. doi:10.1111/1467-9213.00309

Boyd, D., & Ellison, N. (2007). Social network sites: Definition, history, and scholarship. *Journal of Computer-Mediated Communication*, *13*(1), 1–11. doi:10.1111/j.1083-6101.2007.00393.x

Bradu, P., Biswas, A., Nair, C., Sreevalsakumar, S., Patil, M., Kannampuzha, S., Mukherjee, A. G., Wanjari, U. R., Renu, K., Vellingiri, B., & Gopalakrishnan, A. V. (2022). Recent advances in green technology and Industrial Revolution 4.0 for a sustainable future. *Environmental Science and Pollution Research International*, 1–32. doi:10.100711356-022-20024-4 PMID:35397034

Brodny, J., & Tutak, M. (2022). Analyzing the Level of Digitalization among the Enterprises of the European Union Member States and Their Impact on Economic Growth. *Journal of Open Innovation*, *8*(2), 70. doi:10.3390/joitmc8020070

Bukhori, A. (1993). *Pengantar Bisnis*. Alfabeta CV.

Burden, D. (2009). Deploying Embodied AI into Virtual Worlds. *Knowledge-Based Systems*, *22*(7), 540–544. doi:10.1016/j.knosys.2008.10.001

Burton-Jones, A., & Hubona, G. S. (2005). Individual differences and usage behavior. *ACM SIGMIS Database*, *36*(2), 58–77. doi:10.1145/1066149.1066155

Byrne, B. M. (1998). *Structural Equation Modeling with LISREL, PRELIS, and SIMPLIS: Basic Concepts, Applications, and Programming*. Lawrence Erlbaum Associates.

Byrne, B. M. (2001). Structural Equation Modeling with AMOS, EQS, and LISREL: Comparative approaches to testing for the factorial validity of a measuring instrument. *International Journal of Testing*, *1*(1), 55–86. doi:10.1207/S15327574IJT0101_4

CADPA. (2021). Report of China's game industry in 2021. *CADPA*. https://www.sohu.com/a/509131223_120840?scm=&spm=smpc.channel_164.tpl-author-box-pc.14.1655048642096nc9B6Jk_324

Cai, Y., Llorca, J., Tulino, A. M., & Molisch, A. F. (2022). Compute-and data-intensive networks: The key to the Metaverse.

Cai, S., Jiao, X. Y., & Song, B. J. (2022). Open Another Gate to Education—Application, Challenge and Prospect of Educational Metaverse, *Modern. Educational Technology*, *1*, 16–26.

Caminiti, S. (2022). No one knows what the metaverse is and that's what's driving all the hype. CNBC. https://www.cnbc.com/2022/01/25/no-one-knows-what-the-metaverse-is-and-thats-driving-all-the-hype.html

Capaldo, C., Flanagan, K., & Littrell, D. (2008). Teacher interview 7377 – introduction to technology in schools. *Philaliteracy*. https://philaliteracy.org/wp-content/uploads/2014/05/2013-What-are-the-different-types-of-technology-you-use-in-your-classroom.pdf

Carpenter, J. M., Moore, M., & Fairhurst, A. E. (2005). Consumer shopping value for retail brands. *Journal of Fashion Marketing and Management*, *9*(1), 43–53. doi:10.1108/13612020510586398

Caulfield, B. (2021). *What is the Metaverse?* The Official NVIDIA Blog.

Celik, H. (2016). Customer online shopping anxiety within the Unified Theory of Acceptance and Use Technology (UTAUT) framework. *Asia Pacific Journal of Marketing and Logistics*, *28*(2). doi:10.1108/APJML-05-2015-0077

Çelik, R. (2022). Metaverse Nedir? Kavramsal Değerlendirme ve Genel Bakış [What Is The Metaverse? Conceptual Evaluation And Overview]. *Balkan and Near Eastern Journal of Social Sciences*, *8*(1), 67–74.

Chahkoutahi, F., & Khashei, M. (2017). A seasonal direct optimal hybrid model of computational intelligence and soft computing techniques for electricity load forecasting. *Energy*, *140*, 988–1004. doi:10.1016/j.energy.2017.09.009

Chalmers, D. J. (2003). The Matrix as metaphysics. *Science Fiction and Philosophy: From Time Travel to Superintelligence*.

Chalmers, D. J. (2017). The Virtual and the Real. *Disputatio*, *9*(46), 309–352. doi:10.1515/disp-2017-0009

Chan, H. K., Lettice, F., & Durowoju, O. A. (2012). Decision-making for supply chain integration: supply chain integration. Springer.

Chandra, Y. (2022). Non-fungible token-enabled entrepreneurship: A conceptual framework. *Journal of Business Venturing Insights*, *18*, e00323. doi:10.1016/j.jbvi.2022.e00323

Chang, C. (2011). The Effect of the Number of Product Subcategories on Perceived Variety and Shopping Experience in an Online Store. *Journal of Interactive Marketing*, *25*(3), 159–168. doi:10.1016/j.intmar.2011.04.001

Chang, S. C., Chang, H. H., & Lu, M. T. (2021). Evaluating industry 4.0 technology application in SMES: Using a Hybrid MCDM Approach. *Mathematics*, *9*(4), 414. doi:10.3390/math9040414

Chang, S., & Tung, F. (2008). An empirical investigation of students' behavioral intentions to use the online learning course websites. *British Journal of Educational Technology*, *39*(1), 71–83.

Chao, C.-M. (2019). Factors determining the behavioral intention to use mobile learning: An application and extension of the UTAUT model. *Frontiers in Psychology*, *10*, 1652. doi:10.3389/fpsyg.2019.01652 PMID:31379679

Charamba, K. (2022). *Beyond the Corporate Responsibility to Respect in the Dawn of a Metaverse*. doi:10.2139srn.4043254

Chaveesuk, S., & Chaiyasoonthorn, W. (2022). COVID-19 in Emerging Countries and Students' Intention to Use Cloud Classroom: Evidence from Thailand. *Education Research International*. doi:10.1155/2022/6909120

Chen, C. C. (2013). The exploration on network behaviors by using the models of Theory of planned behaviors (TPB), Technology acceptance model (TAM) and C-TAM-TPB. *African Journal of Business Management*, *7*(30), 2976–2984. doi:10.5897/AJBM11.1966

Cheng, C., Guohua, W., & Zhihong, Z. (2020). Research on the influence of price discount level on consumers' purchase intention. *Business economics research*, (23), 76-79.

Chengju, J. (2021). New characteristics and problems of teenagers' entertainment idol worship. *Modern youth*, (04), 51-53.

Cheng, R., Wu, N., Chen, S., & Han, B. (2022, March). Reality Check of Metaverse: A First Look at Commercial Social Virtual Reality Platforms. In *IEEE Conference on Virtual Reality and 3D User Interfaces Abstracts and Workshops (VRW),* (pp. 141-148). IEEE. 10.1109/VRW55335.2022.00040

Chen, J., Chen, S., Liu, Q., & Shen, M. I. (2021). Applying blockchain technology to reshape the service models of supply chain finance for SMEs in China. *The Singapore Economic Review*, 1–18. doi:10.1142/S0217590821480015

Chen, T.-H., & Chang, R.-C. (2021). Using machine learning to evaluate the influence of FinTech patents: The case of Taiwan's financial industry. *Journal of Computational and Applied Mathematics*, *390*, 113215. doi:10.1016/j.cam.2020.113215

Cheong, B. C. (2022). Avatars in the metaverse: potential legal issues and remedies. *International Cybersecurity Law Review*, 1-28.

Chia, A. (2022). The metaverse, but not the way you think: Game engines and automation beyond game development. *Critical Studies in Media Communication*, *39*(3), 1–10. doi:10.1080/15295036.2022.2080850

Childers, T. L., Carr, C. L., Peck, J., & Carson, S. (2001). Hedonic and utilitarian motivations for online retail shopping behavior. *Journal of Retailing*, *77*(4), 511–535. doi:10.1016/S0022-4359(01)00056-2

China Government Network. (2021). *Report on the Work of the Chinese Government in 2021.* http://www.gov.cn/premier/2021-03/12/content_5592671.htm

Chin, B. M. (2006). Regulating Your Second Life-Defamation in Virtual Worlds. *Brook. L. Rev.*, *72*, 1303.

Ching, K. H., Teoh, A. P., & Amran, A. (2020, November). A Conceptual Model of Technology Factors to InsurTech Adoption by Value Chain Activities. In *2020 IEEE Conference on e-Learning, e-Management and e-Services (IC3e),* (pp. 88-92). IEEE. 10.1109/IC3e50159.2020.9288465

Chin, S. L., & Goh, Y. N. (2017). Consumer Purchase Intention Toward Online Grocery Shopping: View from Malaysia. *Global Business and Management Research*, 9.

Chintagunta, P. K., Chu, J., & Cebollada, J. (2012). Quantifying Transaction Costs in Online/Off-line Grocery Channel Choice. *Marketing Science*, *31*(1), 96–114. doi:10.1287/mksc.1110.0678

Chin, W. W., Marcolin, B. L., & Newsted, P. R. (2003). A partial least squares latent variable modeling approach for measuring interaction effects: Results from a Monte Carlo simulation study and an electronic-mail emotion/adoption study. *Information Systems Research*, *14*(2), 189–217. doi:10.1287/isre.14.2.189.16018

Chittum, M. (2022). Morgan Stanley Sees $8 Trillion Metaverse Market — In China Alone. Blockworks. https://blockworks.co/morgan-stanley-sees-8-trillion-metaverse-market-eventually/

Chohan, U. W. (2021). *Non-Fungible Tokens: Blockchains, Scarcity, and Value.* Social Science Research Network. doi:10.2139/ssrn.3822743

Choijil, E., Méndez, C. E., Wong, W.-K., Vieito, J. P., & Batmunkh, M.-U. (2022). Thirty years of herd behavior in financial markets: A bibliometric analysis. *Research in International Business and Finance*, *59*, 101506. doi:10.1016/j.ribaf.2021.101506

Choi, S.-W., Lee, S.-M., Koh, J.-E., Kim, H.-J., & Kim, J.-S. (2021). A Study on the elements of business model innovation of non-fungible token blockchain game: Based on'PlayDapp'case, an in-game digital asset distribution platform. *Journal of Korea Game Society*, *21*(2), 123–138. doi:10.7583/JKGS.2021.21.2.123

Chong, A. Y.-L. (2013). Predicting m-commerce adoption determinants: A neural network approach. *Expert Systems with Applications*, *40*(2), 523–530. doi:10.1016/j.eswa.2012.07.068

Chu, H. C., & Hwang, G. J. (2008). A Delphi-based approach to developing expert systems with the cooperation of multiple experts. *Expert Systems with Applications*, *34*(4), 2826–2840. doi:10.1016/j.eswa.2007.05.034 PMID:32288332

Cialdini, R. B., & Goldstein, N. J. (2004). Social influence: Compliance and conformity. *Annual Review of Psychology*, *55*(1), 591–621. doi:10.1146/annurev.psych.55.090902.142015 PMID:14744228

Cipriani, M., & Guarino, A. (2014). Estimating a structural model of herd behavior in financial markets. *The American Economic Review*, *104*(1), 224–251. doi:10.1257/aer.104.1.224

Clemens, E. S., Powell, W. W., McIlwaine, K., & Okamoto, D. (1995). Careers in print: Books, journals, and scholarly reputations. *American Journal of Sociology*, *101*(2), 433–494. doi:10.1086/230730

Clement J. (2022b). Main reasons for global internet user interest in the metaverse 2021. *Statista*. https://www.statista.com/statistics/1296982/metaverse-interest-reasons-worldwide/

Clement, J. (2021, November 23). *Video gaming market size worldwide 2020-2025*. Statista. https://www.statista.com/statistics/292056/video-game-market-value-worldwide/

Clement, J. (2022a). Leading business sectors worldwide that have already invested in the metaverse as of March 2022. *Statista*. https://www.statista.com/statistics/1302091/global-business-sectors-investing-in-the-metaverse/

CNBC. (2022). Walmart is quietly preparing to enter the metaverse. *CNBC*. https://www.cnbc.com/2022/01/16/walmart-is-quietly-preparing-to-enter-the-metaverse.html.

CNNIC. (2021). The 47th Statistical Report on China's Internet Development was released. *China Broadcasting*, (04), 38

Coburn, J. Q., Freeman, I., & Salmon, J. L. (2017). A Review of the Capabilities of Current Low-Cost Virtual Reality Technology and Its Potential to Enhance the Design Process. *Journal of Computing and Information Science in Engineering*, *17*(3), 031013. doi:10.1115/1.4036921

Collins, B. (2021). The Metaverse: How to Build a Massive Virtual World. *Forbes Magazine*. https://www. forbes. com/sites/barrycollins/2021/09/25/the-metaverse-how-tobuild-a-massive-virtual-world

Collins, C. (2008). Looking to the future: Higher education in the Metaverse. *EDUCAUSE Review*, *43*(5), 51–63.

Comegys, C., Hannula, M., & Váisánen, J. (2009). Effects of Consumer Trust and Risk on Online Purchase Decision-Making: A Comparison of Finnish and United States Students. *International Journal of Management*, *26*(2), 295–308.

Čopič Pucihar, K., & Kljun, M. (2018). ART for art: augmented reality taxonomy for art and cultural heritage. In *Augmented reality art,* (pp. 73–94). Springer. doi:10.1007/978-3-319-69932-5_3

Corbin, J., & Strauss, A. (2008). *Basics of qualitative research: Techniques and procedures for developing grounded theory,* (3rd ed.). SAGE. doi:10.4135/9781452230153

Cornelius, K. (2021). Betraying blockchain: Accountability, transparency and document standards for non-fungible tokens (nfts). *Information*, *12*(9), 358. doi:10.3390/info12090358

Dahan, N. A., Al-Razgan, M., Al-Laith, A., Alsoufi, M. A., Al-Asaly, M. S., & Alfakih, T. (2022). Metaverse Framework: A Case Study on E-Learning Environment (ELEM). *Electronics (Basel)*, *11*(10), 1616. doi:10.3390/electronics11101616

Dalle, J. (2010). *The Relationship Between PU and PEOU Towards the Behavior Intention in New Student Placement (NSP) System of Senior High School in Banjarmasin, South Kalimantan, Indonesia.*

Damar, M. (2021). Metaverse Shape of Your Life for Future: A bibliometric snapshot. *Journal of Metaverse*, *1*(1), 1–8.

Dantas, T. E., De-Souza, E. D., Destro, I. R., Hammes, G., Rodriguez, C. M. T., & Soares, S. R. (2021). How the combination of Circular Economy and Industry 4.0 can contribute towards achieving the Sustainable Development Goals. *Sustainable Production and Consumption*, *26*, 213–227. doi:10.1016/j.spc.2020.10.005

Davis, A., Murphy, J., Owens, D., Khazanchi, D., & Zigurs, I. (2009). Avatars, people, and virtual worlds: Foundations for research in Metaverse s. *Journal of the Association for Information Systems*, *10*(2), 1. doi:10.17705/1jais.00183

Davis, F. D. (1985). *A technology acceptance model for empirically testing new end-user information systems: theory and results*. Massachusetts Institute Of Technology.

Davis, F. D. (1989). Perceived usefulness, perceived ease of use, and user acceptance of information technology. *Management Information Systems Quarterly*, *13*(3), 319–340. doi:10.2307/249008

Davis, F. D., Bagozzi, R. P., & Warshaw, P. R. (1989). User acceptance of computer technology: A comparison of two theoretical models. *Management Science*, *35*(8), 982–1003. doi:10.1287/mnsc.35.8.982

Davis, S., & Wiedenbeck, S. (2001). The mediating effects of intrinsic motivation, ease of use and usefulness perceptions on performance in first-time and subsequent computer users. *Interacting with Computers, 13*(5), 549–580. doi:10.1016/S0953-5438(01)00034-0

Day, C. D. Andrea. (2022, January 12). *Investors are paying millions for virtual land in the metaverse.* CNBC. https://www.cnbc.com/2022/01/12/investors-are-paying-millions-for-virtual-land-in-the-metaverse.html

Decentraland, M. (2021). Property. https:// Metaverse.properties/buy-indecentraland

Dede, C., & Richards, J. (2012). Digital teaching platforms: customizing classroom learning for each student. Teachers college press.

Demirbağ, İ. (2020). Üç boyutlu sanal dünyalar [Three-dimensional virtual worlds]. *Açıköğretim Uygulamaları ve Araştırmaları Dergisi, 6*(4), 97–112.

Dengfeng, C. & Zhenpeng, Y. (2021). Can I make a reservation after a good call? Research on the impact of platform and anchor reputation on consumers' willingness to participate in online live shopping. *Xinjiang Agricultural Reclamation Economy*, (06), 82-92.

Deutschmann, M., Panichi, L., & Molka-Danielsen, J. (2009). Designing oral participation in second life – a comparative study of two language proficiency courses. *ReCALL, 21*(2), 206–226. doi:10.1017/S0958344009000196

Dewey. (2021). Research on the impact of Daren webcast e-commerce on consumers' purchase intention. *Internet Weekly*, (18), 62-64

Dewi, M. V. K., & Darma, G. S. (2019). The Role of Marketing & Competitive Intelligence In Industrial Revolution 4.0. *Jurnal Manajemen Bisnis, 16*(1), 1–12. doi:10.38043/jmb.v16i1.2014

Di Pietro, R., & Cresci, S. (2021). Metaverse: Security and Privacy Issues. In *Third IEEE International Conference on Trust, Privacy and Security in Intelligent Systems and Applications (TPS-ISA).* IEEE. 281-288.

Diep, N. A., Cocquyt, C., Zhu, C., & Vanwing, T. (2016). Predicting adult learners' online participation: Effects of altruism, performance expectancy, and social capital. *Computers & Education, 101*, 84–101. doi:10.1016/j.compedu.2016.06.002

Díez, J. L. (2021). Metaverse: Year One. Mark Zuckerberg's video keynote on Meta in the context of previous and prospective studies on metaverses. *Pensar la publicidad: revista internacional de investigaciones publicitarias, 15*(2), 299-303.

Dilmegani, C. (2021). Data Cleaning in 2021: What it is, Steps to Clean Data & Tools. *Clean (Weinheim).*

Dimock, M. (2019). Defining generations: Where Millennials End and Generation Z Begins. *Pew Research Center.* https://www.pewresearch.org/fact-tank/2019/01/17/where-millennials-end-and-generation-z-begins/

Dincelli, E., & Yayla, A. (2022). Immersive virtual reality in the age of the Metaverse: A hybrid-narrative review based on the technology affordance perspective. *The Journal of Strategic Information Systems, 31*(2), 101717. doi:10.1016/j.jsis.2022.101717

Dinh, T. N., & Thai, M. T. (2018). AI and Blockchain: A disruptive integration. *Computer, 51*(9), 48–53. doi:10.1109/MC.2018.3620971

Diong, I. H., & Toh, E. B. H. (2021). The Influences of Reference Groups Towards the Usage of E-Wallet Payment Systems. In *Handbook of Research on Social Impacts of E-Payment and Blockchain Technology* (pp. 428–455). IGI Global.

Dionisio, J. D. N., III, W. G. B., & Gilbert, R. (2013). 3D virtual worlds and the metaverse: Current status and future possibilities. *ACM Computing Surveys (CSUR), 45*(3), 1-38.

Dionisio, J. D. N., III, W. G. B., & Gilbert, R. (2013). 3D virtual worlds and the metaverse: Current status and future possibilities. *ACM Computing Surveys*, *45*(3), 1-38.

Dionisio, J. D., Burns, W. G. III, & Gilbert, R. (2013, June). 3D Virtual Worlds and the Metaverse: Current Status and Future Possibilities. *ACM Computing Surveys*, *45*(3), 38. doi:10.1145/2480741.2480751

Dirsehan, T., & Can, C. (2020). Examination of trust and sustainability concerns in autonomous vehicle adoption. *Technology in Society*, *63*, 101361. doi:10.1016/j.techsoc.2020.101361

Doan, A. P., Johnson, R. J., Rasmussen, M. W., Snyder, C. L., Sterling, J. B., & Yeargin, D. G. (2021). NFTs: Key US Legal Considerations for an Emerging Asset Class. *Journal of Taxation of Investments*, *38*(4), 63–69.

Dorall, A. (2020). Can An Adult Legally Be In A Relationship With A 17-Year-Old In Malaysia? *TRP*. https://www.therakyatpost.com/2020/09/27/can-an-adult-legally-be-in-a-relationship-with-a-17-year-old-in-malaysia/

Dowling, M. (2022). Is non-fungible token pricing driven by cryptocurrencies? *Finance Research Letters*, *44*, 102097. doi:10.1016/j.frl.2021.102097

Duan, C.(2019). Research on the Quality Evaluation System for Talent Training in Higher Vocational Education in China . *Huainan Vocational and Technical College Journal*, (3), 46-47.

Duan, H., Li, J., Fan, S., Lin, Z., Wu, X., & Cai, W. (2021). Metaverse for social good: A university campus prototype. In *Proceedings of the 29th ACM International Conference on Multimedia,* (pp.153-161). 10.1145/3474085.3479238

Dunner, P. M., Lusch, R. F., & Carver, J. R. (2014). Retailing. (8th Ed). South Western.

Dursun, T., & Yetimova, S. (2022). Güncel İletişim Teknolojilerinin Sunduğu Dijital İş İmkânlarının (Digital Business) Öğrenciler Tarafından Nasıl Algılandığı Üzerine Bir Saha Araştırması [Field Research On How Students Perceive The Digital Business Offered By Current Communication Technologies]. *Third Sector Social Economic Review*, *57*(1), 366–391.

Dutilleux, M., & Chang, K. M. (2022). Future Addiction Concerned for Human-Being. [IMJST]. *International Multilingual Journal of Science and Technology*, *7*(2), 4724–4732.

Dwivedi, Y. K., Hughes, L., Baabdullah, A. M., Ribeiro-Navarrete, S., Giannakis, M., Al-Debei, M. M., Dennehy, D., Metri, B., Buhalis, D., Cheung, C. M. K., Conboy, K., Doyle, R., Dubey, R., Dutot, V., Felix, R., Goyal, D. P., Gustafsson, A., Hinsch, C., Jebabli, I., & Wamba, S. F. (2022). Metaverse beyond the hype: Multidisciplinary perspectives on emerging challenges, opportunities, and agenda for research, practice and policy. *International Journal of Information Management*, *66*, 102542. doi:10.1016/j.ijinfomgt.2022.102542

Eberendu, A. C. (2015). Negative impacts of technology in Nigerian society. *International journal of business and management review 3*(2), 23 – 29.

Education, C. F. I. (2018). R-Squared - Definition, Interpretation, and How to Calculate. *Corporate Finance Institute.* https://corporatefinanceinstitute.com/resources/knowledge/other/r-squared/

Egliston, B., & Carter, M. (2021). Critical questions for Facebook's virtual reality: data, power and the metaverse. *Internet Policy Review, 10*(4).

Eid, M. I. (2011). Customer satisfaction, brand trust and variety seeking as determinants of brand loyalty. *Journal of Electronic Commerce Research*, *12*(1). doi:10.5897/ajbm11.2380

Ejsmont, K. (2021). The Impact of Industry 4.0 on Employees—Insights from Australia. *Sustainability*, *13*(6), 3095. doi:10.3390u13063095

El Beheiry, M., Doutreligne, S., Caporal.,, C., Ostertag, C., Dahan, M., & Masson, J. B. (2019). Virtual reality: Beyond visualization. *Journal of Molecular Biology*, *431*(7), 1315–1321. doi:10.1016/j.jmb.2019.01.033 PMID:30738026

Ellis, R. (2009). *The Study of Second Language Acquisition,* (2nd ed.). Oxford University Press.

Elmassah, S., & Hassanein, E. A. (2022). Digitalization and subjective wellbeing in Europe. Digital Policy, Regulation, and Governance.

Enright, D. S., & McCloskey, M. L. (1985). Yes, Talking!: Organizing the Classroom to Promote Second Language Acquisition. *TESOL Quarterly*, *19*(3), 431–453. http://doi.org.proxy.seattleu.edu/10.2307/3586272Gass.

Eom, S. B., & Ashill, N. (2016). The determinants of students' perceived learning outcomes and satisfaction in university online education: An update. *Decision Sciences Journal of Innovative Education*, *14*(2), 185–215. doi:10.1111/dsji.12097

Eriksson, T., Bigi, A., & Bonera, M. (2020). Think with me, or think for me? On the future role of artificial intelligence in marketing strategy formulation. *The TQM Journal*, *32*(4), 795–814. doi:10.1108/TQM-12-2019-0303

Eri, Y., Aminul Islam, M., & Ku Daud, K. A. (2011). Factors that Influence Customers' Buying Intention on Shopping Online. *International Journal of Marketing Studies*, *3*(1). doi:10.5539/ijms.v3n1p128

Esteves, J., & Curto, J. (2013). A Risk and Benefits Behavioral Model to Assess Intentions to Adopt Big Data. *Journal of Intelligence Studies in Business*, *3*(3), 37–46. doi:10.37380/jisib.v3i3.74

Ethereum. (n.d.). *Find an Ethereum Wallet*. Ethereum.Org. https://ethereum.org

Ethereum. (n.d.). *Non-fungible tokens (NFT)*. Ethereum.Org. https://ethereum.org

Ethereum. (n.d.). *What is Ethereum?* Ethereum.Org. https://ethereum.org

Etikan, I., Musa, S. A., & Alkassim, R. S. (2016). Comparison of Convenience Sampling and Purposive Sampling. *American Journal of Theoretical and Applied Statistics*, *5*(1), 1. C:\Users\salma\Desktop\10.11648\j.ajtas.20160501.11 doi:10.11648/j.ajtas.20160501.11

Fairfield, J. A. (2010). Castles in the Air: Greg Lastowka's Virtual Justice.

Fang, L., Yu, H., & Huang, Y. (2018). The role of investor sentiment in the long-term correlation between U.S. stock and bond markets. *International Review of Economics & Finance*, *58*, 127–139. doi:10.1016/j.iref.2018.03.005

Fangyuan, G. (2020). Research on the current situation, causes and correction of moral anomie in college students' webcast. *Popular literature and art*, (05): 179-180

Featherman, M. S., & Pavlou, P. A. (2003). Predicting e-services adoption: A perceived risk facets perspective. *Human-Computer Studies*, *59*(4), 451–474. doi:10.1016/S1071-5819(03)00111-3

Fei, Z. & Chaoyan, H. (2020). Design of home intelligent remote controller based on interactive design. *Industrial design*, (06), 116-117

Felsberger, A., Qaiser, F. H., Choudhary, A., & Reiner, G. (2022). The impact of Industry 4.0 on the reconciliation of dynamic capabilities: Evidence from the European manufacturing industries. *Production Planning and Control*, *33*(2-3), 277–300. doi:10.1080/09537287.2020.1810765

Fengqi, Z., Yutong, X., & Juan, L. (2021). Research on interface usability based on live broadcast marketing platform. *Science and technology information, 19*(19), 29-31 doi:.1672-3791.2108-5042-1191. doi:10.16661/j.cnki

Feng, Y., Lai, K. H., & Zhu, Q. (2022). Green supply chain innovation: Emergence, adoption, and challenges. *International Journal of Production Economics, 248*, 108497. doi:10.1016/j.ijpe.2022.108497

Fernandez, C. B., & Hui, P. (2022). Life, the Metaverse and Everything: An Overview of Privacy, Ethics, and Governance in Metaverse.

Fernández, A., Peralta, D., Benítez, J. M., & Herrera, F. (2014). E-learning and educational data mining in cloud computing: An overview. *International Journal of Learning Technology, 9*(1), 25–52. doi:10.1504/IJLT.2014.062447

Ferreira, R., Pereira, R., Bianchi, I. S., & da Silva, M. M. (2021). Decision factors for remote work adoption: Advantages, disadvantages, driving forces and challenges. *Journal of Open Innovation, 7*(1), 70. doi:10.3390/joitmc7010070

Fishbein, M., & Ajzen, I. (1975). *Belief, attitude, intention and behavior: an introduction to theory and research.* Addison-Wesley.

Fly, Y. J., & Grünberg, L. (2022). What Will China's Metaverse Look Like? *The-diplomat.* https://thediplomat.com/2022/03/what-will-chinas-metaverse-look-like/

Folger, J. (2022, February 15). *Metaverse.* Investopedia. https://www.investopedia.com/metaverse-definition-5206578

Formanek, R. (2003). Why they collect: Collectors reveal their motivations. In *Interpreting objects and collections.* Routledge.

Fornell, C., & Lacker, D. F. (1981). Evaluating structural equation models with unobservable variables and measurement error. *JMR, Journal of Marketing Research, 18*(1), 39–50. doi:10.1177/002224378101800104

Foundation Team. (2021). *Everything you need to know about the metaverse. Foundation.* https://foundation.app/blog/enter-the-metaverse

Frank, A. G., Dalenogare, L. S., & Ayala, N. F. (2019). Industry 4.0 technologies: Implementation patterns in manufacturing companies. *International Journal of Production Economics, 210*, 15–26. doi:10.1016/j.ijpe.2019.01.004

Fuqiang, Y. & Penghui, H. (2020). Simulacrum, body and emotion: an analysis of webcast in consumer society. *China Youth Research*, (7), 5-12 + 32

Gadekallu, T. R., Huynh-The, T., Wang, W., Yenduri, G., Ranaweera, P., Pham, Q. V., & Liyanage, M. (2022). Blockchain for the Metaverse: A Review.

Gao, L., & Simonson, I. (2015). The positive effect of assortment size on purchase likelihood: The moderating influence of decision order. *Journal of Consumer Psychology, 26*(4), 542–549. doi:10.1016/j.jcps.2015.12.002

Garon, J. (2022). Legal implications of a ubiquitous metaverse and a Web3 future.

Garton, S. (2002). Learner initiative in the language classroom. *ELT Journal, 56*(1), 47–55. doi:10.1093/elt/56.1.47

Gaubert, J. (2021). Seoul to become the first city to enter the metaverse. What will it look like? *Euronews.* https://www.euronews.com/next/2021/11/10/seoul-to-become-the first-city-to-enter-the-metaverse-what-will-it-look-like

Gefen, Karahanna, & Straub. (2003). Trust and TAM in Online Shopping: An Integrated Model. *MIS Quarterly, 27*(1), 51. doi:10.2307/30036519

Gelernter, D. (1991). *Mirror Worlds: Or: The Day Software Puts the Universe in a Shoebox...How It Will Happen and What It Will Mean.* Oxford University Press., doi:10.1093/oso/9780195068122.001.0001

Georgantzinos, S. K., Giannopoulos, G. I., & Bakalis, P. A. (2021). Additive manufacturing for effective smart structures: The idea of 6D printing. *Journal of Composites Science, 5*(5), 119. doi:10.3390/jcs5050119

Ge, R., Zheng, Z., Tian, X., & Liao, L. (2021). Human–robot interaction: When investors adjust the usage of robo-advisors in peer-to-peer lending. *Information Systems Research, 32*(3), 774–785. doi:10.1287/isre.2021.1009

Ghazali, E. M., Mutum, D. S., Chong, J. H., & Nguyen, B. (2018). Do consumers want mobile commerce? A closer look at M-shopping and technology adoption in Malaysia. *Asia Pacific Journal of Marketing and Logistics, 30*(4), 1064–1086. doi:10.1108/APJML-05-2017-0093

Gibbons, L. J. (2008). Law and the emotive avatar. *Vand. J. Ent. & Tech. L., 11*, 899.

Giese, J. L., & Cote, J. A. (2000). Defining Consumer Satisfaction. *Academy of Marketing Science Review*, 1–24.

Gil, G. (2002). Two complementary modes of foreign language classroom interaction. *ELT Journal, 56*(3), 273–279. doi:10.1093/elt/56.3.273

Gillieron, L. (2021). Facebook wants to lean into the metaverse. Here's what it is and how it will work. NPR. https://www.npr.org/2021/10/28/1050280500/what-metaverse-isa nd-how-it-will-work

Giovanis, A. N., & Athanasopoulou, P. (2014). Gaining customer loyalty in the e-tailing marketplace: The role of e-service quality, e-satisfaction and e-trust. *International Journal of Technology Marketing, 9*(3), 288. doi:10.1504/IJTMKT.2014.063857

Glencross, M., Chalmers, A., Lin, M., Otaduy, M., & Gutierrez, D. (n.d.). *Exploiting Perception in High-Fidelity Virtual Environments.*

GlobalData. (2020). COVID-19 Accelerates e-commerce Growth in Malaysia. *GlobalData*. https://www.globaldata.com/covid-19-accelerates-e-commerce-g rowth-malaysia-says-globaldata/

Goh, T. T. (2011). Exploring Gender Differences in SMS-Based Mobile Library Search System Adoption. *Journal of Educational Technology & Society, 14*(4), 192–206.

Goleman, D. (1994). *Emotional intelligence*. Bantam.

Gor, K. (2015). Factors influencing the adoption of online tax filing systems in Nairobi, Kenya. *The Strategic Journal of Business and Change Management, 2*(77), 906–920.

Grabner-Kräuter, S., & Kaluscha, E. A. (2003). Empirical research in on-line trust: A review and critical assessment. *International Journal of Human-Computer Studies, 58*(6), 783–812. C:\Users\salma\Desktop\10.1016\s1071-5819(03)00043-0 doi:10.1016/S1071-5819(03)00043-0

Grider, D. (2021). The Metaverse Web 3.0 Virtual Cloud Economies. *Grayscale Research*. https://grayscale.com/wp-content/uploads/2021/11/Grayscale_M etaverse_Report_Nov2021.pdf.

Grider, D., & Maximo, M. (2021). *The metaverse: Web 3.0 virtual cloud economies*. Grayscale Research.

Guadamuz, A. (2021). The treachery of images: Non-fungible tokens and copyright. *Journal Of Intellectual Property Law and Practice, 16*(12), 1367–1385. doi:10.1093/jiplp/jpab152

Guangyan, H., Xiaofang, Z., & Wanzhen, X. (2020). Investigation and Research on the current situation of Qinghai college students using webcast platform. *Journal of Qinghai Normal University normal college for nationalities, 31*(01): 83-88. doi:. 63-1060/g4. 2020.01.018. doi:10.13780/j.cnki

Guergov, S., & Radwan, N. (2021). Blockchain Convergence: Analysis of Issues Affecting IoT, AI and Blockchain. *International Journal of Computations, Information and Manufacturing (IJCIM), 1*(1).

Guo J. & Fuli, S. (2019). Explore the impact of webcast platform on traditional culture brand communication. *Think tank era,* (43), 259-260

Guoming, Y. & Jiayi, Y. (2020). Understanding live broadcasting: social reconstruction according to communication logic -- an analysis of the value and influence of live broadcasting from the perspective of media. *Journalist,* (08), 12-19

Guo, X., Ling, K. C., & Liu, M. (2012). Evaluating Factors Influencing Consumer Satisfaction towards Online Shopping in China. *Asian Social Science, 8*(13). doi:10.5539/ass.v8n13p40

Gupta, A., Mishra, P., Pandey, C., Singh, U., Sahu, C., & Keshri, A. (2019). Descriptive statistics and normality tests for statistical data. *Annals of Cardiac Anaesthesia, 22*(1), 67. doi:10.4103/aca.ACA_157_18 PMID:30648682

Guzel, M., Sezen, B., & Alniacik, U. (2020). Drivers and consequences of customer participation into value co-creation: A field experiment. *Journal of Product and Brand Management, 30*(7), 1047–1061. doi:10.1108/JPBM-04-2020-2847

Hackl, C., Lueth, D., di Bartolo, T., Arkontaky, J., & Siu, Y. (2022). *Navigating the metaverse: A guide to limitless possibilities in a Web 3.0 world.* https://search.ebscohost.com/login.aspx?direct=true&scope=site&db=nlebk&db=nlabk&AN=3273058

Ha, H.-Y., & Janda, S. (2014). The effect of customized information on online purchase intentions. *Internet Research, 24*(4), 496–519. doi:10.1108/IntR-06-2013-0107

Hair, J. F., Anderson, R. E., Tatham, R. L., & Black, W. C. (1998). *Multivariate data analysis,* (5th ed.). Prentice-Hall.

Hair, J. F., Black, W. C., Babin, B. J., & Anderson, R. E. (2010). *Multivariate Data Analysis,* (7th ed.). Pearson Prentice Hall.

Hair, J. F., Ringle, C. M., & Sarstedt, M. (2011). PLS-SEM: Indeed a Silver Bullet. *Journal of Marketing Theory and Practice, 19*(2), 139–152. doi:10.2753/MTP1069-6679190202

Hamari, J., & Lehdonvirta, V. (2010). Game design as marketing: How game mechanics create demand for virtual goods. *Journal of Business Science and Applied Management, 5*(1), 17.

Hamdan, N. H., Kassim, S. H., & Lai, P. C. (2021). The COVID-19 Pandemic crisis on micro-entrepreneurs in Malaysia: Impact and mitigation approaches. *Journal of Global Business and Social Entrepreneurship, 7*(20), 52–64.

Hamid, A., Razak, F. Z. A., Bakar, A. A., & Abdullah, W. S. W. (2016). The Effects of Perceived Usefulness and Perceived Ease of Use on Continuance Intention to Use E-Government. *Procedia Economics and Finance, 35,* 644–649. doi:10.1016/S2212-5671(16)00079-4

Hamidi, S. R., Aziz, A. A., Shuhidan, S. M., Aziz, A. A., & Mokhsin, M. (2018, March). SMEs maturity model assessment of IR4. 0 digital transformation. In *International Conference on Kansei Engineering & Emotion Research,* (pp. 721-732). Springer, Singapore.

Han, Y., Niyato, D., Leung, C., Miao, C., & Kim, D. I. (2021). A dynamic resource allocation framework for synchronizing metaverse with IoT service and data.

Han, D. I. D., Bergs, Y., & Moorhouse, N. (2022). Virtual reality consumer experience escapes: Preparing for the Metaverse. *Virtual Reality (Waltham Cross), 26*(4), 1–16. doi:10.100710055-022-00641-7

Han, H.-C. (2020). *Sandrine*. From Visual Culture in the Immersive Metaverse to Visual Cognition in Education. doi:10.4018/978-1-7998-3250-8.ch004

Han, J., & Conti, D. (2020). The use of UTAUT and post acceptance models to investigate the attitude towards a telepresence robot in an educational setting. *Robotics, 9*(2), 34. doi:10.3390/robotics9020034

Han, J., Heo, J., & You, E. (2021). Analysis of Metaverse Platform as a New Play Culture: Focusing on Roblox and ZEPETO. *2nd International Conference on Human-Centered Artificial Intelligence, Computing4Human* .

Han, J., Heo, J., & You, E. (2021). Analysis of Metaverse Platform as a New Play Culture: Focusing on Roblox and ZEPETO. *2nd International Conference on Human-Centered Artificial Intelligence.*

Harapan, H., Itoh, N., Yufika, A., Winardi, W., Keam, S., Te, H., Megawati, D., Hayati, Z., Wagner, A. L., & Mudatsir, M. (2020). Coronavirus disease 2019 (COVID-19): A literature review. *Journal of Infection and Public Health, 13*(5), 667–673. doi:10.1016/j.jiph.2020.03.019 PMID:32340833

Hargens, L. L. (2000). Using the literature: Reference networks, reference contexts, and the social structure of scholarship. *American Sociological Review, 65*(6), 846–865. doi:10.2307/2657516

Haridasan, A. C., & Fernando, A. G. (2018). Online or in-store: Unravelling consumer's channel choice motives. *Journal of Research in Interactive Marketing, 12*(2), 215–230. doi:10.1108/JRIM-07-2017-0060

Harrison, R., Scheela, W., Lai, P. C., & Vivekarajah, S. (2017). Beyond institutional voids and the middle income trap? The emerging business angel market in Malaysia. *Asia Pacific Journal of Management*, 1–27.

Harris, S. (2009). Lessons From e-learning: transforming teaching. *Proceedings of the International Conference on e-Learning.*

Hartney, E. (2022). What Is Peer Pressure? Types, Examples, and How to Deal With Peer Pressure. *Very Well Mind.* https://www.verywellmind.com/what-is-peer-pressure-22246

Harwani, Y., & Safitri. (2017). Security and Ease of Use Effect on Customers' Satisfaction Shopping in Tokopedia. *Journal of Resources Development and Management.* https://www.semanticscholar.org/paper/Security-and-Ease-of-Use-Effect-on-Customers%E2%80%99-in-Harwani-Safitri/c54eb79cd78c575b23640394fec36b96909bfe2d

Harwood, T. (2011). Convergence of online gaming and e-commerce. In *Virtual worlds and e-commerce: Technologies and applications for building customer relationships,* (pp. 61–89). IGI Global. doi:10.4018/978-1-61692-808-7.ch004

Haslhofer, B., Stütz, R., Romiti, M., & King, R. (2021). *GraphSense: A General-Purpose Cryptoasset Analytics Platform.* doi:10.48550/arxiv.2102.13613

Hassan, M. (2020). Online teaching challenges during COVID-19 pandemic. *International Journal of Information and Education Technology (IJIET).* doi:10.18178/ijiet.2021.11.1.1487

Hassoun, A., Aït-Kaddour, A., Abu-Mahfouz, A. M., Rathod, N. B., Bader, F., Barba, F. J., Biancolillo, A., Cropotova, J., Galanakis, C. M., Jambrak, A. R., Lorenzo, J. M., Måge, I., Ozogul, F., & Regenstein, J. (2022). The fourth industrial revolution in the food industry—Part I: Industry 4.0 technologies. *Critical Reviews in Food Science and Nutrition*, 1–17. doi:10.1080/10408398.2022.2034735 PMID:35114860

He, W., Chen, J. (2011). Brief Discussion on the Implementation of Positive Psychological Education for Contemporary College Students. *Educational Exploration*, (03), 131-133.

Heart, T. (2010). Who is Out There? Exploring the Effects of Trust and Perceived Risk on SaaS Adoption Intentions. *The Data Base for Advances in Information Systems*, *41*(3), 49–67. doi:10.1145/1851175.1851179

Heimerl, F., Lohmann, S., Lange, S., & Ertl, T. (2014). Word cloud explorer: Text analytics based on word clouds. *Proceedings of the Annual Hawaii International Conference on System Sciences*, 1833–1842. 10.1109/HICSS.2014.231

Henseler, J., Hubona, G., & Ray, P. A. (2016). Using PLS path modeling in new technology research: Updated guidelines. *Industrial Management & Data Systems*, *116*(1), 2–20. doi:10.1108/IMDS-09-2015-0382

Henseler, J., Ringle, C. M., & Sarstedt, M. (2015). A new criterion for assessing discriminant validity in variance-based structural equation modeling. *Journal of the Academy of Marketing Science*, *43*(1), 115–135. doi:10.100711747-014-0403-8

Henseler, J., Ringle, C. M., & Sinkovics, R. R. (2009). The use of partial least squares path modeling in international marketing. *Advances in International Marketing*, *20*, 277–319. doi:10.1108/S1474-7979(2009)0000020014

Henslin, J. M., Possamai, A. M., Possamai-Inesedy, A. L., Marjoribanks, T., & Elder, K. (2015). *Sociology: A down to earth approach*. Pearson Higher Education AU.

Hernandez, B., Jimenez, J., & Jose Martin, M. (2009). The impact of self-efficacy, ease of use and usefulness on e-purchasing: An analysis of experienced e-shoppers. *Interacting with Computers*, *21*(1-2), 146–156. doi:10.1016/j.intcom.2008.11.001

Higgins, S., Xiao, Z., Katsipataki, M., (2012). The impact of digital technology on learning: a summary for the education endowment foundation.

Hirsh Pasek, K., Zosh, J., Hadani, H. S., Golinkoff, R. M., Clark, K., Donohue, C., & Wartella, E. (2022). *A whole new world: Education meets the Metaverse*. Policy.

Hirsh-Pasek, K., Zosh, J., Hadani, H. S., Golinkoff, R. M., Clark, K., Donohue, C., & Wartella, E. (2022). *A whole new world: Education meets the metaverse*. Policy.

Ho, R. C. (2019). *The outcome expectations of promocode in mobile shopping apps: An integrative behavioral and social cognitive perspective*, 74–79.

Ho, R. C. (2021). Chatbot for Online Customer Service: Customer Engagement in the Era of Artificial Intelligence. In Impact of Globalization and Advanced Technologies on Online Business Models, (pp. 16–31). IGI Global.

Ho, R. C., & Amin, M. (2019). What drives the adoption of smart travel planning apps? The relationship between experiential consumption and mobile app acceptance. *KnE Social Sciences*, 22–41.

Ho, R. C., & Chua, H. K. (2015). The influence of mobile learning on learner's absorptive capacity: a case of bring-your-own-device (BYOD) learning environment. In Taylor's 7th Teaching and Learning Conference 2014 Proceedings (pp. 471-479). Springer, Singapore.

Hoang, M., Alija Bihorac, O., & Rouces, J. (2019). Aspect-Based Sentiment Analysis Using BERT. Linköping University Electronic Press, 187–196.

Hoffman, D. L., & Novak, T. P. (1996). Marketing in Hypermedia Computer-Mediated Environments: Conceptual Foundations *Journal of Marketing*, *60*(3), 50–68. doi:10.1177/002224299606000304

Hollensen, S., Kotler, P., & Opresnik, M. O. (2022). Metaverse–the new marketing universe. *The Journal of Business Strategy*. doi:10.1108/JBS-01-2022-0014

Holly, L., Wong, B., Agrawal, A., Awah, I., Azelmat, M., Kickbusch, L., & Ndili, N. (2022). Opportunities and threats for adolescent well- being provided by digital transformations. *NCBI.* www.ncbi.nlm.nih.gov › pmc › articles › PMC7366938

Hongzhao, F. (2018). Research on Online Live Broadcasting Marketing Strategy and Mode in the Era of Fans Economy. *Modern Economic Information*, (11), 336-337+339.

Hood, B. M., & Bloom, P. (2008). Children prefer certain individuals over perfect duplicates. *Cognition*, *106*(1), 455–462. doi:10.1016/j.cognition.2007.01.012 PMID:17335793

Ho, R. C., & Amin, M. (2022). Exploring the role of commitment in potential absorptive capacity and its impact on new financial product knowledge: A social media banking perspective. *Journal of Financial Services Marketing*, 1–14. doi:10.105741264-022-00168-7

Ho, R. C., Amin, M., Ryu, K., & Ali, F. (2021). Integrative model for the adoption of tour itineraries from smart travel apps. *Journal of Hospitality and Tourism Technology*, *12*(2), 372–388. doi:10.1108/JHTT-09-2019-0112

Ho, R. C., & Song, B. L. (2021). Immersive live streaming experience in satisfying the learners' need for self-directed learning. *Interactive Technology and Smart Education*, *19*(2), 145–160. doi:10.1108/ITSE-12-2020-0242

Horan, B., Gardner, M., & Scott, J. (2015), MiRTLE: a mixed reality teaching & learning environment, Hybrid Learning and Education, In *International Conference on Hybrid Learning and Education,* (pp. 54-65). Springer, Berlin, Heidelberg.

Howell, J. (2022). Key Technologies That Power The Metaverse. *Blockchains*. https://101blockchains.com/technologies-powering-metaverse/

Hsu, C. L., Chang, K. C., & Chen, M. C. (2012). The impact of website quality on customer satisfaction and purchase intention: Perceived playfulness and perceived flow as mediators. *Information Systems and e-Business Management*, *10*(4), 549–570. doi:10.100710257-011-0181-5

Hsu, J. S. C., Huang, H. H., & Linden, L. P. (2011). Computer-mediated Counter-Arguments and Individual Learning. *Journal of Educational Technology & Society*, *14*(4), 111–123.

Hu, M. B., & Liu, B. Q. (2019). The New Era's Value Guidance and Realization Strategy for the Quality of Higher Vocational Education Talents Training. *Vocational Education Forum, 6*(66), 17-22.

Huang, S., Deng, F., & Xiao, J. (2021). Research on impulse purchase decision of audience on webcast platform -- from the perspective of dual path influence. *Financial science,* (05): 119-132

Huang, N. X. (2019). Thoughts on Improving the Quality of Talents in Post-secondary Vocational Education. *Shanxi Youth*, *11*, 188.

Huddleston, P., Whipple, J., Nye Mattick, R., & Jung Lee, S. (2009). Customer satisfaction in food retailing: Comparing specialty and conventional grocery stores. *International Journal of Retail & Distribution Management*, *37*(1), 63–80. doi:10.1108/09590550910927162

Hughes, C. E., Stapleton, C. B., Hughes, D. E., & Smith, E. M. (2005). Mixed reality in education, entertainment, and training. *IEEE Computer Graphics and Applications*, *25*(6), 24–30. doi:10.1109/MCG.2005.139 PMID:16315474

Hui, L. (2017). Empirical Study on Influencing Factors of webcast use attitude from the perspective of TAM theory. *South China University of technology.*

Huijts, N. M. A., Molin, E. J. E., & Steg, L. (2012). Psychological factors influencing sustainable energy technology acceptance: A review-based comprehensive framework. *Renewable & Sustainable Energy Reviews*, *16*(1), 525–531. doi:10.1016/j.rser.2011.08.018

Hui, W. (2021). sahelan An empirical analysis of the impact of additional comments on e-commerce shopping platforms on consumers' Purchase Intention — Based on the survey data of colleges and universities in Xinjiang. *Journal of Xinjiang Radio and Television University*, *25*(02), 45–51.

Hu, L., & Bentler, P. M. (1999). Cutoff criteria for fit indexes in covariance structure analysis: Conventional criteria versus new alternatives. *Structural Equation Modeling*, *6*(1), 1–55. doi:10.1080/10705519909540118

Humayun Kabir, M. (2018). Did investors herd during the financial crisis? Evidence from the US financial industry. *International Review of Finance*, *18*(1), 59–90. doi:10.1111/irfi.12140

Hung, M. C., Yang, S. T., & Hsieh, T. C. (2012). An examination of the determinants of mobile shopping continuance. *International Journal of Electronic Business Management*, *10*(1), 29–37. http://203.72.2.146/bitstream/987654321/27104/1/An+Examination.pdf

Hussain, S., Fangwei, Z., Siddiqi, A., Ali, Z., & Shabbir, M. (2018). Structural Equation Model for Evaluating Factors Affecting Quality of Social Infrastructure Projects. *Sustainability*, *10*(5), 1415. https://www.forbes.com/sites/bernardmarr/2022/03/21/a-short-history-of-the-metaverse/?sh=503572be5968. doi:10.3390u10051415

Huynh-The, T., Pham, Q. V., Pham, X. Q., Nguyen, T. T., Han, Z., & Kim, D. S. (2022). Artificial Intelligence for the Metaverse: A Survey.

Hwang, G. J., & Chien, S. Y. (2022). Definition, roles, and potential research issues of the metaverse in education: An artificial intelligence perspective. *Computers and Education: Artificial Intelligence*, 100082.

Hwang, H. (2017). The "Edufication" of Gaming. *LinkedIn*. https://www.linkedin.com/pulse/edufication-gaming-hans-hwang

Ibáñez, M. B., & Delgado-Kloos, C. (2018). Augmented reality for STEM learning: A systematic review. *Computers & Education*, *123*, 109–123. doi:10.1016/j.compedu.2018.05.002

Icrowdnewswire. (2021). In-game advertising market to reach usd 18.41 billion by 2027, is going to boom with rapidfire inc., playwire media llc, atlas alpha inc, engage. ISP News.

Idow, P.A, Idowu, A.O. & Adawgouno. (2004). A comparative survey of Information and communication technologies in High Educational Institutions in Africa: a case study from Nigeria and Mozambi. *Journal of Information Technology impact, 4*(6), 67-74.

iiMedia Research (2022). Report on user behavior analysis of China' s meta universe industry in 2021. iiMedia Research. Retrieved from https://report.iimedia.cn/repo3-0/43053.html?acPlatCode=IIMReport&acFrom=recomBar_1020&iimediaId=85146

Im, I., Hong, S., & Kang, M. S. (2011). An international comparison of technology adoption: Testing the UTAUT model. *Information & Management*, *48*(1), 1–8. doi:10.1016/j.im.2010.09.001

industry and Information Technology Bureau. (2022). Measures for Promoting the Innovation and Development of the Metaverse in Guangzhou Huangpu District and Guangzhou Development Zone. *People's Government of Huangpu District.* http://www.hp.gov.cn/gzjg/qzfgwhgzbm/qgyhxxhj/xxgk/content/post_8171935.html

IPFS. (n.d.). How IPFS Works. *IPFS*. https://ipfs.io/#how

Iranmanesh, M., Min, C. L., Senali, M. G., Nikbin, D., & Foroughi, B. (2022). Determinants of switching intention from web-based stores to retail apps: Habit as a moderator. *Journal of Retailing and Consumer Services*, *66*, 102957. doi:10.1016/j.jretconser.2022.102957

Jagtiani, J., & Lemieux, C. (2019). The roles of alternative data and machine learning in fintech lending: Evidence from the LendingClub consumer platform. *Financial Management, 48*(4), 1009–1029. doi:10.1111/fima.12295

Jalal, R. N.-U.-D., Alon, I., & Paltrinieri, A. (2021). A bibliometric review of cryptocurrencies as a financial asset. *Technology Analysis and Strategic Management, 1*(1), 1–16. doi:10.1080/09537325.2021.1939001

James, M. X., Hu, Z., & Leonce, T. E. (2019). Predictors of organic tea purchase intentions by Chinese consumers: Attitudes, subjective norms and demographic factors. *Journal of Agribusiness in Developing and Emerging Economies, 9*(3), 202–219. doi:10.1108/JADEE-03-2018-0038

Jamwal, A., Agrawal, R., Sharma, M., & Giallanza, A. (2021). Industry 4.0 technologies for manufacturing sustainability: A systematic review and future research directions. *Applied Sciences (Basel, Switzerland), 11*(12), 5725. doi:10.3390/app11125725

Jang, J., & Choi, J., Jeon, H., & Kang, J. (. (2019). Understanding US travellers' motives to choose Airbnb: A comparison of business and leisure travellers. *International Journal of Tourism Sciences, 19*(3), 192–209. doi:10.1080/15980634.2019.1664006

Javaid, M., Haleem, A., Singh, R. P., & Suman, R. (2021). Substantial capabilities of robotics in enhancing industry 4.0 implementation. *Cognitive Robotics, 1*, 58–75. doi:10.1016/j.cogr.2021.06.001

Javaid, M., Haleem, A., Singh, R. P., & Suman, R. (2022). Artificial intelligence applications for industry 4.0: A literature-based study. *Journal of Industrial Integration and Management, 7*(01), 83–111. doi:10.1142/S2424862221300040

Jenkins, J. R. (2008). Taiwanese Private University EFL Students' Reticence in Speaking Language. *Taiwan Journal of TESOL, 5*(1), 61–93.

Jen, W., Lu, T., & Liu, P. T. (2009). An integrated analysis of technology acceptance behaviour models: Comparison of three major models. *MIS REVIEW: An International Journal, 15*(1), 89–121.

Jeon, H. J., Youn, H. C., Ko, S. M., & Kim, T. H. (2022). Blockchain and AI Meet in the Metaverse. *Advances in the Convergence of Blockchain and Artificial Intelligence, 73.* 10.5772/intechopen.99114

Jiang, C., Ying, W., & Lilong, Z. (2021). Research on the influence of e-commerce online Red self presentation on consumers' purchase behavior and its mechanism. *Journal of Chongqing Industrial and Commercial University,* 1-14. https://kns.cnki.net/kcms/detail/50.1154.c.20210506.1637.004.html

Jiang, J., Hsu, M., Klein, G., & Lin, B. (2000). E-commerce user behaviour model: An empirical study. *Human Systems Management, 19*(4), 265–276. doi:10.3233/HSM-2000-19406

Jia, Y. L., & Yang, Y. L. (2020). The Development of a System for Assessing the Quality of Talent Training in Higher Vocational Education. *Shandong Chemical Industry, 49*, 174–175.

Jie, L. (2021). Influence of online live broadcasting on Fujian tea marketing. *Fujian tea, 43*(10), 42-43

Jie, R. & Peng, Z. (2019). College Students' irrational herd consumption -- from the perspective of social psychology. *Knowledge economy*, (18), 60-61 doi:. zsjj. 2019.18.031. doi:10.15880/j.cnki

Jin, C. C., Seong, L. C., & Khin, A. A. (2019). Factors affecting the consumer acceptance towards fintech products and services in Malaysia. *International Journal of Asian Social Science, 9*(1), 59–65. doi:10.18488/journal.1.2019.91.59.65

Jing, H. (2019). Research on differentiated innovation of webcast platform. *China newspaper industry,* (06), 12-13. doi:. cni. 2019.06.005. doi:10.13854/j.cnki

Jingmei, W. (2020). Analysis on business development model of webcast platform. *News research guide, 11*(09), 214-215

Johnson, J. (2022a). Metaverse - statistics & facts. *Statista.* https://www.statista.com/topics/8652/metaverse/#dossierContents__outerWrapper

Johnson, J. (2022b). Countries with the highest number of internet users 2022. *Statista.* https://www.statista.com/statistics/262966/number-of-internet-users-in-selected-countries/

Johnson, J. (2022c). Benefits of the metaverse worldwide 2021. Statista. Retrieved from https://www.statista.com/statistics/1285117/metaverse-benefits/

Joshua, J. (2017). Information Bodies: Computational Anxiety in Neal Stephenson's Snow Crash. *Interdisciplinary Literary Studies, 19*(1), 17–47. doi:10.5325/intelitestud.19.1.0017

Jun, F., Chen, T., & Qing, Z. (2021). The impact of the host's interaction strategy on the audience's willingness to reward in different types of live broadcast scenes *Nankai Management Review*, *24*(06), 195–204.

Jungherr, A., & Schlarb, D. B. (2022). The Extended Reach of Game Engine Companies: How Companies Like Epic Games and Unity Technologies Provide Platforms for Extended Reality Applications and the Metaverse. *Social Media+ Society, 8*(2), 20563051221107641.

Jung, W. K., Kim, D. R., Lee, H., Lee, T. H., Yang, I., Youn, B. D., Zontar, D., Brockmann, M., Brecher, C., & Ahn, S. H. (2021). Appropriate smart factory for SMEs: Concept, application and perspective. *International Journal of Precision Engineering and Manufacturing*, *22*(1), 201–215. doi:10.100712541-020-00445-2

Junjun, C., Huiwen, L., Ruijing, S., & Xiaolu, Z. J. (2020). Investigation and Research on the use of College Students' webcast platform -- Taking an agricultural university as an example. *Rural economy and science and technology, 31*(1), 361-362

Ju, P. (2021). Research on the Impact of Live Broadcast on College Students' Values and Guiding Countermeasures.. *News Research Guide*, *12*(09), 58–59.

Ju, P. (2021). Research on the Influence of Webcasting on College Students' Values and Guiding Countermeasures.. *Journal of News Research*, *12*(09), 58–59.

Kahle, J. H., Marcon, É., Ghezzi, A., & Frank, A. G. (2020). Smart Products value creation in SMEs innovation ecosystems. *Technological Forecasting and Social Change*, *156*, 120024. doi:10.1016/j.techfore.2020.120024

Kai, S., Liu, L., & Liu, C. (2022). Impulsive purchase intention of live broadcast e-commerce consumers from an emotional perspective.. *China's Circulation Economy*, *36*(01), 33–42. doi:10.14089/j.cnki.cn11-3664/f.2022.01.004

Kalia, P., Arora, D. R., & Kumalo, S. (2016). E-service quality, consumer satisfaction and future purchase intentions in e-retail. *e-Service Journal*, *10*(1), 24. doi:10.2979/eservicej.10.1.02

Kalsoom, T., Ahmed, S., Rafi-ul-Shan, P. M., Azmat, M., Akhtar, P., Pervez, Z., Imran, M. A., & Ur-Rehman, M. (2021). Impact of IoT on Manufacturing Industry 4.0: A new triangular systematic review. *Sustainability*, *13*(22), 12506. doi:10.3390u132212506

Kalsoom, T., Ramzan, N., Ahmed, S., & Ur-Rehman, M. (2020). Advances in sensor technologies in the era of smart factory and industry 4.0. *Sensors (Basel)*, *20*(23), 6783. doi:10.339020236783 PMID:33261021

Kamal, S. A., Shafiq, M., & Kakria, P. (2020). Investigating acceptance of telemedicine services through an extended technology acceptance model (TAM). *Technology in Society*, *60*, 101212. doi:10.1016/j.techsoc.2019.101212

Kamińska, D., Sapiński, T., Wiak, S., Tikk, T., Haamer, R. E., Avots, E., & Anbarjafari, G. (2019). Virtual reality and its applications in education: Survey. *Information (Switzerland)*. doi:10.3390/info10100318

Kang, H. (2013). The prevention and handling of the missing data. *Korean Journal of Anesthesiology*, *64*(5), 402. doi:10.4097/kjae.2013.64.5.402 PMID:23741561

Kaplan, A. D., Cruit, J., Endsley, M., Beers, S. M., Sawyer, B. D., & Hancock, P. A. (2021). The effects of virtual reality, augmented reality, and mixed reality as training enhancement methods: A meta-analysis. *Human Factors*, *63*(4), 706–726. doi:10.1177/0018720820904229 PMID:32091937

Karandikar, N., Chakravorty, A., & Rong, C. (2021). Blockchain based transaction system with fungible and non-fungible tokens for a community-based energy infrastructure. *Sensors (Basel)*, *21*(11), 3822. doi:10.339021113822 PMID:34073110

Katrina, B., & Benedict, L. (2019). *Why Are Malaysians Shopping Online and What's Their Shopping Behaviour Like?* Janio. https://janio.asia/articles/why-are-malaysians-shopping-online/

Katz, E., Blumer, J. G., & Gurevitch, M. (1974). Uses and gratifications research. *Public Opinion Quarterly*, *37*(4), 509–524. doi:10.1086/268109

Kaur, P., Stoltzfus, J., & Yellapu, V. (2018). Descriptive statistics. *International Journal of Academic Medicine*, *4*(1), 60. doi:10.4103/IJAM.IJAM_7_18

Kaushik, M. (2022). Artificial Intelligence (Ai). In Intelligent System Algorithms and Applications in Science and Technology, (pp. 119-133). Apple Academic Press.

Kayyali, A. (Feb 6, 2022). How the Metaverse will Alter the World's Direction. *Insidetelecom*. https://www.insidetelecom.com/how-the-metaverse-will-alter-the-worlds-direction/

Keleko, A. T., Kamsu-Foguem, B., Ngouna, R. H., & Tongne, A. (2022). Artificial intelligence and real-time predictive maintenance in industry 4.0: a bibliometric analysis. *AI and Ethics*, 1-25.

Kemp, J., & Livingstone, D. (2006). Putting a Second Life "metaverse" skin on learning management systems. *Proceedings of the First Second Life Education Workshop, Part of the 2006 Second Life Community Convention, 20.*

Khairani, N. A., & Rajagukguk, J. (2019, December). Development of Moodle E-Learning Media in Industrial Revolution 4.0 Era. In *4th Annual International Seminar on Transformative Education and Educational Leadership (AISTEEL 2019),* (pp. 559-565). Atlantis Press.

Khan, S. A. R., Godil, D. I., Jabbour, C. J. C., Shujaat, S., Razzaq, A., & Yu, Z. (2021). Green data analytics, blockchain technology for sustainable development, and sustainable supply chain practices: Evidence from small and medium enterprises. *Annals of Operations Research*, 1–25. doi:10.100710479-021-04275-x

Khan, S. N., Loukil, F., Ghedira-Guegan, C., Benkhelifa, E., & Bani-Hani, A. (2021). Blockchain smart contracts: Applications, challenges, and future trends. *Peer-to-Peer Networking and Applications*, *14*(5), 2901–2925. doi:10.100712083-021-01127-0 PMID:33897937

Kharpal, A. (Feb 14, 2022). China's tech giants push toward an $8 trillion metaverse opportunity. Tech drivers. *CNBC*. https://www.cnbc.com/2022/02/14/china-metaverse-tech-giants-latest-moves-regulatory-action.html

Khazaei, H. (2020). Integrating cognitive antecedents to UTAUT model to explain adoption of blockchain technology among Malaysian SMEs. *JOIV: International Journal on Informatics Visualization*, *4*(2), 85–90. doi:10.30630/joiv.4.2.362

Kian, T. P., Loong, A. C. W., & Fong, S. W. L. (2018). Customer Purchase Intention on Online Grocery Shopping. *International Journal of Academic Research in Business and Social Sciences*, *8*(12). doi:10.6007/IJARBSS/v8-i12/5260

Kim, D. J., Ferrin, D. L., & Rao, H. R. (2009). Trust and Satisfaction, Two Stepping Stones for Successful E-Commerce Relationships: A Longitudinal Exploration. *Information Systems Research*, *20*(2), 237–257. doi:10.1287/isre.1080.0188

Kim, D., & Chang, H. (2007). Key functional characteristics in designing and operating health information websites for user satisfaction: An application of the extended technology acceptance model. *International Journal of Medical Informatics*, *76*(11-12), 790–800. doi:10.1016/j.ijmedinf.2006.09.001 PMID:17049917

Kim, H., Kwon, Y. T., Lim, H. R., Kim, J. H., Kim, Y. S., & Yeo, W. H. (2021). Recent advances in wearable sensors and integrated functional devices for virtual and augmented reality applications. *Advanced Functional Materials*, *31*(39), 2005692. doi:10.1002/adfm.202005692

Kim, J. (2021). Advertising in the Metaverse: Research Agenda. *Journal of Interactive Advertising*, *21*(3), 141–144. do i:10.1080/15252019.2021.2001273

Kim, J., Jin, B., & Swinney, J. L. (2009). The role of etail quality, e-satisfaction and e-trust in online loyalty development process. *Journal of Retailing and Consumer Services*, *16*(4), 239–247. doi:10.1016/j.jretconser.2008.11.019

Kim, S., Williams, R., & Lee, Y. (2004). Attitude Toward Online Shopping and Retail Website Quality. *Journal of International Consumer Marketing*, *16*(1), 89–111. doi:10.1300/J046v16n01_06

Kingaby, S. A. (2022). The Stock Market API. *Data-Driven Alexa Skills*, 387–404. doi:10.1007/978-1-4842-7449-1_16

Kiong, L. V. (2022). *Metaverse Made Easy: A Beginner's Guide to the Metaverse: Everything you need to know about Metaverse*. NFT and GameFi.

Kline, R. B. (2011). *Principles and practice of structural equation modelling,* (3rd ed.). Guilford Press.

Klopfer, E. (2008). *Augmented learning: Research and design of mobile educational games*. MIT press. doi:10.7551/mitpress/9780262113151.001.0001

Knox, J. (2022). The Metaverse, or the Serious Business of Tech Frontiers. *Postdigital Science and Education*, *4*(2), 207–215. doi:10.100742438-022-00300-9

Kocur, M., Henze, N., & Schwind, V. (2021). *The Extent of the Proteus Effect as a Behavioral Measure for Assessing User Experience in Virtual Reality*, 4.

Ko, T., Lee, J., & Ryu, D. (2018). Blockchain technology and manufacturing industry: Real-time transparency and cost savings. *Sustainability*, *10*(11), 4274. doi:10.3390u10114274

Koul, S., & Eydgahi, A. (2018). Utilizing technology acceptance model (TAM) for driverless car technology adoption. *Journal of Technology Management & Innovation*, *13*(4), 37–46. doi:10.4067/S0718-27242018000400037

Kozinets, R. V. (2022). Immersive netnography: A novel method for service experience research in virtual reality, augmented reality and metaverse contexts. *Journal of Service Management*. doi:10.1108/JOSM-12-2021-0481

Kraus, S., Kanbach, D. K., Krysta, P. M., Steinhoff, M. M., & Tomini, N. (2022). Facebook and the creation of the Metaverse: Radical business model innovation or incremental transformation? *International Journal of Entrepreneurial Behaviour & Research*, *28*(9), 52–77. doi:10.1108/IJEBR-12-2021-0984

Kress, B. C., & Peroz, C. (2020, February). Optical architectures for displays and sensing in augmented, virtual, and mixed reality (AR, VR, MR). *Proceedings of the Society for Photo-Instrumentation Engineers*, *11310*, 1131001.

Krueger, M. W., Gionfriddo, T., & Katrin, H. (1985). Videoplace - An Artifical Reality. *Chi '85 Proceedings*, (pp. 35-40).

Kumar, A., Pujari, P., & Gupta, N. (2021). Artificial Intelligence: Technology 4.0 as a solution for healthcare workers during COVID-19 pandemic. *Acta Universitatis Bohemiae Meridionales*, *24*(1), 19–35. doi:10.32725/acta.2021.002

Kumar, S., Lim, W. M., Sivarajah, U., & Kaur, J. (2022). Artificial intelligence and blockchain integration in business: Trends from a bibliometric-content analysis. *Information Systems Frontiers*, 1–26. doi:10.100710796-022-10279-0 PMID:35431617

Kye, B., Han, N., Kim, E., Park, Y., & Jo, S. (2021). Educational applications of metaverse: Possibilities and limitations. *Journal of Educational Evaluation for Health Professions*, *18*, 32. doi:10.3352/jeehp.2021.18.32 PMID:34897242

Laeeq, K. (2022). *Metaverse: Why, How and What.* https://www.researchgate.net/publication/358505001_Metaverse_Why_How_and_

Laeeq, K. (2022). *Metaverse: Why*. How, and What.

Lai, P. C. (2020). Intention to use a drug reminder app: a case study of diabetics and high blood pressure patients *SAGE Research Methods Cases*. doi:10.4135/9781529744767

Lai, P. C., & Tong, D. L. (2022). An Artificial Intelligence-Based Approach to Model User Behavior on the Adoption of E-Payment. Handbook of Research on Social Impacts of E-Payment and Blockchain, 1-15.

Lai, P. C., & Tong, D. L. (2022). An Artificial Intelligence-Based Approach to Model User Behavior on the Adoption of E-Payment. Handbook of Research on Social Impacts of E-Payment and Blockchain. IGI Global.

Lai, P. C., Toh, E. B. H. & Alkhrabsheh, A. A. (2020), Empirical Study of Single Platform E-Payment in South East Asia. *Strategies and Tools for Managing Connected Consumers*, 252-278

Lai, P. C., Toh, E. B. H. & Alkhrabsheh, A. A. (2020). Empirical Study of Single Platform E-Payment in South East Asia. *Strategies and Tools for Managing Connected Consumers,* 252-278

Lai, C. P., & Zainal, A. A. (2015). Consumers' intention to use a single platform e-payment system: A study among Malaysian internet and mobile banking users. *Journal of Internet Banking and Commerce*, *20*(1), 1–13.

Lai, E., & Wang, Z. (2012). An empirical research on factors affecting customer purchasing behavior tendency during online shopping. *IEEE International Conference on Computer Science and Automation Engineering*. 10.1109/ICSESS.2012.6269534

Lai, P. C. (2016). *Smart Healthcare @ the palm of our hand.* ResearchAsia.

Lai, P. C. (2016). *SMART Healthcare @ the palm of your hand.* ResearchAsia.

Lai, P. C. (2016). *SMART HEALTHCARE @ the palm of your hand.* ResearchAsia.

Lai, P. C. (2018). Research, Innovation and Development Strategic Planning for Intellectual Property Management. *Economic Alternatives*, *3*(12), 303–310.

Lai, P. C. (2018). Research, Innovation and Development Strategic Planning for Intellectual Property Management. *Economic Alternatives.*, *12*(3), 303–310.

Lai, P. C. (2019). *Factors That Influence the tourists' or Potential Tourists' Intention to Visit and the Contribution to the Corporate Social Responsibility Strategy for Eco-Tourism. International Journal of Tourism and Hospitality Management in the Digital Age*. IGI-Global.

Lai, P. C., & Liew, E. J. (2021). Towards a Cashless Society: The Effects of Perceived Convenience and Security on Gamified Mobile Payment Platform Adoption. *AJIS. Australasian Journal of Information Systems*, *25*. doi:10.3127/ajis.v25i0.2809

Lai, P. C., & Liew, E. J. Y. (2021). *Towards a Cashless Society: The Effects of Perceived Convenience and Security on Gamified E-Payment Platform Adoption*. Australasian.

Lai, P. C., & Lim, C. S. (2019). *The Effects of Efficiency, Design and Enjoyment on Single Platform E-payment Research in Business and Management*, *2*(6), 19–34.

Lai, P. C., & Lim, C. S. (2019). The Effects of Efficiency, Design and Enjoyment on Single Platform E-payment. Research in Business and Management, *6*(2), 19–34.

Lai, P. C., & Tong, D. L. (2022). *An Artificial Intelligence-Based Approach to Model User Behavior on the Adoption of E-Payment. Handbook of Research on Social Impacts of E-Payment and Blockchain*. IGI Global.

Lai, P. C., & Zainal, A. A. (2015). Perceived Risk as an Extension to TAM Model: Users' Intention To Use A Single Platform E-Payment. *Australian Journal of Basic and Applied Sciences*, *9*(2), 323–330.

Lan, L., Jingwen, L., & Yinuo, W. (2019). Investigation on the impact of webcast marketing on College Students' consumption. *China business theory,* (13), 87-90 doi:. issn2096-0298.2019.13.087. doi:10.19699/j.cnki

Lan, Z. & Huicui, Q. (2021). Research on the impact of webcast based on content marketing on consumers' purchase intention. *Mall modernization*, (14), 121-123 doi:. scxdh. 2021.14.042. doi:10.14013/j.cnki

Lanier, J. (1992). Virtual reality: The promise of the future. *Interactive Learning International*, *8*(4), 275–279.

Lanzini, F., Ubacht, J., & De Greeff, J. (2021). Blockchain adoptioin factors for SMEs in supply chain management. *Journal of Supply Chain Management Science*, *2*(1-2), 47–68.

Lastowka, G. (2010). *Virtual justice*. Yale University Press.

Latikka, R., Turja, T., & Oksanen, A. (2019). Self-efficacy and acceptance of robots. *Computers in Human Behavior*, *93*, 157–163. doi:10.1016/j.chb.2018.12.017

Lee, J. Y. (2021). *A Study on Metaverse Hype for Sustainable Growth. 10*(3), 72–80.

Lee, J. Y. (2021). A Study on Metaverse Hype for Sustainable Growth. *International Journal of Advanced Smart Convergence*, 72-80.

Lee, K. W. (2000). English Teachers' Barriers to the Use of Computer-Assisted Language Learning. *The Internet TESL Journal, 6*(12).

Lee, L. H., Braud, T., Zhou, P., Wang, L., Xu, D., Lin, Z., & Hui, P. (2021). All one needs to know about metaverse: A complete survey on technological singularity, virtual ecosystem, and research agenda.

Lee, L. H., Braud, T., Zhou, P., Wang, L., Xu, D., Lin, Z., & Hui, P. (2021). All one needs to know about Metaverse: A complete survey on technological singularity, virtual ecosystem, and research agenda.

Lee, L.-H., Braud, T., Wang, L., Xu, D., Kumar, A., Bermejo, C., & Hui, P. (2021). All One Needs to Know About MetaverseL A Complete Survey on Technological Singularity, Virtual Ecosystem, and Research Agenda. *Journal of Latex Class Files*, 1-66.

Lee, L.H., Braud, T., Zhou, P., Wang, L., Xu, D., Lin, Z., Kumar, A., Bermejo, C. & Hui, P. (2021). All one needs to know about metaverse: A complete survey on technological singularity, virtual ecosystem, and research agenda.

Lee, O. (2022). When Worlds Collide: Challenges and Opportunities in Virtual Reality. *Embodied: The Stanford Undergraduate Journal of Feminist, Gender, and Sexuality Studies, 1*(1).

Lee, B.-K. (2021). The Metaverse World and Our Future. *Review of Korea Contents Association, 19*(1), 13–17. https://www.koreascience.or.kr/article/JAKO202119759273785.pdf

Leedy, P. D., Ormrod, J. E., & Johnson, L. R. (2010). Practical research: planning and design (9th ed.). New York Pearson Education, Inc.

Lee, H. (2019). Real-time manufacturing modeling and simulation framework using augmented reality and stochastic network analysis. *Virtual Reality (Waltham Cross), 23*(1), 85–99. doi:10.100710055-018-0343-6

Lee, H., Woo, D., & Yu, S. (2022). Virtual Reality Metaverse System Supplementing Remote Education Methods: Based on Aircraft Maintenance Simulation. *Applied Sciences (Basel, Switzerland), 12*(5), 2667. doi:10.3390/app12052667

Lee, J. (2021). *Blockchain 3.0 Era and Future of Cryptocurrency; Future Info Graphics*. Research Center for New Industry Strategy.

Lee, M. K. O., Shi, N., Cheung, C. M. K., Lim, K. H., & Sia, C. L. (2011). Consumer's decision to shop online: The moderating role of positive informational social influence. *Information & Management, 48*(6), 185–191. doi:10.1016/j.im.2010.08.005

Leenes, R. (2007, August). Privacy in the Metaverse. In *IFIP International Summer School on the Future of Identity in the Information Society,* (pp. 95–112). Springer.

Lee, S. G., Trimi, S., Byun, W. K., & Kang, M. (2011). Innovation and imitation effects in Metaverse service adoption. *Service Business, 5*(2), 155–172. doi:10.100711628-011-0108-8

Lee, T., & Jun, J. (2007). The role of contextual marketing offer in Mobile Commerce acceptance: Comparison between Mobile Commerce users and nonusers. *International Journal of Mobile Communications, 5*(3), 339. doi:10.1504/IJMC.2007.012398

Lee, Y. H., Hsieh, Y. C., & Hsu, C. N. (2011). Adding innovation diffusion theory to the technology acceptance model: Supporting employees' intentions to use e-learning systems. *Journal of Educational Technology & Society, 14*(4), 124–137.

Lee, Y., & Kozar, K. A. (2008). An empirical investigation of anti-spyware software adoption: A multitheoretical perspective. *Information & Management, 45*(2), 109–119. doi:10.1016/j.im.2008.01.002

Lehdonvirta, V. (2009). Virtual item sales as a revenue model: Identifying attributes that drive purchase decisions. *Electronic Commerce Research, 9*(1–2), 97–113. doi:10.100710660-009-9028-2

Lei, C., & Feng, L. (2017). Research on consumers' willingness to continue to use in the socialized e-commerce environment -- from the perspective of continuous trust. *Science, technology and economy, 30*(06), 61-65 doi:. cn32-1276n. 2017.06.013. doi:10.14059/j.cnki

Leng, J., Ruan, G., Jiang, P., Xu, K., Liu, Q., Zhou, X., & Liu, C. (2020). Blockchain-empowered sustainable manufacturing and product lifecycle management in industry 4.0: A survey. *Renewable & Sustainable Energy Reviews, 132*, 110112. doi:10.1016/j.rser.2020.110112

Lewis, S. (2008). Where young adults intend to get news in five years. *Newspaper Research Journal, 29*(4), 36–5. http://findarticles.com/p/ articles/mi_qa3677/ is_200810/ ai _n39229321/. doi:10.1177/073953290802900404

Li, S. X., & Jiang, L. J. (2019). How to enhance colleges' connotation construction in the New Era. *Chinese University Technology*, 70–72.

Li, T. (2019). *Variance Inflation Factor Definition*. Investopedia. https://www.investopedia.com/terms/v/variance-inflation-factor.asp

Liang, B., Jianchao, Z., & Zhijie, D. (2011). Three-dimensional reflection on the ideological and political education of college students in the era of webcasting. *Jiangsu Higher Education,* (11), 95-98. doi: . doi:10.13236/j.cnki.jshe.2021.11.016

Lian, X. (2021). The influence of "Internet popularity" on college students and its countermeasures.. *Journal of Zunyi Normal University, 23*(05), 122–125.

Li, D. Y., Li, J. F., & Yan, D. (2019). The Talent Training Model of Higher Vocational Education in China: Issues and Reforms. *Hebei Vocational Education, 1*, 30–34.

Lidan, C. & Zhiwei, X. (2015). Zhou Liang's Commentary from the perspective of online collective action. *Journalist,* (01), 74-78. doi:10.16057/j.cnki.31-1171/g2.2015.01.011

Li, J. (2017). On the Optimization Control of Visual Elements in UI Design.. *Art Science and Technology, 30*(11), 120–121.

Lik-Hang, L., Braud, T., Zhou, P., Wang, L., Xu, D., Lin, Z., & Hui, P. (2021). All One Needs to Know about Metaverse: A Complete Survey on Technological Singularity, VIrtual Ecosystem, and Research Agenda. *Journal of LATEX Class*, 1-66.

Li, L., & Buhalis, D. (2006). E-commerce in China: The case of travel. *International Journal of Information Management, 26*(2), 153–166. doi:10.1016/j.ijinfomgt.2005.11.007

Lili, L., Xinyue, C., & Shi, C. (2021). Research on e-commerce live broadcasting mode. *China's collective economy,* (s33), 65-68

Lim, C. H., Lim, S., How, B. S., Ng, W. P. Q., Ngan, S. L., Leong, W. D., & Lam, H. L. (2021). A review of industry 4.0 revolution potential in a sustainable and renewable palm oil industry: HAZOP approach. *Renewable & Sustainable Energy Reviews, 135*, 110223. doi:10.1016/j.rser.2020.110223

Lim, W. M., & Ting, D. H. (2012). E-shopping: An Analysis of the Technology Acceptance Model. *Modern Applied Science, 6*(4). doi:10.5539/mas.v6n4p49

Ling, H. (2017). Saving Campus Loans - In depth Scanning of Campus Network Consumption Loans. *China Credit,* (06): 16-27

Ling, L., Meng, Q., Hong, W., Yanzhao, W., Hao, H., Darong, L. (2020). Analysis on the influence of "comparison psychology" on College Students' consumption psychology and behavior -- Taking the research of Shengli University as an example. *Value engineering, 39*(15), 38-40 doi:. cn13-1085/n.2020.15.018. doi:10.14018/j.cnki

Lingrong, L., & Lan, W. (2021). Research on the impact of consumer purchase decisions from the perspective of online celebrity live broadcast with goods. *Modern Commerce,* (28), 48-50. doi:10.14097/j.cnki.5392/202128.015

Lingrong, L., & Lan, W. (2021). Research on the impact of consumers' purchase decision from the perspective of online live broadcasting. *Modern business*, (28), 48-50 doi:. 5392/2021.28.015. doi:10.14097/j.cnki

Linhao, Z. (2013). *Sentiment analysis on Twitter with stock price and significant keyword correlation*. Texas ScholarWorks. https://repositories.lib.utexas.edu/handle/2152/20057

Lin, Y., Wang, M., & Ma, T. (2020). The impact of online shopping on physical retail — a survey based on the business center of Shanghai Southern Mall.. *World Geographic Research*, *29*(03), 568–578.

Liu, X., & Liu,W. K. (2019). The Development of Higher Vocational Education in the Context of Recruitment Expansion: Challenges and Countermeasures. *Education and occupation, 14*, 5-11.

Liu, Z. B. (2018). Strategy for Improving the Quality of Higher Vocational Education Talent Training . *Chinese and international entrepreneurs, 31,* 165.

Liu, H., Bowman, M., Adams, R., Hurliman, J., & Lake, D. (2010). Scaling Virtual Worlds: Simulation Requirements and Challenges. *Winter Simulation Conference*, (pp. 778-790). 10.1109/WSC.2010.5679112

Liu, L., & Fang, C. C. (2020). Accelerating the Social Media Process: The Impact of Internet Celebrity Word-of-Mouth Communication and Relationship Quality on Consumer Information Sharing.. *International Journal of Human Resource Studies*, *10*(1), 201222–201222. doi:10.5296/ijhrs.v10i1.16043

Liu, X. J., & Wen, Y. F. (2022). The New Era's Implications for Talent Development in Higher Vocational Colleges. *Education and Career*, *1*, 58–63.

Liu, X., & Qian, J. N. (2020). Major Construction and Industrial Development in Vocational Education: Aligning Logic and Theoretical Framework. *Higher Engineering Education Research*, *2*, 142–147.

Li, Y. (2016). Empirical Study of Influential Factors of Online Customers' Repurchase Intention. *IBusiness*, *08*(03), 48–60. doi:10.4236/ib.2016.83006

Li, Y. Y. (2003). The "four services" should govern higher vocational education reform. *Shandong Youth Political College Journal*, *4*, 76–77.

Li, Y., Xu, Z., & Xu, F. (2018). Perceived control and purchase intention in online shopping: The mediating role of self-efficacy. *Social Behavior and Personality*, *46*(1), 99–105. doi:10.2224bp.6377

Li, Z. (2021). *Guide online live broadcast of college students with socialist core values*. Guizhou Normal University. doi:10.27048/d.cnki.ggzsu.2021.000042

Li, Z. H. (2018). Theoretical foundations and methods for my country's industrial development's "three major changes". *Reform*, *9*, 91–101.

Li, Z., Chen, S., & Zhou, B. (2021). Electric vehicle peer-to-peer energy trading model based on smes and blockchain. *IEEE Transactions on Applied Superconductivity*, *31*(8), 1–4. doi:10.1109/TASC.2021.3091074

Loper, E., & Bird, S. (2002). *NLTK: The Natural Language Toolkit*. http://nltk.sf.net/

Louro, D., Fraga, T., & Pontuschka, M. (2009). Metaverse: Building Affective Systems and Its Digital Morphologies in Virtual Environments. *Journal of Virtual Worlds Research*, *2*(5). doi:10.4101/jvwr.v2i5.950

Lucas, T., Alexander, S., Firestone, I. J., & Baltes, B. B. (2006). Self-efficacy and independence from social influence: Discovery of an efficacy–difficulty effect. *Social Influence*, *1*(1), 58–80. doi:10.1080/15534510500291662

Lumpur, K. (2020). Malaysia's online retail sales up 28.9% in April. *The Star*. Www.thestar.com.my. https://www.thestar.com.my/business/business-news/2020/06/11/malaysia039s-online-retail-sales-up-289--in-april

Lune, H., & Berg, B. L. (2017). *Qualitative research methods for the social sciences*. Pearson.

Luo, S., & Choi, T. M. (2022). Operational research for technology-driven supply chains in the industry 4.0 Era: Recent development and future studies. *Asia-Pacific Journal of Operational Research*, *39*(01), 2040021. doi:10.1142/S0217595920400217

Luo, X., Li, H., Zhang, J., & Shim, J. P. (2010). Examining multi-dimensional trust and multifaceted risk in initial acceptance of emerging technologies: An empirical study of mobile banking services. *Decision Support Systems*, *49*(2), 222–234. doi:10.1016/j.dss.2010.02.008

Lu, S. (2016). The Impact of Social Utility Software on College Students' Interpersonal Relations — Taking Live Online Software as an Example.. *Digital Media Research*, *33*(10), 42–46.

Lu, S. S. (2019). The Development of a System for Assessing the Quality of Higher Vocational Talent Training. *Henan Education Education*, *1*, 53–54.

Lu, Y. (2017). Industry 4.0: A survey on technologies, applications and open research issues. *Journal of Industrial Information Integration*, *6*, 1–10. doi:10.1016/j.jii.2017.04.005

Lu, Y., Yang, S., Chau, P. Y., & Cao, Y. (2011). Dynamics between the trust transfer process and intention to use mobile payment services: A cross-environment perspective. *Information & Management*, *48*(8), 393–403. doi:10.1016/j.im.2011.09.006

Lynn, M. (1991). Scarcity effects on value: A quantitative review of the commodity theory literature. *Psychology and Marketing*, *8*(1), 43–57. doi:10.1002/mar.4220080105

Ma, W. (2022). China's tech giants in the metaverse. *CNBC*. https://www.cnbc.com/2022/02/14/china-metaverse-tech-giants-latest-moves-regulatory-action.html

MacCallum, K., & Parsons, D. (2019, September). Teacher perspectives on mobile augmented reality: The potential of metaverse for learning. In *World Conference on Mobile and Contextual Learning*, (pp. 21-28).

Macionis, J. J. (2006). *Society: The Basics* (8th ed.). Pearson/Prentice Hall.

Malaysian Communications and Multimedia Commission. (2018). E-Commerce Consumers Survey 2018. *MCMC*. https://www.mcmc.gov.my/skmmgovmy/media/General/pdf/ECS-2018.pdf

Malik, A. A., Masood, T., & Bilberg, A. (2020). Virtual reality in manufacturing: Immersive and collaborative artificial-reality in design of human-robot workspace. *International Journal of Computer Integrated Manufacturing*, *33*(1), 22–37. doi:10.1080/0951192X.2019.1690685

Mandic, V. (2020). Model-based manufacturing system supported by virtual technologies in an Industry 4.0 context. In *Proceedings of 5th International Conference on the Industry 4.0 Model for Advanced Manufacturing*, (pp. 215-226). Springer, Cham. 10.1007/978-3-030-46212-3_15

Manis, K. T., & Choi, D. (2019). The virtual reality hardware acceptance model (VR-HAM): Extending and individuating the technology acceptance model (TAM) for virtual reality hardware. *Journal of Business Research*, *100*, 503–513. doi:10.1016/j.jbusres.2018.10.021

Maohua, C. (2021). Research on the induction and strategy of capital logic on Contemporary College Students' consumption behavior. *Journal of Mianyang Normal University,40*(12): 15-20. doi:.cn51-1670/g.2021.12.003. doi:10.16276/j.cnki

Mao, J., Chiu, C., Owens, B. P., Brown, J. A., & Liao, J. (2019). Growing followers: Exploring the effects of leader humility on follower self-expansion, self-efficacy, and performance. *Journal of Management Studies, 56*(2), 343–371. doi:10.1111/joms.12395

Maouchi, Y., Charfeddine, L., & El Montasser, G. (2022). Understanding digital bubbles amidst the COVID-19 pandemic: Evidence from DeFi and NFTs. *Finance Research Letters, 47*(1), 102584. doi:10.1016/j.frl.2021.102584

Marczewski, A. (2013). *Gamification: a simple introduction.* Andrzej Marczewski.

Mares, R. (Ed.). (2011). *The UN guiding principles on business and human rights: foundations and implementation.* Martinus Nijhoff Publishers.

Margaret, S., Barbara, A., & Elisabeth, P. (2003). *Total Physical Response.* TPR.

Marini, A., Nafisah, S., Sekaringtyas, T., Safitri, D., Lestari, I., Suntari, Y., Umasih, Sudrajat, A., & Iskandar, R. (2022). Mobile Augmented Reality Learning Media with Metaverse to Improve Student Learning Outcomes in Science Class. *International Journal of Interactive Mobile Technologies, 16*(7), 99–115. doi:10.3991/ijim.v16i07.25727

Martín, G. F. (2018). Social and psychological impact of musical collective creative processes in virtual environments; Te Avatar Orchestra Metaverse in Second Life. Musica/Tecnologia Music. *Technology*, 75.

Maryanti, N., Rohana, R., & Kristiawan, M. (2020). The principal's strategy in preparing students ready to face the industrial revolution 4.0. *International Journal of Educational Research, 2*(1), 54–69.

Maselli, A., & Slater, M. (2013). The building blocks of the full body ownership illusion. *Frontiers in Human Neuroscience, 7.* https://www.frontiersin.org/article/10.3389/fnhum.2013.00083. doi:10.3389/fnhum.2013.00083 PMID:23519597

Mastos, T. D., Nizamis, A., Terzi, S., Gkortzis, D., Papadopoulos, A., Tsagkalidis, N., Ioannidis, D., Votis, K., & Tzovaras, D. (2021). Introducing an application of an industry 4.0 solution for circular supply chain management. *Journal of Cleaner Production, 300*, 126886. doi:10.1016/j.jclepro.2021.126886

Matsumoto, D. (2006). Culture and nonverbal behavior. In V. Manusov & M. L. Patterson (Eds.), *The SAGE handbook of nonverbal communication,* (pp. 219–235). Sage Publications, Inc., doi:10.4135/9781412976152.n12

Maurer, M. (2022, January 7). Accounting Firms Scoop Up Virtual Land in the Metaverse. *The Wall Street Journal.* https://www.wsj.com/articles/accounting-firms-scoop-up-virtual-land-in-the-metaverse-11641599590

Mayer-Schonberger, V., & Crowley, J. (2006). Napster's Second Life: The Regulatory Challenges of Virtual Worlds. *Nw. UL Rev., 100,* 1775.

McCloskey, D. (2003). Evaluating electronic commerce acceptance with the technology acceptance model. *Journal of Computer Information Systems, 44*(2).

Meng, J. Z. (2021). Chinese Vocational Education has a distinct value and aim . *Latitude theoretical, 6,* 23-28

Mengmeng, D. (2020). A survey on college students' online live broadcast purchase behavior. *Marketing (Theory and Practice).*

Merugula, S., Dinesh, G., Kathiravan, M., Das, G., Nandankar, P., & Karanam, S. R. (2021, March). Study of Blockchain Technology in Empowering the SME. In *2021 International Conference on Artificial Intelligence and Smart Systems (ICAIS),* (pp. 758-765). IEEE. 10.1109/ICAIS50930.2021.9395831

Meta Platforms and Technologies | Meta. (n.d.). Meta Platforms and Technologies. https://about.facebook.com/technologies/

Meyer, K. A. (2005). Planning for cost-efficiencies in online learning. *Planning for Higher Education, 33*, 19–30.

Mihalyi, C. (nd). Adolescence. *Britannica.* https://www.britannica.com/science/adolescence

Miles, M. B., & Huberman, A. M. (1994). *Qualitative data analysis: An expanded sourcebook,* (2nd ed.). Sage Publications, Inc.

Mileva, G. (2022). 50+ Metaverse Statistics-- Market Size & Growth. *Influencer Marketing*. https://influencermarketinghub.com/metaverse-stats/

Milgram, P., Takemura, H., Utsumi, A., & Kishino, F. (1995). Augmented reality: A class of displays on the reality-virtuality continuum. In *Telemanipulator and telepresence technologies* (Vol. 2351, pp. 282–292). International Society for Optics and Photonics. doi:10.1117/12.197321

Ming, T. R., Norowi, N. M., Wirza, R., & Kamaruddin, A. (2021). Designing a Collaborative Virtual Conference Application: Challenges. *Requirements and Guidelines. Future Internet, 13*(10), 253. doi:10.3390/fi13100253

Miraz, M. H., Hasan, M. T., Rekabder, M. S., & Akhter, R. (2022). Trust, transaction transparency, volatility, facilitating condition, performance expectancy towards cryptocurrency adoption through intention to use. *Journal of Management Information and Decision Sciences, 25*, 1–20.

Mittal, S., Khan, M. A., Romero, D., & Wuest, T. (2018). A critical review of smart manufacturing & Industry 4.0 maturity models: Implications for small and medium-sized enterprises (SMEs). *Journal of Manufacturing Systems, 49*, 194–214. doi:10.1016/j.jmsy.2018.10.005

Mofokeng, T. E. (2021). The impact of online shopping attributes on customer satisfaction and loyalty: Moderating effects of e-commerce experience. *Cogent Business & Management, 8*(1), 1968206. doi:10.1080/23311975.2021.1968206

Mohamed Yusoff, A. F., Hashim, A., Muhamad, N., & Wan Hamat, W. N. (2021). Application of fuzzy delphi technique to identify the elements for designing and developing the e-PBM PI-Poli module. [AJUE]. *Asian Journal of University Education, 7*(1), 292–304. doi:10.24191/ajue.v17i1.12625

Mohamed, N., Hussein, R., Zamzuri, N. H. A., & Haghshenas, H. (2014). Insights into individual's online shopping continuance intention. *Industrial Management & Data Systems, 114*(9), 1453–1476. doi:10.1108/IMDS-07-2014-0201

Mohanta, B. K., Jena, D., Satapathy, U., & Patnaik, S. (2020). Survey on IoT security: Challenges and solution using machine learning, artificial intelligence and blockchain technology. *Internet of Things, 11*, 100227. doi:10.1016/j.iot.2020.100227

Mohd Salleh, N. H., Selvaduray, M., Jeevan, J., Ngah, A. H., & Zailani, S. (2021). Adaptation of industrial revolution 4.0 in a seaport system. *Sustainability, 13*(19), 10667. doi:10.3390u131910667

Mohd. Ridhuan, M. J., Zaharah, H., Nurul Rabihah, M. N., Ahmad Arifin, S., & Norlidah, A. (2014). *Application of Fuzzy Delphi Method in Educational Research. Design and Developmental Research.* Pearson Malaysia Sdn. Bhd.

Mok, C. (2022). Metaverse is the future of social network;. *CNBC*. https://www.cnbc.com/2022/02/14/china-metaverse-tech-giants-latest-moves-regulatory-action.html

Moon, J. W., & Kim, Y. G. (2001). Extending the TAM for a World-Wide-Web context. *Information & Management, 38*(4), 217–230. doi:10.1016/S0378-7206(00)00061-6

Morgan, J. P. (2022). Opportunities in the metaverse. *JP Morgan*. https://www.jpmorgan.com/content/dam/jpm/treasury-services/documents/opportunities-in-the-metaverse.pdf

Morrar, R., Arman, H., & Mousa, S. (2017). The fourth industrial revolution (Industry 4.0): A social innovation perspective. *Technology Innovation Management Review*, *7*(11), 12–20. doi:10.22215/timreview/1117

Moy, C., & Gadgil, A. (2022). Opportunities in the Metaverse: How Businesses Can Explore the Metaverse and Navigate the Hype vs. Reality. *Erişim Tarihi, 23*(02). https://www. jpmorgan. com/content/dam/jpm/treasury-services /documents/opportunities-in-the-metaverse. pdf .

Moy, C., & Gadgil, A. (2022). Opportunities in the Metaverse: How Businesses Can Explore the Metaverse and Navigate the Hype vs. Reality. *Opportunities in the metaverse, 23*(02).

Mubin, S. A., Thiruchelvam, V., & Andrew, Y. W. (2020). Extended Reality: How They Incorporated for ASD Intervention. In *2020 8th International Conference on Information Technology and Multimedia, ICIMU 2020*. doi:10.1109/ICIMU49871.2020.9243332

Müller, J. M., Kiel, D., & Voigt, K. I. (2018). What drives the implementation of Industry 4.0? The role of opportunities and challenges in the context of sustainability. *Sustainability*, *10*(1), 247. doi:10.3390u10010247

Mumtaz, H., Aminul Islam, M., Ku Ariffin, K. H., & Karim, A. (2011). Customers Satisfaction on Online Shopping in Malaysia. *International Journal of Business and Management*, *6*(10). C:\Users\salma\Desktop\10.5539\ijbm.v6n10p162 doi:10.5539/ijbm.v6n10p162

Murray, A. (2022). High Fidelity invests in Second Life to expand the virtual world. *ZD Net*. https://www.zdnet.com/article/high-fidelity-invests-in-second-life-to-expand-virtual-world/.

Murray, J. H. (2020). Virtual/reality: How to tell the difference. *Journal of Visual Culture*, *19*(1), 11–27. doi:10.1177/1470412920906253

Museum Of Applied Arts and Sciences. *(2017).* Museum of applied arts and sciences. Retrieved 20th *NSW Government.* https://maas.museum/about/

Mustafa, I.E. (2011). Determinants of e---commerce customer satisfaction, trust, and loyalty in Saudi Arabia. *Journal of Electronic Commerce Research, 12*(1), 78---93.

Mustapha, U. F., Alhassan, A. W., Jiang, D. N., & Li, G. L. (2021). Sustainable aquaculture development: A review on the roles of cloud computing, internet of things and artificial intelligence (CIA). *Reviews in Aquaculture*, *13*(4), 2076–2091. doi:10.1111/raq.12559

Mystakidis, S. (2022). Metaverse. *Encyclopedia*, 486-497. doi:10.3390/

Mystakidis, S. (2022). Metaverse. *Metaverse. Encyclopedia*, *2*(1), 486–497. doi:10.3390/encyclopedia2010031

Mystakidis, S., Christopoulos, A., & Pellas, N. (2021). A systematic mapping review of augmented reality applications to support STEM learning in higher education. *Education and Information Technologies*, 1–45.

Nadlifatin, R., Miraja, B. A., Persada, S. F., Belgiawan, P. F., Redi, A. P., & Lin, S.-C. (2020). The Measurement of University Students' Intention to Use Blended Learning System through Technology Acceptance Model (TAM) and Theory of Planned Behavior (TPB) at Developed and Developing Regions: Lessons Learned from Taiwan and Indonesia. *International Journal of Emerging Technologies in Learning*, *15*(09), 219–230. doi:10.3991/ijet.v15i09.11517

Naheed F. (2005) Interactive Approach in English Language Learning and Its Impact on Developing Language Skills, 94-101.

Najork, M. (2017). *Web Crawler Architecture*. doi:10.1007/978-1-4899-7993-3457-3

Nakamoto, S. (2009). *Bitcoin: A Peer-to-Peer Electronic Cash System, 9.*

Narin, N. G. (2021). A Content Analysis of the Metaverse Articles. *Journal of Metaverse*, *1*(1), 17–24.

Ndukwe, E. (2011). The telecommunication revolution in Nigeria. *A Convocation Lecture.*

Negroponte, N. (1996). *Being digital* (1. Vintage Books ed). *NFT basics*. https://nftschool.dev/concepts/non-fungible-tokens/

Nevelsteen, K. J. (2018). Virtual world, defined from a technological perspective and applied to video games, mixed reality, and the Metaverse. *Computer Animation and Virtual Worlds*, *29*(1), e1752. doi:10.1002/cav.1752

Nica, E., Stan, C. I., Luțan, A. G., & Oașa, R. Ș. (2021). Internet of things-based real-time production logistics, sustainable industrial value creation, and artificial intelligence-driven big data analytics in cyber-physical smart manufacturing systems. *Economics, Management, and Financial Markets*, *16*(1), 52–63. doi:10.22381/emfm16120215

Nichols, G. (2018). *Augmented and virtual reality mean business: Everything you need to know An executive guide to the technology and market drivers behind the hype in AR.* VR, and MR.

Niewiadomski, P., Stachowiak, A., & Pawlak, N. (2019). Knowledge on IT tools based on AI maturity–Industry 4.0 perspective. *Procedia Manufacturing*, *39*, 574–582. doi:10.1016/j.promfg.2020.01.421

Nigerian Communication Commission. (2019). Study on young children and digital technology: a survey across Nigeria. *NCC.* www.ncc.gov.ng › Technical Regulation › Research & Development

Ning, H., Wang, H., Lin, Y., Wang, W., Dhelim, S., Farha, F., & Daneshmand, M. (2021). A Survey on Metaverse: the State-of-the-art, Technologies, Applications, and Challenges.

Ning, H., Wang, H., Lin, Y., Wang, W., Dhelim, S., Farha, F., & Daneshmand, M. (n.d.). A Survey on Metaverse: The State-of-the-art, Technologies, Applications, and Challenges. 1-34.

Ning, Y. & Yuqi, G. (2020). E-commerce's marketing strategy based on webcast platform. *Farm staff*, (22), 172 + 212

Ning, H., Wang, H., Lin, Y., Wang, W., Dhelim, S., Farha, F., & Daneshmand, M. (2021). *A Survey on Metaverse: the State-of-the-art.* Technologies, Applications, and Challenges.

Nižetić, S., Šolić, P., González-de, D. L. D. I., & Patrono, L. (2020). Internet of Things (IoT): Opportunities, issues, and challenges towards a smart and sustainable future. *Journal of Cleaner Production*, *274*, 122877. doi:10.1016/j.jclepro.2020.122877 PMID:32834567

NVIDIA. (2021). *GTC Spring 2021 Keynote with NVIDIA CEO Jensen Huan.* YouTube. https://www.youtube.com/watch?v=eAn_oiZwUXA

Nwosu, N. (2019). *Developmental Psychology and Education.* Published by Agatha Series Ltd.

O'Leary, D. E. (2017). Configuring blockchain architectures for transaction information in blockchain consortiums: The case of accounting and supply chain systems. *Intelligent Systems in Accounting, Finance & Management*, *24*(4), 138–147. doi:10.1002/isaf.1417

Odofin, T. (2014). the influence of digital technology on secondary school students academic engagementbehaviour in nigeria. *The Educational Psychologist, 14*(1). https://www.journals.ezenwaohaetorc.org/index.php/ncep/article/viewfile/1847/1879

OECD (2016). Trends Shaping Education 2016. *OECD Publishing Paris*. doi:. doi:10.1787/trends_edu-2016-en

Oh, J. H. (2021). A Study on Factors Affecting the Intention to Use the Metaverse by Applying the Extended Technology Acceptance Model (ETAM): Focused on the Virtual World Metaverse. *The Journal of the Korea Contents Association*, *21*(10), 204–216.

Ojedokon, A. A & Owolabi, E.O (2005).Internet access competence and the use of internet for teaching and research activities by university of Botswana Academic staff. *Africa Journal of Library, archives and information science, 13*(1), 43-53

Okoli, C., & Pawlowski, S. D. (2004). The Delphi method as a research tool: An example, design considerations and applications. *Information & Management*, *42*(1), 15–29. doi:10.1016/j.im.2003.11.002

Okumus, B., Ali, F., Bilgihan, A., & Ozturk, A. B. (2018). Psychological factors influencing customers' acceptance of smartphone diet apps when ordering food at restaurants. *International Journal of Hospitality Management*, *72*, 67–77. doi:10.1016/j.ijhm.2018.01.001

Olajide M. S., Afolabi, O.P. & Ajayi, A.A. (2020). *Impact of digital technology on adolescents in Nigeria.* https://www.researchgate.net/publication/343796926

Olaniyi, A. (2019). Type and Cronbach's Alpha Analysis in an Airport Perception Study. *Scholar Journal of Applied Sciences and Research, 2*(4). https://www.innovationinfo.org/articles/SJASR/SJASR-4-223.pdf

Oliveira, T. A., Oliver, M., & Ramalhinho, H. (2020). Challenges for connecting citizens and smart cities: ICT, e-governance and blockchain. *Sustainability*, *12*(7), 2926. doi:10.3390u12072926

Olweus, D. (1996). Bullying at school: Knowledge base and an effective intervention program. In C. Ferris & T. Grisso (Eds.), *Understanding aggressive behaviour in children*. New York Academy of Sciences. doi:10.1111/j.1749-6632.1996.tb32527.x

Openkov, M. Y., & Tetenkov, N. B. (2021). Digital Empires And The Annexation Of The Metaverse. 86-89.

OpenSea. (n.d.). *the largest NFT marketplace*. Opensea. https://opensea.io/

Oranje, A., Mislevy, B., Bauer, M. I., & Jackson, G. T. (2019). Summative game-based assessment. In D. Ifenthaler & Y. J. Kim (Eds.), *Game-based assessment revisited,* (pp. 37–65). Springer International Publishing., doi:10.1007/978-3-030-15569-8_3

Osazee-Odia, O. U. (2016). Reflections on smartphones: A new paradigm of modernity in Nigeria. *IISTE New Media and Communication Journal*, *56*(1), 39–52.

Owoyele, J. W., & Toyobo, O. M. (2008). Parental will, peer pressure, academic ability and school subjects selection by students in senior secondary schools. *Social Sciences*, *3*(8), 583–586. http://docsdrive.com/pdfs/medwelljournals/sscience/2008/583-586.pdf

Oztemel, E., & Gursev, S. (2020). Literature review of Industry 4.0 and related technologies. *Journal of Intelligent Manufacturing*, *31*(1), 127–182. doi:10.100710845-018-1433-8

Page-Thomas, K. (2006). Measuring task-specific perceptions of the world wide web. *Behaviour & Information Technology*, *25*(6), 469–477. doi:10.1080/01449290500347962

Pallant, J. (2005). SPSS survival guide. Allen & Unwin.

Pan, J. S. (2019). Thoughts on the Expansion of One Million Students in Higher Vocational Colleges . *Education and occupation, 14,* 12-16.

Pandrangi, V. C., Gaston, B., Appelbaum, N. P., Albuquerque, F. C. Jr, Levy, M. M., & Larson, R. A. (2019). The application of virtual reality in patient education. *Annals of Vascular Surgery*, *59*, 184–189. doi:10.1016/j.avsg.2019.01.015 PMID:31009725

Pan, X., Pan, X., Song, M., Ai, B., & Ming, Y. (2020). Blockchain technology and enterprise operational capabilities: An empirical test. *International Journal of Information Management*, *52*, 101946. doi:10.1016/j.ijinfomgt.2019.05.002

Pan, Z., Cheok, A. D., Yang, H., Zhu, J., & Shi, J. (2006). Virtual reality and mixed reality for virtual learning environments. *Computers & Graphics*, *30*(1), 20–28. doi:10.1016/j.cag.2005.10.004

Papagiannidis, S., Bourlakis, M., & Li, F. (2008). Making real money in virtual worlds: MMORPGs and emerging business opportunities, challenges and ethical implications in metaverses. *Technological Forecasting and Social Change*, *75*(5), 610–622. doi:10.1016/j.techfore.2007.04.007

Parente, M., Figueira, G., Amorim, P., & Marques, A. (2020). Production scheduling in the context of Industry 4.0: Review and trends. *International Journal of Production Research*, *58*(17), 5401–5431. doi:10.1080/00207543.2020.1718794

Park, A., Kietzmann, J., Pitt, L., & Dabirian, A. (2022). The evolution of nonfungible tokens: Complexity and novelty of NFT use-cases. *IT Professional*, *24*(1), 9–14. doi:10.1109/MITP.2021.3136055

Park, S. M., & Kim, Y. G. (2022). A Metaverse: Taxonomy, components, applications, and open challenges. *IEEE Access : Practical Innovations, Open Solutions*, *10*, 4209–4251. doi:10.1109/ACCESS.2021.3140175

Park, S., & Kang, Y. J. (2021). A Study on the intentions of early users of metaverse platforms using the Technology Acceptance Model. *Journal of Digital Convergence*, *19*(10), 275–285.

Park, S., & Kim, S. (2022). Identifying World Types to Deliver Gameful Experiences for Sustainable Learning in the Metaverse. *Sustainability*, *14*(3), 1361. doi:10.3390u14031361

Park, Y. S., Konge, L., & Artino, A. R. Jr. (2019). The Positivism Paradigm of Research. *Academic Medicine*, *95*(5), 690–694. doi:10.1097/ACM.0000000000003093 PMID:31789841

Parsons, D., & Stockdale, R. (2010). Cloud as context: Virtual world learning with open wonderland. In *Proceedings of the 9th World Conference on Mobile and Contextual Learning,* Malta, (pp. 123-130).

Paschek, D., Luminosu, C. T., & Ocakci, E. (2022). Industry 5.0 Challenges and Perspectives for Manufacturing Systems in the Society 5.0. *Sustainability and Innovation in Manufacturing Enterprises*, 17-63.

Pascual-Ezama, D., Scandroglio, B., & de Lian, B. G. G. (2014). Can we predict individual investors' behavior in stock markets? A psychological approach. *Universitas Psychologica*, *13*(1), 25–36. doi:10.11144/Javeriana.UPSY13-1.cwpi

Paulins, V. A., & Geistfeld, L. V. (2003). The effect of consumer perceptions of store attributes on apparel store preference. *Journal of Fashion Marketing and Management*, *7*(4), 371–385. doi:10.1108/13612020310496967

Pavlou, P. A., & Fygenson, M. (2006). Understanding and predicting electronic commerce adoption: An extension of the theory of planned behavior. *Management Information Systems Quarterly*, *30*(1), 115–143. doi:10.2307/25148720

Pellas, N., Dengel, A., & Christopoulos, A. (2020). A scoping review of immersive virtual reality in STEM education. *IEEE Transactions on Learning Technologies*, *13*(4), 748–761. doi:10.1109/TLT.2020.3019405

Pellas, N., Mystakidis, S., & Kazanidis, I. (2021). Immersive Virtual Reality in K-12 and Higher Education: A systematic review of the last decade scientific literature. *Virtual Reality (Waltham Cross)*, *25*(3), 835–861. doi:10.100710055-020-00489-9

Peñarroja, V., Sánchez, J., Gamero, N., Orengo, V., & Zornoza, A. M. (2019). The influence of organisational facilitating conditions and technology acceptance factors on the effectiveness of virtual communities of practice. *Behaviour & Information Technology*, *38*(8), 845–857. doi:10.1080/0144929X.2018.1564070

Pereira, N., Lima, A. C., Lanceros-Mendez, S., & Martins, P. (2020). Magnetoelectrics: Three centuries of research heading towards the 4.0 industrial revolution. *Materials (Basel)*, *13*(18), 4033. doi:10.3390/ma13184033 PMID:32932903

Peres, R. S., Jia, X., Lee, J., Sun, K., Colombo, A. W., & Barata, J. (2020). Industrial artificial intelligence in industry 4.0-systematic review, challenges and outlook. *IEEE Access : Practical Innovations, Open Solutions*, *8*, 220121–220139. doi:10.1109/ACCESS.2020.3042874

Ping, L. (2021). Research on behavior problems of higher vocational students from the perspective of positive psychology-- consumer psychological comparison, etc. *The age of wealth*, (08), 229-230

Pisano, G. P. (2015). You Need an Innovation Strategy. *Harvard Business Review*.

Popkova, E. G., Ragulina, Y. V., & Bogoviz, A. V. (Eds.). (2019). *Industry 4.0: Industrial revolution of the 21st century*, (Vol. 169, p. 249). Springer.

Prasad, Ch. J. S., & Aryasri, A. R. (2009). Determinants of Shopper Behaviour in E-tailing: An Empirical Analysis. *Paradigm*, *13*(1), 73–83.

Premkumar, G., & Bhattacherjee, A. (2008). Explaining information technology usage: A test of competing models. *Omega*, *36*(1), 64–75. doi:10.1016/j.omega.2005.12.002

Prensky, M. (2001). Digital Natives, Digital Immigrants. *On the Horizon*, *9*(5), 1–6. doi:10.1108/10748120110424816

Preston, J. (2021). Facebook, the Metaverse and the monetisation of higher education. *Impact of Social Sciences Blog*.

Putit, L., Muhammad, N. S., & Aziz, Z. D. A. (2017). *Retailing*. Oxford Revision Series.

Qin, Y. & Renjun, L. (2020). Research on the design of Taobao live broadcast interface based on user characteristics. *Packaging engineering*, *41*(08), 219-222 doi:.1001-3563.2020.08.032. doi:10.19554/j.cnki

Rafee, H. (2019). Cover Story: Retail market remains steady amid changes in e-commerce and demography. *The Edge Markets*. https://www.theedgemarkets.com/article/cover-story-retail-market-remains-steady-amid-changes-ecommerce-and-demography

Rafiola, R., Setyosari, P., Radjah, C., & Ramli, M. (2020). The effect of learning motivation, self-efficacy, and blended learning on students' achievement in the industrial revolution 4.0. [iJET]. *International Journal of Emerging Technologies in Learning*, *15*(8), 71–82. doi:10.3991/ijet.v15i08.12525

Rafli, D. P. A. D. (2022). NFT Become a Copyright Solution. *Journal of Digital Law and Policy*, *1*(2), 43–52.

Ragulina, Y. V., Alekseev, A. N., Strizhkina, I. V., & Tumanov, A. I. (2019). Methodology of criterial evaluation of consequences of the industrial revolution of the 21st century. In *Industry 4.0: Industrial Revolution of the 21st Century*, (pp. 235–244). Springer. doi:10.1007/978-3-319-94310-7_24

Rahardja, U., Aini, Q., Graha, Y. I., & Tangkaw, M. R. (2019, December). Gamification framework design of management education and development in industrial revolution 4.0. *Journal of Physics: Conference Series*, *1364*(1), 012035. doi:10.1088/1742-6596/1364/1/012035

Raja, G. B. (2021). Impact of Internet of Things, Artificial Intelligence, and Blockchain Technology in Industry 4.0. In *Internet of Things, Artificial Intelligence and Blockchain Technology*, (pp. 157–178). Springer. doi:10.1007/978-3-030-74150-1_8

Rakshit, S., Islam, N., Mondal, S., & Paul, T. (2022). Influence of blockchain technology in SME internationalization: Evidence from high-tech SMEs in India. *Technovation, 115*, 102518. doi:10.1016/j.technovation.2022.102518

Ramadania, S., & Braridwan, Z. (2019). The influence of perceived usefulness, ease of use, attitude, self-efficacy, and subjective norms toward intention to use online shopping. *International Business and Accounting Research Journal, 3*(1), 1–14.

Raman, P. (2019). Understanding female consumers' intention to shop online. *Asia Pacific Journal of Marketing and Logistics, 31*(4), 1138–1160. doi:10.1108/APJML-10-2018-0396

Ramesh, U. V., Harini, A., Gowri, C. S. D., Durga, K. V., Druvitha, P., & Kumar, K. S. (2022). International Journal of Research Publication and Review. Metaverse. *Future of the Internet., 3*(2), 93–97.

Ramírez-Correa, P., Rondán-Cataluña, F. J., Arenas-Gaitán, J., & Martín-Velicia, F. (2019). Analysing the acceptation of online games in mobile devices: An application of UTAUT2. *Journal of Retailing and Consumer Services, 50*, 85–93. doi:10.1016/j.jretconser.2019.04.018

Raut, R. D., Mangla, S. K., Narwane, V. S., Gardas, B. B., Priyadarshinee, P., & Narkhede, B. E. (2019). Linking big data analytics and operational sustainability practices for sustainable business management. *Journal of Cleaner Production, 224*, 10–24. doi:10.1016/j.jclepro.2019.03.181

Reed, P., Hughes, A., & Phillips, G. (2013). Rapid recovery in sub-optimal readers in Wales through a self- placed computer-based reading programme. *British Journal of Special Education, 40*(4), 162–166. doi:10.1111/1467-8578.12040

Revels, J., Tojib, D., & Tsarenko, Y. (2010). Understanding consumer intention to use mobile services. *Australasian Marketing Journal, 18*(2), 74–80. doi:10.1016/j.ausmj.2010.02.002

Riegler, A., Riener, A., & Holzmann, C. (2021). Augmented reality for future mobility: Insights from a literature review and hci workshop. *i-com, 20*(3), 295-318.

Rillig, M. C., Gould, K. A., Maeder, M., Kim, S. W., Dueñas, J. F., Pinek, L., Lehmann, A., & Bielcik, M. (2022). Opportunities and Risks of the "Metaverse" For Biodiversity and the Environment. *Environmental Science & Technology, 56*(8), 4721–4723. doi:10.1021/acs.est.2c01562 PMID:35380430

Ringle, C. M., Wende, Sven, & Baker, J. M. (2017). SmartPLS 4. *SmartPLS*. https://www.smartpls.com/documentation/algorithms-and-techniques/bootstrapping

Risch, M. (2009). Virtual rule of law. *West Virginia Law Review, 112*(1), 1–52.

Rita, P., Oliveira, T., & Farisa, A. (2019). The impact of e-service quality and customer satisfaction on customer behavior in online shopping. *Heliyon, 5*(10), e02690. doi:10.1016/j.heliyon.2019.e02690 PMID:31720459

Ritter, M. (2021). Cryptocurrencies rally on Facebook's "Meta" rebrand - CNN. *CNN Business*.

Ritter, M. (2021). *Cryptocurrencies Rally on Facebook's 'Meta' Rebrand - CNN*. CNN Business.

Robbins, L. (1932). *An Essay on the Nature and Significance of Economic Science*. doi:10.2307/2342397

Robertson, H. (2021). *Jefferies says the metaverse is as big an investment opportunity as the early internet*. Insider.

Robertson, H. (2021). *Jefferies Says the Metaverse Is as Big an Investment Opportunity as the Early Internet*. Insider.

Robinson, J. (2017, June 23). The Sci-Fi Guru Who Predicted Google Earth Explains Silicon Valley's Latest Obsession. *Vanity Fair.* https://www.vanityfair.com/news/2017/06/neal-stephenson-metaverse-snow-crash-silicon-valley-virtual-reality

Roblox Corporation. (2020, November 19). Registration statement on form S-1. *SEC.gov.* https://www.sec.gov/Archives/edgar/data/0001315098/000119312520298230/d87104ds1.htm

Rocha, D. M., Brasil, L. M., Lamas, J. M., Luz, G. V., & Bacelar, S. S. (2020). Evidence of the benefits, advantages and potentialities of the structured radiological report: An integrative review. *Artificial Intelligence in Medicine, 102,* 101770. doi:10.1016/j.artmed.2019.101770 PMID:31980107

Rogers, E. M. (2003). Diffusion of Innovations, 5th Edition. Simon and Schuster.

Romero, D., Stahre, J., Wuest, T., Noran, O., Bernus, P., Fast-Berglund, Å., & Gorecky, D. (2016, October). Towards an operator 4.0 typology: a human-centric perspective on the fourth industrial revolution technologies. In proceedings of the international conference on computers and industrial engineering (CIE46), Tianjin, China, (pp. 29-31).

Rong, W. & Juan, L. (2021). Research on the impact of webcast on the mental health of media art college students. *Western radio and television, 42*(05), 54-56

Rönkkö, M., & Cho, E. (2020). An Updated Guideline for Assessing Discriminant Validity. *Organizational Research Methods, 109442812096861.* doi:10.1177/1094428120968614

Rospigliosi, P. A. (2022). Metaverse or Simulacra? Roblox, Minecraft, Meta and the turn to virtual reality for education, socialisation and work. *Interactive Learning Environments, 30*(1), 1–3. doi:10.1080/10494820.2022.2022899

Roy, A., Boninsegni, M. F., Kumar, S., Peronard, J. P., & Reimer, T. (2020). Transformative Outcomes of Consumer Well-Being in the Era of IR 4.0: Opportunities and Threats of Physical, Biological and Digital Technologies Across Sectors. *Consumer Interests Annual, 66.*

Ruedinger, B. M. (2022). *Social Rejection, Avatar Creation, & Self-esteem.* University Of Oklahoma.

Ruimeng, S. (2022). Research on the Impact of Online Red Economy on College Students' Consumption Behavior. *Time honored Brand Marketing*, (19), 64-66.

Ruonan, T. (2021). The influence of e-commerce consumption on college students and its countermeasures. *China management informatization, 24* (23), 77-80

Ryan, G. V., Callaghan, S., Rafferty, A., Higgins, M. F., Mangina, E., & McAuliffe, F. (2022). Learning Outcomes of Immersive Technologies in Health Care Student Education: Systematic Review of the Literature. *Journal of Medical Internet Research, 24*(2), e30082. doi:10.2196/30082 PMID:35103607

Rymarczyk, J. (2020). Technologies, opportunities and challenges of the industrial revolution 4.0: Theoretical considerations. *Entrepreneurial Business and Economics Review, 8*(1), 185–198. doi:10.15678/EBER.2020.080110

Rymarczyk, J. (2021). The impact of industrial revolution 4.0 on international trade. *Entrepreneurial Business and Economics Review, 9*(1), 105–117. doi:10.15678/EBER.2021.090107

Safeena, R., Date, H., Hundewale, N., & Kammani, A. (2013). Combination of TAM and TPB in internet banking adoption. *International Journal of Computer Theory and Engineering, 5*(1), 146–150. doi:10.7763/IJCTE.2013.V5.665

Sağlık, Z. Y., & Yıldız, M. (2021). Türkiye'de Dil Öğretiminde Web 2.0 Araçlarının Kullanımına Yönelik Yapılan Çalışmaların Sistematik İncelemesi [A Systematic Review of Studies on the Use of Web 2.0 Tools in Language Teaching in Turkey] [JRES]. *Eğitim ve Toplum Araştırmaları Dergisi*, *8*(2), 418–442. doi:10.51725/etad.1011687

Sahai, P., Sharma, M., & Singh, V. K. (2020). Effect of Perceived Quality, Convenience, and Product Variety on Customer Satisfaction in Teleshopping. *Management and Economics Research Journal*, *6*(3), 1–8. doi:10.18639/MERJ.2020.9900021

San Martín, H., & Herrero, Á. (2012). Influence of the user's psychological factors on the online purchase intention in rural tourism: Integrating innovativeness to the UTAUT framework. *Tourism Management*, *33*(2), 341–350. doi:10.1016/j.tourman.2011.04.003

Sanchez, M., Exposito, E., & Aguilar, J. (2020). Industry 4.0: Survey from a system integration perspective. *International Journal of Computer Integrated Manufacturing*, *33*(10-11), 1017–1041. doi:10.1080/0951192X.2020.1775295

Sandbox. (2020). The Sandbox Whitepaper. *The Sandbox*. https://installers.sandbox. game/The_Sandbox_Whitepaper_2020.pdf [Accessed:2021-01-30]

Sang-Min, P., & Young-Gab, K. (2021). A Metaverse: Taxonomy, Components, Applications, and Open Challenges. IEEE, 4209-4251.

Sarstedt, M., & Mooi, E. (2014). Regression Analysis. *Springer Texts in Business and Economics*, 193–233.

Schaar, L., & Kampakis, S. (2022). Non-fungible tokens as an alternative investment: Evidence from cryptopunks. *The Journal of The British Blockchain Association*, *5*(2), 2516–3949. doi:10.31585/jbba-5-1-(2)2022

Schaupp, L.C., and Belanger, F. (2005). A conjoint analysis of online customer satisfaction. *Journal of Electronic Commerce Research, 6*(2), 95---111

Schuemie, M., Van Der Straaten, P., Krijn, M., & Van Der Mast, C. (2001). Research on Presence in Virtual Reality: A Survey. *Cyberpsychology & Behavior*, *4*(2), 183–201. doi:10.1089/109493101300117884 PMID:11710246

Schwoerer, C. E., May, D. R., Hollensbe, E. C., & Mencl, J. (2005). General and specific self-efficacy in the context of a training intervention to enhance performance expectancy. *Human Resource Development Quarterly*, *16*(1), 111–129. doi:10.1002/hrdq.1126

Seigneur, J. M., & Choukou, M. A. (2022). How should Metaverse augment humans with disabilities? In *13th Augmented Human International Conference Proceedings*. ACM. 10.1145/3532525.3532534

Selamat, Z., Jaffar, N., & Ong, H. B. (2009). (PDF) Technology Acceptance in Malaysian Banking Industry. *European Journal of Economics, Finance and Administrative Sciences*.

Seo, J., Kim, K., Park, M., & Park, M. (2018). *An Analysis of Economic Impact on IOT Industry under GDPR. Mobile Information Systems. 1-6.*

Shahroom, A. A., & Hussin, N. (2018). Industrial revolution 4.0 and education. *International Journal of Academic Research in Business & Social Sciences*, *8*(9), 314–319. doi:10.6007/IJARBSS/v8-i9/4593

Shaikh, I. M., Qureshi, M. A., Noordin, K., Shaikh, J. M., Khan, A., & Shahbaz, M. S. (2020). Acceptance of Islamic financial technology (FinTech) banking services by Malaysian users: An extension of technology acceptance model. *Foresight*, *22*(3), 367–383. doi:10.1108/FS-12-2019-0105

Shannon, C. E., & Weaver, W. (2003). *The mathematical theory of communication*. University of Illinois Press.

Shen, B. Q., Tan, W.M., Guo, J.Z., Zhao, L.S., & Qin. P. (2021). How to Promote User Purchase in Metaverse? A Systematic Literature Review on Consumer Behavior Research and Virtual Commerce Application Design. *applied sciences*, 11, 11087

Sherwani, F., Asad, M. M., & Ibrahim, B. S. K. K. (2020, March). Collaborative robots and industrial revolution 4.0 (ir 4.0). In *2020 International Conference on Emerging Trends in Smart Technologies (ICETST),* (pp. 1-5). IEEE. 10.1109/ICETST49965.2020.9080724

Shin, D. H. (2010). MVNO services: Policy implications for promoting MVNO diffusion. *Telecommunications Policy, 34*(10), 616–632. doi:10.1016/j.telpol.2010.07.001

Shraddha, C. (2018) Uses and Gratifications Theory. *Businesstopia*, https://www.businesstopia.net/mass-communication/uses-gratifications-theory

Shuang, Y. (2021). The influence of Pan entertainment on College Students' Ideological and political education and its countermeasures. *Foreign trade and economic cooperation*, (12), 96-99

Shue, M. & Yuyi, W. (2014). Research on business model of technology trading platform from the perspective of value network. *Scientific and technological progress and countermeasures, 31*(06), 1-5

Shuo, G., Juan, L., Jiaxin, Y., & Yu, H. (2019). An Analysis of the Current Situation of the Development of College Students' Live Broadcasting and the Countermeasures -- Taking Henan University as an Example. *Science and Technology Communication, 11* (07), 131-134+191. doi:10.16607/j.cnki.1674-6708.2019.07.066

Shuqin, B. (2021). An empirical study on College Students' webcast behavior and psychology. *News knowledge*, (10), 65-70

Siddiqui, K. (2013). Heuristics for sample size determination in multivariate statistical techniques. *World Applied Sciences Journal, 27*(2), 285–287.

Silva, H., Resende, R., & Breternitz, M. (2018). Mixed reality application to support infrastructure maintenance. In *International Young Engineers Forum (YEF-ECE),* (pp. 50-54). IEEE. 10.1109/YEF-ECE.2018.8368938

Silva, A. J., Cortez, P., Pereira, C., & Pilastri, A. (2021). Business analytics in industry 4.0: A systematic review. *Expert Systems: International Journal of Knowledge Engineering and Neural Networks, 38*(7), e12741. doi:10.1111/exsy.12741

Singh, P., & Keswani, S. (2016). A Study of Adoption Behavior for Online Shopping: An Extension of Tam Model. *International Journal Advances in Social Science and Humanities, 4*(7).

Singh, D. P. (2015). Integration of TAM, TPB, and self-image to study online purchase intentions in an emerging economy. *International Journal of Online Marketing, 5*(1), 20–37. doi:10.4018/IJOM.2015010102

Sivadas, E., & Baker-Prewitt, J. L. (2000). An examination of the relationship between service quality, customer satisfaction, and store loyalty. *International Journal of Retail & Distribution Management, 28*(2), 73–82. doi:10.1108/09590550010315223

Siyue, Q. & Dongju, W. (2021). Research on the Current Situation and Development Countermeasures of Direct Broadcasting with Goods to Assist Agriculture -- Taking Taobao Direct Broadcasting as an Example.. *China Agricultural Accounting*, (10), 88-89. doi:10.13575/j.cnki.319.2021.10.034

Skarbez, R., Neyret, S., Brooks, F. P., Slater, M., & Whitton, M. C. (2017). A Psychophysical Experiment Regarding Components of the Plausibility Illusion. *IEEE Transactions on Visualization and Computer Graphics, 23*(4), 1369–1378. doi:10.1109/TVCG.2017.2657158 PMID:28129171

Slater, M. (2003). A note on presence terminology. *Presence Connect, 3*(3), 1–5.

Slater, M. (2009). Place illusion and plausibility can lead to realistic behaviour in immersive virtual environments. *Philosophical Transactions of the Royal Society of London. Series B, Biological Sciences, 364*(1535), 3549–3557. doi:10.1098/rstb.2009.0138 PMID:19884149

Slater, M., Pérez Marcos, D., Ehrsson, H., & Sanchez-Vives, M. (2009). Inducing illusory ownership of a virtual body. *Frontiers in Neuroscience, 3*(2), 214–220. https://www.frontiersin.org/article/10.3389/neuro.01.029.2009. doi:10.3389/neuro.01.029.2009 PMID:20011144

Slater, M., & Sanchez-Vives, M. V. (2016). Enhancing our lives with immersive virtual reality. *Frontiers in Robotics and AI, 3*, 74. doi:10.3389/frobt.2016.00074

Slater-Robins, M. (2022). Salesforce employees are reportedly unhappy about its NFT plans. *Techradar,* (22). https://www.techradar.com/news/salesforce-employees-are-reportedly-unhappy-about-its-nft-plans

Smart, J. M., Cascio, J., & Paffendorf, J. (2007). *Metaverse Roadmap Overview.* https://www.metaverseroadmap.org/overview/

Smart, J. M., Cascio, J., & Paffendorf, J. (2007). *Metaverse Roadmap Overview.* Acceleration Studies Foundation.

Soliman, M. S. M., Karia, N., Moeinzadeh, S., Islam, M. S., & Mahmud, I. (2019). Modelling intention to use ERP systems among higher education institutions in Egypt: UTAUT perspective. *International Journal of Supply Chain Management, 8*(2), 429–440.

Sonal, K., & Thomas, S. (2019). (PDF) Online grocery retailing – exploring local grocers beliefs. *International Journal of Retail & Distribution Management, 47*(12).

Sonvilla-Weiss, S. (2008). *(In)visible: Learning to act in the metaverse.* Springer.

Soon, K. W. K., Lee, C. A., & Boursier, P. (2016). A study of the determinants affecting adoption of big data using integrated Technology Acceptance Model (TAM) and diffusion of innovation (DOI) in Malaysia. *International journal of applied business and economic research,* 14(1), 17-47.

Sourin, A. (2017). Case study: shared virtual and augmented environments for creative applications. In *Research and development in the academy, creative industries and applications,* (pp. 49–64). Springer. doi:10.1007/978-3-319-54081-8_5

Sparkes, M. (2021). What is a metaverse. *New Scientist, 251*(3348), 1–18. doi:10.1016/S0262-4079(21)01450-0

Sreeram, A., Kesharwani, A., & Desai, S. (2017). Factors affecting satisfaction and loyalty in online grocery shopping: An integrated model. *Journal of Indian Business Research, 9*(2), 107–132. doi:10.1108/JIBR-01-2016-0001

Stam, E., & Wennberg, K. (2009). The roles of R&D in new firm growth. *Small Business Economics, 33*(1), 77–89. doi:10.100711187-009-9183-9

Stăncioiu, A. (2017). The Fourth Industrial Revolution 'Industry 4.0'. *Fiabilitate Şi Durabilitate, 1*(19), 74–78.

Stephenson, N. (2003). *Snow crash: A novel.* Spectra.

Stephenson, N. (2008). *Snow crash.* Bantam Books.

Sterne, J., & Razlogova, E. (2019). Machine learning in context, or learning from landr: Artificial intelligence and the platformization of music mastering. *Social Media + Society, 5*(2), 2056305119847525. doi:10.1177/2056305119847525

Strikwerda, L. (2014). Should Virtual Cybercrime by Regulated by Means of Criminal Law: Philosophical, Legal-Economic, Pragmatic and Constitutional Dimension. *Information & Communications Technology Law*, *23*(1), 31–60. doi:10.1080/13600834.2014.891870

Studio, R. (2021). Anything you can imagine, make it now! *Roblox*. https://www.roblox. com/create

Sudan, R., Petrov, O., & Gupta, G. (2022). Can the metaverse offer benefits for developing countries? *Worldbank*. https://blogs.worldbank.org/digital-development/can-metaverse-offer-benefits-developing-countries

Sudarmo. (1996). *Prinsip Dasar Manajemen*. BPFE-Yogyakarta.

Sudiwedani, A., & Darma, G. S. (2020). Analysis of the effect of knowledge, attitude, and skill related to the preparation of doctors in facing industrial revolution 4.0. *Bali Medical Journal*, *9*(2), 524–530. doi:10.15562/bmj.v9i2.1895

Suh, W., & Ahn, S. (2022). Utilizing the Metaverse for Learner-Centered Constructivist Education in the Post-Pandemic Era: An Analysis of Elementary School Students. *Journal of Intelligence*, *10*(1), 17. doi:10.3390/jintelligence10010017 PMID:35324573

Sung, H.-N., Jeong, D.-Y., Jeong, Y.-S., & Shin, J.-I. (2015). The relationship among self-efficacy, social influence, performance expectancy, effort expectancy, and behavioral intention in mobile learning service. *International Journal of U-and e-Service. Science and Technology*, *8*(9), 197–206.

Sun, Y. C. (2021). Evaluation of regional higher vocational colleges' social service performance——Based on an empirical analysis of higher vocational colleges in Guangdong Province. *Heilongjiang Journal of Higher Education*, *3*(93), 115–120.

Sun, Y. S. (2020). Internal Logic Analysis of Xi Jinping's Talent Development. *Human Resource Development*, *19*, 10–11.

Suzor, N. (2010). The role of the rule of law in virtual communities. *Berkeley Technology Law Journal*, *25*, 1817.

Suzuki, S. N., Kanematsu, H., Barry, D. M., Ogawa, N., Yajima, K., Nakahira, K. T., Shirai, T., Kawaguchi, M., Kobayashi, T., & Yoshitake, M. (2020). Virtual experiments in metaverse and their applications to collaborative projects: The framework and its significance. *Procedia Computer Science*, *176*, 2125–2132. doi:10.1016/j.procs.2020.09.249

Swan, M. (2015). *Blockchain: Blueprint for a New Economy*. O'Reilly Media.

Tabachnick, B. G., & Fidell, L. S. (2007). *Using Multivariate Statistics*. Pearson Education Inc.

Taherdoost, H. (2016). Validity and Reliability of the Research Instrument; How to Test the Validation of a Questionnaire/Survey in a Research. *SSRN Electronic Journal*, *5*(3). doi:10.2139/ssrn.3205040

Tambare, P., Meshram, C., Lee, C. C., Ramteke, R. J., & Imoize, A. L. (2021). Performance measurement system and quality management in data-driven Industry 4.0: A review. *Sensors (Basel)*, *22*(1), 224. doi:10.339022010224 PMID:35009767

Tang, C. S., & Veelenturf, L. P. (2019). The strategic role of logistics in the industry 4.0 era. *Transportation Research Part E, Logistics and Transportation Review*, *129*, 1–11. doi:10.1016/j.tre.2019.06.004

Tang, J. C., Venolia, G., & Inkpen, K. (2016). *Meerkat and Periscope:I Stream, You Stream, Apps Stream for LiveStreams*. Human Factors In Computing Systems. doi:10.1145/2858036.2858374

Tang, L.-L., & Nguyen, H. T. H. (2013). Common causes of trust, satisfaction and TAM in online shopping: An integrated model. *Journal of Quality*, *20*(5), 483–501.

Tarhini, A., Alalwan, A. A., Al-Qirim, N., Algharabat, R., & Masa'deh, R. (2018). An analysis of the factors influencing the adoption of online shopping. [IJTD]. *International Journal of Technology Diffusion*, *9*(3), 68–87. doi:10.4018/IJTD.2018070105

Taye, C. & Samuel, O. (2019). From news breakers to news followers: the influence of Facebook on the coverage of the January 2010 crisis in Jos. *Bingham journal of humanities, social and management sciences, 1,* 7-9.

Taylor, S., & Todd, P. (1995). Assessing it usage: The role of prior experience. *Management Information Systems Quarterly*, *19*(4), 561–570. doi:10.2307/249633

Technavio. (2020). The increasing demand for AR and VR technology to boost growth. *Technavio.* https://www.businesswire.com/news/home/20200903005356/en

Teo, T. S. H. (2001). Demographic and motivation variables associated with Internet usage activities. *Internet Research*, *11*(2), 125–137. doi:10.1108/10662240110695089

The Economist (2022). Building a metaverse with Chinese characteristics. *The Economist.* https://www.economist.com/china/2022/02/04/building-a-metaverse-with-chinese-characteristics

The Ministry of Education, et al. (2020). *The Action Plan for Improving the Quality of Vocational Education (2020-2023), 7.*

Thomala, L. L. (2022a). Level of understanding of metaverse in China 2021. *Statista.* https://www.statista.com/statistics/1287390/china-familiarity-with-metaverse/

Thomala, L. L. (2022b). Expectations of metaverse adoptions in China 2021. *Statista.* https://www.statista.com/statistics/1287403/china-most-in-demand-applications-of-metaverse-2021/

Thomala, L. L. (2022c). Major concerns of metaverse in China 2021. *Statista.* https://www.statista.com/statistics/1287437/china-worries-about-the-popularity-of-metaverse/

Tian, L. & Yongqi, L. (2021). Analysis of Taobao online marketing strategy. *Market weekly, 34*(08), 72-74

Tian, X., Guo, R., & Wang, B. (2021). Research on the purchase intention of clothing consumers in Taobao live broadcast based on perceived risk theory. *Journal of Beijing Institute of Clothing Technology*, *41*(01), 61–66. doi:10.16454/j.cnki.issn.1001-0564.2021.01.010

Tims, M., Bakker, A. B., & Derks, D. (2014). Daily job crafting and the self-efficacy–performance relationship. *Journal of Managerial Psychology*, *29*(5), 490–507. doi:10.1108/JMP-05-2012-0148

Törhönen, M., Sjöblom, M., & Hamari J. (2018). Likes and Views:Investigating Internet Video Content Creators Perceptionsof Popularity. New York:The2nd International GamiFIN Conference.

Tran, V. Le, & Leirvik, T. (2020). Efficiency in the markets of crypto-currencies. *Finance Research Letters, 35*(November), 101382. doi:10.1016/j.frl.2019.101382

Treiblmaier, H., & Špan, Ž. (2022). Will blockchain really impact your business model? Empirical evidence from Slovenian SMEs. *Economic and Business Review*, *24*(2), 132–140. doi:10.15458/2335-4216.1302

Tsang, Y. P., & Lee, C. K. M. (2022). Artificial intelligence in industrial design: A semi-automated literature survey. *Engineering Applications of Artificial Intelligence*, *112*, 104884. doi:10.1016/j.engappai.2022.104884

Tseng, M. L., Tran, T. P. T., Ha, H. M., Bui, T. D., & Lim, M. K. (2021). Sustainable industrial and operation engineering trends and challenges Toward Industry 4.0: A data driven analysis. *Journal of Industrial and Production Engineering*, *38*(8), 581–598. doi:10.1080/21681015.2021.1950227

Tsinghua University New Media Research Center. (2021). *2020-2021 Metaverse Development Research Report.* https://xw.qq.com/cmsid/20210920A0095N00

Türk, G. D., & Darı, A. B. (2022). Metaverse'de bireyin toplumsallaşma süreci [The Socializatıon Process Of The Individual In The Metaverse]. *Stratejik ve Sosyal Araştırmalar Dergisi*, *6*(1), 277–297.

Turkle, S. (2007). *Evocative Objects, Things we Think With.* MIT Press.

Um, T., Kim, H., Kim, H., Lee, J., Koo, C., & Chung, N. (2022). Travel Incheon as a Metaverse: Smart Tourism Cities Development Case in Korea. In *ENTER22 e-Tourism Conference,* (pp. 226–231). Springer. doi:10.1007/978-3-030-94751-4_20

Uslu, M. (2013). Relationship between degrees of self-esteem and peer pressure in high school adolescents. *International Journal of Academic Research Part B,* 117122.https://www.researchgate.net/publication/314524207

Uzuegbunam, C. (2019). A child-cantered study of teens' digital lifeworld's from a Nigerian perspective. https://blogs.lse.ac.uk/parenting4digitalfuture/2019/10/16/a-child-centred-study-of-teens-digital-lifeworlds-from-a-nige
rian-perspective/

Vabalas, A., Gowen, E., Poliakoff, E., & Casson, A. J. (2019). Machine learning algorithm validation with a limited sample size. *PLoS One*, *14*(11), e0224365. doi:10.1371/journal.pone.0224365 PMID:31697686

Valaei, N., Rezaei, S., Ho, R. C., & Okumus, F. (2019). Beyond structural equation modelling in tourism research: Fuzzy Set/qualitative comparative analysis (fs/QCA) and data envelopment analysis (DEA). In Quantitative Tourism Research in Asia, (pp. 297–309). Springer.

Valaskova, K., Machova, V., & Lewis, E. (2022). Virtual Marketplace Dynamics Data, Spatial Analytics, and Customer Engagement Tools in a Real-Time Interoperable Decentralized Metaverse. Linguistic and Philosophical Investigations, 21.

Valeonti, F., Bikakis, A., Terras, M., Speed, C., Hudson-Smith, A., & Chalkias, K. (2021). Crypto collectibles, museum funding and OpenGLAM: Challenges, opportunities and the potential of Non-Fungible Tokens (NFTs). *Applied Sciences (Basel, Switzerland)*, *11*(21), 9931. doi:10.3390/app11219931

Van Der Merwe, D. (2021). The Metaverse as Virtual Heterotopia. *World Conference on Research in Social Sciences.* 10.33422/3rd.socialsciencesconf.2021.10.61

van Tonder, R., Trockman, A., & le Goues, C. (2019). A panel data set of cryptocurrency development activity on GitHub. *IEEE International Working Conference on Mining Software Repositories,* 186–190. 10.1109/MSR.2019.00037

Vasic, N., Kilibarda, M., & Kaurin, T. (2019). The Influence of Online Shopping Determinants on Customer Satisfaction in the Serbian Market. *Journal of Theoretical and Applied Electronic Commerce Research, 14*(2), 0. doi:10.4067/S0718-18762019000200107

Venkatesh, V., & Davis, F. D. (2000). A theoretical extension of the technology acceptance model: Four longitudinal field studies. *Management Science*, *46*(2), 186–204. doi:10.1287/mnsc.46.2.186.11926

Venkatesh, V., & Morris, M. (2000). Why don't men ever stop to ask for direction? Gender, social influence and their role in technology acceptance and usage behavior. *Management Information Systems Quarterly*, *24*(1), 115–139. doi:10.2307/3250981

Venkatesh, V., Thong, J. Y., & Xu, X. (2016). Unified theory of acceptance and use of technology: A synthesis and the road ahead. *Journal of the Association for Information Systems*, *17*(5), 328–376. doi:10.17705/1jais.00428

Versprille, A. (2022, January 6). NFT Market Surpassed $40 Billion in 2021, New Estimate Shows. *Bloomberg*. https://www.bloomberg.com/news/articles/2022-01-06/nft-market-surpassed-40-billion-in-2021-new-estimate-shows?leadSource =uverify%20wall

Wahab, H. K. A., & Tao, M. (2019). The Influence of Internet Celebrity on Purchase Decision and Materialism: The Mediating Role of Para-social Relationships and Identification.. *European Journal of Business and Management*, *11*, 15.

Wall Street Journal. (2022). Accounting Firms Scoop Up Virtual Land in the Metaverse. *The Wall Street Journal*. https://www.wsj.com/articles/accounting-firms-scoop-up-virtual-land-in-the-Metaverse-11641599590/.

Walsh, S. (2002). 'Construction or obstruction: Teacher talk and learner involvement in the EFL classroom'. *Language Teaching Research*, *6*(1), 3–23. doi:10.1191/1362168802lr095oa

Wang, P. (2022). Research on the Management of Epidemic Prevention and Control in Colleges and Universities.. *China Higher Education*, (09), 47-49

Wang, Q., Li, R., Wang, Q., & Chen, S. (2021). Non-fungible token (NFT): Overview, evaluation, opportunities and challenges.

Wang, W., Zhang, J., & Huang, H. (2020). Research on Optimization of New Media Platform for Ideological and Political Education of College Students.. *Higher Education Forum*, (12), 91-93

Wang, Y., Su, Z., Zhang, N., Liu, D., Xing, R., Luan, T. H., & Shen, X. (2022). A survey on metaverse: Fundamentals, security, and privacy.

Wang, Y., Su, Z., Zhang, N., Liu, D., Xing, R., Luan, T. H., & Shen, X. (2022). A survey on Metaverse: Fundamentals, security, and privacy. doi:10.36227/techrxiv.19255058.v2

Wang, H., Tao, D., Yu, N., & Qu, X. (2020). Understanding consumer acceptance of healthcare wearable devices: An integrated model of UTAUT and TTF. *International Journal of Medical Informatics*, *139*, 104156. doi:10.1016/j.ijmedinf.2020.104156 PMID:32387819

Wang, M. (2018). On Consumer Knowledge and Consumer Behavior — Taking College Students as an Example.. *Value Engineering*, *37*(17), 233–235. doi:10.14018/j.cnki.cn13-1085/n.2018.17.103

Wang, Q. Y., & Castro, C. D. (2010, June). Class Interaction and Language Output. *English Language Teaching*, *3*(2). doi:10.5539/elt.v3n2p175

Wang, R., Lin, Z., & Luo, H. (2019). Blockchain, bank credit and SME financing. *Quality & Quantity*, *53*(3), 1127–1140. doi:10.100711135-018-0806-6

Wang, Y.-S., Yeh, C.-H., & Liao, Y.-W. (2013). What drives purchase intention in the context of online content services? The moderating role of ethical self-efficacy for online piracy. *International Journal of Information Management*, *33*(1), 199–208. doi:10.1016/j.ijinfomgt.2012.09.004

Wang, Y., Seo, J. H., & Song, W.-K. (2019). Factors That Influence Consumer Satisfaction with Mobile Payment: The China Mobile Payment Market. *International Journal of Contents*, *15*(4), 82–88.

Wang, Z. B., & Bi, Y. R. (2019). Consider the software technology major at Huai'an Vocational Information Technology College as an example. *Think Tank Times*, *45*, 176–177.

Warner, D. S. (2022). *The Metaverse with Chinese Characteristics: A Discussion of the Metaverse through the Lens of Confucianism and Daoism* [Doctoral dissertation]. University of Pittsburgh.

Weisberg, M. (2011). Student attitudes and behaviors towards digital textbooks. *Publishing Research Quarterly*, *27*(2), 188–196. doi:10.100712109-011-9217-4

Weisheng, Z. (2022). Key Problems and Solutions of Music Works in Online Live Broadcasting. *Friends of Editors*, (05), 83-87. doi:10.13786/j.cnki.cn14-1066/g2.2022.5.013

Wen, C. & Kaiyan, Z. (2021). Research on the impact of online live broadcasting on consumers' purchase intention. *Communication and copyright*, (07), 55-58. doi:. 45-1390/g2. 2021.07.019. doi:10.16852/j.cnki

Wen, Z. & Xuguang, C. (2019). Research on College Students' motivation and behavior of watching webcast. *Youth exploration*, (02), 78-86. doi:. issn1004-3780.2019.02.008. doi:10.13583/j.cnki

Whitty & Young. (2012). Rape, pillage, murder, and all manner of ills: Are there some possibilities for action that should not be permissible even within gamespace? In K. Poels & S. Malliet (Eds.), *Moral Issues in Digital Game Play*, (p. 2). ACCO Academic.

Wikipedia. (n.d.). COVID-19 pandemic. *Wikipedia*. https://en.wikipedia.org/wiki/COVID-19_pandemic

Williams, C. (2007). Research Methods. *Journal of Business & Economics Research*, *5*(3). doi:10.19030/jber.v5i3.2532

Williams, D. M., & Rhodes, R. E. (2016). The confounded self-efficacy construct: Conceptual analysis and recommendations for future research. *Health Psychology Review*, *10*(2), 113–128. doi:10.1080/17437199.2014.941998 PMID:25117692

Williams, J. A. (2001). Classroom conversations: Opportunities to learn for ESL students in mainstream classrooms. *The Reading Teacher*, *54*(8), 750–757. http://login.proxy.seattleu.edu/login?url=http://search.proquest.com/docview/203275419?accountid=28598

Wilson, N., & Christella, R. (2019). An Empirical Research of Factors Affecting Customer Satisfaction: A Case of the Indonesian E-Commerce Industry. *DeReMa (Development Research of Management). Jurnal Manajemen*, *14*(1), 21. doi:10.19166/derema.v14i1.1108

Wood, L. (2021). Global Gaming Industry to Cross $314 Billion by 2026 - Microsoft, Nintendo, Twitch, and Activision Have All Reached New Heights in Player Investment, Helped by COVID-19. *Research and Markets*. (https://finance.yahoo.com/news/global-gaming-industry-cross-314-183000310.html)

Wood, L. (2021). Global Gaming Industry to Cross $314 Billion by 2026 - Microsoft, Nintendo, Twitch, and Activision Have All Reached New Heights in Player Investment, Helped by COVID-19. *Research and Markets*. https://finance.yahoo.com/news/global-gaming-industry-cross-314-183000310.html

Wood, R., Atkins, P., & Tabernero, C. (2000). Self-efficacy and strategy on complex tasks. *Applied Psychology*, *49*(3), 430–446. doi:10.1111/1464-0597.00024

World Health Organization. (2020). Listings of WHO's response to COVID-19. *WHO*. https://www.who.int/news/item/29-06-2020-covidtimeline

World Health Organization. (2020). Orientation program me on adolescent health for health-care providers. *WHO*. https://www.who.int/health-topics/adolescent-health#tab=tab_1

Wortley, D. J., & Lai, P. C. (2017) The Impact of Disruptive Enabling Technologies on Creative Education, *3rd International Conference on Creative Education.*

Wortley, D. J., & Lai, P. C. (2017) The Impact of Disruptive Enabling Technologies on Creative Education. *3rd International Conference on Creative Education.*

Wright, J. (2016, October 21). Nosedive. Black Mirror.

Wright, O., Standish, J., & Weiss, E. (2011). *How consumer companies can innovate to grow?* Accenture.

Wu, J., Yang, C., & Wenbing, X. (2020). Research on College Students' daily reasonable consumption. *Guangxi quality supervision guide*, (01), 220-221

Wu, H. Q. (2020). Conducting research on the social and environmental impacts of my country's industrial growth. *Journal of Zhongnan University of Economics and Law*, *5*, 52–63.

Wu, J. H., & Wang, S. C. (2005). What drives mobile commerce? An empirical evaluation of the revised technology acceptance model. *Information & Management*, *42*(5), 719–729. doi:10.1016/j.im.2004.07.001

Wu, L. (2013). The antecedents of customer satisfaction and its link to complaint intentions in online shopping: An integration of justice, technology, and trust. *International Journal of Information Management*, *33*(1), 166–176. doi:10.1016/j.ijinfomgt.2012.09.001

Xiaoya, A. I. (2018). Development status and prospective research of webcast platform in China. *New media research*, 4,(4), 7-10 doi:. issn2096-0360.2018.04.003. doi:10.16604/j.cnki

Xin, L. V., & Lu, D. (2021). Analysis on business development model of webcast platform. *Marketing*, (37), 29-30

Xingli, D. (2007). Discussion on the Morbid Star Chasing of Teenagers. *Contemporary Youth Research,* (08), 70-75

Xingnan, J. (2020). Data Research and Evaluation on the Status Quo of College Students' E-commerce Platform Shopping. *North Economic and Trade Journal,* (06), 48-50

Xinhuanet. (2017). *Report of the 19th National Congress of the Communist Party of China.* http://www.news.cn/politics/19cpcnc

Xiong, G., Mengmeng, L., & Wan, L. (2019). Study on Influencing Factors of College Students' willingness to continue using webcast platform. *Science and technology communication,11*(07):173-176.

Xu, H., Zhao, M., Zhuang, J., Chen, W., & Yang, J. (2021). The influence of "special price economy" on College Students' rational consumption behavior *Collection of award-winning papers of the national statistical modeling competition for college students in 2021,* China Statistical Education Society, (p. 46). doi:. 2021.041766.10.26914/c.cnkihy

Xu, J. (2021). Research on the Development of a System for Evaluating the Quality of Higher Vocational Education Talent Training Using "People's Satisfaction". *Journal of Hunan Industrial Vocational and Technical College*, *2*(21), 13–19.

Xu, J. D., Benbasat, I., & Cenfetelli, R. T. (2013). Integrating Service Quality with System And information Quality: An Empirical Test in the E-Service Context. *Management Information Systems Quarterly*, *37*(3), 777–794. doi:10.25300/MISQ/2013/37.3.05

Xu, Y., & Lv, L. (2021, August). A New Method of Free Combat Teaching Based on Artificial Intelligence. *Journal of Physics: Conference Series, 1992*(4), 042010. doi:10.1088/1742-6596/1992/4/042010

Yadav, R. & Moon, S. (2022, March 11). Opinion: Is the pandemic ending soon? *WHO*. https://www.who.int/philippines/news/detail/11-03-2022-opinion-is-the-pandemic-ending-soon

Yang, H. H., & Su, C. H. (2017). Learner Behaviour in a MOOC Practice-oriented Course: In Empirical Study Integrating TAM and TPB. *International Review of Research in Open and Distributed Learning*, *18*(5), 35–63. doi:10.19173/irrodl.v18i5.2991

Yang, J. (2015). On China's Vocational Education Modernization Vision. *Higher Vocational Education Exploration*, *5*(15), 1–6.

Yang, X., Gao, M., Wu, Y., & Jin, X. (2018). Performance evaluation and herd behavior in a laboratory financial market. *Journal of Behavioral and Experimental Economics*, *75*, 45–54. doi:10.1016/j.socec.2018.05.001

Yao, W. M. (2019). The new era's building of high-quality teachers. *Journal of Chinese Education,* 103-104.

Yao, G., & Yao, C. (2021). Research on the Development of Rural E-Commerce under the Network Live Broadcast Mode.. *Tropical Agricultural Engineering*, *45*(05), 89–91.

Yee, N. (n.d.). *Motivations of Play in MMORPGs*, *46*.

Yee, N., Bailenson, J. N., & Ducheneaut, N. (2009). The Proteus Effect: Implications of Transformed Digital Self-Representation on Online and Offline Behavior. *Communication Research*, *36*(2), 285–312. doi:10.1177/0093650208330254

Yeong, Y.-C., Kalid, K. S., Savita, K., Ahmad, M., & Zaffar, M. (2022). Sustainable cryptocurrency adoption assessment among IT enthusiasts and cryptocurrency social communities. *Sustainable Energy Technologies and Assessments*, *52*, 102085. doi:10.1016/j.seta.2022.102085

Yi, F. (2021). Research on online live broadcasting e-commerce model under the background of new media marketing. *China media science and technology,* (11): 109-111 doi:.11-4653/n.2021.11.033. doi:10.19483/j.cnki

Yi, M. Y., Jackson, J. D., Park, J. S., & Probst, J. C. (2006). Understanding information technology acceptance by individual professionals: Toward an integrative view. *Information & Management*, *43*(3), 350–363. doi:10.1016/j.im.2005.08.006

Ying, X. (2019). Reflections on the "loss" of social networking in the live webcast from the perspective of strangers.. *Journal of Chongqing University of Science and Technology (Social Science Edition),* (04), 30-33+40. doi:10.19406/j.cnki.cqkjxyxbskb.2019.04.008

Yip, W. Y. (2019). Singapore is the world's second most digitally competitive country, after the US. *The Straits Times*. https://www.straitstimes.com/tech/singapore-is-the-worlds-second-most-digitally-competitive-country-after-the-us

Yoder, M. (2022). An" OpenSea" of Infringement: The Intellectual Property Implications of NFTs. *The University of Cincinnati Intellectual Property and Computer Law Journal.*, *6*(2), 4.

Yong, Z. (2018). The Tianping "Autonomous" situation: Contemporary representation of the construction of the relationship between live broadcasting and social interaction -- a re examination of merowitz's situation theory. International Press.

Yong'an, Z. & Tao, W. X. (2017). Profit model, profit change and driving factors of webcast platform -- an exploratory case study based on the era of happy gathering. *China Science and Technology Forum,* (12), 182-192 doi:.fstc.2017.12.022. doi:10.13580/j.cnki

Yonhap News Agency. (2021). Seoul to offer new concept administrative services via metaverse platform. *Korea Herald*. http://www.koreaherald.com/view.php?ud=20211103000692

Younus, A. M., Tarazi, R., Younis, H., & Abumandil, M. (2022). The Role of Behavioural Intentions in Implementation of Bitcoin Digital Currency Factors in Terms of Usage and Acceptance in New Zealand: Cyber Security and Social Influence. *ECS Transactions*, *107*(1), 10847–10856. doi:10.1149/10701.10847ecst

Yuan, Z. (2021). From "618" to "double 11": e-commerce festival marketing experience and Its Enlightenment. *Business economics research*, (14), 102-105

Yusuf Dauda, S., & Lee, J. (2015). Technology adoption: A conjoint analysis of consumers' preference on future online banking services. *Information Systems*, *53*, 1–15. doi:10.1016/j.is.2015.04.006

Zallio, M., & Clarkson, P. (2022). Inclusive Metaverse. How businesses can maximize opportunities to deliver an accessible, inclusive, safe Metaverse that guarantees equity and diversity.

Zarmpou, T., Saprikis, V., Markos, A., & Vlachopoulou, M. (2012). Modeling users' acceptance of mobile services. *Electronic Commerce Research*, *12*(2), 225–248. doi:10.100710660-012-9092-x

Zhang X. Q., Head K. (2009) Dealing with learner reticence in the speaking class. *ELT Journal*, *64*(1). doi:10.1093/elt/ccp0181

Zhang, F. Q., & Xu, F. K. (2018) . On the Establishment of a Quality Assurance System Throughout the Process of Developing Talent in Higher Vocational Education . *Education and occupation, 2,* 46-51.

Zhang, W. L., & Fan, M. M. (2019). The meaning, fundamental principles, and promotion path for the development of high-quality higher vocational education in the modern era . *Education and occupation, 221,* 25-32.

Zhang, Y., Weng, Q., & Zhu, N. (2018). The relationships between electronic banking adoption and its antecedents: A meta-analytic study of the role of national culture. *International Journal of Information Management*, *40*, 76–87. doi:10.1016/j.ijinfomgt.2018.01.015

Zhao, C. (2012, November 17). Classroom Interaction and Second Language Acquisition: *The more interactions the better?* doi:10.3968/j.sll.1923156320130701.3085

Zhao, X. & Jingli, F. (2021). The negative influence of online celebrity culture on College Students' values and its countermeasures. *Contemporary educational theory and practice*, 13 (06), 68-74. doi:.1674-5884.2021.06.012. doi:10.13582/j.cnki

Zhao, Q. (2011). 10 Scientific Problems in Virtual Reality. *Communications of the ACM*, *54*(2), 116–119. doi:10.1145/1897816.1897847

Zhao, Q., Chen, C., Cheng, H., & Wang, J.-L. (2017). Determinants of Live Streamers' Continuance Broadcasting Intentions on Twitch:A Self-determination Theory Perspective.. *Telematics and Informatics*, *35*(2), 406–420. doi:10.1016/j.tele.2017.12.018

Zhao, Y., Jiang, J., Chen, Y., Liu, R., Yang, Y., Xue, X., & Chen, S. (2022). *Metaverse: Perspectives from graphics, interactions and visualization*. Visual Informatics.

Zhong, Bi. L. (2022). Preserve a century-old legacy and advance the quality of higher education. Chongqing University of Technology, Chongqing, 1(10), 3-5.

Zhong, B. L. (2018). Ascend to a new era and embrace new possibilities to overcome new obstacles. *Higher Education in China*, *1*, 1.

Zhong, R. Y., Xu, X., Klotz, E., & Newman, S. T. (2017). Intelligent manufacturing in the context of industry 4.0: A review. *Engineering, 3*(5), 616–630. doi:10.1016/J.ENG.2017.05.015

Zhou, K., Liu, T., & Zhou, L. (2015, August). Industry 4.0: Towards future industrial opportunities and challenges. In *2015 12th International conference on fuzzy systems and knowledge discovery (FSKD),* (pp. 2147-2152). IEEE.

Zhou, T., Lu, Y., & Wang, B. (2010). Integrating TTF and UTAUT to explain mobile banking user adoption. *Computers in Human Behavior, 26*(4), 760–767. doi:10.1016/j.chb.2010.01.013

Zuckerberg, M. (2021). *Connect 2021 Keynoate: Our Vision for the Metaverse.* Facebook.

Žukauskas, P., Vveinhardt, J., & Andriukaitienė, R. (2018). Research Ethics. *Management Culture and Corporate Social Responsibility.* doi:10.5772/intechopen.70629

Zutter, N. (2016, October 24). *Trying Too Hard: Black Mirror, "Nose-dive."* Tor.Com. https://www.tor.com/2016/10/24/black-mirror-season-3-nosedive-television-review/

About the Contributors

PC Lai is a professor supervising Ph.D. students with "UNIRAZAK" and facilitating postgraduate programs with various universities as well as conducting corporate training with industry. She also spoke and judged at local and international conferences, competitions, and events including winning numerous ICT local and international awards. She has more than 15 years of experience in the Asia Pacific region in various industries, mainly telecommunications and Finance. Her works also involved corporate strategy, e-business, marketing, and transforming business processes between the PEOPLE (employees), PRODUCTS, PARTNERS and CUSTOMERS by enhancing customer satisfaction levels and ultimately resulting in enhanced Business Relationships, reduced costs, processes, and increased revenue. She was also the Chief Strategy Information Officer (CSIO) with a publicly listed company. She is also affiliated with a few technology transfer offices globally besides advisors with a few associations.

Vinesh Thiruchelvam is a professional Engineer with over 25 years' experience in engineering education/academics, ICT, operations & maintenance, facility management, design and project management in various industries ranging from building services/property, oil & gas, district cooling plants, co-generation power plants and the port industry. I have managed deliveries of international projects in the Maldives, India, Russia, Iran, UAE, Qatar, Saudi Arabia, Oman, Vietnam, Brunei and Malaysia for the property sector, ports, oil & gas and power plant industries. I was the principal engineer on Malaysia's first Waste to Energy technology transfer program 2002 to 2004 under Ministry Housing & Local Government, head of a port's Operations & Maintenance Department from 1998 to 2002 where I managed the facility management, operations & maintenance programs and technical teams. Having moved to the Middle East in 2005, I focused my specialization mainly in the building services/property sector primarily in designing sustainable and energy efficient buildings, rail and airports. I am currently the Chief Innovation & Enterprise Officer for the APIIT Education Group after stints as Dean and Deputy Vice Chancellor. Industry Projects Completed in Value as Key Accountable Design Engineering Personnel – RM6.78 Billion (1998-2012). Industry consultancy & research grants (Industry) Obtained – RM132.7 million (2008-2012). Research grants (University) received – RM1.96 million (2014 to date).

Tinfah Chung is a professor at ELM Graduate Business School, HELP University with industry experience in the following areas of Regulatory, Corporate, Consulting and Academia. Dr Chung specializes in quantitative finance and economics, having taught in two other private universities. He has served as an independent member of the Rating Committee of RAM Ratings and Market Participants Committee of Bursa Malaysia. His career in economic research has spanned working for Bank Negara Malaysia, several investment banks and academia. In addition, his experience and expertise in economic research has been applied to fund management (SOCSO) and surveillance of the equity market (Securities Commission). He has undertaken consulting work for the World Bank, Government of Malaysia, Securities Commission and RAM Holdings. He has a PhD in finance from Universiti Putra Malaysia, with a Masters in Economics from Yale University, USA and a B Econs from University Malaya. His publications have been accepted in Global Finance Journal, Japan and the World Economy, International Banking and Finance and Pertanika.

Sonali Vyas is serving as an academician and researcher for around a decade. Currently, she is working as Assistant Professor (Selection Grade) at University of Petroleum and Energy Studies, Uttarakhand. She has been awarded as "National Distinguished Educator Award 2021", instituted by the "International Institute of organized Research (I2OR) which is a registered MSME with the Ministry of Micro, Small and Medium Enterprises, Government of India". Also awarded as "Best Academician of the Year Award (Female)" in "Global Education and Corporate Leadership Awards (GECL-2018)". Her research interest includes Blockchain, Database Virtualization, Data Mining, and Big Data Analytics. She has authored the ample number of research papers, articles and chapters in refereed journals/ conference proceedings and books. She authored a book on "Smart Health Systems", Springer. She is also an editor of "Pervasive Computing: A Networking Perspective and Future Directions", Springer Nature and "Smart Farming Technologies for Sustainable Agricultural Development, IGI Global". She acted as a guest editor in a special issue of "Machine Learning and Software Systems" in "Journal of Statistics & Management Systems (JSMS)" (Thomson Reuters)". She is also a member of Editorial Board and Reviewer Board in many referred National and International journals. She has also been a member of Organizing Committee, National Advisory Board and Technical Program Committee at many International and National conferences. She has also Chaired Sessions in various reputed International and National Conferences. She is a professional member of CSI, IEEE, ACM-India, IFERP, IAENG, ISOC, SCRS and IJERT.

Desmond Okocha, PhD, is a Social Scientist with specialisation in management and mass communication. He has over 15 years experience in consulting, Research and lecturing. He obtained his B.A degree in Management from the United Kingdom, holds a M.A and PhD in Journalism and Mass Communication, both from India. Additionally, has PGDs in Education Management and Leadership and another in Logistics and Supply Chain Management. He was the pioneer National Knowledge Management and Communication Coordinator for the International Fund for Agricultural Development project in the Niger Delta. He is presently, a Senior Lecturer, Department of Mass Communication, Bingham University, Nigeria. He is the Founder of Institute for Leadership and Development Communication, Nigeria. As an international voice, Dr. Okocha is a frequent speaker in conferences across continents. In 2018, he was invited to speak at Harvard University, USA, Vienna University, Austria, and at the MIRDEC-8th, International Academic Conference on Social Sciences, Portugal.

Yuyang Wang is a research assistant professor at Institute of Automation, Chinese Academy of Sciences. Her main research interests are in consumer behavior, pan media communication, big data analysis, new coming technologies research, digital economy, and business intelligence in Sustainable Development Goals (SDGs). Dr Wang earned a Bachelor of Science and Engineering degree in Management, Master degree of Law, and a Doctorate degree of Business Administration. She has participated in editing three books, attended one international conference, and published articles as the first author on various topics ranging from culture and construction of personal core value system, campus network management and online guidance, network option and ideological education to the enlightenment of new coming technologies to higher education. She has also presided over two research projects, participated in three National Fund projects and three other research projects as principal investigator, and won two prize of social science achievement. She was also once employed as the Deputy Director of the Public Opinion Center of a Public University, engaged in network of public opinion and communication, E-government and Digital government construction, college students' mental health education, the Belt and Road cultural communication.

Hatice Büber is currently a faculty member of Software Engineering Department at Faculty of Engineering, Kırklareli University, Turkey. She completed her BA in Computer Education and Instructional Technology, in 2007 at Gazi University, Turkey. She did her MS in Information Science, in 2011 at the University of Pittsburgh, the USA. She obtained her Ph.D in Technology Enhanced Learning (Computer Science), in 2019 at the Durham University, UK. Her areas of Interests include but not limited to technology use in education, cyberfeminism, gamification and human-computer interaction.

Siti Azreena Mubin is a senior lecturer at Asia Pacific University of Technology & Innovation (APU) with over 13 years of experience in teaching Multimedia and Computer Science modules. Obtained her PhD in Software Engineering and Master of Computer Science in Multimedia from University of Putra Malaysia (UPM). She is a certified trainer by Human Resources Development Fund (HRDF) and certified by Malaysia Board of Technologists (MBOT) as a Professional Technologists (Ts.). She had various experienced in web and iOS mobile development especially with multimedia content applications. Her research passion and experience is on gamification & multimedia technologies including spatial computing such as virtual reality, augmented reality and mixed reality. She is also accountable as Extended Reality (XR) subject matter of expert in APU and appointed as research leader in her cluster. She is actively involved in research and publications as she acknowledges that academic research is an integral part of teaching. Student-centered learning and visualization aids are several of her approach to enhance students' understanding. She believes learning and gaining knowledge should be fun and exciting.

Evelyn Toh is currently a lecturer / program leader in the Department of Marketing Strategy & Innovation (DMSI) at Sunway University Business School (SUBS), Malaysia. Her work experience spans across in various industries - retail, trading and manufacturing as well as in Higher Education Institutions (HEIs). Teaching and supervising students' thesis (PG and UG) remains her passion as she has been teaching in HEIs for the past 18 years. Apart from teaching, she is also a journal reviewer for the Journal of Applied Structural Equation Modeling (JASEM), Journal of Cleaner Production, Asia Pacific Journal of Marketing and Logistics as well as IGI Global Publications. Currently, she teaches Marketing Channels for Undergraduates and Consumer Behaviour for Post graduate students. Her research areas include marketing channels, Omni channels, metaverse in retail, retailing as well as consumer behaviour.

Mitali Chugh has 22 years of experience in academics. Presently she is Assistant Professor (Senior Scale), School of Computer Science, at the University of Petroleum and Energy Studies. She has research publications in various international journals and conferences with SCI/SCOPUS/UGC indexing. Her research interests include software process improvement, software quality, knowledge management, and technical debt.

Ree Chan Ho is an associate professor at the Faculty of Business and Law, Taylor's University, Malaysia. He has vast academic and administrative experience in tertiary education institutes. His academic portfolio includes dean, principal lecturer, program director, stream coordinator, program leader, etc. Before his academic career, he worked in the capacity of business analyst, project manager, and regional manager in the development of business enterprise systems. He has published in numerous indexed journals and received several best paper awards at international conferences. His current research interests include business innovation and technology, information and knowledge management, online business, and data analytics.

Tugba Mutlu is currently an Assistant Professor at Department of Office Management and Executive Assistant Training, Yozgat Bozok University, Turkey. She completed her BA in Mathematics, in 2007 at Kafkas University, Turkey. She did her MA in Mathematics Education, in 2008 at Erciyes University, Turkey and an MS in Informaion Science, in 2011 at the University of Pittsburgh, the USA. She obtained her Ph.D in Instructional Technologies, in 2016 at the University of Sheffield, UK. Her areas of Interests include but not limited to technology use in education, technology management, human-computer interaction.

Glaret Sinnappan has 21 years lecturing, research and training experiences in the field of Computing. She obtained her PhD form University Malaysia in 2012 in the field of Information System. Master's degree in information technology (IT) from University Putra Malaysia (UPM) and), B.A (Hons) from University Malaysia. Her expertise is in the field of Information system, Cognitive Computing and Artificial Intelligence.

Bee Lian Song is a senior lecturer at Asia Pacific University of Technology and Innovation, Malaysia. Dr. Song's teaching and research areas are in marketing and management. Her research has been published in Thomson Reuters Social Sciences Citation Indexed (SSCI) and Emerging Sources Citation Indexed (ESCI), Australian Business Dean's Council (ABDC), Scopus and Excellent Research Australia (ERA) ranked journals. She has received several best paper research awards at international conferences. Her research interests include consumer behaviour, digital marketing, sustainable marketing and corporate social responsibility.

Sia Chin Chin was admitted as Barrister in the Gray's Inn, UK in 2003 after reading law at Cardiff University (Bar Vocational Course) in 2002 and University of Sheffield (LLB). In 2007, She was conferred Erasmus Mundus Bursary to complete her European Master in Law and Economics in University of Paul Cezanne (France), Ghent University (Belgium) and University of Bologna (Italy). She was conferred PhD in 2020 from the Taylor's University (Malaysia). She is a member of International Bar Association Probono (IBA Probono), Young International Arbitration Group (YIAG), Society of Legal Scholar (SLS) and International Association of Law School (IALS). Prior to her academic career as a law faculty member, Chin Chin has practised and been appointed as legal advisors in Malaysia, Saudi Arabia and Italy in the areas of international commercial laws and scientific grants in research contracts for the past 10 years. She holds numerous committee appointments, including Committee Member of Animal Ethics in Teaching and Research, Student Disciplinary Committee and Conferences. Her primary areas of research are in commercial law, public health focusing on food law and legal education.

Eik Den Yeoh has more than 17 years of experiences in digital and technology; He started as a software engineer and expanded his knowledge into artificial intelligence, big data analytics, and robotic process automation along his career. Which intensively covered end to end integration for most transformation needs from business challenge into practical solutions. He graduated from Asia Pacific University (APU) with MSc in Finance & Investment in Apr 2018 and currently continue pursuing a Doctor in Business Admin with HELP University. This improves his grasp of the distinction between functional and technical capability in relation to the requirements of the majority of business transformations and enables him to combine business and technology more effectively. In his free time, he pursues research on new trends, particularly those that have developed in the previous year, such as the blockchain industry, in which He began to invest time and money to learn more about NFT, DeFi, GameFi, and Blockchain. This will help him gain a deeper understanding of how these technologies can be applied to new concepts. He is always learning and prepared to anticipate upcoming technologies that might have an impact on the business sector, such as AI or technologies like Blockchain and Web 3.0.

Liu Fengmei is a teacher at Guangdong Polytechnic of Science and Technology. Mainly engaged in higher vocational education and student management research and is a PhD student with Universiti Tun Abdul Razak

Salman Faiz is currently a Post graduate student for the MSc in Marketing by Lancaster University. Amongst his area of interest include digital marketing, Marketing in the metaverse world, Omnichannels, Retailing and Consumer Behaviour. During his studies, he has been actively involved in clubs and societies as well as student affairs.

Tay Liang Han attends Tunku Abdul Rahman University. He began his studies in Diploma in Information Technology in 2018. He has enrolled in a bachelor's degree program in Information Security program. He is interested in researching in Information Technology field.

Terhile is a PhD Candidate in the Department of Mass Communication, Bingham University, Nigeria

Index

A

B

C

D

E

F

G

H

I

L

M

N

O

P

Q

R

S

T

U

V

W

Printed in the United States
by Baker & Taylor Publisher Services